Heteromagnetic Microelectronics

Alexander A. Ignatiev • Alexander V. Lyashenko

Heteromagnetic Microelectronics

Microsystems of Active Type

Springer

Professor Alexander A. Ignatiev
Saratov State University
Department of Physics
Astrakhanskaya 83
Russia 410026
ket07@rambler.ru

Professor Alexander V. Lyashenko
Open Society "Tantal"
50 years of October, 110
Russia 410040
secretery@oao-tantal.ru

ISBN 978-1-4419-6001-6 e-ISBN 978-1-4419-6002-3
DOI 10.1007/978-1-4419-6002-3
Springer New York Dordrecht Heidelberg London

Library of Congress Control Number: 2010924185

Printed on acid-free paper

Springer is part of Springer Science+Business Media (www.springer.com)

Foreword

The book proposed to the readers' attention represents an attempt to state and systematize extensive material of our experimental and theoretical investigations of heteromagnetic interactions in ferrite semiconductor structures of the active type carried out at the department of general physics, Saratov State University named after N.G. Chernyshevskiy and in the Design office of critical technologies (DO CT) of SRI–Tantal Corp. of the Holding company "Tantal" in recent years. The novelty and complexity of the physical phenomena determined the high-technology character of our investigations at the joint of some leads – semiconductor microelectronics, microcircuitry, radio engineering, radiophysics, physics of magnetic phenomena, magnetoelectronics.

Accumulation of extensive theoretical and experimental material on magnetoelectronics of the microwave and EHF-ranges, investigations on bigyrotropic microelectronics in ferrite films and structures on their basis, decisive experiments confirming the multifunctionality of interactions in ferrite semiconductor structures of the active type have determined the new lead being promising.

The results of our physical investigations of multifunctional, multiparametric interactions in ferrite semiconductor structures of the active type – (oscillators, converters, amplifiers, frequency synthesizers, and sensors) in the radio-wave range are discussed in the book.

Performance of such a great volume of investigations became possible by joining the efforts of leading experts and scientists of Saratov State University, leading industrial enterprises of Russia in the spheres of semiconductor microelectronics manufacturing, development of microcontrollers, radioelectronic systems, and ferrites.

The obtained results were discussed at scientific and technical meetings held regularly at Tantal® and SRI–Tantal® with the participation of leading experts and reported in our publications of accounts, papers, reports in the topical collected books "Heteromagnetic microelectronics" (INNS 1810–9594), which have been published on a regular basis (one to two times a year) at the Publishing house of Saratov State University since 2004.[1]

[1] Electronic variants are located on the specialized site: http://www.oao-tantal.ru

The book is intended for experts partly familiar with the fundamentals of this lead and working in the field of designing new radio components, microsystems, multiparameter sensors (including intellectual ones), systems of navigation, communication, location, crack detection, monitoring of the environment and complex objects and mechanisms. It will also be useful for professors, students, postgraduate students to whom these problems are close.

In preparation of the book great help was rendered by many researchers, students, and postgraduate students. The materials of some chapters have been obtained together with the following persons: the principal engineer A.S. Stolyarov, j.r.a. D.V. Tugushov and the engineer A.G. Peredumov (Chaps. 1–3), Dr. L.S. Sotov (Chaps. 5, 6, 10, 11, 14), Dr. A.L. Khvalin, the programming engineer V.V. Pleshkov and j.r.a. V.N. Samoldanov (Chap. 7), Dr. S.V. Ovchinnikov (Chaps. 8 and 9), Dr. M.N. Kulikov and Dr. V.A. Kostyakov (Chap. 11), Dr. V.V.Gurzo, the programming engineer V.V.Pleshkov and Dr M.N. Kulikov (Chap. 12), j.r.a. A.V. Vasiliev and j.r.a. V.N. Samoldanov (Chap. 13), and Dr V.V. Gurzo (Chap. 15).

Invaluable help in primary editing and computer design of the book was rendered by Mrs. O.G. Danke and T.N. Sirotinina.

The formation of this lead and its development would be impossible without the considerable help of Dr V.G. Dmitriev and, especially, without the design of first operational laboratory models of high power by N.I. Odintsov, head of the teaching and scientific laboratory "Telecommunications, communication facilities and information processing" at the department of general physics of Saratov State University named after N.G. Chernyshevskiy.

A special role in intensification of our works and investigations on heteromagnetic microelectronics belongs to the authorities and a large number of services and structural divisions of Tantal®, because this lead would not be realized in such a short time interval without their support.

Alexander Ignatiev
Saratov City
September 2006

Reviewers
Full member of the Russian Academy of Natural Sciences N.I. Sinicyn
Doctor of Science S.G. Souchkov

Translators
Dr. E.A. Ignatiev and Dr. S.L. Shmakov

Introduction

Heteromagnetic microelectronics (*heteros* – Greek. – "another," "a different") is a different magnetoelectronics of active devices in relation to the traditional magnetoelectronics of passive devices, which has been intensively developed in 1980s–1990s [1]. This is a new direction of multipurpose, multiple parametric microfield interactions in ferrite-semiconductor structures, magnetotransistors, devices, and microsystems of active type.

Extensive research and development of last decades in the field of solid-state electronics and microelectronics, magnetonics have been directed on advance to the ranges of millimetric and submillimetric waves, design of a new element base for processing, recording, storage, protection of information, creation of vector-type sensors with expanded capabilities, functional properties and sensitivity, development of new means of location, navigation, communication, defectoscopy; systems and means of control of medical, biologic, and psychophysical state of the human, information transfer, etc. A special place in these researches is occupied by the works of academician Yu. V. Guliaev's scientific school to expand integration and functional properties, to attach intellectual capabilities of microsystems, computer facilities, research of the physics of spinwave electronics, spin-wave resonance, and also spinotronics (*spin-transport electronics*) studying transport of an electron's spin and magnetic moment in various junctions like "ferromagnetic metal – ultrathin nonmagnetic separating layer – ferromagnetic metal." They are also directed on creation of magneto-photon crystals, operated microwave devices and high-frequency switches, memory devices, sensors of magnetic field, new nano-sized means for high-density recording, media of information transfer, and design of an optical computer with the use of the photon-crystal base [2–11].

There was an impact for the development of heteromagnetic microelectronics – first of all, the results of theoretical researches of excitation and propagation of various types of electromagnetic waves in bigyrotropic layered film structures containing magnetic semiconductors in the centimetric and millimetric ranges of radiowaves [1]. A new type of structures – magnetic semiconductors [12] combining the properties of semiconductors, on the one hand, and magnetic crystals, on the other hand – should provide, under certain conditions, energetic compensation of the distribution losses of waves and oscillations in such structures, and design of ferreed amplifiers and generators. Other approaches concerned energy interactions of various kinds of waves in magneto-arranged structures [13–16]. However, such amplifiers and laboratory breadboard models mainly had so-called electronic amplification (the level of signal on their output was below the input one and partially compensated the losses introduced by the amplifier).

The basic, principal restriction consists in the fact that in magnetic semiconductors at their doping with atoms and ions of various elements the structure and properties of the base semiconductor and the magnetic properties of the ferrite structures worsened. Therefore, at the first stage of research, compound FSCS combining the high qualities of semiconductor and ferrite could be more effective. The first researches on bulk and film ferrites in the modes of mixed signals were carried out by this way. However, the weak sagging of high-frequency magnetic fields from the semiconductor subsystem into the surrounding space provided no effective interactions with the ferrite subsystem.

In the mode of microwave oscillation generation by various types of transistors and transistor-based oscillators, effective interaction of the high-frequency magnetic fields of the transistor and the FMCR was achieved. Just the first experiments have shown that the generating interactions in compound FSCS were multipurpose ones, covered regular (spectrally pure), noise-type and noise signals close to white noise, signals in the form of evenly spaced frequency spectrum like operated synthesizers of frequencies. Control over the power and spectral characteristics, change of the central frequency of signals in FSCS were carried out by the voltage and current of the semiconductor subsystem and the bias field value in the ferrite subsystem. The spectra of the formed signals overlapped multioctave frequency ranges with signals of one kind, namely, harmonics in the common modes, parametrical harmonics and parametrical subharmonics with synchronous control over the central frequencies (phases) of all the components and equidistance – the frequency distances of all the spectral components – in the parametrical modes. The interactions were effective and provided a high technical efficiency of the used FSCS.

FMCR in such devices plays the role of a multiconnected, generally nonlinear, oscillatory contour. The set of characteristic frequencies, precessions of magnetization vectors in such contours can reach 5–7 and more [17–20].

In the domain, nonlinear mode, the end of the magnetization vector of each characteristic frequency describes an ellipse trajectory. "Turn-on" and "turn-out" of this or that contour into FSCS, and their set by characteristic frequencies are carried out due to a choice of the static parameters, namely, magnetizations, fields of anisotropy, the shape and sizes of ferrite, its orientation relative to the high-frequency magnetic fields of excitation in the semiconductor subsystem and the external bias field, as well as due to the dynamic parameters, namely, the level of high-frequency power on the input of the structure or the power developed in the structure, which can switch the ferrite subsystem into a nonlinear mode and various parametrical modes – multiplication, divisions, frequency modulation of signals of the basic (fundamental) frequency [20, 21].

Experimental researches of new kinds of multipurpose interactions were paid special attention. It is due to the complexity and variety of the effects observable in active-type FSCS in the multioctave frequency ranges, as well as due to the necessity of formation of correct physical models of the processes for theoretical research.

Experimental researches have been done on various types of transistors (bipolar, field ones) on low (few mW) and high (few W) levels of continuous and pulse power. Experiment covered the widest variations of the parameters of various FMCR types and their orientations in an external magnetic field and their locations relative to the electrodes (currents) in the semiconductor system – in a magnetotransistor.

Modeling of physical processes in the investigated structures was made on the basis of several approaches, namely, the electrodynamic one, the method of equivalent circuits, the method of harmonious balance with the use of modern CADs, and included modeling and working off of coupling elements of various types and designs, modeling of the parameters and characteristics of field and bipolar transistors and magnetotransistor of low, medium, and high levels of continuous and pulse power in the VHF, UHF, MWF, EHF, and HHF ranges[1] [22]. Theoretical calculations of thermal loadings in such devices for continuous and pulse powers, time of thermal readiness in various topological models were conducted. The mechanico-climatic and thermal stability of magnetotransistors to various factors was estimated. Some applied aspects of heteromagnetic microelectronics and the possibilities of design of multipurpose operated generators, amplifiers, including low-noise ones with the operated central frequency, multipurpose synthesizers of an evenly spaced frequency spectrum, sensors of magnetic induction vector and its deviations, multiparameter sensors of the displacement vector and the related dynamic physical quantities (linear and angular speeds, forces, pressure, moments of forces, moments of momentum, uninertial forces at simple and complex trajectories of movement) are considered as examples. The multipurpose properties of HMG are most effectively realized at management of their parameters and characteristics by specialized microprocessors.

Researches and development in this direction will produce new types of heteromagnetic MIC and SSI, intellectual microsystems of various purposes, including self-diagnostics, increase of survivability and working resource, distinction of objects' portraits, formation of various kinds of signals for information protection, monitoring of Earth's magnetosphere, prevention of earthquakes and tsunami, etc. Interesting directions may be new communication systems of passive location by the magnetic component of electromagnetic radiation, Earth's magnetic induction, including new types of magnetovision, operative tomography in mobile conditions, etc.

Of special value are developments of "know-how" of heteromagnetic components on gallium arsenide, silicon, other materials of semiconductor microelectronics, modern CADs for transistors and digital microcircuits to control their parameters and to process data from response signals on various kinds and types of electromagnetic, magnetic, and mechanical influences upon HMT.

---

[1] According to the international rules of a radio communication division into ranges of frequencies is entered: VHF – (30–300 MHz), UHF – (0.3–3.0 GHz), MWF – (3.0–30.0 GHz), EHF – (30.0–300.0 GHz), HHF – (300.0–3,000.0 GHz).

Increase of the integration level of heteromagnetic transistors, designing of CHIPS, MIC, and SSI on their basis determines the necessity of creation of standard libraries and databases of equivalent parameters and algorithms of their calculations for CADs. This applies to FMCR made of magnetic materials at low and high power levels, to various types of microstrip coupling elements, to various types of magnetotransistors, and to the dependences of their parameters and characteristics on temperature and mechanical influences.

The advantages of heteromagnetic microsystems of active type consist in the following:

- The common element base, system of designing, manufacturing techniques
- An increased reliability and working resource
- Ecological safety at manufacturing, and essential reduction of the required accessories
- Depreciation
- An increased adaptability to manufacture
- Formation of final-kind signals, including signals with complex spectra and PSD
- Multifunctionality at formation of various signals (regular – spectral pure, noise-type, noise ones, including signals with a uniform PSD in a wide range of frequencies as white noise, signals like those generated by synthesizers of evenly spaced frequency spectra with operated frequency distances between spectral components, etc.)
- Super broadbandness (overlapping of multioctave frequency ranges at parametrical multiplication and division of signals of the basic frequency)
- Multiparametric vector quantities (magnetic induction and its variable components, and, hence, the vector of electric field and polarization, vector mechanical quantities such as displacement, linear and angular speeds, accelerations, forces, pulses, moments of forces and moment, noninertial forces at simple and complex trajectories of movement)
- A raised noise immunity
- Capabilities of maintenance of complex circuit solutions on processing and formation of signals by one crystal (CHIP)
- New directions of research and applications of heteromagnetic interactions can be: Gunn's operated multipurpose magnetodiodes and IMPATTD in the radiowave range
- Operated multipurpose magnetolasers for the optical range
- Magnetic aerials (distributed systems of sensors with an operated mobile spatial aperture)
- Systems of signal and message coding
- Intellectual microsystems, nonlinear locators to recognize images and the organization level of radio-electronic systems
- Vector control systems of the medical and biologic parameters and the psychophysical state of the human
- Portable systems of magnetovision and magnetic tomography
- Systems of multipurpose vector multiparameter control and testing of complex radio-electronic modules and electronic systems

Heteromagnetic microelectronics satisfies the criteria of critical, break-through technologies for which characteristics are: the common element base and technologies, an essential expansion of functionalities and parameters, overlapping of multioctave frequency ranges, a high technical efficiency, a lower cost of development, power inputs, ecological loads on the environment, a lower cost of products, effective replacement of complex and complexificated modules and blocks by a microsystem, and CHIP with expanded capabilities and higher parameters.

This pioneer research in this direction has been carried out in brief terms.[2] The extensive experimental material, development of various models of processes in the radiorange (from 0.03 to 300 GHz) and submillimetric range of frequencies (below 1,000 GHz), development, manufacturing, and research of the parameters of a big number of laboratory and preproduction models of heteromagnetic devices and microsystems of various types, development of author's programs of designing of elements and units of heteromagnetic devices of various types, operating radioelectronic systems were the result of creativity of our scientific personnel headed by the authors of the book.

It is necessary to emphasize the role of comprehensive scientific and technical examinations of our obtained results on heteromagnetic microelectronics, which were done in 2000–2004 by the employees and leading experts of the Fryazino and Saratov branches of IRE of the Russian Academy of Science, specialized research institutes, industrial enterprises and scientific centres, research-and-production associations, and some high schools of Russia. At these examinations not only the used concepts, maintenance and treatments of effects and objects of research were specified and corrected, but also the terms were specified, perspective research directions and expected results were estimated, and new directions were formulated.

Of great value for intense researches in the new area were:

- Modern information equipment and Internet access to world know-hows in this and adjacent areas of science and engineering
- Equipment (since 2002) of our educational process and research of students and postgraduate students at the chair of general physics of Saratov State University (SSU) by licensed CADs of analog and digital devices "MWO-2002" (AWR®, USA[3]), which are deeply debugged and have a high methodical level
- Formation of a research team on the basis of leading experts and teachers from SSU, Tantal Corp., Tantal Institute; students, postgraduate students, postdoctorals

[2] The effects are theoretically predicted during 1990–1992, found out experimentally on 03.11.1995 by initiative research in Saratov State University on small (up to 1 mW) and average (up to 100 mW) power levels by using ferrite microresonators and various types of field and bipolar transistors. The most significant and first experimental results at high power levels (up to several W) were received between 2000 and 2001 on powerful bipolar transistors. The most intensive research, development, and tests of experimental samples were executed between 2000 and 2005.

[3] Serial number is 4283, number of license is 2590, registered and applied in SSU.

- Creation of the necessary infrastructure and corresponding services, modern maintenance with domestic and foreign devices and equipment, such as a vector analyzer of circuits N (N5250A) (Agilent Technologies®) and a station of precision positioning station (Cascade Microtech®), etc.

No research materials are included in the book about:

- Heteromagnetic technologies of monolithic devices and microsystems
- Magnetodiode modes (Gunn's multipurpose diodes and IMPATTD)
- Multipurpose operated magneto-lasers
- Heteromagnetic modules of high integration
- Means of signal coding and testing of complex electronic circuits
- Design of distributed systems of heteromagnetic sensors

These are independent directions to be described in books in the near future, and the intensity of similar works will be determined by interested young researchers, engineers-developers and their colleagues (teachers, instructors, heads), the most important factor being the demand of results by various segments of the market.

Contents

Part III Calculation of Parameters of Heteromagnetic Structures

Part IV Applied Aspects

Abbreviation

AEM	Axis of easy magnetization
AFC	Amplitude–frequency characteristics
AM	Amplitude modulation
APLC	Automatic phase-lock control
AUM	Axis of unfavorable magnetization
BHF	Barium hexaferrite
BMT	Bipolar magnetotransistor
CAD	Computer-aided design
CE	Coefficient of efficiency
CE*	Control element
CHIP	Crystal microcircuit
CVS	Constant-voltage source
DAC	Digital-to-analog converter
DCA	Direct-current amplifier
DVDR	Divisor with variable division ratio
EC	Electromagnetic compatibility
EEF	Effecting external factors
EHF	Extremely high frequencies (30.0–300.0 GHz)
ES FS	Evenly spaced[4] frequency spectrum
ES NT FS	Evenly spaced noise-type frequency spectrum
FD	Frequency divider
FEMT	Field-effect magnetotransistor
FET	Field-effect transistor
FFT	Fast Fourier transform
FM	Ferrite microresonator
FM*	Frequency modulation
FMCR	Ferrite microresonator
FMR	Ferromagnetic resonance
FN	Flat noise
FSCS	Ferrite-semiconductor structure

[4] With equal frequency distances between spectral components.

FSS	Ferrite-semiconductor structure
FTS	Ferrite-transistor structure
GCR	Generator of clock rate
GFC	Gain-frequency characteristic
HF	High frequency
HHF	Hyper-high frequency (300.0–3,000.0 GHz)
HMG	Heteromagnetic generator
HMS	Heteromagnetic sensor
HMT	Heteromagnetic transistor
HPL	High power level
IC	Impulse counter
IMPATTD	Impact avalanche transit time (diode)
LPF	Low pass filter
LPL	Low power level
MC	Monolithic chip
MCG	Magnet (magnet field)-controlled generator
MDM	Metal–dielectric–metal
MECE	Magnetoelectronic coupling element
MES	Magnetoelectronic system
MFMS	Multifunctional frequency magnetosynthesizer
MIC	Monolithic integrated circuits
MOS	Metal–oxide–semiconductor
MPL	Mean power level
MPMS	Moving part of magnetic system
MSW	Magnetostatic wave
MWF	Microwave frequency (3.0–30.0 GHz)
NFMR	Nonlinear ferromagnetic resonance
NPSD	Spectral noise power density
NS	Noise signal
OS	Operating storage
PAA	Phased antenna array
PD	Phase discriminator
PFC	Phase–frequency characteristics
PLG	Photolithography
PM	Phase modulation
PS	Pseudonoise signal
PS*	Permanent storage
PSD	Power spectral density
PSK	Phase-shift keying
PSK PS	Phase-shift-keyed pseudonoise signal
PWFT	Pseudorandom working frequency tuning
RPS	Reflection phase shifter
RQG	Reference quarts generator
SCS	Semiconductor structure
SLC	Super large chip

SPS	Spectral-pure signal
STF	Stripline transmitting filter
SWR	Standing wave ratio
SWR*	Spin-wave resonance
TC	Technical conditions
TEC	Temperature expansion coefficient
UHF	Ultrahigh frequencies (0.03–0.3 GHz)
VAC	Volt–ampere characteristics
VCG	Voltage-controlled generator
VHF	Very high frequencies (0.3–3.0 GHz)
VLSIC	Very large scale integrated circuits
YIG	Yttrium-iron garnet

Symbols

A	Amplitude, dimension
a	Dimension, acceleration, thermal conductivity
B	Dimension
b	Dimension
C	Capacitance
c (index)	Collector
D	Parameter
d	Diameter, middle width of domain
E	EMF source, error function, elasticity modulus (Young's), magnetic energy density, degaussing field density
e (index)	Emitter
F	Tuning out frequency, free energy density, energy of crystallographic anisotropy (F_{an})
f	Frequency
f (index)	Ferrite
G	Conductivity, coupling oscillation coefficient, thermodynamic potential
g	Free fall acceleration, parameter
H	Magnetic intensity, intensity of anisotropy field (H_{A}), saturation field intensity (H_S)
h	High-frequency magnetic field, threshold field, distance, thickness, parameter
I	Current
i	Current number ($i = 1, 2, \ldots$)
K	Gain, screening number (K_{S}), diode feature coefficient
k	Boltzmann's constant, coefficient, wave number (vector $\overline{k}$), dynamic coefficient
L	Inductance, a component of dissipation matrix, impulse moment, Lagrange function
l	Length
M	Magnetization, mutual induction
m	Mass, high-frequency magnetization ($\widetilde{m}$), surface density of effective mass of domain wall (m'_{Y})

N	Form factor, moment of force
n	Concentration of charge carriers, attenuation
P	Power, power of phase noise (P_φ), extraneous shifting force (P), static force (P_{st}), pulse
p	Concentration of charge carriers, nonlinear capacitance
Q	Total charge, unloaded Q(quality), Q-quality
q	Electron charge, density of surface heat evolution, parameter
R	Active part of radiation resistance, resistance, thermal resistance (R_T)
r	Coefficient, radius, resistance
S	Spectral power density, matrix, scattering parameter, area, coefficient of magnetic screening (S_H), screening number of high-frequency fields (S_{HF}), sensitivity to velocity changes (S_v), sensitivity to pressure changes (S_p)
s.w. (index)	Spin waves
T	Temperature, period, kinetic energy density of magnetization vector precession
t	Time, coefficient
U	Voltage
V	Voltage, volume
W	Power, Volume density of heat evolution, energy of warming
X, Y, Z	Spatial coordinates
x, y, z	Spatial coordinates
Y	Matrix
α	Differential transconductance, coefficient, increment of vibrational amplitude, coefficients of convective heat transfer and heat exchange
β	Phase constant, coefficient
γ	K factor of dissipation, gyromagnetic ratio of electron
Δ	Delta
δ	Parameter of mistiming, static flexure (δ_{st}), skin layer thickness
η	Efficiency, coordinate
θ	Angle
λ	Characteristic constant
μ	Magnetic permeability, tensor ($\overleftrightarrow{\mu}$)
ν	Frequency
ξ	Error function, coordinate
Σ	Summation sign
σ	Breaking point, maximum voltage
τ	Time, pulse duration, heating time (τ_H)
γ	Angle, potential
φ	Function
Ψ	Generalized phase
ψ	Angle
ω	Circular frequency

Part I
Experimental Investigation of the Properties of Oscillating Heteromagnetic Structures at Low, Medium, and High Power Levels

The results of our experimental investigation of the properties of various FSS at low (mW), medium (up to 100 mW), and high (W) power levels are given. Generalized models of the phenomena, the properties of structures with FMCR of various types, and various orientations in an external magnetic field are discussed.

Chapter 1
Spectra of Regular and Noise Signals

1.1 General Remarks: Generalized Models

Heteromagnetic microelectronics or the magnetoelectronics of active microsystems investigates multifunction interactions in FSS, devices, MC, VLSIC, containing one or several FMCR[1] in their saturated (single-domain) or unsaturated (multi-domain) states in the interelectrode gaps of a transistor or diode, or in the gaps between the main electrodes (which are energized) and additional electrodes [23].

Such an integrated magneto-semiconductor device can contain a part of a magnetotransistor as a control element and provides formation of various signals, control of their power, and spectral characteristics [23]. The signals are taken from the main or additional electrodes on the frequencies of FMCR resonance, harmonic, or subharmonic components. FMCR can have various magnetic parameters, can be homogeneous volume or film ones, multilayer epitaxial film ones with specified laws of the transverse and longitudinal gradients of their magnetic parameters, can have various resonant frequencies, and be in linear or nonlinear states. The resonant FMCR frequencies and their modulation spectra are determined by several parameters: saturation magnetization $4\pi M_s$; fields of crystallographic anisotropy of the first – H_{A1} and second – H_{A2} orders; the shape and values of the sample and its demagnetization factors $\overleftrightarrow{N} M_S$ and $\overleftrightarrow{N}_{A1,2}\ H_{A1,2}$ ($\overleftrightarrow{N}$ and $N_{A1,2}$ are the tensors of demagnetization factors); value and direction of a bias field – $\overline{H}_0$; gradients of transverse (along the axis $0X$) magnetization over FMCR thickness – $\Delta_x M_{si}$ ($i = 1, 2 \ldots$ – the number of a ferrite layer with homogeneous magnetization); the transverse gradients of the anisotropy fields – $\Delta_x H_{A1}$ and $\Delta_x H_{A2}$; the gradients of magnetization and anisotropy fields on the FMCR area – $\Delta_{y,z} M_{si}$ and $\Delta_{y,z} H_{A1,2}$, and the half-width of the ferromagnetic resonance line –ΔH, the threshold high-frequency power level – h_{thr}, modulation characteristics of high-frequency magnetic fields – $\tilde{h}_{\text{mod}}$ and high-frequency magnetization – $\tilde{m}$, their values in relation to magnetization of the FMCR material and the internal (effective) magnetic field

[1] The diameter of YIG spheres is 0.4–0.5 mm, the thickness of ferrite films is 10–45 = μm, and the dimensions of film structures are about $1.5 \times 1.5\,\text{mm}^2$, the domain size is about $10\,\mu$m.

A.A. Ignatiev and A.V. Lyashenko, *Heteromagnetic Microelectronics: Microsystems of Active Type*, DOI 10.1007/978-1-4419-6002-3_1,

$\overline{H}_{0i}$ ($\overline{H}_0$, $4\pi M_s$, $H_{A1,2}$ $\overleftrightarrow{N} M_S$, $\overleftrightarrow{N}_A$ $H_{A1,2}$), the kind of polarization – (linear, circular, elliptic), and vector orientation of high-frequency $\bar{\bar{h}}$ and constant $\overline{H}_0$ magnetic fields.

FMCR are placed in the area(s) with localization of high-frequency magnetic fields. FMCR plays the role of a multifunctional, multicircuit, multiconnected, nonlinear element, which parameters are controlled in the linear mode by the bias field, and in nonlinear modes by the level of a high-frequency magnetic field (power) developed in the device, or by the level of input or output signals. Control of the power and spectral characteristics is carried out on both the central frequency and the frequencies of harmonic and subharmonic components. For uniaxial ferrites the number of characteristic FMCR frequencies is five, for antiferromagnetics their number is even more [17–21].

Physical processes in such a device are determined by the tensor of high-frequency magnetic conductivity and its components: in the saturated (equilibrium) tensor state – $\overleftrightarrow{\mu}$; in the unsaturated (nonequilibrium) state – $\overleftrightarrow{\mu}_{\mathrm{n}}$; in the transition state – $\overleftrightarrow{\mu}_{\mathrm{t}}$.

The particular design of a device and its functional applicability (generator, amplifier, converter, mixer, oscillator, superheterodyne, frequency synthesizer, sensor of the magnetic-field vector, sensor of the displacement vector, or other vector physical value: speed, acceleration, force, pulse, momentum of force, moment of pulse, noninertial forces) determine the FMCR location in the semiconductor, circuits of internal positive or negative feedbacks, circuits of input and output signal filtration, control of the bias field by the impedance of magnetoactive transitions, input, output of the device – active and reactive components within the limits of one–two orders of magnitude.

In Figs. 1.1–1.21, some suggested multifunction integrated heteromagnetic devices: magnetotransistors, magnetodiodes, magnetolasers, magnetomixers, etc. are sketched.

Symbols $\rightleftarrows$, $\longleftrightarrow$ designate the nonlinear and linear FMCR modes and the place of their location. A bipolar magnetotransistor with FMCR in the nonlinear modes is shown in Fig. 1.1. Here and below transistors with main electrodes (emitter – E, base – B, collector – C) and additional ones for connection with FMCR (E_f, B_f, and C_f) are given. FMCR are located on the main electrodes and in the

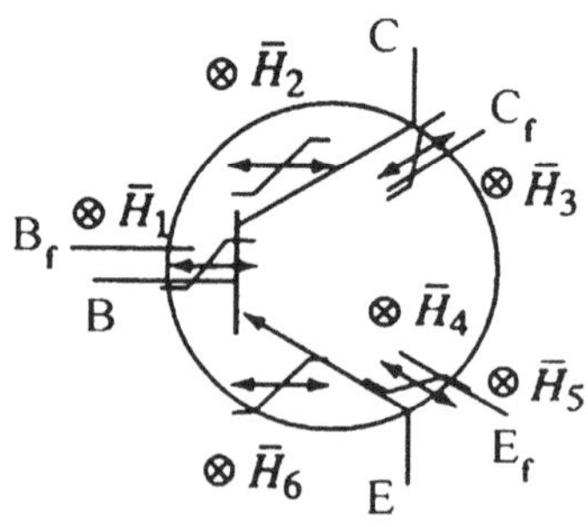

Fig. 1.1 A bipolar magnetotransistor with FMCR in the nonlinear mode

interelectrode gaps and are in various bias fields at the ends of the main electrodes ($\overline{H}_1$, $\overline{H}_3$, $\overline{H}_5$) and interelectrode gaps ($\overline{H}_2$, $\overline{H}_4$, $\overline{H}_6$). Figure 1.2 presents a bipolar magnetotransistor with FMCR in the nonlinear modes with a general bias field $\overline{H}_0$. Figure 1.3 shows a bipolar magnetotransistor with FMCR in the nonlinear self-resonance modes [17, 37]. Figure 1.4 shows a bipolar magnetotransistor with FMCR in the linear modes and various bias fields at the ends of the main electrodes ($\overline{H}_1, \overline{H}_3, \overline{H}_5$) and interelectrode gaps ($\overline{H}_2, \overline{H}_4, \overline{H}_6$). Figure 1.5 shows a bipolar magnetotransistor with FMCR in the linear modes with a common bias field $\overline{H}_0$. Figure 1.6 shows a bipolar magnetotransistor with FMCR in the linear self-resonance modes.

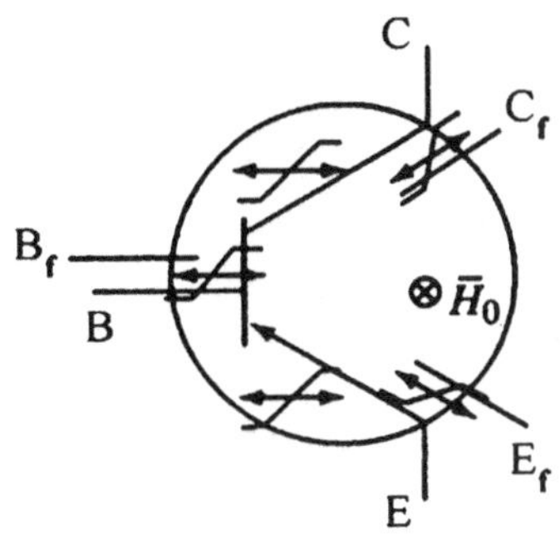

Fig. 1.2 A bipolar magnetotransistor with FMCR in the nonlinear modes with a general bias field $\overline{H}_0$

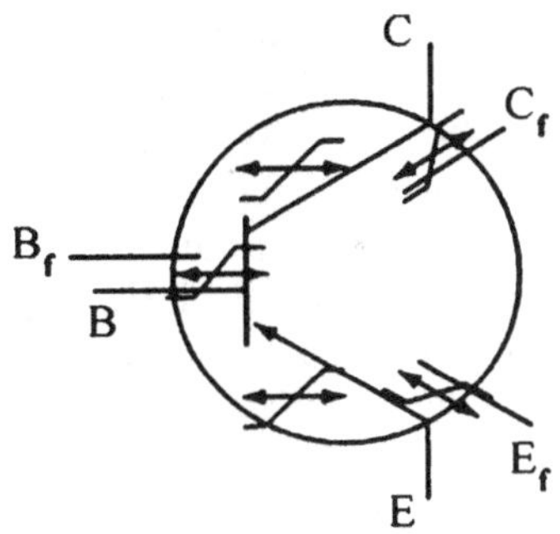

Fig. 1.3 A bipolar magnetotransistor with FMCR in the nonlinear self-resonance mode

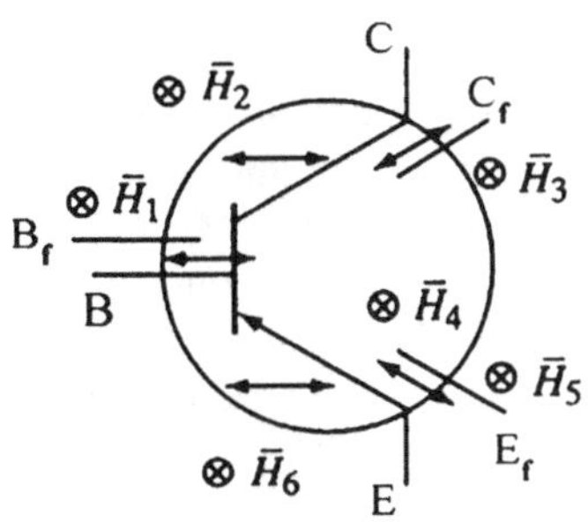

Fig. 1.4 A bipolar magnetotransistor with FMCR in the linear mode and various bias fields at the ends of the main electrodes ($\overline{H}_1$, $\overline{H}_3$, $\overline{H}_5$) and interelectrode gaps ($\overline{H}_2$, $\overline{H}_4$, $\overline{H}_6$)

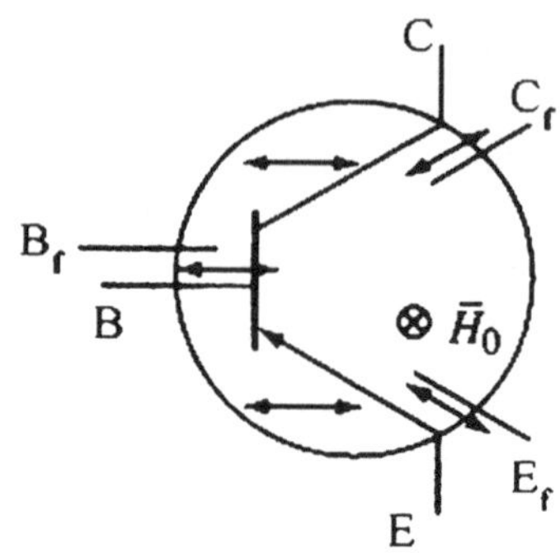

Fig. 1.5 A bipolar magnetotransistor with FMCR in the linear modes with a common bias field $\overline{H}_0$

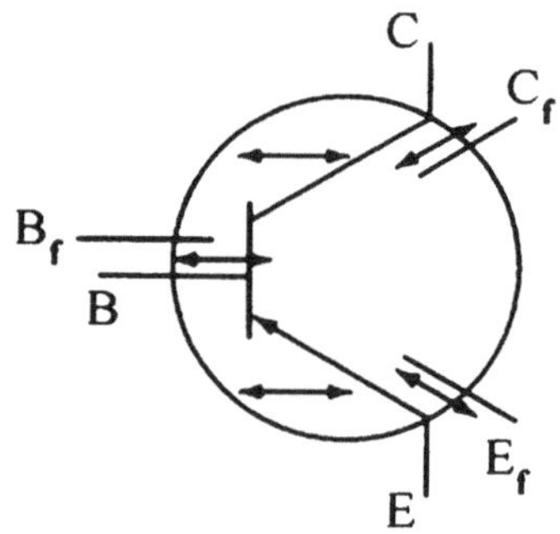

Fig. 1.6 A bipolar magnetotransistor with FMCR in the linear self-resonance mode

Figure 1.7 shows a field transistor with FMCR in the nonlinear modes (the main electrodes: source – S, shutter – Sh, drain – D, and additional electrodes S_f, Sh_f, D_f). FMCR are located in a transistor and in various bias fields at the ends of the main electrodes ($\overline{H_2}, \overline{H_3}, \overline{H_5}$) and interelectrode gaps ($\overline{H_1}, \overline{H_4}, \overline{H_6}$). Figure 1.8 shows a field magnetotransistor with FMCR in the nonlinear modes and a common bias field $\overline{H}_0$. Figure 1.9 shows a field magnetotransistor with FMCR in the nonlinear self-resonance modes. Figure 1.10 shows a field magnetotransistor with FMCR in the linear modes and various bias fields at the ends of the main electrodes ($\overline{H_2}, \overline{H_3}, \overline{H_5}$) and interelectrode gaps ($\overline{H_1}, \overline{H_4}, \overline{H_6}$). Figure 1.11 shows a field magnetotransistor with FMCR in the linear modes and a common bias field $\overline{H}_0$. Figure 1.12 shows a field magnetotransistor with FMCR in the linear self-resonance modes.

In Figs. 1.13–1.15, you see a Hann magnetodiode or magneto-avalanche diode (AD). A magneto-semiconductor laser is shown in Figs. 1.16–1.18. Figures 1.19–1.21 show a magnetomixer. In Figs. 1.13–1.21, "OC_f" means an ohmic contact for connection with FMCR. Figure 1.22 shows a multifunction oscillator on a bipolar transistor with a grounded base, with FMCR included into the emitter-base transition, in the linear mode with the magnetic field H_0 control, i.e., an MCO.

In Figs. 1.22–1.24, sketches of some controlled devices on bipolar magnetotransistors with FMCR in various modes are shown: p – the transient mode (PиC.1.25) and n – the unsaturated one (Fig. 1.24).

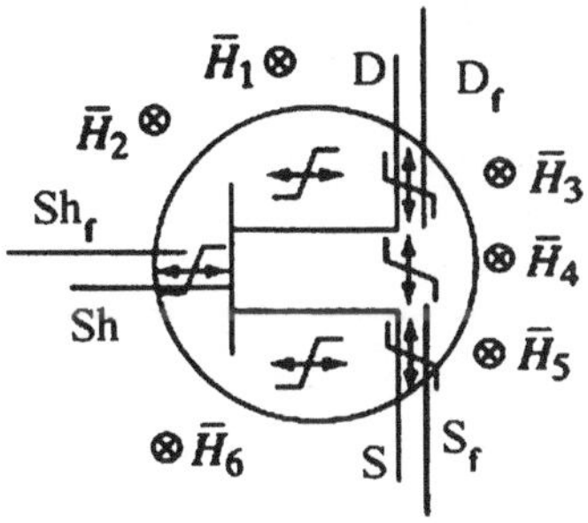

Fig. 1.7 A field transistor with FMCR in the nonlinear mode

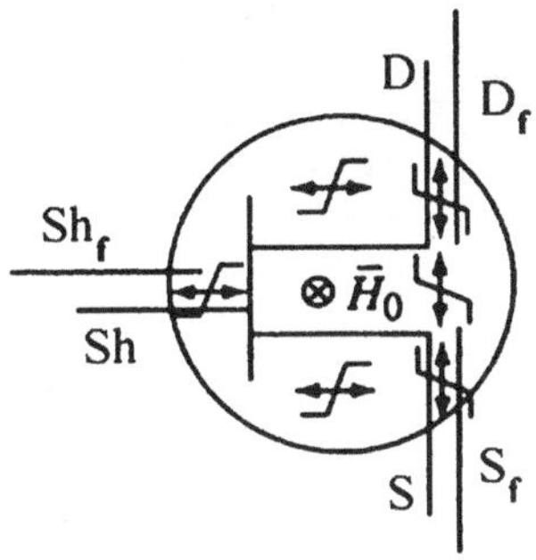

Fig. 1.8 A field magnetotransistor with FMCR in the nonlinear mode and a common bias field $\overline{H}_0$

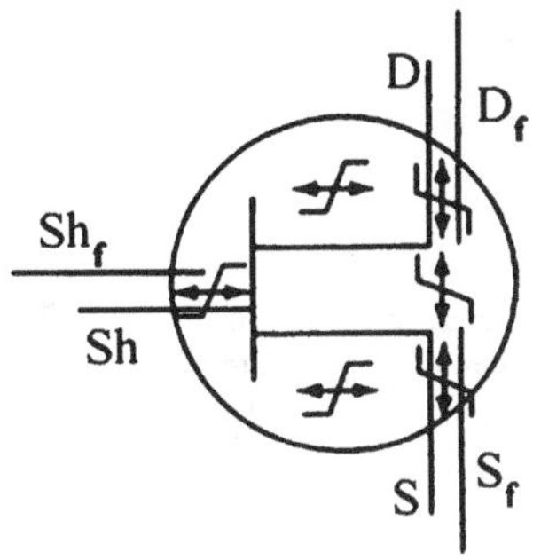

Fig. 1.9 A field magnetotransistor with FMCR in the nonlinear self-resonance mode

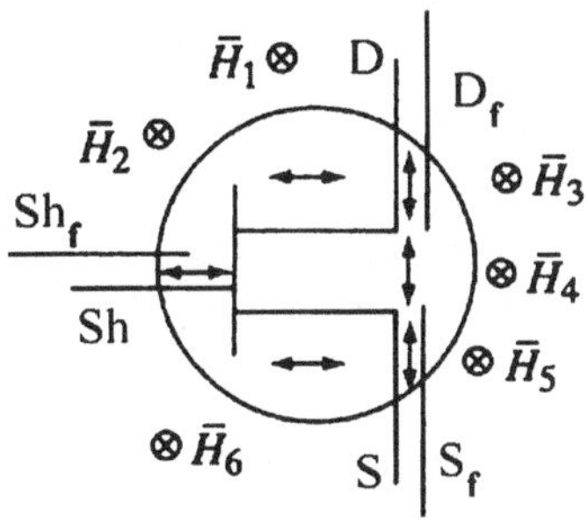

Fig. 1.10 A field magnetotransistor with FMCR in the linear mode and various bias fields at the ends of the main electrodes ($\overline{H_2}$, $\overline{H_3}$, $\overline{H_5}$) and interelectrode gaps ($\overline{H_1}$, $\overline{H_4}$, $\overline{H_6}$)

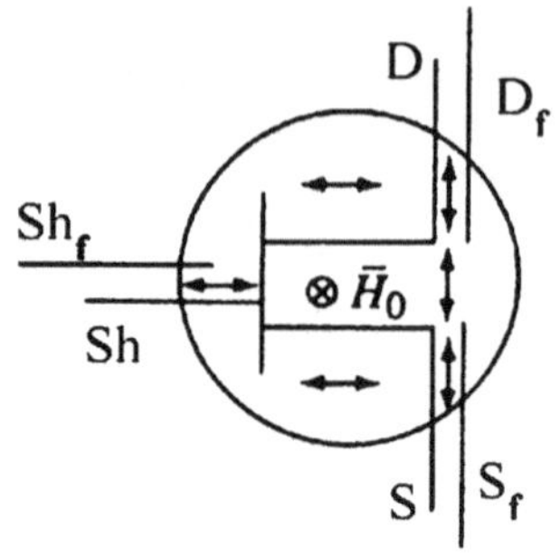

Fig. 1.11 A field magnetotransistor with FMCR in the linear mode and a common bias field $\overline{H}_0$

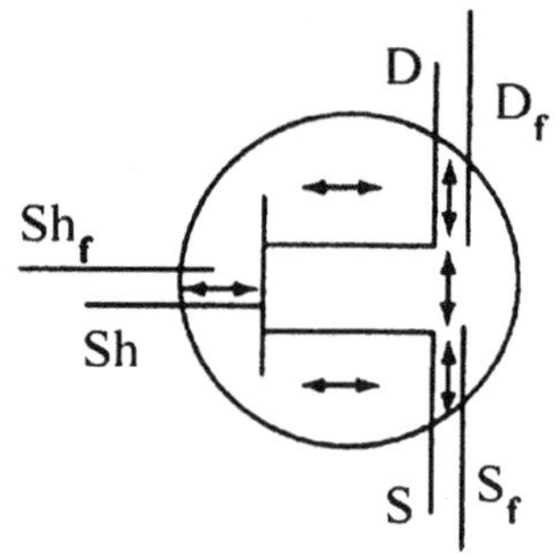

Fig. 1.12 A field magnetotransistor with FMCR in the linear self-resonance mode

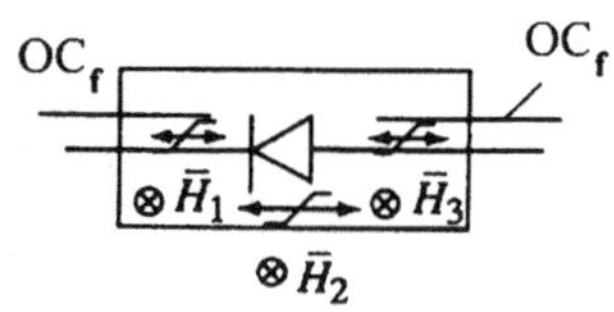

Fig. 1.13 A Hann magnetodiode or magneto-avalanche diode (AD)

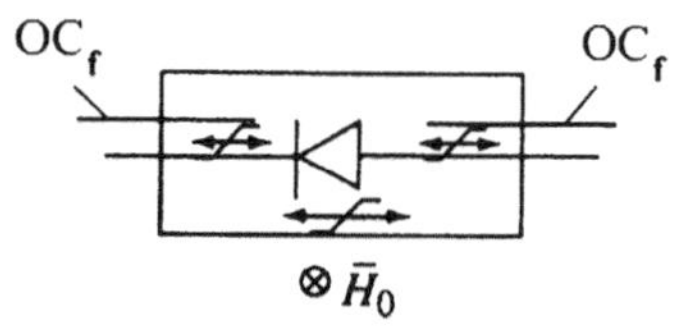

Fig. 1.14 A Hann magnetodiode or magneto-avalanche diode (AD)

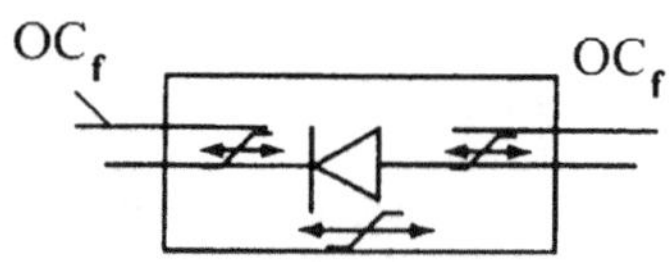

Fig. 1.15 A Hann magnetodiode or magneto-avalanche diode (AD)

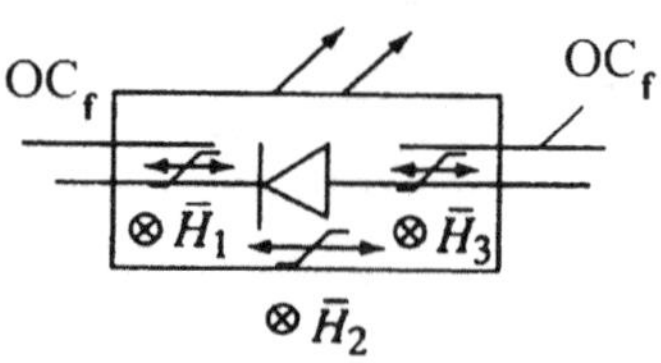

Fig. 1.16 A magneto-semiconductor laser

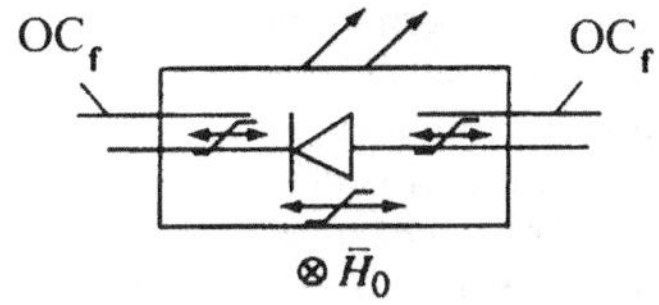

Fig. 1.17 A magneto-semiconductor laser

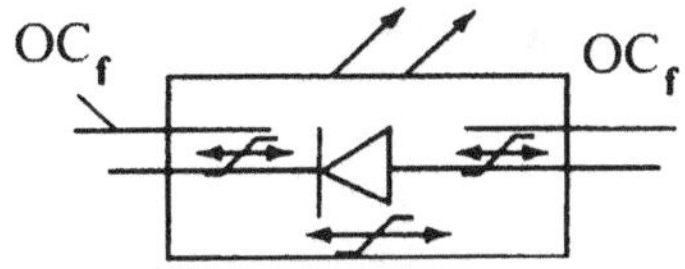

Fig. 1.18 A magneto-semiconductor laser

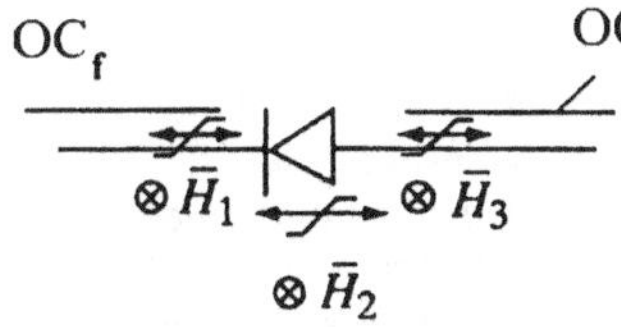

Fig. 1.19 A magnetomixer

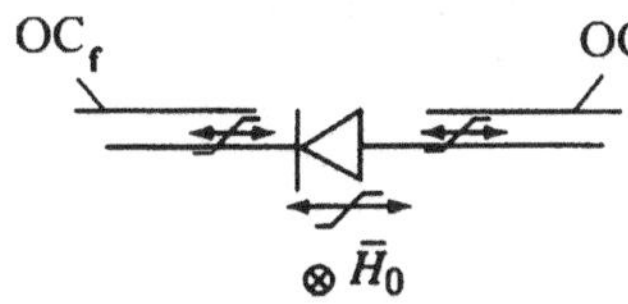

Fig. 1.20 A magnetomixer

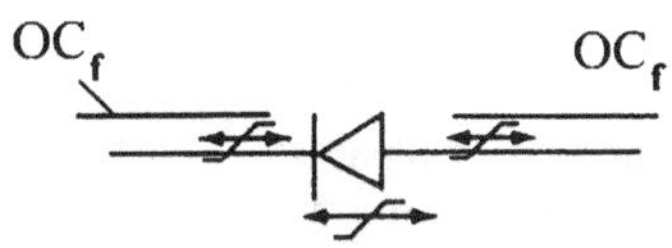

Fig. 1.21 A magnetomixer

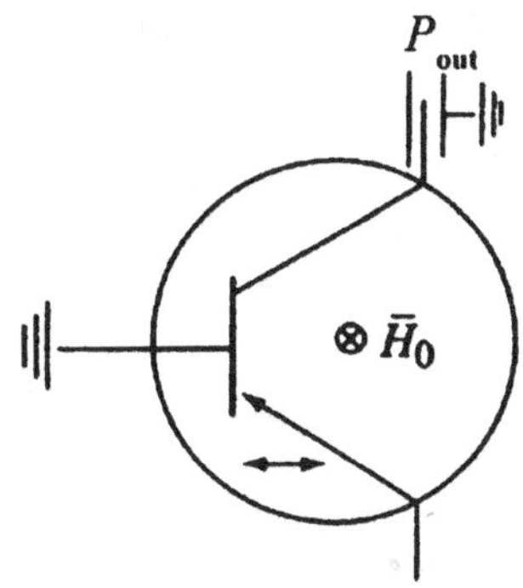

Fig. 1.22 An oscillator on a bipolar magnetotransistor with a common FMCR base in the saturated linear mode, controlled by a magnetic field $\overline{H}_0$

In Fig 1.22, an oscillator on a bipolar magnetotransistor with a common FMCR base in the saturated linear mode, controlled by a magnetic field $\overline{H}_0$, is shown.

In Fig 1.23, a multifunction frequency synthesizer (an equidistant frequency spectrum oscillator) on a bipolar magnetotransistor with FMCR in the nonlinear, transient mode with controlled parameters (equidistance, the central frequencies and phases of all the harmonic and subharmonic components, control of the noise level of spectral components and transition into the pseudonoise and noise signals with a nonuniform and uniform (white noise) spectral density of noise power) is shown.

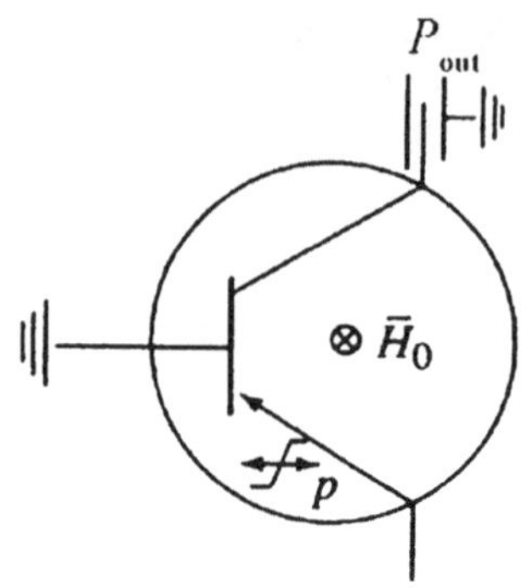

Fig. 1.23 A multifunction frequency synthesizer (an equidistant frequency spectrum oscillator) on a bipolar magnetotransistor with FMCR in the nonlinear, transient mode with controlled parameters

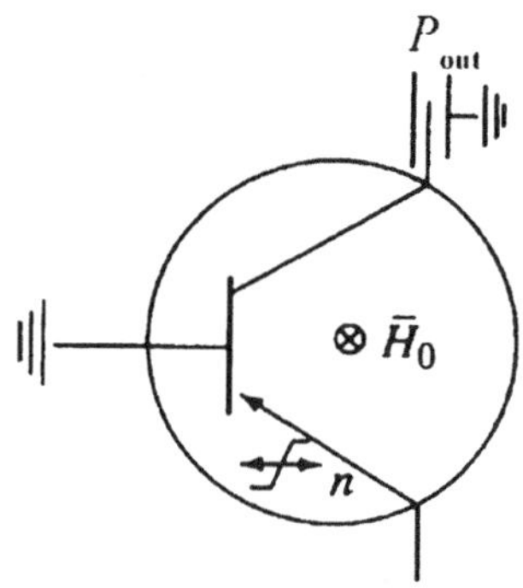

Fig. 1.24 A bipolar magnetotransistor with FMCR in the interelectrode emitter-base space in the nonlinear, unsaturated mode – a multiplier and divider of frequency

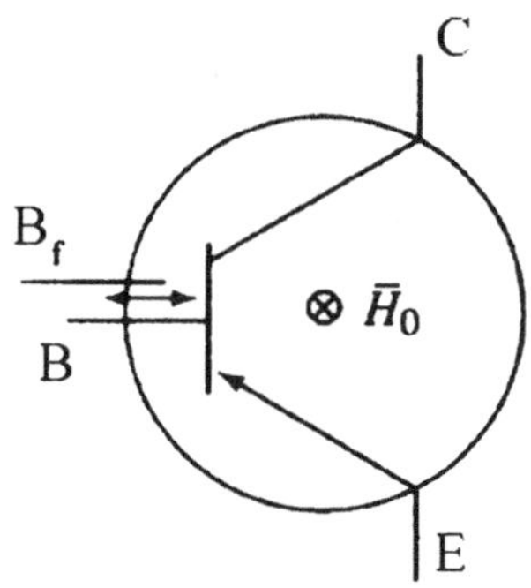

Fig. 1.25 A bipolar magnetotransistor with controlled selectivity and a reduced input noise factor

Figure 1.24 shows a bipolar magnetotransistor with FMCR in the interelectrode emitter-base space in the nonlinear, unsaturated mode – a multiplier and divider of frequency generated by the device and transformable around some reference frequency, with control of the central frequencies and phases by a bias field $\overline{H}_0$.

Figure 1.25 shows a bipolar magnetotransistor with controlled selectivity and a reduced input noise factor. Figure 1.26 shows a bipolar magnetotransistor with controlled output selectivity. Figure 1.27 shows a powerful bipolar magnetotransistor with fine tuning of FMCR by magnetic fields $\overline{H_1}, \overline{H_2}, \overline{H_3}$ of impedances on all the electrodes in the linear saturated mode.

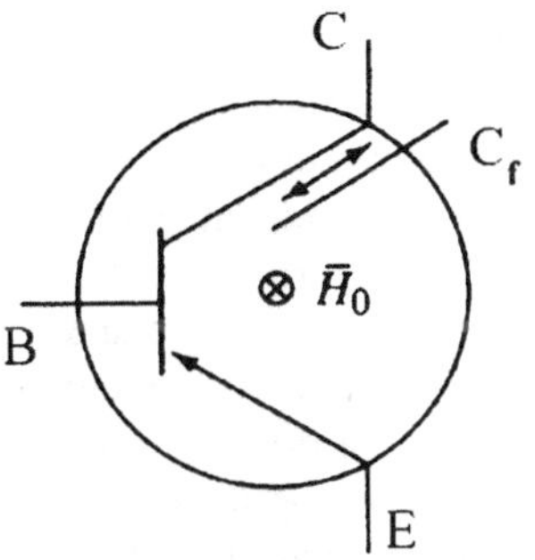

Fig. 1.26 A bipolar magnetotransistor with controlled output selectivity

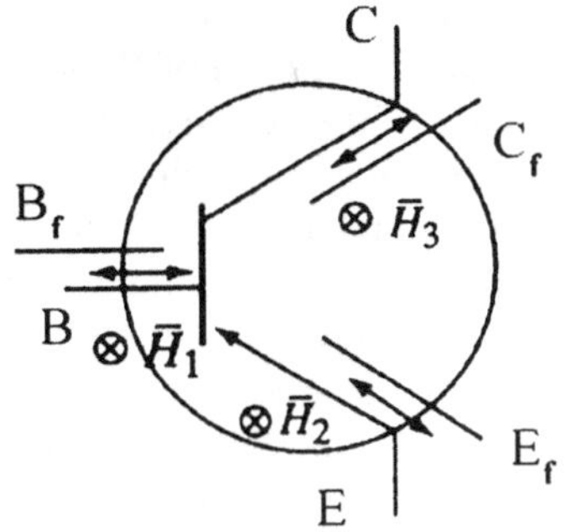

Fig. 1.27 A powerful bipolar magnetotransistor with fine tuning of FMCR by magnetic fields $\overline{H_1}, \overline{H_2}, \overline{H_3}$ of impedances on all the electrodes in the linear saturated mode

The basic advantages of the suggested heteromagnetic small-size devices and microsystems in comparison with the known ones are as follows:

- The usage of one element base, one CHIP to form signals and spectra of the final level of various kinds (regular, pseudonoise, noise ones, as multifunction frequency synthesizers do in the oscillator modes)
- The design of amplifiers with controlled selectivity and a reduced noise factor
- The design of miniature and subminiature vector magnetometric sensors and gyromagnetic microsystems
- The design of miniature vector magnetic sensors of mechanical vector static and dynamic quantities of forward, rotary, and complex movements
- Registration of the magnetic induction vector and its deviations by amplitude and frequency
- Control of the total impedance of devices
- A high technical efficiency
- Simplification of the design and miniaturization of devices
- Multichannel control of the power, spectral, and noise characteristics, the central frequencies and phases, passband of signals by feed of the transistor subsystem, the magnetic subsystem and dynamic control of the level of the high-frequency input power brought to the device, the high-frequency power level in the device and on its output
- Signal formation on the reference frequency, harmonics, and subharmonics and their synchronous frequency and phase changes
- Formation of the direct amplification modes and the signal superheterodyning modes

With the known, traditional engineering solutions and ways of device design in the microwave range many of the specified requirements and properties are contradictory and mutually exclude each other. For example, formation of signals of the final shape, enhancement of multifunctionality and opportunities of the application of one CHIP as various microsystems and devices, mass-dimension reduction, decrease of power consumption and increase of technical efficiency, increase of endurance and reliability, improvement of manufacturability and decrease of environment load, depreciation down to the record values, and expansion of the totality of parameters, properties, application ranges.

Now transistor oscillators are assembled under various schemes containing one or several transistors, in which one or several dielectric [24] and ferrite [25] resonators in the form of YIG spheres included in the external (outside the transistor) circuits of the devices are applied as frequency-stabilizing elements.

Control of noise signal parameters in a transistor oscillator by means of application of dynamic chaos in the external circuit (outside the transistor) of feedback of FMCR (filter) in the nonlinear mode is examined in [26, 52]. Under this scheme a noise generator in the microwave range has been created, in which a YIG filter in the saturated (single-domain) state is used.

Ferrite microresonators were also used in such complex integrated devices as microwave frequency synthesizers, in which YIG filters in the saturated (single-domain) linear states [27, 28] were used for their designated purpose.

Complex integrated devices (high-frequency amplifiers, oscillators, pulse microwave transmitters) contain various devices: transistors, dielectric resonators, a topology of microstrip lines, capacities, and resonators [29–32, 44].

For registration of the induction vector of constant and variable magnetic fields, magnetosensitive devices [33, 34] and magnetotransistors [35, 36] are used.

For formation of various signals in the active modes, various semiconductor and microelectronic circuits, including MC and VLSIC, and devices made by the monolithic technology in the radio and optical ranges are traditionally used. The high level of integration allows a large number of various elements to be placed into small volumes, signals of various complexity levels to be formed and processed in real time. The basic part of the active devices of various power levels and frequency ranges (active oscillators, amplifiers, receivers, transmitters, location stations, frequency synthesizers, vector sensors, etc.) consists of a large number of component elements with point-to-point wiring, and planar elements in integrated circuits as well. These devices, modules are rather bulky and labor-intensive. Their endurance is limited. Their mass-dimensional parameters and cost are rather significant and cannot be reduced by some orders of magnitude. The technical efficiency of the modules of the devices on the known elements and accessories is either negligibly low or limited and cannot be essentially increased, and their power consumption cannot be lowered.

At transition to the microelectronic element base, MC, VLSIC, and microcontrollers, microprocessors the mass – dimensions of the devices and modules sharply decrease but the cost and labor-intensity remain high. Besides, reduction of environmental contamination and the CO_2 emission level in the atmosphere are problematic.

The typical microelectronic multifunction heteromagnetic device contains a transistor, a diode with negative resistance, and a ferrite microresonator at least [23].

On one of the CVS surfaces or under the surface of conductive conductors, coverings, contacts, or on the boundary of layers with various conductivity, or inside a layer (layers) in the area or several areas with high-frequency magnetic fields one or several FMCR with identical or various sizes and shapes, magnetic parameters, which can vary over thickness and area under certain laws, have intended gradients of their magnetic parameters (magnetization, dissipation, anisotropy fields, etc.) over thickness, surface, or in the volume of the microresonator (s), which are in one or various states, namely, the domain (unsaturated), transient, or single-domain (saturated) ones are introduced.

The microresonator can have conductivity of carriers of this or that kind (sign). In the device, control of the central frequency (phase) of input and output signals, their power, spectral, noise parameters, and characteristics is realized by choice of the value and direction of the constant component of the bias field concerning the crystallographic axes of the microresonator, and its variable component – the frequency, phase, kind of polarization of a high-frequency magnetic field for input (external) signals, internal signals, their intensity, the frequency and parameters of modulation of a high-frequency field, and voltages and currents applied to the semiconductor part of the device (the transistor, diode, laser).

FMCR as an oscillatory circuit at increased power levels are described in [37,38].

In all these devices FMCR were included into their external circuits.

The most multifunction, multiparameter modes of FMCR are provided by change of the bias field or of the HF magnetic field level (power).

The known types and structures of ferrites allow the frequency range from the radio-wave one (on ferro- and ferrimagnetics – yttrium-iron garnets $Y_3Fe_5O_5$, spinels $MgAl_2O_3$ barium hexaferrite $Ba_3Fe_2O_5$) up to the optical one (on antiferromagnetics: αFe_2O_3 – hematite, $FeBO_3$ – ferric borate, $NiCO_3$ – nickel carbonate) to be covered. The presence of nonlinear effects depends on the level of the threshold high-frequency power (about 0.5–1.0 mW on a frequency of 1 GHz) and determines the modes of parametrical multiplication, division, frequency modulation in FMCR. Besides, FMCR can be made of materials with various magnetic parameters. New kinds of structures, namely, epitaxial films, including multilayer ones with intended transverse gradients of the magnetic parameters (saturation magnetizations), provide a decrease of the resonator size down to 5–50 μm and expand the functionalities of HMT.

FMCR is usually included into the antinode of high-frequency magnetic fields of this or that topology of microstrip lines, of one or several loop coils, in the interelectrode spaces of FSS, exciting oscillations of the magnetization vector. The parameters and characteristics of FMCR, its radiation resistance $Z = R + jB$, are controlled by the bias field and HF power [1].

At change of the bias field in FMCR in the linear, equilibrium, single-domain states, a single-circuit interaction (in terms of the equivalent circuit) is realized in FSS and the central frequencies are changed by the magnetic field or by voltages of the transistor or diode feed.

At nonlinear, non-uniform multidomain states, multicircuit interactions in the FMCR, and processes of parametric multiplication and division, parametric frequency modulation of the reference frequency signal are realized. This provides the formation of deterministic signals like those generated by multifunction frequency synthesizers, stochastic signals – (narrow-band and broadband pseudonoise signals), and white noise.

In Fig. 1.28 ($\vec{M}_1$, $\vec{M}_2$ are the magnetizations of ferrites, $\overleftrightarrow{\mu}_1$, $\overleftrightarrow{\mu}_2$ the tensors of magnetic conductivity, $\overline{H}_0$ the external magnetic field), several variants of heteromagnetic transistors of n–p–n type with FMCR are schematically presented: (a) as spheres; (b) hemispheres; and (c) a film placed in the n–p transition. In Fig. 1.29, heteromagnetic diodes with FMCR are shown: (a) as a hemisphere and (b) as a film in the transition.

In Fig. 1.30, FMCR of the passage type, which can play the role of a multifunction controlled transformer of impedances (generally, a nonlinear one), is shown.

In Fig. 1.31, a variant of the bipolar magnetotransistor with possible inclusions of FMCR on its input and output is given. Figure 1.32 shows a field magnetotransistor.

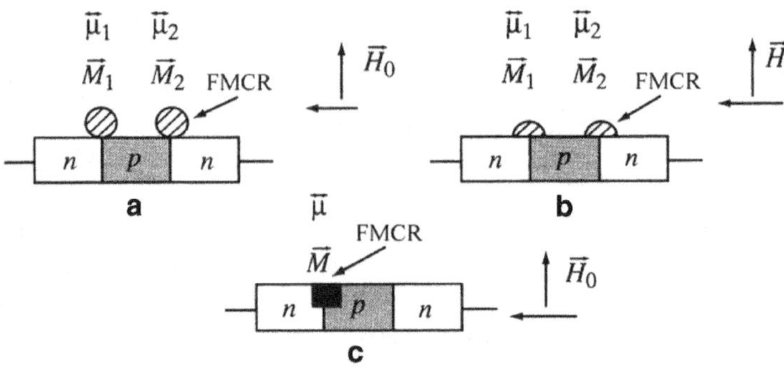

Fig. 1.28 Heteromagnetic transistors of n–p–n type with FMCR are schematically presented: (**a**) as spheres; (**b**) hemispheres; (**c**) a film placed in the n–p transition

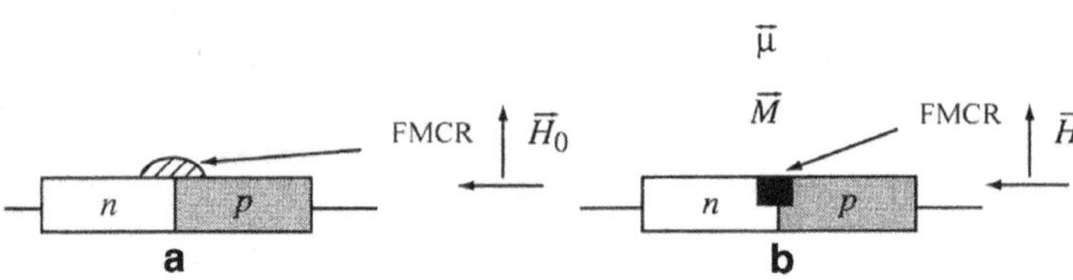

Fig. 1.29 Heteromagnetic diodes with FMCR are shown: (**a**) as a hemisphere; (**b**) as a film in the transition

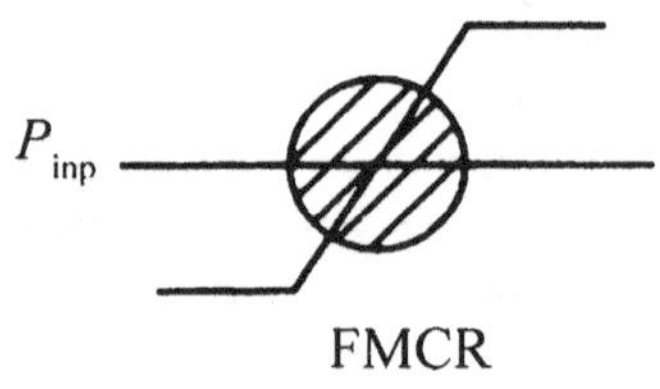

Fig. 1.30 FMCR of the transition type

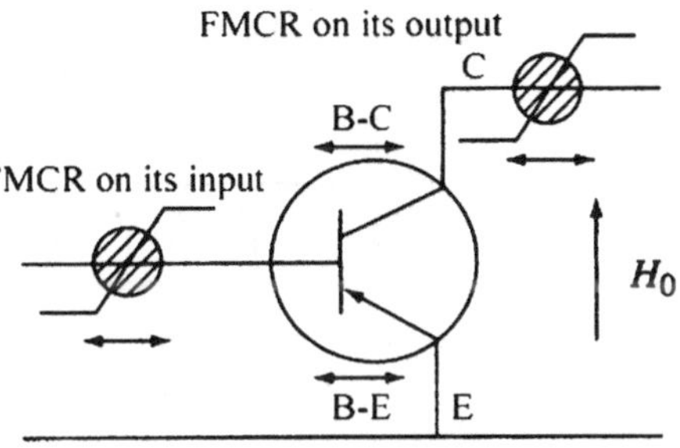

Fig. 1.31 A variant of the bipolar magnetotransistor with possible inclusions of FMCR on its input and output

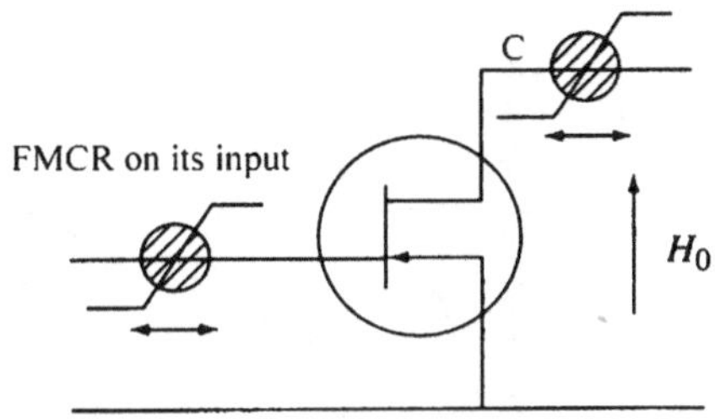

Fig. 1.32 A field magnetotransistor

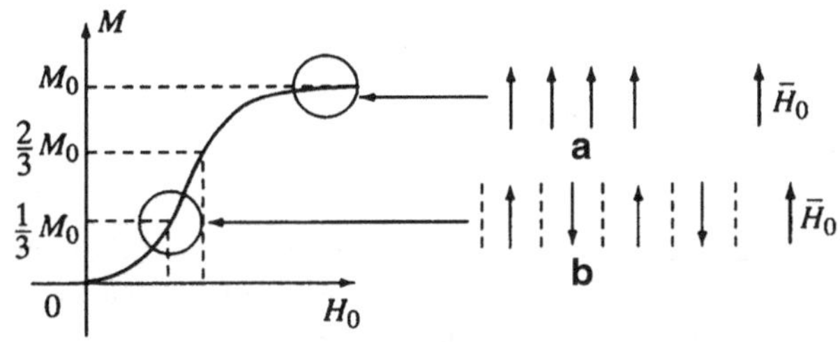

Fig. 1.33 The dependence of the ferrite magnetization M on the applied field H_0, and schematic images of its equilibrium saturated state ((**a**) single-domain) and unsaturated ((**b**) multidomain) states

In Fig. 1.33, the dependence of the ferrite magnetization M on the applied field H_0, and schematic images of its equilibrium saturated state ((a) single-domain) and unsaturated ((b) multidomain) states are presented.

In Fig. 1.34, the characteristic frequencies of an unsaturated ferrite of the cubic crystallographic structure are given: $\nu_{\mathrm{p}1,2}$ are the frequencies along the domain boundaries, $\nu_{\mathrm{t}1,2}$ the frequencies across the domain boundaries, and ν_{d} is the oscillation frequency of the interdomain boundaries: $\nu_{\mathrm{d}} << \nu_{\mathrm{p}1,2}, \nu_{\mathrm{t}1,2}$. In FSS, the harmonic or pseudonoise components of oscillations can be kept with each of these intrinsic frequencies or with their whole set.

For the linear states of FMCR the inequalities $\overline{h} << \overline{H}_{0i}$; $\overline{m} << \overline{M}$ should be satisfied, where $\overline{h}$ is the HF magnetic component, $\overline{H}_{0i}$ the internal magnetic field, $\overline{m}$ the HF magnetization of the ferrite, and $\overline{M}$ is the saturation magnetization.

For the nonlinear states of FMCR it is necessary that $\overline{h} \gtrsim \overline{H}_{0i}$ or $\overline{m} >> \overline{M}$.

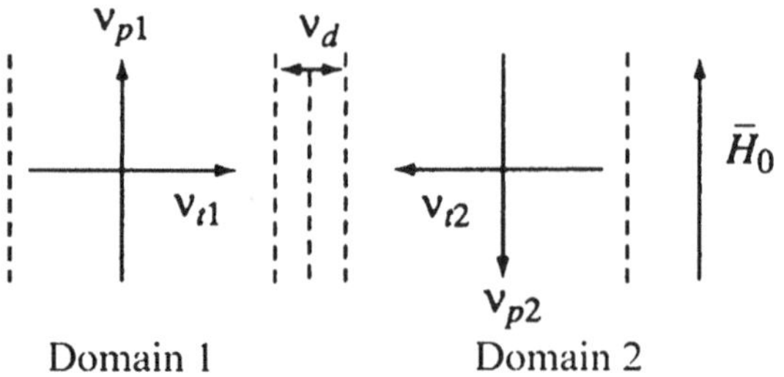

Fig. 1.34 Frequencies of an unsaturated ferrite of the cubic crystallographic structure are given: $\nu_{p1,2}$ are the frequencies along the domain boundaries; $\nu_{t1,2}$ the frequencies across the domain boundaries; and ν_d is the oscillation frequency of the interdomain boundaries: $\nu_d << \nu_{p1,2}, \nu_{t1,2}$

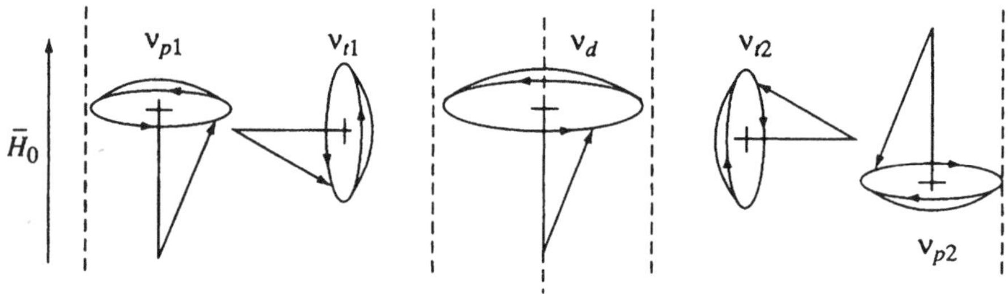

Fig. 1.35 A nonlinear model of an unsaturated ferrite

In Fig. 1.35, a nonlinear model of an unsaturated ferrite, when movement of the magnetization vectors occurs on the segments of the elliptic surfaces (the length of the vector and the speed of its movement change), is shown.

Generalized equivalent circuits of the field and bipolar magnetotransistors with FMCR, included into the corresponding output electrodes of the transistors' feed circuits, are given in Figs. 1.36 and 1.37, where the designations from [39–41] are used. FMCR in the output circuits are marked with the letters A, B, C, and in the interelectrode spaces – with figures 1, 2, 3 to connect transitions 1–3, 3–2, 1–2.

The circuit of the field magnetotransistor (Fig. 1.36) is made on the base Materk model, the circuit of the bipolar magnetotransistor (Fig. 1.37) is on the Humel–Poon model. Special cases of the particular magnetotransistors will be reflected in the corresponding arrangements of FMCR.

1.2 Regimes of Low and Middle Power Levels

In Fig. 1.38a, an oscillation FSS used for investigation of the physical mechanisms of multifunction heteromagnetic interactions on low and medium levels of continuous power in the VHF, UHF, microwave, and the near part of the EHF frequency

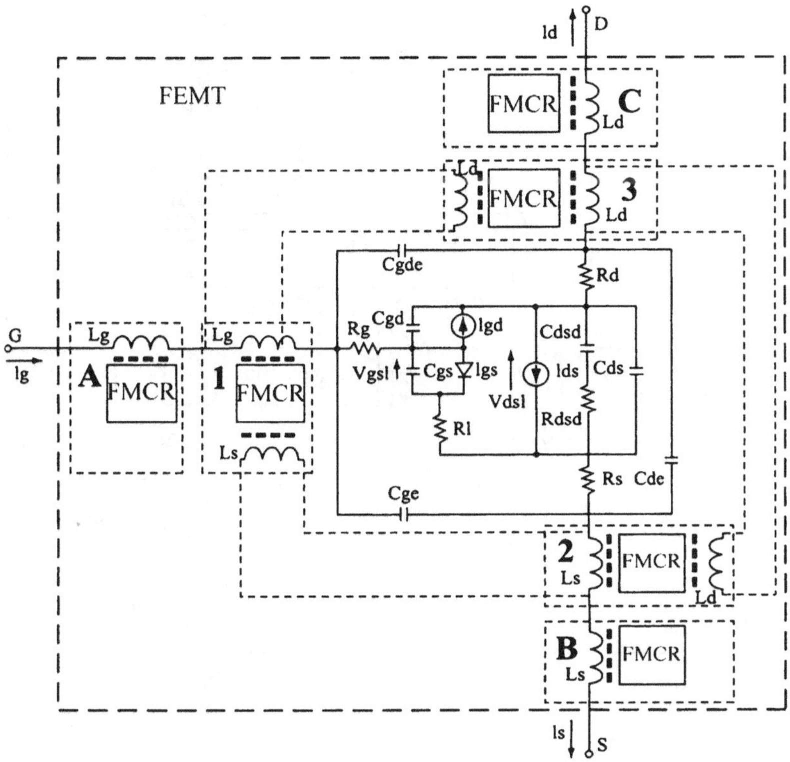

Fig. 1.36 The field magnetotransistors with FMCR

ranges (from 10 MHz up to 40 GHz) is schematically shown. The domain structure of ferrite with the characteristic frequencies ν_p, ν, and ν_d in one domain is shown as a fragment in Fig. 1.38b.

The basic kinds of signals formed by oscillation FSS in the modes of parametric multiplication and division[2] of the reference (fundamental) frequency signal (Fig. 1.39), parametric multiplication and parametric frequency modulation (Figs. 1.40 and 1.41), selective noisiness of separate areas in the spectrum (Fig. 1.42), selective increase of the power level (Fig. 1.43), and generation in the multioctave frequency range of superbroadband white noise (Fig. 1.44) are presented.

Let us examine in detail the multifunction modes. We shall introduce the following designations: harmonic components $\nu_{n-1} = n\nu_0, n = 1, 2, 3$ ($n = 1$, $\nu_0 = \nu_0$ – is the reference frequency; $n = 2$, $\nu_1 = 2\nu_0$ the first harmonic, etc.); subharmonic components $\nu_{-m+1} = \nu_0/m, m = 1, 2, 3, \ldots$ ($m = 1$, $\nu_0 = \nu_0$ – is the reference frequency; $m = 2$, $\nu_{-1} = \nu_0/2$ the first subharmonic, etc.).

[2] Parametric division of the signal of the fundamental frequency ν_0 was observed in the certain types, constructions of FSS in the operating modes.

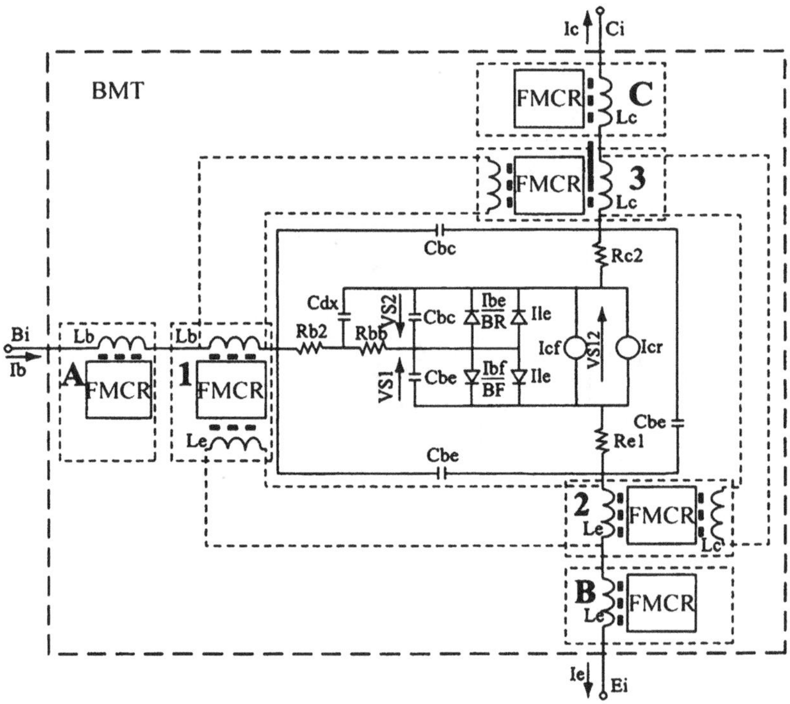

Fig. 1.37 The bipolar magnetotransistors with FMCR

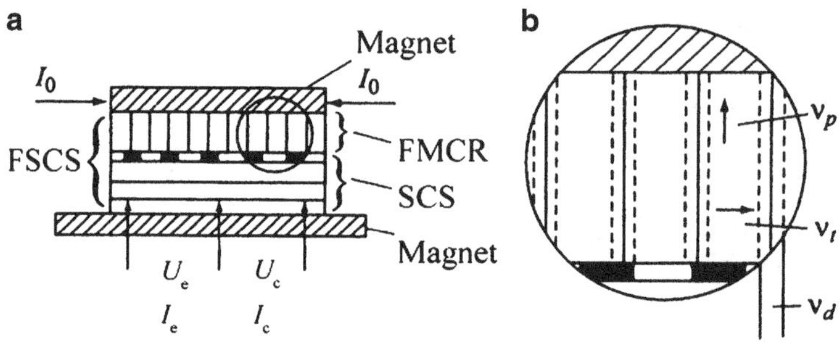

Fig. 1.38 (**a**) An oscillation FSS; (**b**) the domain structure of ferrite with the characteristic frequencies ν_p, ν, and ν_d in one domain

Figure 1.39 shows:

- A signal on the reference (fundamental) frequency $\nu_0 = 1$ GHz, for which the spectrum linewidth is $(\Delta\nu_{-3\text{dB}})_0$
- Signals of the harmonic components with $n = 21$ and $n = 41$, for which $(\Delta\nu_{-3\text{dB}})_0 = (\Delta\nu_{-3\text{dB}})_{20} = (\Delta\nu_{-3\text{dB}})_{40} = \text{const}$
- A signal of the subharmonic component at the subharmonic number $m = 71$, for which $(\Delta\nu_{-3\text{dB}})_0 = (\Delta\nu_{-3\text{dB}})_{-70} = \text{const}$.

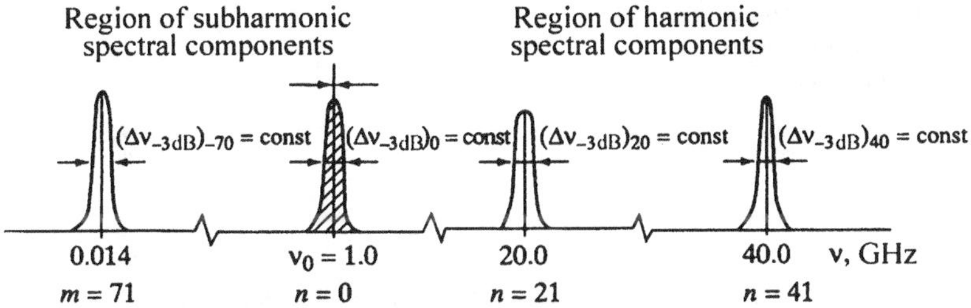

Fig. 1.39 The basic kinds of signals formed by oscillation FSS in the modes of parametric multiplication and division (parametric division of the signal of the fundamental frequency ν_0 was observed in the certain types, constructions of FSS in the operating modes) of the reference (fundamental) frequency signal

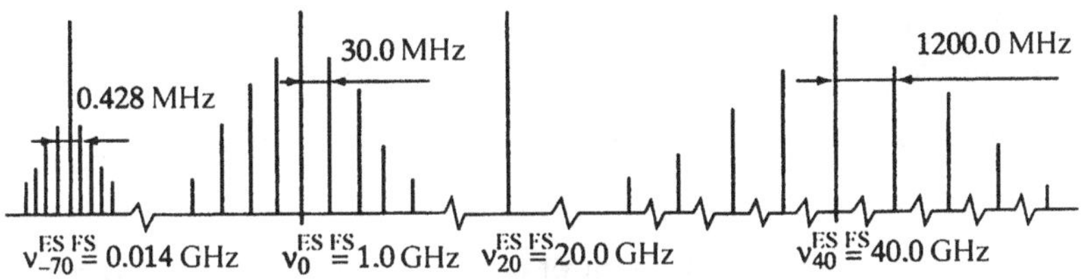

Fig. 1.40 Signal spectra in the mode of generation of equidistant frequency spectra

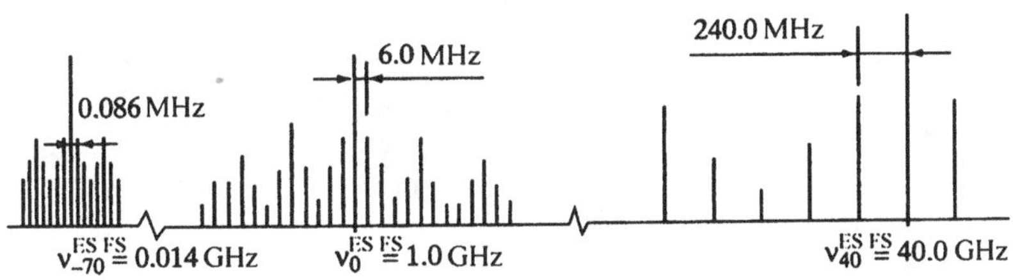

Fig. 1.41 Signal spectra in the mode of generation of equidistant frequency spectra

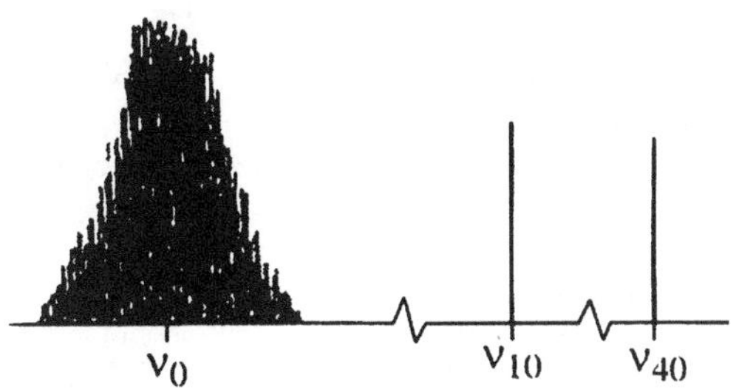

Fig. 1.42 Selective noisiness of the components in certain frequency regions

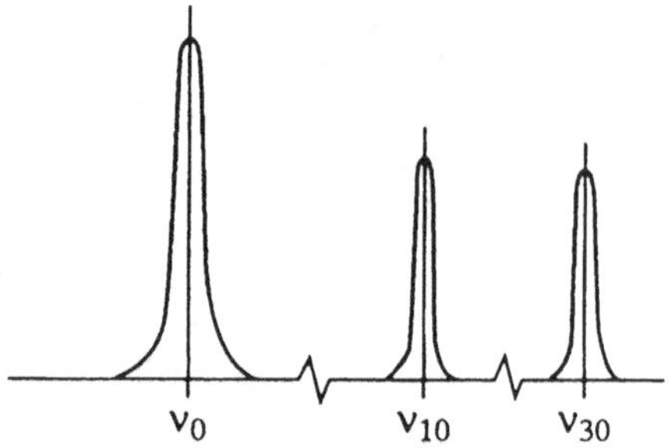

Fig. 1.43 Selective change of the power level of certain spectral components or their groups

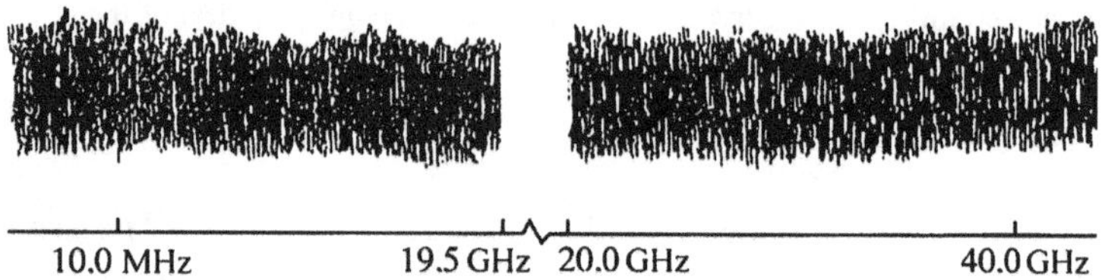

Fig. 1.44 A signal of FSS generation, close to white noise

Parametric multiplication and division of a signal[3] were observed experimentally at integral power levels 40–80 mW on a FSS made on the basis of a KT938A transistor and an FMCR with magnetization $4\pi M_s = 410$ G. In the region of the harmonic components of frequencies ($\nu > 1$ GHz) parametric multiplication was highly effective, right up to $\nu \geq 40$ GHz and for all the spectral components with $n = 2, 3, \ldots$ the amplitudes of signals did not vary essentially, and in the range of the subharmonic components ($\nu < 1$ GHz) at parametric division of the signal ν_0 the conversion losses were essentially superior and amounted to minus (40–45) dBmW of the harmonic component level. The latter is connected with experimental presence of a low-frequency filter in the pilot model of FSS. The processes were parametric ones with $\Delta\nu_{-3\text{dB}} = \text{const}$ for the harmonics with $n > 1$ and subharmonics with $m > 1$.

With change of the bias field H_0 the central frequencies (phase) of all the spectral components of parametric signals on ν_0, ν_{n-1}, and ν_{-m+1} were synchronously changed, and the change steepness by the magnetic field for harmonics was $\beta_{n-1} = \Delta\nu_{n-1}/\Delta H_0 = n\Delta\nu_0/\Delta H_0$, where $n = 1, 2, 3, \ldots$, and for subharmonics $\alpha_{-m+1} = \Delta\nu_{-m+1}/H_0 = \Delta\nu_0/mH_0$, where $m = 1, 2, 3, \ldots$, and $(\Delta\nu_{-3\text{dB}})_{n-1} \cong (\Delta\nu_{-3\text{dB}})_{-m+1} \cong (\Delta\nu_{-3\text{dB}})_0$.

Figures 1.40 and 1.41 show signal spectra in the mode of generation of equidistant frequency spectra[4] in the vicinity of each spectrum component (Fig. 1.39):

[3] Constancy of the width and form of the spectral components both for harmonics ($n = 2, 3, \ldots$) and subharmonics ($m = 2, 3, \ldots$) tells about the parametric character of the process of the signal processing of the reference frequency ν_0: $(\Delta\nu_{-3\text{dB}})_0 = (\Delta\nu_{-3\text{dB}})_{n-1} = (\Delta\nu_{-3\text{dB}})_{-m+1} = \text{const}$.

[4] With equal frequency distances between the spectral components.

$\nu_0^{\mathrm{ES\,FS}}$, $\nu_{n-1}^{\mathrm{ES\,FS}}$, $\nu_{-m+1}^{\mathrm{ES\,FS}}$, moreover the frequency distances for the harmonic components are $\Delta\nu_{n-1}^{\mathrm{ES\,FS}} = n\Delta\nu_0^{\mathrm{ES\,FS}}$, and for the subharmonic ones are $\Delta\nu_{-m+1}^{\mathrm{ES\,FS}} = \Delta\nu_0^{\mathrm{ES\,FS}}/m(\Delta\nu_0^{\mathrm{ES\,FS}}$ being the frequency distances between the spectral components in the vicinity of the reference frequency ν_0). At change of the magnetic field H_0 the groups of signals of the equidistant frequency spectra (Figs. 1.40 and 1.41) were changed synchronously with the steepness β_{m-1} or α_{-m+1}. At certain values of the field H_0 the equidistance of frequency distances changed stepwise, and in the frequency regions of each component (Fig. 1.40) splitting of the frequency distances into the spectral components of next orders (Fig. 1.41 where frequency spectra of the second order in the regions of the fundamental frequency ν_0^{s}, harmonic ν_{40}^{s} with number $n = 41$ and subharmonic ν_{-70}^{s} with number $m = 71$ are shown) was observed.

At change of the HF power level, equidistant spectra of higher orders (Fig. 1.41) were also observed.

In Figs. 1.42 and 1.43, oscillator operating modes of FSS are shown, where selective noisiness of the components in certain frequency regions (ν_0 in Fig. 1.42) and selective change of the power level of certain spectral components or their groups (Fig. 1.43) took place.

An interesting operating mode of oscillating FSS is the mode of noise generation with a uniform PSD in a wide multioctave frequency range. In Fig. 1.44, a signal of FSS generation, close to white noise, in the frequency band of 10 MHz–40 GHz at an integral power level of 4.5 mW is shown.

Diagrams in the power–frequency coordinates for some domestic and foreign types of transistors, structures, and heteromagnetic multifunction oscillators on their basis are given in Fig. 1.45.

In Fig. 1.46, a generalized dependence of the resonant frequencies ν_{res} on saturation magnetization $4\pi M_{\mathrm{s}}$ for various ferrites is shown: garnets (KG), spinels (KS), and barium hexaferrite (KB), which can be used in FSS.

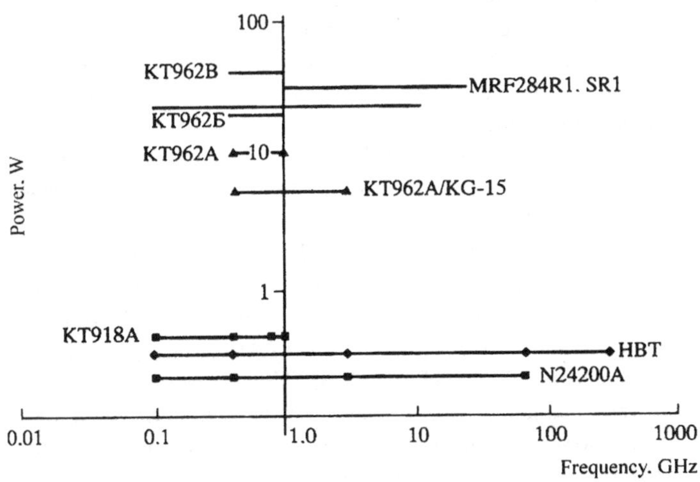

Fig. 1.45 Diagrams in the power–frequency coordinates

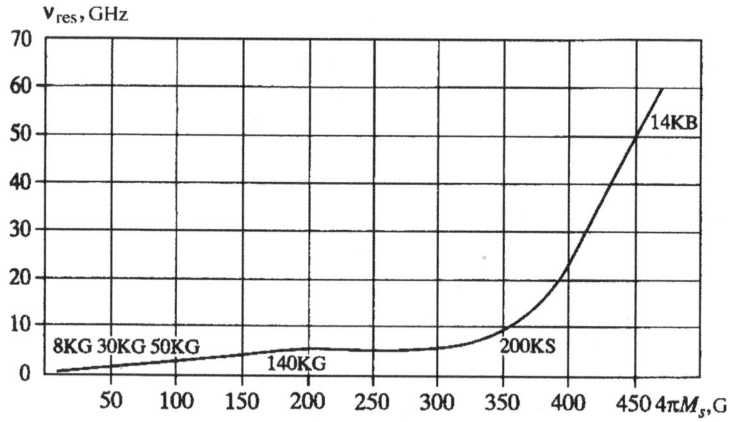

Fig. 1.46 A dependence of the resonant frequencies ν_{res} on saturation magnetization $4\pi M_s$

1.3 Regimes of High Power Level

1.3.1 Control by Magnetic Field and High-Frequency Signals Power

The multifunction properties of a heteromagnetic oscillator of high power level, controlled by a magnetic field, are shown in Fig. 1.47. The oscillator is made according to the circuit with a common base on a bipolar powerful KT962B transistor with FMCR – as a ferrite sphere having its saturation magnetization $4\pi M_s = 190$ G, a transistor feed voltage: $U_c = 4$ V and $U_e = 3$ V at an average output power $P_{out} \cong$ 0.5 W and an efficiency $\eta \cong 50\%$ for signals on the reference frequency ν_0 (solid line), the first ν_1 (dotted line) and second ν_2 harmonics (dash-and-dot line). In Fig. 1.47a, the dependencies of the spectrum linewidth $(\Delta\nu_{-3dB})_{0,1,2}$ on the bias field H_0, and in Fig. 1.47b that on its base $(\Delta\nu_{-60dB})_{0,1,2}$ are shown. Oscillograms 1–9 in Fig. 1.47a show the most typical spectra of signals formed on the frequency ν_0 at the corresponding values of field H_0.

With the field $H_0 \cong 0$ corresponding to self-resonance, generation of rather broadband pseudonoise signals (oscillogram 1 in Fig. 1.47a) was observed. With increase of the harmonic number the spectrum linewidth increased.

At increase of the bias field up to the values $H_0 \cong (40–60)$ Oe synchronous change of the central frequencies $\nu_{0,1,2}$ and narrowing of the spectrum lines $(\Delta\nu_{-3dB})_{0,1,2}$ by an order of magnitude for all the spectral components of the signal was observed, and narrowing at the level $(\Delta\nu_{-60dB})_{0,1,2}$ was more effective for higher harmonics ν_1 and ν_2 (Fig. 1.47b). The signal from a pseudonoise one at $H_0 \cong 0$ at self-resonance (oscillogram 1) was switched into a spectrally pure one at $H_0 \cong (40–60)$ Oe (oscillogram 2). At these values of magnetic field H_0 in an MCO parametric processes of multiplication of the reference frequency signal $(\Delta\nu_0 \cong \Delta\nu_1 \cong \Delta\nu_2$ const) were observed.

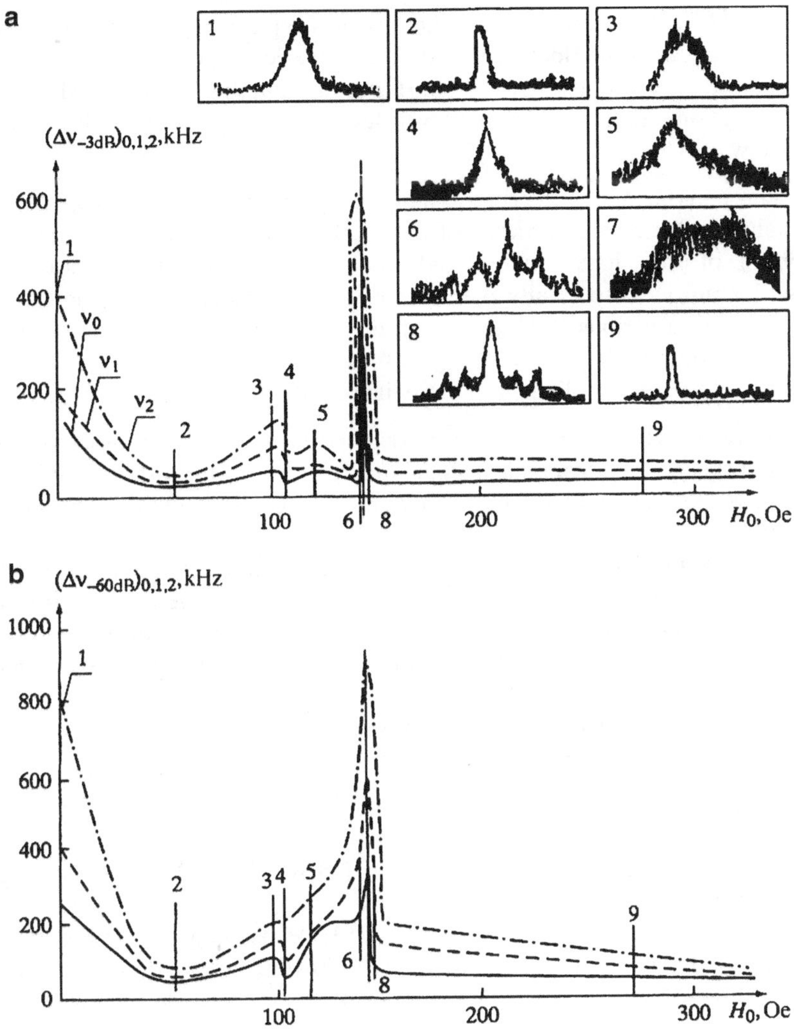

Fig. 1.47 The multifunction properties of spectral line of a heteromagnetic oscillator of high power level, controlled by a magnetic field: (**a**) on the level –3 dB, (**b**) on the level –60 dB

At further increase of the magnetic field up to $H_0 \cong 100$ Oe transition to a pseudonoise signal (oscillogram 3) was observed. At the field $H_0 \cong 105$ Oe, transition to a narrow-band pseudonoise signal (oscillogram 4) was observed. At $H_0 \cong 120$ Oe (oscillogram 5), the base of pseudonoise signals of all the spectral components was widened. At a magnetic field $H_0 \cong 130$ Oe, broadband pseudonoise signals (oscillogram 6) were observed. At $H_0 \cong 139$ Oe, a broadband noise signal which surpassed (oscillogram 7) the initial spectral lines at $H_0 = 0$ Oe (oscillogram 1) by spectrum width for all the components $\nu_{0,1,2}$ was observed.

At $H_0 \cong 148$ Oe, the mode of noisy equidistant frequency spectra (oscillogram 8) was observed. At further increase of the magnetic field, spectrally pure lines with $\Delta\nu_1 = \text{const}$, $\Delta\nu_2 = \text{const}$, $\Delta\nu_3 = \text{const}$ (oscillogram 9) were observed.

Thus, on one FSS at a high power level at change of the bias field H_0 observed was:

- Synchronous change of the central frequencies for the components ν_0, ν_1, ν_2
- Simultaneous control of the kinds of signals, their spectral and noise characteristics
- Control of the signal quality and transition from the initial noise mode (oscillogram 1) to spectrally pure signals (oscillograms 2 and 9) and to noise signals with a various PSD nonuniformity (oscillograms 3–8) and the greatest PSD (oscillogram 7)
- Control of the spectral linewidth more than by 10–40 times

Our investigations have shown that in an oscillating FSS multifunction control of the power and spectral characteristics, signal quality, PSD of various kinds of signals, the magnetic field H_0, voltage of the transistor feed U_c, U_e, and the HF power level is possible.

In Fig. 1.48, the dependences of the power and spectral characteristics of regular and pseudonoise signals $(\Delta\nu_{-3\text{dB}})_{0,1,2}$, $(\Delta\nu_{-60\text{dB}})_{0,1,2}$ are given; oscillograms 1–7 in Fig. 1.48a and oscillograms 1–5 in Fig. 1.48b – at various values of the magnetic field H_0 and output power levels P_{out} for the reference frequency ν_0 are marked with figures in sections.

It is obvious that for the given type of oscillating FSS at certain output power levels observed is generation of:

- Spectrally pure signals ($P_{\text{out}} = 0.1$ W, $H_0 = 253$ Oe, oscillogram 1 in Fig. 1.48a)
- The most broadband noise signals ($P_{\text{out}} = 0.5$ W, $H_0 \cong (253\text{–}260)$ Oe, oscillograms 2, 3 in Fig. 1.48a, b)
- Noise signals ($P_{\text{out}} \cong 1.0$ W, $H_0 \cong (250\text{–}253)$ Oe, oscillogram 4 in Fig. 1.48a, b)
- Noise signals ($P_{\text{out}} \cong (1.0\text{–}2.8)$ W, $H_0 \cong (250\text{–}260)$ Oe, oscillograms 5, 6, 7 in Fig. 1.48a and oscillograms 4, 5, in Fig. 1.4b)

These data show that control of PSD by 10–50 times from the medium up to high (Ws) power levels is possible in oscillating FSS.

The main kinds of signal spectra, which were observed in the self-oscillating modes in powerful FSS made on KT962B transistors and on KG-15, KG-30, KG-65 ferrites, are shown in the oscillograms of Fig. 1.49. The signals were generated by FSS simultaneously on the reference frequency ν_0, harmonic components ν_{n-1} and, in certain types of structures, on the subharmonic ν_{-m+1} components, which were synchronously changed by frequency (phase) at change of the transistor feed voltage (U_c, U_e), the bias field H_0, or the HF power level P_{out}.

Figure 1.49a presents a spectrally pure signal with a spectral linewidth $(\Delta\nu_{-3\text{dB}})_0 = 10$ kHz.

Figure 1.49b, c shows signal spectra in the modes of pedestal noisiness near the carrier frequency in the Doppler frequency range of tuning outs (up to 100 kHz).

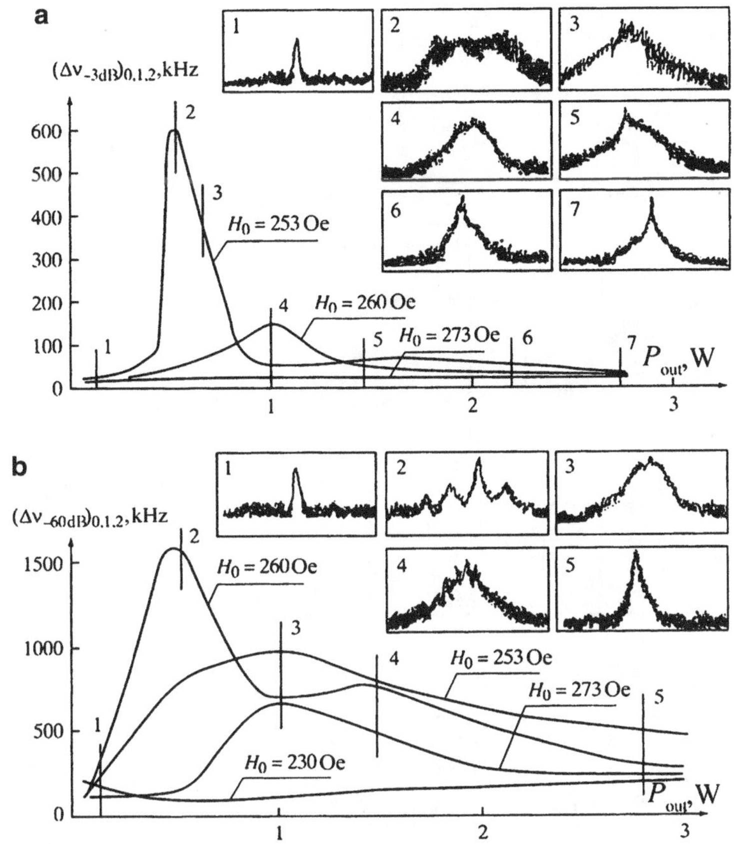

Fig. 1.48 The dependences of the power and spectral characteristics of regular and pseudonoise signals: (**a**) on the level –3 dB, (**b**) on the level –60 dB

Figure 1.49d shows a signal spectrum in the mode of pedestal noisiness in a wide frequency range of tuning outs, including the ranges of Doppler and intermediate frequencies of tuning outs from the carrier frequency.

Figure 1.49e–g shows spectra of pseudonoise broadband signals of various kinds.

In Fig. 1.49h–m, equidistant frequency spectra in various modes of noisiness are shown.

In Fig. 1.49n–q, various spectra of broadband pseudonoise signals are given.

Figure 1.49r shows a spectrum of white noise in the frequency range of 10 MHz–40 GHz with an integral power of 50 mW at an efficiency = 45%.

The considered FSS in the mode of generation provides synchronous control of the central frequencies, spectrum width, their shape and quality of signal–noise level in various ranges of tuning outs due to change of the bias field H_0 and the transistor feed voltage (U_c, U_e).

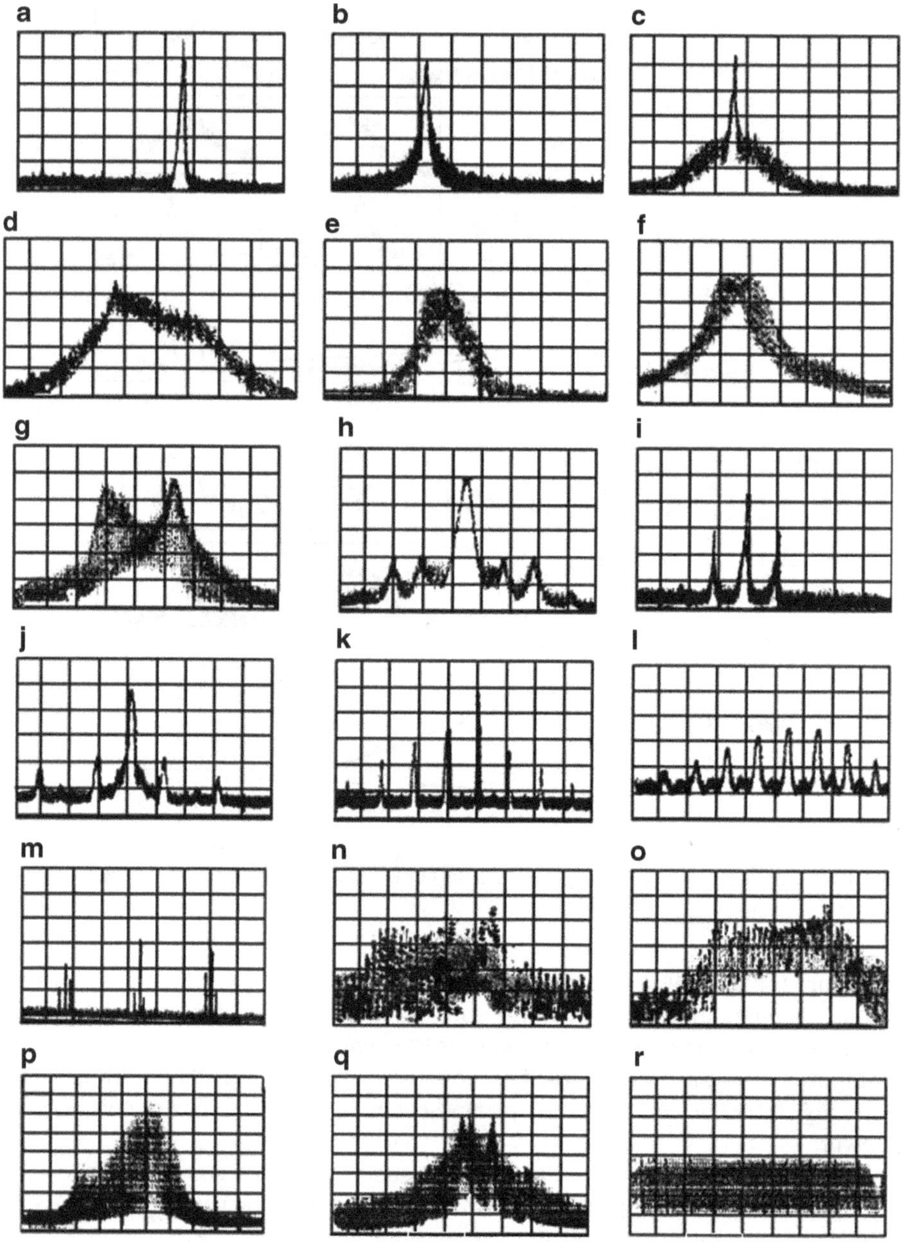

Fig. 1.49 The main kinds of signal spectra in the self-oscillating modes in powerful FSS

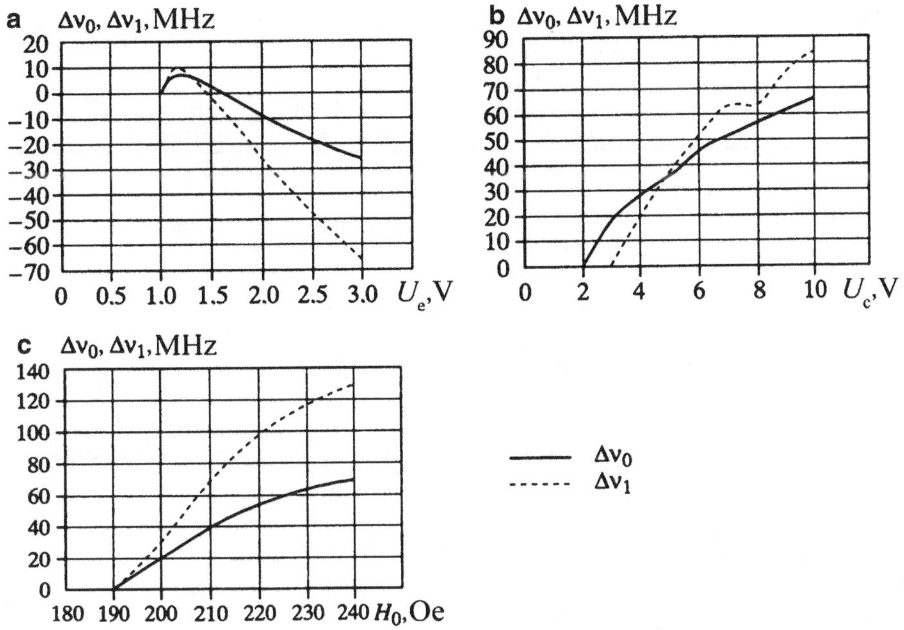

Fig. 1.50 The experimental dependences of the central frequency change of signals in oscillating FSS: (**a**, **b**) of the basic signal, (**c**) of the first harmonic component

In Fig. 1.50, the experimental dependences of the central frequency change of signals in oscillating FSS are shown: (a), (b) are the basic signal $\Delta\nu_0$ and the first harmonic component $\Delta\nu_1$ due to change of the voltage on the emitter U_e and that on the collector U_c; and (c) is the bias fields H_0. The minimum signal spectral linewidth made $\Delta\nu_{-3\text{dB}} = 10\ldots15$ kHz, its widening due to change of the transistor feed is more than by 20–30 times, and due to change of the magnetic field by 3×10^2 times.

The average steepness of reconstruction of the central frequency ν_0 and the first harmonic component ν_1 of the signals due to changes:

- Of the voltage on the emitter $\Delta\nu_0/\Delta U_e \cong -18$ MHz/V, $\Delta\nu_1/\Delta U_e \cong$ –42 MHz/V
- Of the voltage on the collector $\Delta\nu_0/\Delta U_c \cong +8$ MHz/V, $\Delta\nu_1/\Delta U_e \cong +12$ MHz/V
- Of the bias field $\Delta\nu_0/\Delta H_0 = +1.4$ MHz/Oe, $\Delta\nu_1/\Delta H_0 = +2.6$ MHz/Oe

In Table 1.1, the parameters changed most essentially at the use of powerful oscillating FSS are given.

Table 1.1 The parameters changed most essentially at the use of powerful oscillating FSS

Parameters	Conditional oscillating heteromagnetic structure	Conditional prototype	Advantage, its estimation
1. Mass	Tens of g	Tens–hundreds of kg	10^3–10^5 times and more
2. Dimensions	$\sim 1\,\text{mm}^3$	$0.5\,\text{dm}^3$	10^5 times and more
3. Cost price	*S* 50–400	Tens of thousand *S*	10^2–10^3 and more
4. Endurance	10^3–10^4 h and more	100–200 h	10^2–10^3 times and more
5. Technical efficiency	50–60%	–	Essential
6. Range of overlapping of operating frequencies, GHz	Up to 5–7 frequency octaves and more	2 frequency octaves	2.5–3 times and more
7. Multifrequency (multioctave overlapping of frequency range by one kind of signal)	Yes	No	Extremely essential
8. Multifunctionality (spectrally pure, pseudonoise, equidistant spectra of frequencies, white noise)	Main advantage (no analogues)	Limited	Extremely essential
9. Advance into new frequency ranges	On one CHIP structure without financial expenses	Significant financial expenses for research and development, industrial design, and manufacturing	Extremely essential
10. Specific power, W/kg	10^2–10^4	No data	Extremely essential
11. Power spectral density, W/MHz	100–500 and more	100–500	At reduction of mass-dimensions by an order and the efficiency $\approx 50\%$
12. Control of spectral linewidth	10^3–10^4 times and more	No data	10^3–10^4 times and more, extremely essential
13. Synchronous control of frequency (phase) change of all spectral components	Main advantage	No data	Extremely essential

(continued)

Table 1.1 (continued)

Parameters	Conditional oscillating heteromagnetic structure	Conditional prototype	Advantage, its estimation
14. Multimodiness (electric switching from one kind of signal to another, control of the value and nonuniformity of PSD, transition from continuous into pulse mode, signals from spectrally pure up to white noise and frequency spectra with controlled equidistance)	Main advantage	No analogues	Extremely essential
15. Level of continuous power, W	5–10 (can be increased by 10^2–10^3 times)	Tens–hundreds	Extremely essential

1.3.2 *Multifunctional Properties of Powerful Heteromagnetic Oscillators*

Oscillating FSS of the power up to 0.5–1.0 W were experimentally investigated in the UHF and microwave ranges for the following kinds of signals:

- Spectrally pure
- With noisiness of the spectral components in the range of Doppler tuning out frequencies from the central frequency (up to 100–150 kHz from the central frequency)
- With noisiness of the spectral components in the range of the intermediate tuning out frequencies from the central frequency (from 100 to 150 kHz up to units-tens of MHz)
- With broadband uniform and nonuniform noisiness
- Equidistant frequency spectra with various frequency distances
- With noisiness of equidistant frequency spectra

The oscillograms of signals (Fig. 1.51) from the output of the oscillator with heteromagnetic interaction are given below.

In Fig. 1.51a, the mode of generation of a spectrally pure line is presented: the central frequency $\nu_0 = 715$ MHz, the spectral linewidth $\Delta\nu_{-3\mathrm{dB}} = 15$ kHz, the spectral linewidth on the base $\Delta\nu_{-60\mathrm{dB}} = 130$ kHz, the integral power $P_{\mathrm{out}} = 0.54$ W, the transistor feed voltage $U_{\mathrm{e}} = 3$ V, $U_{\mathrm{c}} = 4.5$ V, and the bias field $H_0 = 200$ Oe. The noise level is below minus (60–70) dB.

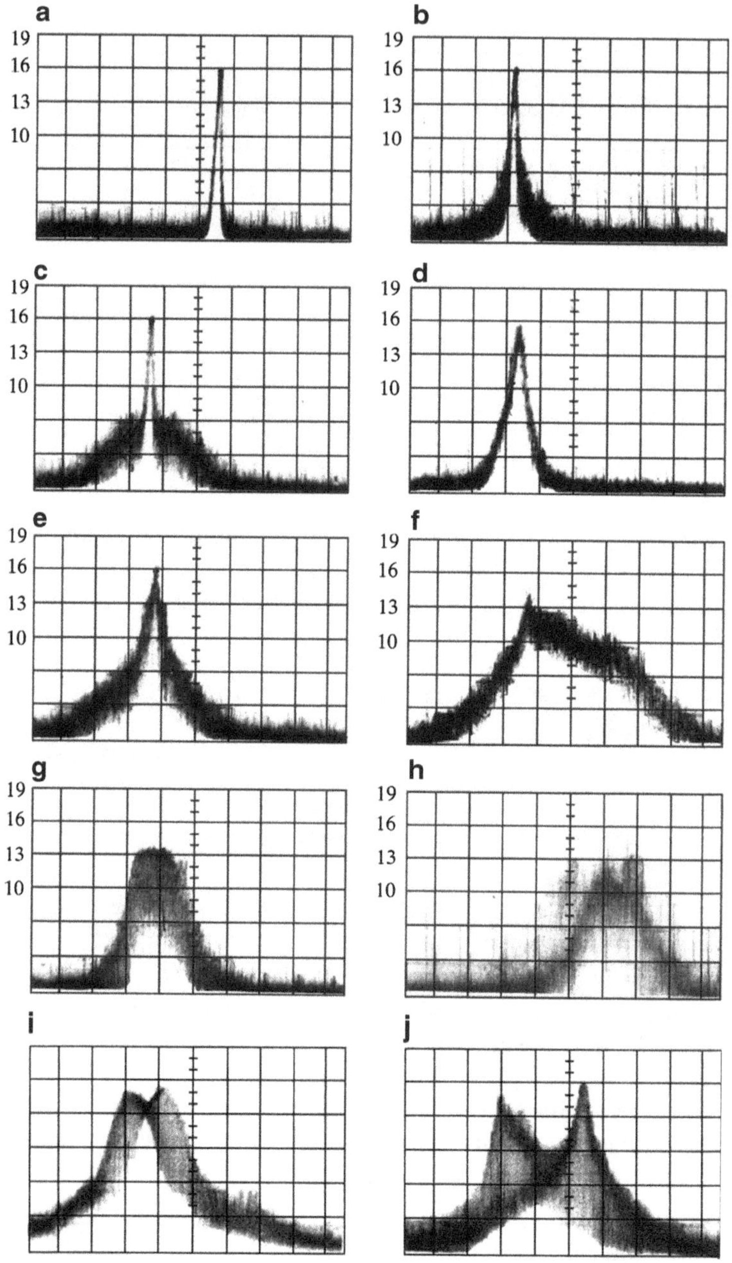

Fig. 1.51 The oscillograms of signals from the output of the oscillator with heteromagnetic interaction

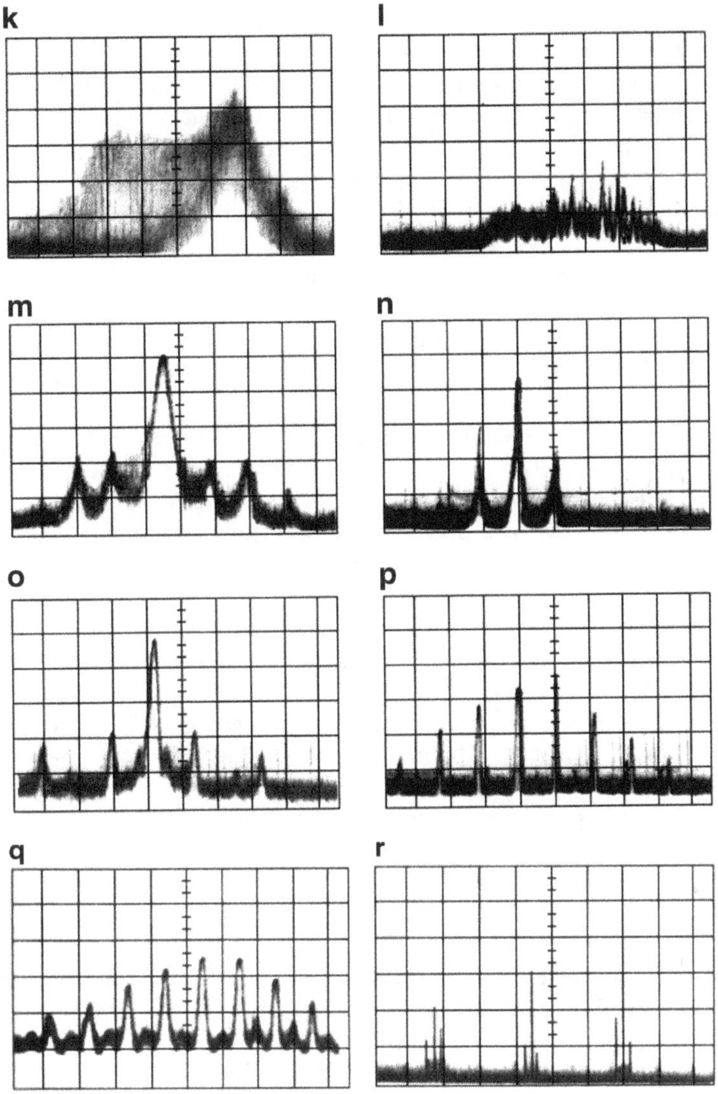

Fig. 1.51 (continued)

In Fig. 1.51b, the case of noisiness of the carrier frequency wings in the Doppler frequency range is given. The mode is: $\nu_0 = 719\,\text{MHz}$, $\Delta\nu_{-3\text{dB}} = 15\,\text{kHz}$, $\Delta\nu_{-60\text{dB}} = 250\,\text{kHz}$, $P_{\text{out}} = 0.69\,\text{W}$, and $H_0 = 213\,\text{Oe}$. The noisiness band in the region of the spectral line is $\Delta\nu_{-30\text{dB}} \cong 50\text{–}60\,\text{kHz}$.

In Fig. 1.51c, the case of pedestal noisiness both in the range of Doppler frequencies and in the near area of intermediate frequencies of the tuning out from the carrier one is shown. The mode is: $\nu_0 = 720\,\text{MHz}$, $\Delta\nu_{-60\text{dB}} = 15\,\text{MHz}$, $\Delta\nu_{-60\text{dB}} = 700\,\text{kHz}$, $\Delta\nu_{-30\text{dB}} \cong 280\,\text{kHz}$, $P_{\text{out}} = 0.68\,\text{mW}$, and $H_0 = 218\,\text{Oe}$.

Figure 1.51d shows the mode of widening of the spectral line $\nu_0 = 723\,\mathrm{MHz}$, $\Delta\nu_{-3\mathrm{dB}} = 20\,\mathrm{kHz}$, $\Delta\nu_{-60\mathrm{dB}} = 500\,\mathrm{kHz}$, $P_{\mathrm{out}} = 0.66\,\mathrm{mW}$, and $H_0 = 228\,\mathrm{Oe}$.

Figure 1.51e shows the case of noisiness of the spectral line wings in the Doppler and intermediate frequency tuning out ranges. The mode is: $\nu_0 = 728\,\mathrm{MHz}$, $\Delta\nu_{-3\mathrm{dB}} = 20\,\mathrm{kHz}$, $\Delta\nu_{-60\mathrm{dB}} = 1\,\mathrm{MHz}$, $P_{\mathrm{out}} = 0.66\,\mathrm{W}$, and $H_0 = 228\,\mathrm{Oe}$.

Figure 1.51f shows the case of broadband noisiness of signal in the Doppler and intermediate tuning out frequency range with nonuniform noise decay in the region of the upper and lower tuning out frequencies. The mode is: $\nu_0 = 736\,\mathrm{MHz}$, $\Delta\nu_{-60\mathrm{dB}} = 1.8\,\mathrm{MHz}$, $P_{\mathrm{out}} = 0.83\,\mathrm{W}$, here and below $U_c = 11.5\,\mathrm{V}$, $H_0 = 232\,\mathrm{Oe}$.

Figure 1.51g shows the case of broadband uniform noisiness of the Doppler frequency range and the near part of the intermediate tuning out frequency range. The mode is: $\nu_0 = 736\,\mathrm{MHz}$, $\Delta\nu_{-3\mathrm{dB}} = 200\,\mathrm{kHz}$, $\Delta\nu_{-60\mathrm{dB}} = 800\,\mathrm{kHz}$, $P_{\mathrm{out}} = 0.86\,\mathrm{W}$, and $H_0 = 224\,\mathrm{Oe}$.

Figure 1.51h shows a spectrum with nonuniform noisiness (two frequency regions with bands of 50–60 kHz in the Doppler tuning out frequency range). The mode is: $\nu_0 = 740\,\mathrm{MHz}$, $\Delta\nu_{-60\mathrm{dB}} = 1\,\mathrm{MHz}$, $P_{\mathrm{out}} = 0.82\,\mathrm{W}$, and $H_0 = 232.5\,\mathrm{Oe}$.

In Fig. 1.51i, the case of a double noise spectrum with tuning on 200 kHz is given. The mode is: $\nu_0 = 723\,\mathrm{MHz}$, $\Delta\nu_{-3\mathrm{dB}} = 300\,\mathrm{kHz}$, $\nu_{-60\mathrm{dB}} = 1.8\,\mathrm{MHz}$, $P_{\mathrm{out}} = 0.84\,\mathrm{W}$, and $H_0 = 231\,\mathrm{Oe}$.

Figure 1.51j shows a double noise spectrum with tuning on 450 kHz. The mode is: $\nu_0 = 737\,\mathrm{MHz}$, $\Delta\nu_{-3\mathrm{dB}} = 50\,\mathrm{kHz}$, $\nu_{-60\mathrm{dB}} = 1.6\,\mathrm{MHz}$, $P_{\mathrm{out}} = 0.84\,\mathrm{W}$, and $H_0 = 231\,\mathrm{Oe}$.

Figure 1.51k shows a spectrum with additional noisiness in the region of the lower tuning out frequencies. The mode is: $\nu_0 = 730\,\mathrm{MHz}$, $\Delta\nu_{-3\mathrm{dB}} = 50\,\mathrm{MHz}$, $\nu_{-60dB} = 1.2\,\mathrm{MHz}$, $P_{\mathrm{out}} = 0.3\,\mathrm{W}$, and $H_0 = 242\,\mathrm{Oe}$.

Figure 1.51l shows the mode of broadband noise: $\nu_0 = 734\,\mathrm{MHz}$, $\Delta\nu_{-60\mathrm{dB}} = 25\,\mathrm{MHz}$, $P_{\mathrm{out}} = 84.5\,\mathrm{mW}$, and $H_0 = 255\,\mathrm{Oe}$.

Figure 1.51m shows the case of noisiness of an equidistant frequency spectrum. The mode is: $\nu_0 = 723\,\mathrm{MHz}$, frequency distances of 200 kHz, $\Delta\nu_{-60\mathrm{dB}} = 1.4\,\mathrm{MHz}$, $P_{\mathrm{out}} = 0.64\,\mathrm{mW}$, and $H_0 = 228\,\mathrm{Oe}$.

Figure 1.51n shows the mode of noisiness of an equidistant frequency spectrum. The mode is: $\nu_0 = 748\,\mathrm{MHz}$, $\Delta\nu_{-3\mathrm{dB}} = 130\,\mathrm{kHz}$, $\Delta\nu_{-60\mathrm{dB}} = 4\,\mathrm{MHz}$, equidistance – 2 MHz, $P_{\mathrm{out}} = 0.2\,\mathrm{W}$, and $H_0 = 246\,\mathrm{Oe}$.

Figure 1.51o shows the case of controlled noisiness of signal in the region of the central frequency pedestal. The mode is: $\nu_0 = 729\,\mathrm{MHz}$, $\Delta\nu_{-3\mathrm{dB}} = 40\,\mathrm{kHz}$, $\Delta\nu_{-60\mathrm{dB}} = 6\,\mathrm{MHz}$, $P_{\mathrm{out}} = 0.46\,\mathrm{W}$, and $H_0 = 256.5\,\mathrm{Oe}$.

Figure 1.51p shows an equidistant frequency spectrum. The mode is: $\nu_0 = 745\,\mathrm{MHz}$, $\Delta\nu_{-3\mathrm{dB}} = 10\,\mathrm{kHz}$, $\Delta\nu_{-60\mathrm{dB}} = 16\,\mathrm{MHz}$, equidistance – 2 MHz, $P_{\mathrm{out}} = 51\,\mathrm{mW}$, and $H_0 = 256.5\,\mathrm{Oe}$.

Figure 1.51q shows the case of noisiness of an equidistant frequency spectrum. The mode is: $\nu_0 = 740\,\mathrm{MHz}$, $\Delta\nu_{-3\mathrm{dB}} = 270\,\mathrm{kHz}$, $\Delta\nu_{-60\mathrm{dB}} = 18\,\mathrm{MHz}$, equidistance – 2 MHz, $P_{\mathrm{out}} = 77\,\mathrm{mW}$, and $H_0 = 55\,\mathrm{Oe}$.

Figure 1.51r shows a frequency spectrum of the second order (generation of an equidistant spectrum in the region of each spectral component of the first-order spectrum). The mode is: $\nu_0 = 430\,\mathrm{MHz}$, $\Delta\nu_{-3\mathrm{dB}} = 13\,\mathrm{kHz}$, $P_{\mathrm{out}} = 0.4\,\mathrm{W}$, and

$H_0 = 243$ Oe, the frequency spectrum of overlapping – 5 MHz, equidistance in the second-order frequency spectrum – 400 kHz.

Thus, our experimental investigations confirm the possibility of multifunction generation of various kinds of spectrally pure, pseudonoise, and noise signals in an oscillator with heteromagnetic interaction at power levels from 0.5 up to 1 W. Various kinds of signals are realized on one CHIP at change of the electric modes of the oscillator feed. The power level, the spectral linewidth, and the kind of noisiness spectrum vary within wide limits.

Let us examine the results of our experimental investigation of ways to control the power and spectral characteristics of noise and pseudonoise signals in heteromagnetic oscillators with a power level up to 3.5 W.

In Fig. 1.52, the dependence of the central frequency of the fundamental component ν_0 on the bias field H_0 at tuning on the maximum level of the output power ($P_{out} = 3.5$ W) due to change of the electrocapacity of the capacitor $C_2 = C_{2opt}$, included in the collector circuit of a bipolar magnetotransistor on the oscillation input is given. At change of the field H_0 together with change of the central frequency ν_0, and of all the harmonic spectral components as well, change of the spectrum shape and the noisiness level within wide limits was observed.

Most typical changes of the spectra parameters of various signals for the corresponding values of the field H_0 are indicated by 1–5 in Fig. 1.52. The critical spectra parameters of these signals for sections 1–5 are given in Table 1.2.

Let us note that in Fig. 1.52 the steepness of the frequency change $\nu_0(H_0)$ has various values and signs. For example, within the limits of change of the magnetic field $H_0 = (0–70)$ Oe, the steepness of the frequency change is $\alpha \cong -14$ kHz/Oe. At change of the field H_0 within the limits of (70–100) Oe the value is $\alpha = +17$ kHz/Oe. In the range of changing $H_0 = (100–175)$ Oe the steepness is $\alpha = -25$ kHz/Oe, and at $H_0 = (175–300)$ Oe – $\alpha = +6$ kHz/Oe.

At change of the capacitor electrocapacity C_2 in the collector circuit of a bipolar magnetotransistor the oscillator tunes out from the mode of the maximal output power and the efficiency (C_{2opt}) into the mode of soft excitation of generation.

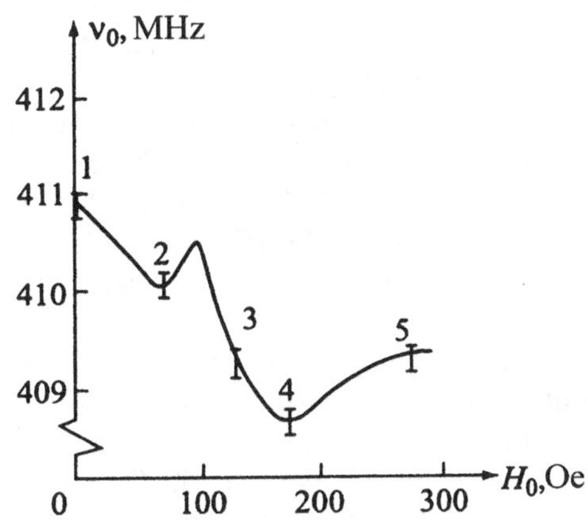

Fig. 1.52 The dependence of the central frequency of the fundamental component ν_0 on the bias field H_0

Table 1.2 The critical spectra parameters of signals for sections 1–5 (Fig. 1.52)

No. of Section	P_{out}, W	U_c, V	U_e, V	I_c, A	H_0, Oe	ν_0, MHz	$\Delta\nu_{-3dB}$, kHz	$\Delta\nu_{-60dB}$, kHz	Note
1	3.5	6	3	0.78	0	411.0	30	150	Spectral line is noisy and widened
2	3.5	6	3	0.78	70	410.0	25	100	Spectrally pure line
3	3.5	6	3	0.78	142	409.0	30	130	Noisy spectral line with the appearance of side tones
4	3.5	6	3	0.78	176	408.6	20	100	Spectrally pure line
5	3.5	6	3	0.78	288	409.3	25	300	Noisy line with side tones

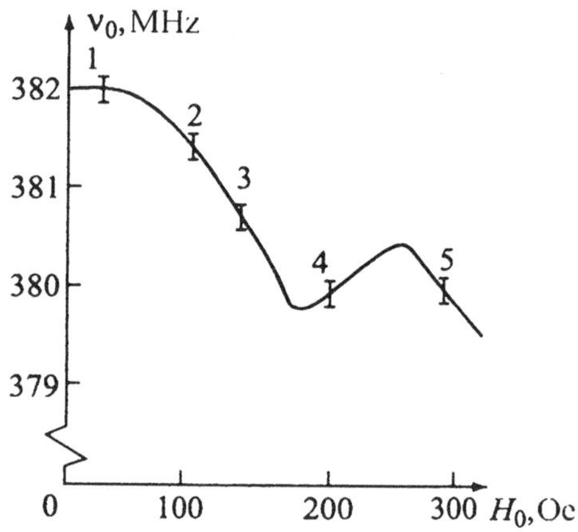

Fig. 1.53 The dependence of change of the central frequency of signal ν_0 on the change of the magnetic field H_0 at $P_{out} = 3$ W

In Fig. 1.53, the dependence of change of the central frequency of signal ν_0 on the change of the magnetic field H_0 at $P_{out} = 3$ W is given. At change of the magnetic field H_0 within the limits of (0–100) Oe the generation frequency ν_0 remains practically constant. At change of the magnetic field H_0 within the limits of (100–180) Oe the steepness of the frequency change is $\alpha = -11$ kHz/Oe. At change of the magnetic field H_0 within the limits of (180–270) Oe the steepness is $\alpha = +5.6$ kHz/Oe, and at $H_0 =$ (270–290) Oe $-\alpha = -15$ kHz/Oe. The change of the critical parameters and the spectra of signals generated by a bipolar magnetotransistor in this case are reflected in Table 1.3.

In Fig. 1.54, the dependence $\nu_0(H_0)$ of generated signals for the capacitor tune-out C_2 in the collector circuit of a bipolar magnetotransistor into the other region on the optimal (C_{2opt}) at $P_{out} = 1.5$ W is given. At change of the magnetic field H_0 within the limits of (0–50) Oe the oscillator frequency and the shape of the signal practically did not change. At change of the field H_0 within the limits

Table 1.3 The change of the critical parameters and the spectra of signals generated by a bipolar magnetotransistor in this case

No. of Section	P_{out}, W	U_c, V	U_e, V	I_c, A	H_0, Oe	ν_0, MHz	$\Delta\nu_{-3dB}$, kHz	$\Delta\nu_{-60dB}$, kHz	Note
1	3	6	3	0.6	20	382.0	20	100	Spectral line is noisy
2	3	6	3	0.6	90	381.4	20	70	Spectrally pure line
3	3	6	3	0.6	130	380.7	50	200	Heavily noisy spectral line
4	3	6	3	0.6	200	380.0	20	70	Spectrally pure line
5	3	6	3	0.6	290	380.0	30	300	Strongly noisy line with side tones

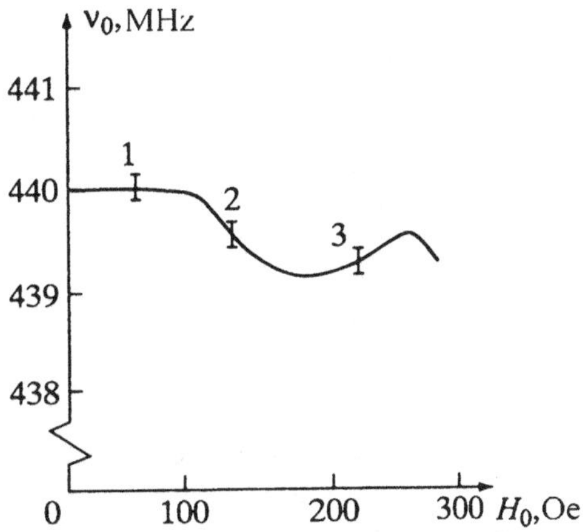

Fig. 1.54 The dependence $\nu_0(H_0)$ of generated signals for the capacitor tune-out C_2 in the collector circuit of a bipolar magnetotransistor

Table 1.4 The critical parameters of the spectra of signals generated by a powerful bipolar magnetotransistor at tuning out from optimum mode

No. of Section	P_{out}, W	U_c, V	U_e, V	I_c, A	H_0, Oe	ν_0, MHz	$\Delta\nu_{-3dB}$, kHz	$\Delta\nu_{-60dB}$, kHz	Note
1	1.5	6	3	0.7	50	440.00	20	60	Spectrally pure line
2	1.5	6	3	0.7	130	439.50	150	200	Noise signal
3	1.5	6	3	0.7	225	439.25	20	60	Spectrally pure line

of (50–180) Oe the steepness is $\alpha = -17.7$ kHz/Oe. At change of the magnetic field H_0 within the limits of (170–260) Oe – $\alpha = +10$ kHz/Oe, and at $H_0 = (260\text{–}320)$ Oe – $\alpha = -11.8$ kHz/Oe.

In Table 1.4, the critical parameters of the spectra of signals generated by a powerful bipolar magnetotransistor in this mode are given.

Experimental dependences of the fundamental frequency $\Delta\nu_0$ and the harmonic components $\Delta\nu_1$, $\Delta\nu_2$ on the magnetic field H_0 at various levels of continuous

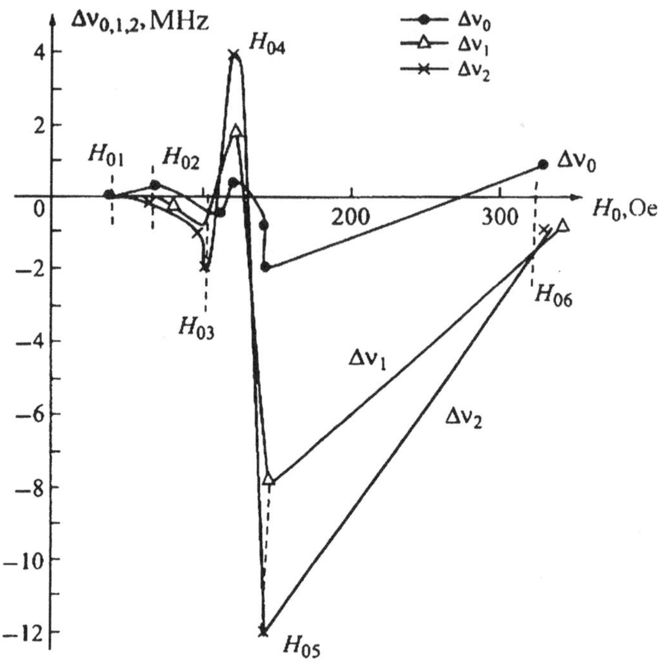

Fig. 1.55 Experimental dependences of the fundamental frequency $\Delta\nu_0$ and the harmonic components $\Delta\nu_1$, $\Delta\nu_2$ on the magnetic field H_0 at various levels of continuous integral output power ($P_{out} = 0.05$ W)

integral output power are given in: Fig. 1.55 – $P_{out} = 0.05$ W; Fig. 1.56 – $P_{out} = 1.0$ W; and Fig. 1.57 – $P_{out} = 3.6$ W. The specified dependences have a number of common regularities.

We shall examine in detail the dependences $\Delta\nu(H_0)$ for the fundamental frequency ν_0 and harmonic components ν_1 and ν_2 at an integral power level $P_{out} = 0.05$ W (Fig. 1.55). In the change range of the magnetic field H_0 from 0 up to ~50 Oe the quantities $\Delta\nu_0$, $\Delta\nu_1, \Delta\nu_2 \cong$ const. At change of the field within the limits of H_{01}–H_{02} a positive steepness of the change of the fundamental frequency $\Delta\nu_0$ is noted. At change of the magnetic field within the limits of H_{02}–H_{03} the steepness of the frequency change for all the spectral components $\Delta\nu_0$, $\Delta\nu_1, \Delta\nu_2$ is approximately identical and $\alpha \cong -20$ kHz/Oe At change of the magnetic field from H_{03} up to H_{04} we have an increase of the steepness of change at increase of the harmonic number: $\alpha_{\nu_0} = +50$ kHz/Oe $\alpha_{\nu_1} = +150$ kHz/Oe and $\alpha_{\nu_2} = +300$ kHz/Oe At change of the magnetic field from H_{04} up to H_{05}: $\alpha_{\nu_0} = -100$ kHz/Oe $\alpha_{\nu_1} = -200$ kHz/Oe and $\alpha_{\nu_2} = -560$ kHz/Oe at change from H_{05} up to H_{06}: $\alpha_{\nu_0} = +17$ kHz/Oe, $\alpha_{\nu_1} = +40$ kHz/Oe, and $\alpha_{\nu_2} = +62$ kHz/Oe.

At increase of the integral output power up to 1 W (Fig. 1.56) and up to 3.6 W (Fig. 1.57) the above tendencies are kept on the whole, and with growth of P_{out} a decrease of the steepness of the frequency change on ν_0, ν_1, ν_2 and values of the maximal frequency deviations $\Delta\nu_{0,1,2}$ depending on change of the magnetic field H_0 takes place.

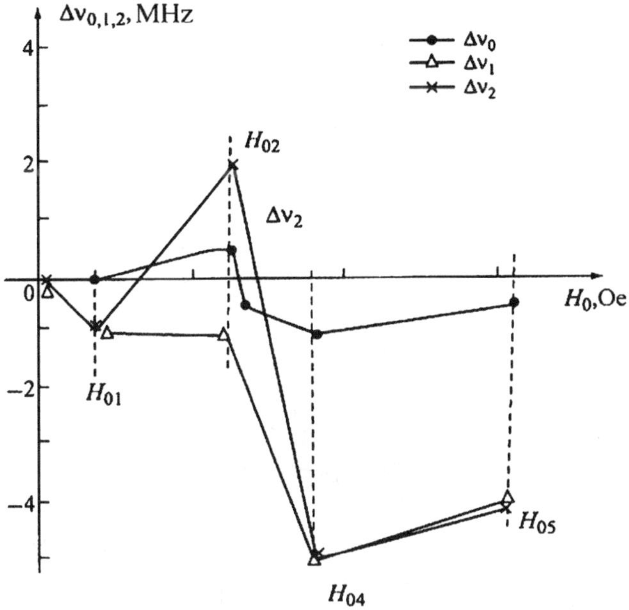

Fig. 1.56 Experimental dependences of the fundamental frequency $\Delta\nu_0$ and the harmonic components $\Delta\nu_1$, $\Delta\nu_2$ on the magnetic field H_0 at various levels of continuous integral output power ($P_{out} = 1.0\,\text{W}$)

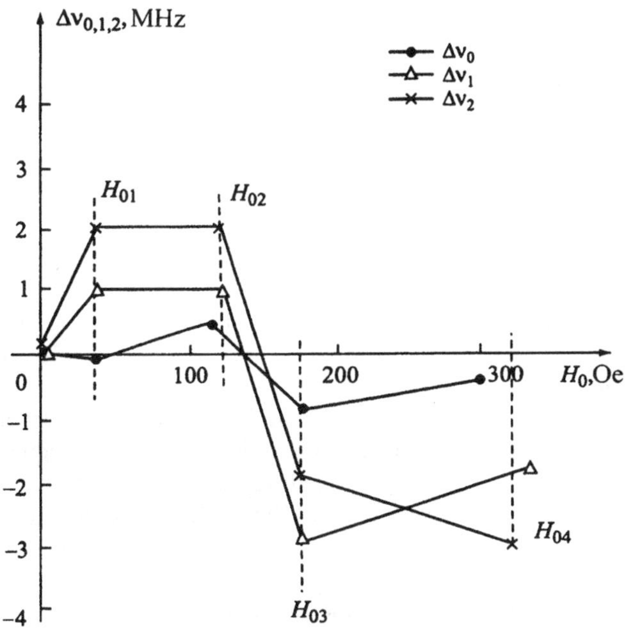

Fig. 1.57 Experimental dependences of the fundamental frequency $\Delta\nu_0$ and the harmonic components $\Delta\nu_1$, $\Delta\nu_2$ on the magnetic field H_0 at various levels of continuous integral output power ($P_{out} = 3.6\,\text{W}$)

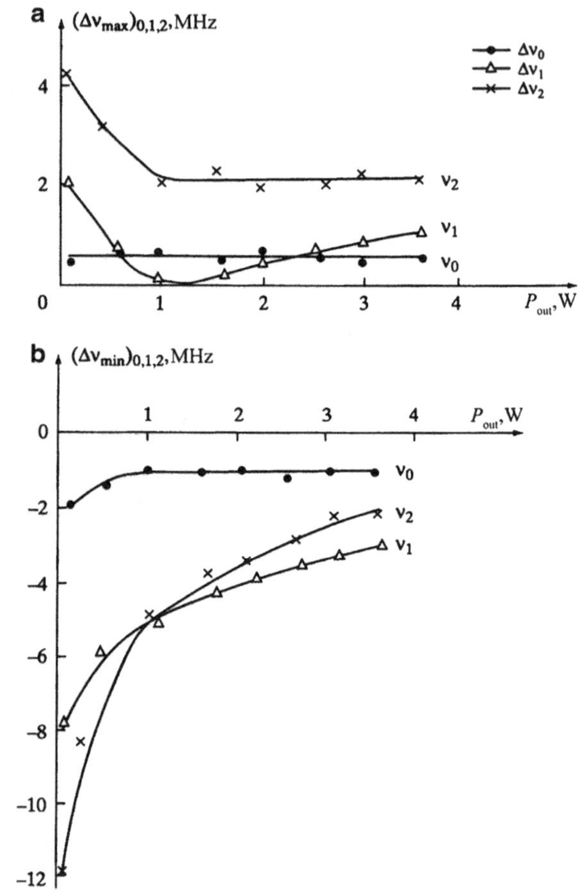

Fig. 1.58 The dependences of the maximal $(\Delta\nu_{max})_{0,1,2}$ (**a**) and minimal $(\Delta\nu_{min})_{0,1,2}$ (**b**) frequency deviation on the level P_{out} of signals generated by a bipolar transistor

In Fig. 1.58, the dependences of the maximal $(\Delta\nu_{max})_{0,1,2}$ (Fig. 1.58a) and minimal $(\Delta\nu_{min})_{0,1,2}$ (Fig. 1.58b) frequency deviations on the level P_{out} of signals generated by a bipolar transistor are shown.

In Fig. 1.59, the dependences of the modulus of the maximal frequency deviation $|\Delta\nu_{max}|_{0,1,2} = |\nu_{max} - \nu_{min}|_{0,1,2}$ on the level P_{out} for the spectral components on the frequencies ν_0, ν_1, and ν_2 are given.

From Figs. 1.58 and 1.59, it follows that at transition to the W levels of the generated power in a bipolar magnetotransistor a decrease of the frequency deviation for all the spectral components of generated signals is observed.

Other parameters influencing change of the central frequency of generation on the fundamental ν_0 and harmonic components ν_1 and ν_2 are the feed voltages U_e and U_c on the transistor.

In Fig. 1.60a, the dependencies of the frequency changes ν_0, ν_1, and ν_2 on the voltage on the emitter U_e at the voltage on the collector $U_c = 3\,V$ are given.

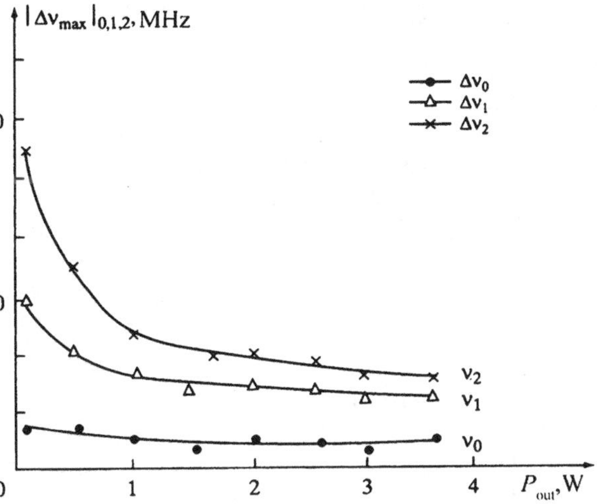

Fig. 1.59 The dependences of the modulus of the maximal frequency deviation $|\Delta\nu_{max}|_{0,1,2} = |\nu_{max} - \nu_{min}|_{0,1,2}$ on the level P_{out} for the spectral components on the frequencies ν_0, ν_1, and ν_2

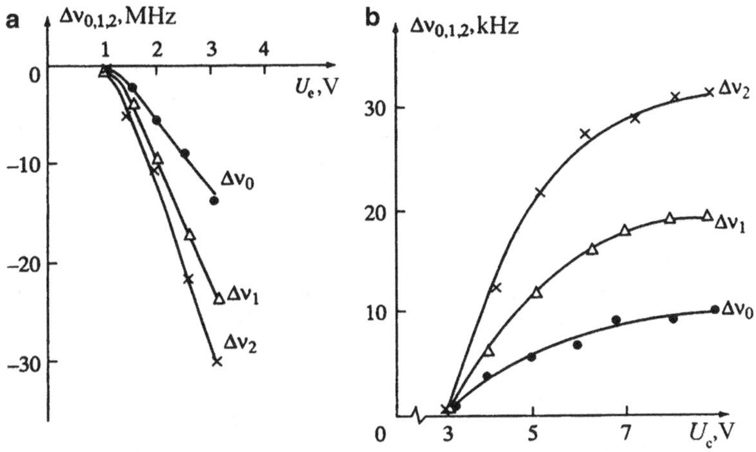

Fig. 1.60 The dependencies of the frequency changes ν_0, ν_1, and ν_2 on the voltage on the emitter U_e at the voltage on the collector $U_c = 3$ V (**a**). The dependencies of the frequency changes $\Delta\nu_0$, $\Delta\nu_1$, $\Delta\nu_2$ on the voltage on the collector U_c at $U_e = 4$ V (**b**)

The highest steepness of change was observed for the second harmonic $\beta_{\nu_2} = -15$ MHz/V, for the first harmonic $\beta_{\nu_1} = -14$ MHz/V, and for the fundamental frequency $\beta_{\nu_0} = -7$ MHz/V.

In Fig. 1.60b, the dependencies of the frequency changes $\Delta\nu_0$, $\Delta\nu_1$, $\Delta\nu_2$ on the voltage on the collec tor U_c at $U_e = 4$ V are given. A stronger nonlinearity of the dependences $\Delta\nu(U_c)$ in comparison with $\Delta\nu(U_e)$, but at a lower steepness, is noted. On the section $U_c = (3–5)$ V we have: $\gamma_{\nu_0} = +3$ MHz/V, $\gamma_{\nu 1} = +6$ MHz/V, and $\gamma_{\nu_2} = +10$ MHz/V.

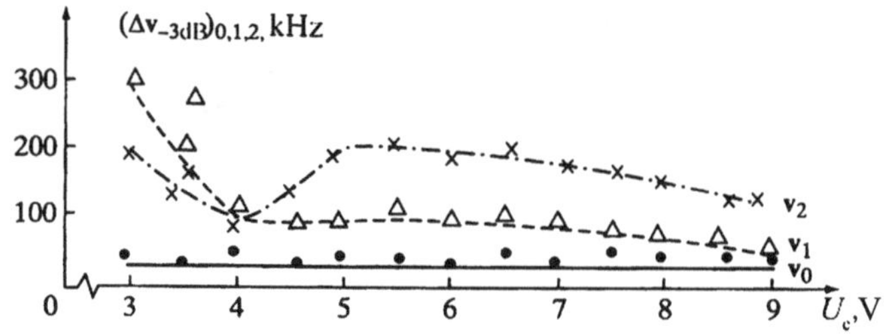

Fig. 1.61 The dependencies of the spectral linewidth $(\Delta\nu_{-3\mathrm{dB}})_{0,1,2}$ of the fundamental frequency ν_0, the first ν_1 and second ν_2 harmonics on the voltage on the collector U_c for $H_0 = 135\,\mathrm{Oe}$

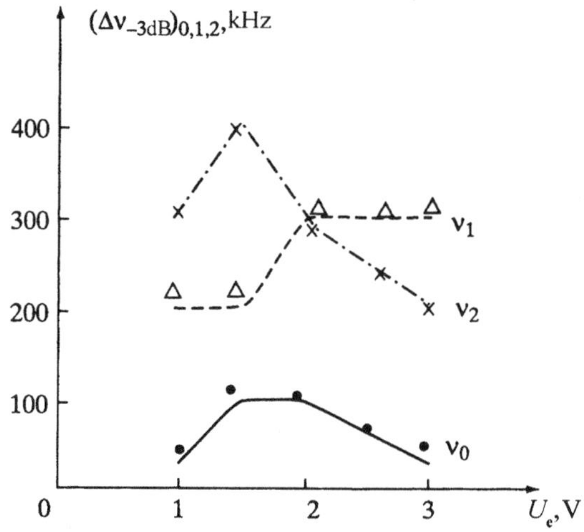

Fig. 1.62 The dependencies of the spectral linewidth $\Delta\nu_{-3\mathrm{dB}}$ for the spectral components of signals ν_0,ν_1, and ν_2 on the voltage on the emitter U_e at $H_0 = 135\,\mathrm{Oe}$

In Fig. 1.61, the dependencies of the spectral linewidth $(\Delta\nu_{-3\mathrm{dB}})_{0,1,2}$ of the fundamental frequency ν_0, the first ν_1 and second ν_2 harmonics on the voltage on the collector U_c for $H_0 = 135$ Oe are shown. It is obvious that the spectral linewidth for ν_0 is practically constant. At $U_c = 4\,\mathrm{V}$ the spectral linewidth for ν_1 and ν_2 is identical. At $U_c > 6\,\mathrm{V}$ the spectral lines ν_1 and ν_2 get narrower.

In Fig. 1.62, the dependencies of the spectral linewidth $\Delta 1_{-3\mathrm{dB}}$ for the spectral components of signals ν_0,ν_1, and ν_2 on the voltage on the emitter U_e at $H_0 =$ 135 Oe are shown. There are modes in which widening of the spectral lines of generated signals by 5 times for ν_0, by 1.5 times for ν_1, and by 2 times for ν_2 is observed.

For detailed studying the physical mechanisms in oscillators with heteromagnetic interactions investigations of the dependence of the oscillator critical parameters on the level of the output power of signals were carried out.

The results of these experiments, which are discussed below, are given for FMCR with a magnetization $4\pi M_s = 360\,\mathrm{G}$ and a half-width of the FMR line $\Delta H = 0.3\,\mathrm{Oe}$.

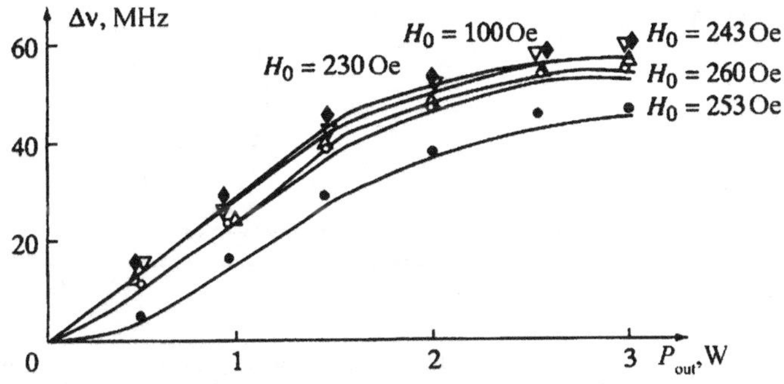

Fig. 1.63 The dependencies of the central frequency drift $\Delta\nu_0$ of generated signals in a bipolar magnetotransistor (B) on the output power level P_{out} at various bias fields H_0

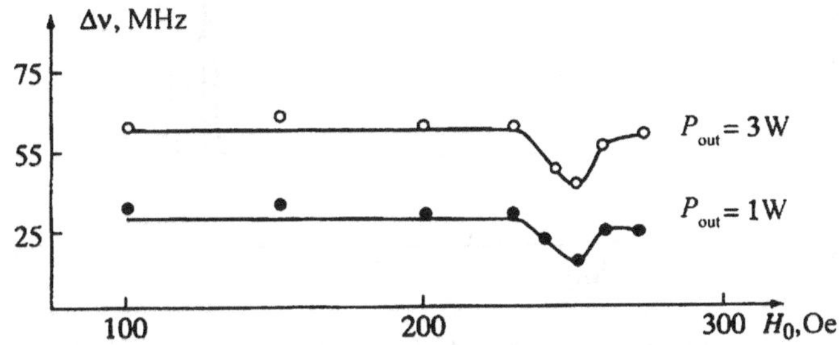

Fig. 1.64 The dependences of the generation frequency drift $\Delta\nu_0$ in B on the field H_0 for two values of the output power $P_{out} = 1$ W and $P_{out} = 3$ W

In Fig. 1.63, the dependencies of the central frequency drift $\Delta\nu_0$ of generated signals in a BMT on the output power level P_{out} at various bias fields H_0 which changed within the limits of (100–260) Oe are presented. The output power level P_{out} of the oscillator changed due to the voltage on the collector of the transistor U_c.

From Fig. 1.63, it is obvious that the tuning out of the central frequency of signal in BMT $\Delta\nu_0$ with increase of the output power level P_{out} increases monotonously at change of the magnetic field H_0 within the limits of (253–100) Oe and has a tendency to saturation at $P_{out} \geq (2.5–3)$ W.

In Fig. 1.64, the dependences of the generation frequency drift $\Delta\nu_0$ in BMT on the field H_0 for two values of the output power $P_{out} = 1$ W and $P_{out} = 3$ W are given.

From Fig. 1.64, it follows that in the region of magnetic fields $H_0 = (100–230)$ Oe the central frequency drift $\Delta\nu_0 = \text{const}$ and with increase of the output power level P_{out} the value $\Delta\nu_0$ increases. In the region of fields $H_0 \approx 250$ Oe a decrease of the frequency change $\Delta\nu_0$ for both $P_{out} = 1$ W and $P_{out} = 3$ W takes place.

Let us examine the experimental dependences illustrating the basic regularities at the central frequency change of an oscillating B and dynamics of change of the shape of the spectral lines of generated signals on the magnetic field H_0 for various values of the integral power level ($P_{out} = \text{const}$).

In Fig. 1.65, the dependences of change of the central frequency $\Delta\nu_0$ of various signals on the magnetic field $H_0 = (0–320)$ Oe at continuous output power

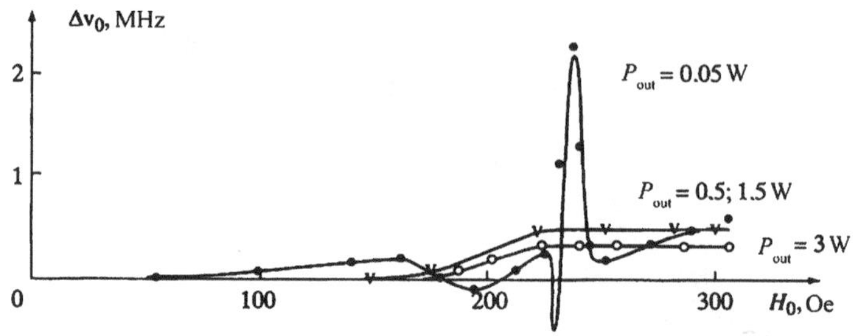

Fig. 1.65 The dependences of change of the central frequency $\Delta\nu_0$ of various signals on the magnetic field H_0 = (0–320) Oe at continuous output power levels $P_{\text{out}} = 0.05; 0.5; 1.5; 3.0$ W

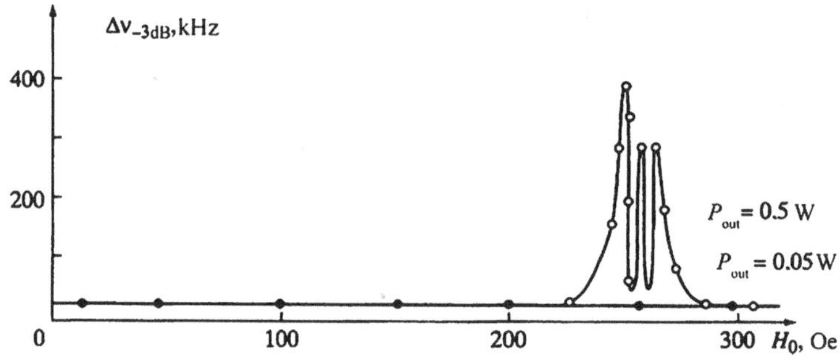

Fig. 1.66 The dependences $\Delta\nu_{-3\text{dB}}$ on the field H_0 for $P_{\text{out}} = 0.05$ and $P_{\text{out}} = 0.5$ W

levels $P_{\text{out}} = 0.05; 0.5; 1.5; 3.0$ W are given. The most complex sign-changing dependences $\Delta\nu_0(H_0)$ were observed at low power levels $P_{\text{out}} = 0.05$ W. In the range of change of the magnetic field H_0 = (0–160) Oe the steepness is $\alpha \cong +1.25$ kHz/Oe. In the range of change of the magnetic field H_0 = (160–185) Oe $-\alpha \cong$ –0.75 kHz/Oe. At H_0 = (225–227) Oe the characteristic steepness of $\Delta\nu 0(H_0)$ is again negative and maximal:$\alpha \cong$ –375 kHz/Oe. In the range of change of the magnetic field H_0 = (227–237) Oe the steepness $\alpha = +0.25$ Hz/Oe, and at $H_0 \cong$ (237–245) Oe the value $\alpha \cong$ –0.25 Hz/Oe. At further increase of the magnetic field H_0 = (245–285) Oe the steepness is considerably low and positive $-\alpha = +7.5$ kHz/Oe.

At increase of the output power level in an oscillating B from 0.05 W up to $P_{\text{out}} = 0.5; 1.5$ W control of the spectral characteristics of generated signals has a different regularity. Control of the frequency change $\Delta\nu_0$ at such power levels was observed at the values of magnetic field beginning with $H_0 > 150$ Oe. In the range of change of the field H_0 = (150–225) Oe, the value $\alpha = +6.7$ kHz/Oe. At $H_0 >$ (225–325) Oe, the value $\Delta\nu$ = const for P_{out} = (0.5–1.0) W.

A similar dependence and behavior regularity of $\Delta\nu(H_0)$ was observed at a higher power level ($P_{\text{out}} = 3$ W).

In Figs. 1.66–1.68, the experimental dependences illustrate the dynamics of change of the shape of the spectral lines of generated signals on the bias field H_0 for powers $P_{\text{out}} \cong$ (0.05–3) W.

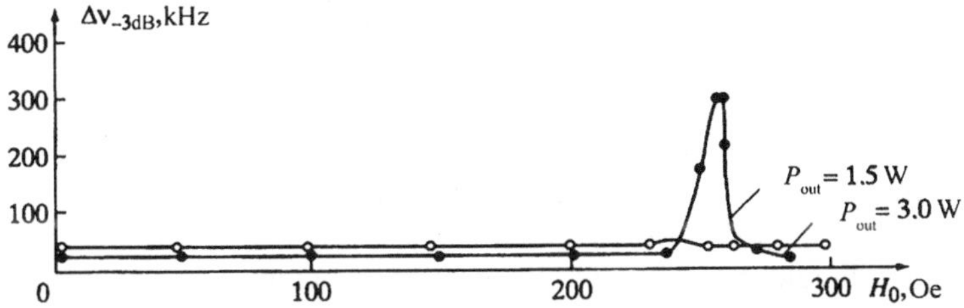

Fig. 1.67 The dependences of the spectral linewidth $\Delta\nu_{-3\text{dB}}$ on the magnetic field H_0 for power levels $P_{\text{out}} = 1.5$ and $P_{\text{out}} = 3$ W

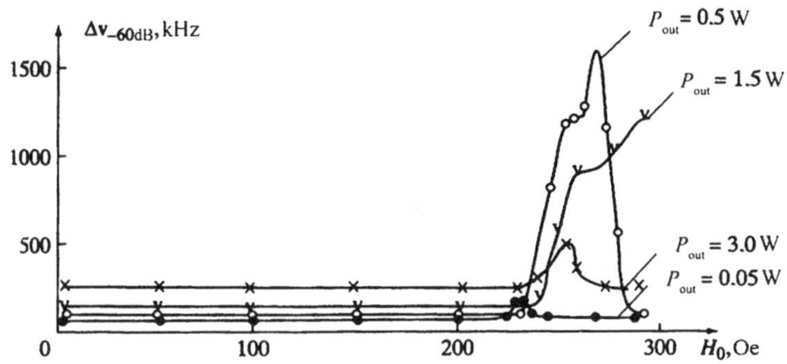

Fig. 1.68 The dependencies of the spectral linewidth by the base ($\Delta\nu_{-60\text{dB}}$) on the magnetic field H_0 for various values of integral power $P_{\text{out}} = 0.05; 0.5; 1.5; 3.0$ W

In Fig. 1.66, the dependences $\Delta\nu_{-3\text{dB}}$ on the field H_0 for $P_{\text{out}} = 0.05$ and $P_{\text{out}} = 0.5$ W are shown.

It is obvious that at a low output power level ($P_{\text{out}} = 0.05$ W) the spectral linewidth practically did not change ($\Delta\nu_{-3\text{dB}} = \text{const}$) within the limits of change of the magnetic field $H_0 = (0–300)$ Oe. At increase of the output power level by an order of magnitude ($P_{\text{out}} \cong 0.5$ W), beginning with certain values of magnetic field ($H_0 \geq 227$ Oe), a significant change of the spectral linewidth $\Delta\nu_{-3\text{dB}}$ takes place. Thus, at change of the field within the limits of $H_0 = (227–285)$ Oe the spectral linewidth changes from the minimum value $(\Delta\nu_{-3\text{dB}})_{\min} \cong 25$ kHz up to the maximum values: $(\Delta\nu_{-3\text{dB}})_{\min} \cong 400$ kHz at $H_0 = 252$ Oe and $(\Delta\nu_{-3\text{dB}})_{\max} \cong 300$ kHz at $H_0 = 258$ and 262 Oe.

At $H_0 > 285$ Oe $(\Delta\nu_{-3\text{dB}})_{\min} \cong 25$ kHz again. We shall note that at $H_0 \cong 255$ Oe and $H_0 \cong 262$ Oe the spectral linewidth $\Delta\nu_{-3\text{dB}} \cong 50$ kHz is close to the minimally observed value ($\Delta\nu_{-3\text{dB}} = 25$ kHz) in the whole range of change of the magnetic field H_0.

From the data of Fig. 1.66, it follows that in the range of change of the magnetic field $H_0 = (225–285)$ Oe an increase of the spectral linewidth $\Delta\nu_{-3\text{dB}}$ by 20 times is observed at $H_0 \cong (225–252)$ Oe. At change of the field within the limits of $H_0 \cong (252–255)$ Oe the spectral linewidth decreases by 18 times ($\Delta\nu_{-3\text{dB}} = 50$ kHz). At further change of the field $H_0 = (255–258)$ Oe the spectral line $\Delta\nu_{-3\text{dB}}$ widens by 6 times. In the range of change of the field $H_0 = (258–262)$ Oe the spectral line changes by 6 times again and reaches a value $\Delta\nu_{-3\mathrm{dB}} = 50\,\mathrm{kHz}$. At $H_0 =$ (260–262) Oe widening of the spectral line by 6 times ($\Delta\nu_{-3\mathrm{dB}} \cong 300\,\mathrm{kHz}$) is observed, and at $H_0 =$ (262–285) Oe the spectral line gets narrow again by 12 times down to a value $\Delta\nu_{-3\mathrm{dB}} = 25\,\mathrm{kHz}$. Thus, in the oscillator with heteromagnetic interaction the spectral linewidth at the continuous power $P_{\mathrm{out}} = 0.5\,\mathrm{W}$ is changed by the magnetic field H_0 by 10–20 times.

In Fig. 1.67, the dependencies of the spectral linewidth $\Delta\nu_{-3\mathrm{dB}}$ on the magnetic field H_0 for power levels $P_{\mathrm{out}} = 1.5$ and $P_{\mathrm{out}} = 3\,\mathrm{W}$ are shown. As well as in the previous case, the most significant widening of the spectral line takes place at $H_0 \cong$ 258 Oe for $P_{\mathrm{out}} = 1.5\,\mathrm{W}$. The line widens more than by an order of magnitude. At $P_{\mathrm{out}} = 3\,\mathrm{W}$ widening of the spectral line was observed and $\Delta\nu_{-3\mathrm{dB}} \cong 40\,\mathrm{kHz}$ in the whole range of change of H_0.

In Fig. 1.68, the dependencies of the spectral linewidth by the base ($\Delta\nu_{-60\mathrm{dB}}$) on the magnetic field H_0 for various values of integral power $P_{\mathrm{out}} = 0.05; 0.5; 1.5;$ 3.0 W are given.

From comparison of Figs. 1.66–1.68, it follows that significant changes of the shape of the spectral line take place when the magnetic field $H_0 \cong$ (230–290) Oe. The most significant widening of the spectral line base falls on the range of magnetic field $H_0 \cong$ (230–270) Oe. It is obvious that the maximum widening of the spectral line pedestal ($\Delta\nu_{-60\mathrm{dB}}$) is observed at $P_{\mathrm{out}} \cong 0.5\,\mathrm{W}$.

In Fig. 1.69, the dependence of the maximum value of the spectral linewidth base $(\Delta\nu_{-60\mathrm{dB}})_{\max}$ on the power level P_{out} is given. It is obvious that the steepness $(\Delta\nu_{-60\mathrm{dB}})_{\max}/P_{\mathrm{out}} \cong 2.8$, MHz/W lays in the range of $P_{\mathrm{out}} =$ (0.05–0.5) W. When $P_{\mathrm{out}} > 0.5\,\mathrm{W}$ we have $(\Delta\nu_{-60\mathrm{dB}})_{\max}/P_{\mathrm{out}} \cong -0.56\,\mathrm{MHz/W}$.

Our investigations show that in oscillating heteromagnetic structures at continuous power levels up to 3 W the following take place:

- A change of the spectral linewidth by several orders of magnitude
- A sign-changing character of the steepness of change of the central frequencies of the spectral components depending on the magnetic field
- Various laws and values of the steepness of change of the central frequency of signals depending on the feed voltage of a semiconductor subsystem

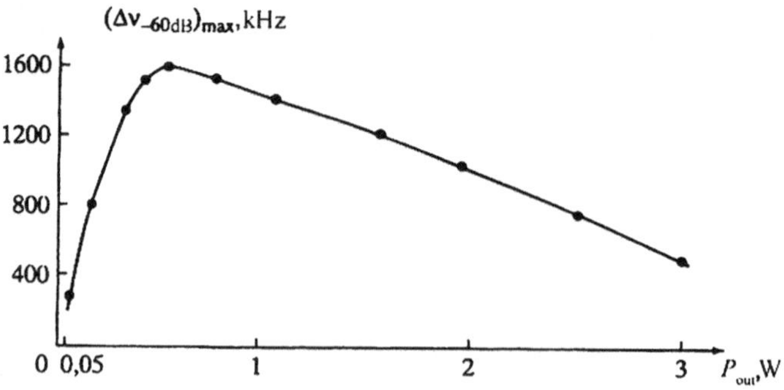

Fig. 1.69 The dependence of the maximum value of the spectral linewidth base $(\Delta\nu_{-60\mathrm{dB}})_{\max}$ on the power level P_{out}

- An increase of the change steepness of signal by frequency and expansion of the limits of change of the spectral line shape of signals at increase of the saturation magnetization of ferrite.

1.4 Signal Spectra of Heteromagnetic Interactions on High Power Levels

The multifunction modes of formation of various kinds of spectra in an oscillator with heteromagnetic interaction at power levels up to 3 W are presented in the corresponding oscillograms (Figs. 1.70–1.102) and in Tables 1.5–1.8.

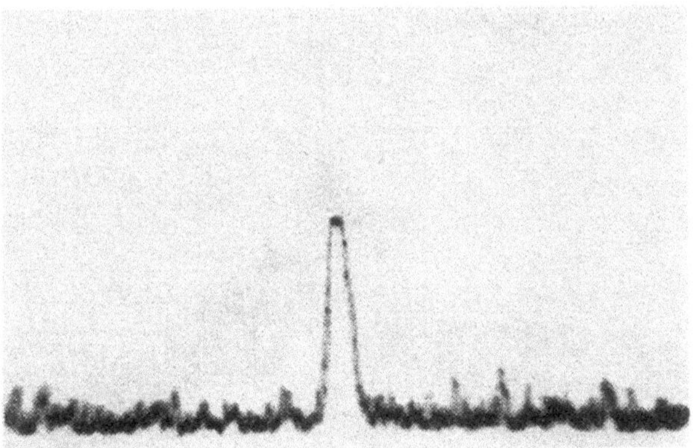

Fig. 1.70 The spectra of the signal at $P_{out} = 0.05\,W$

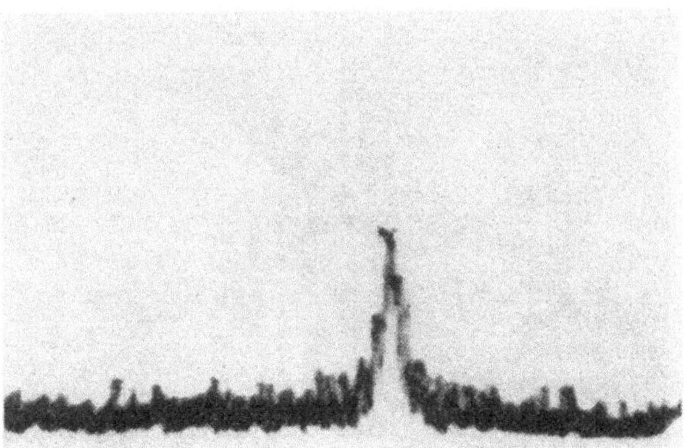

Fig. 1.71 The spectra of the signal at $P_{out} = 0.05\,W$

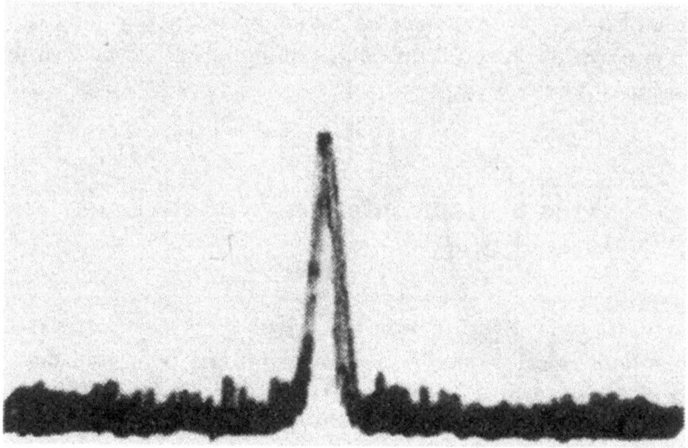

Fig. 1.72 The spectra of pseudonoise signal at $P_{out} = 0.5\,W$

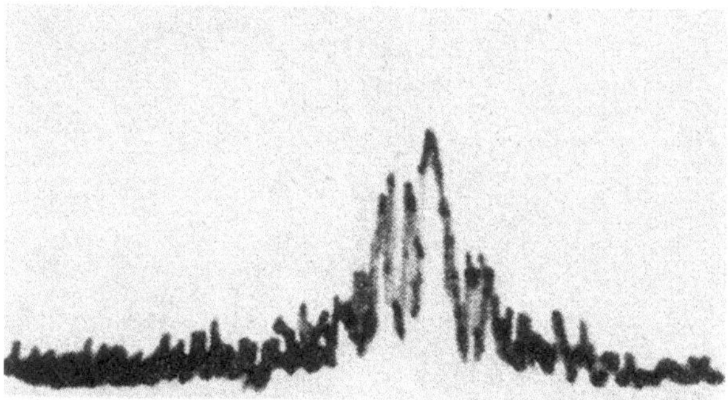

Fig. 1.73 The spectra of pseudonoise signal at $P_{out} = 0.5\,W$

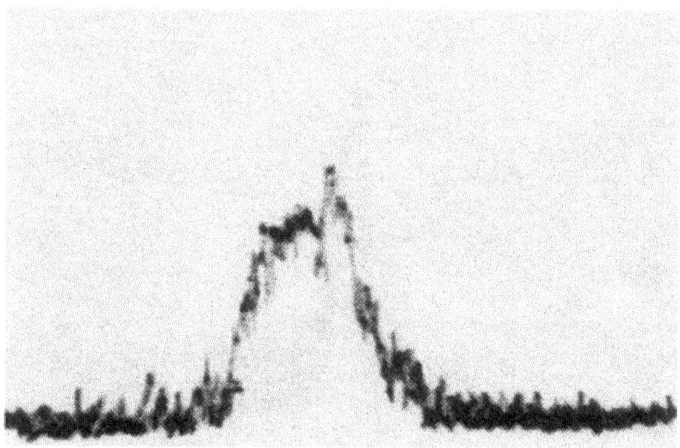

Fig. 1.74 The spectra of pseudonoise signal at $P_{out} = 0.5\,W$

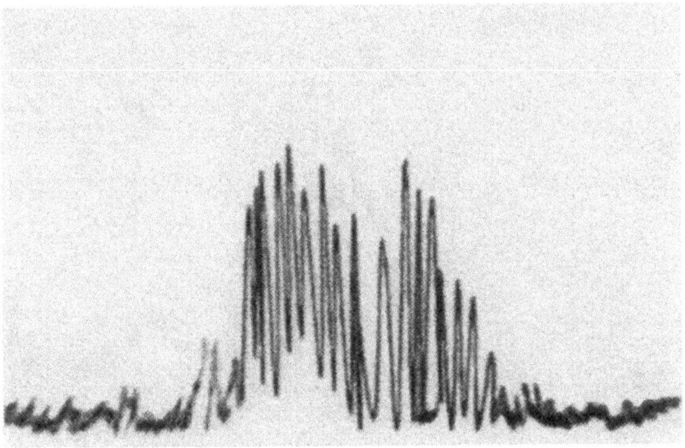

Fig. 1.75 The spectra of pseudonoise signal at $P_{out} = 0.5\,W$

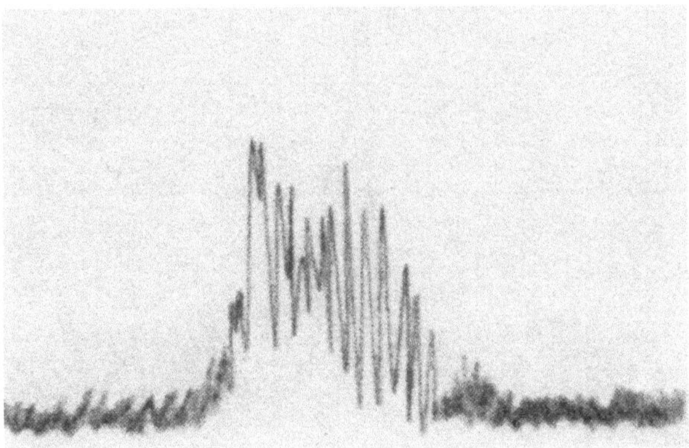

Fig. 1.76 The spectra of pseudonoise signal at $P_{out} = 0.5\,W$

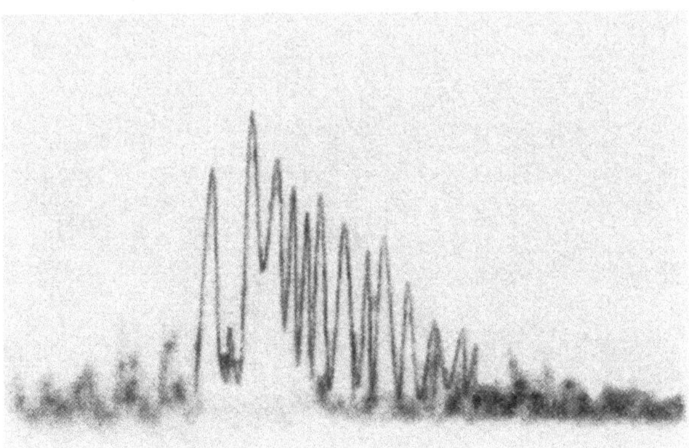

Fig. 1.77 The spectra of pseudonoise signal at $P_{out} = 0.5\,W$

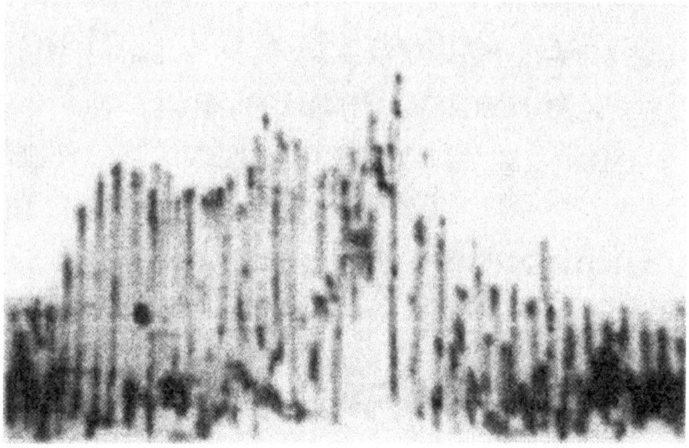

Fig. 1.78 The spectra of pseudonoise signal at $P_{out} = 0.5\,W$

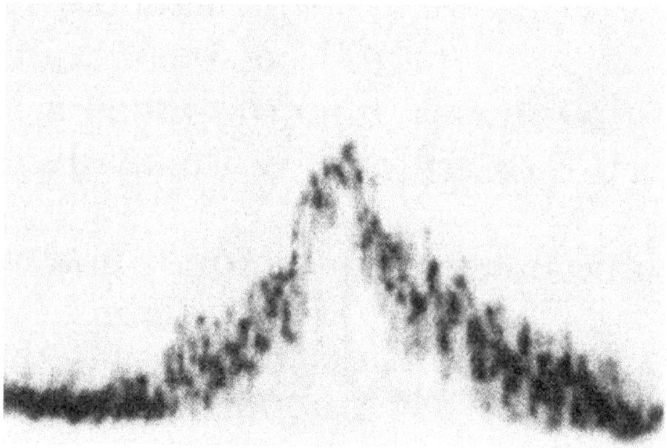

Fig. 1.79 The spectra of pseudonoise signal at $P_{out} = 0.5\,W$

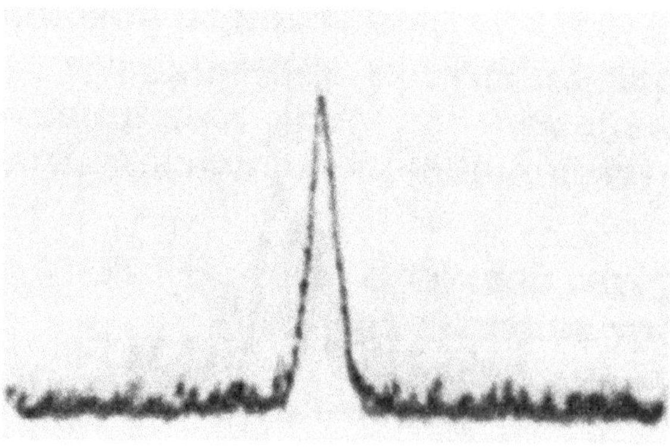

Fig. 1.80 The spectra of the signal at $P_{out} = 1.5\,W$

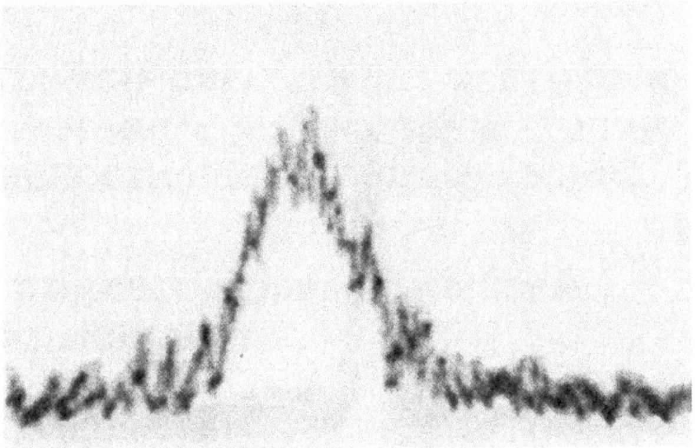

Fig. 1.81 The spectra of the signal at $P_{out} = 1.5\,W$

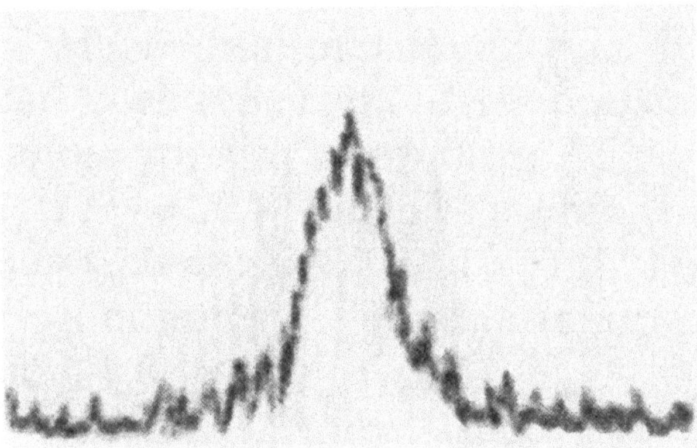

Fig. 1.82 The spectra of the signal with pedestal noisiness

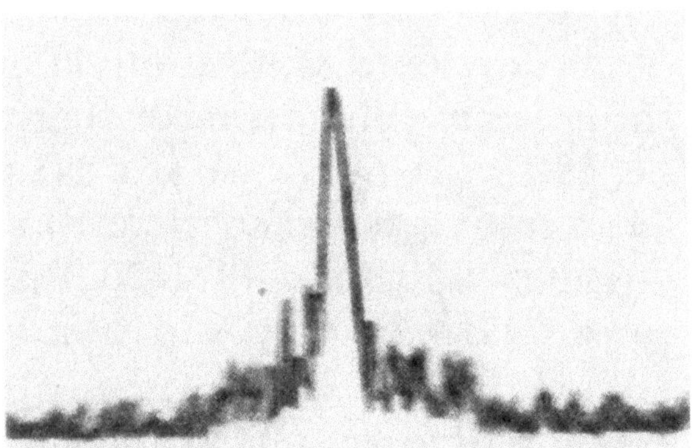

Fig. 1.83 The spectra of the signal with pedestal noisiness

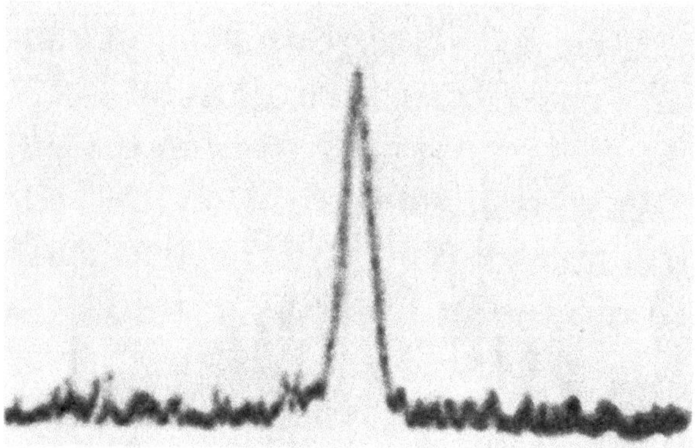

Fig. 1.84 The spectra of the signal at a field $H_{01} = 0\,\text{Oe}$

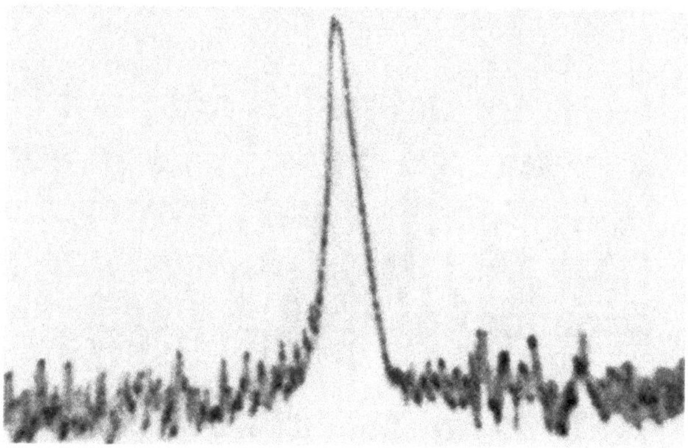

Fig. 1.85 The spectra of the signal at a field $H_{02} = 100\,\text{Oe}$

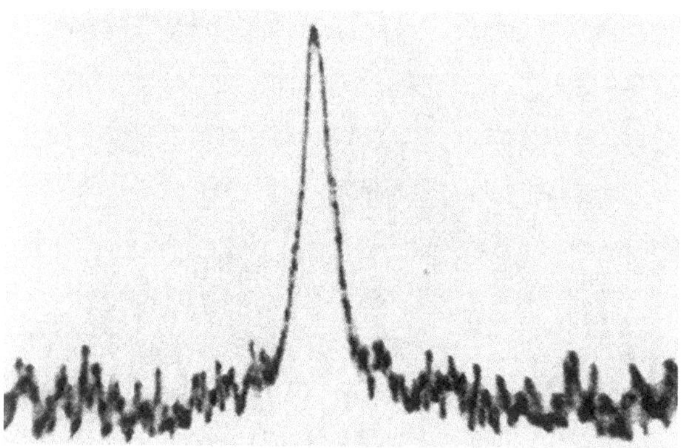

Fig. 1.86 The spectra of the signal at a field $H_{03} = 230\,\text{Oe}$

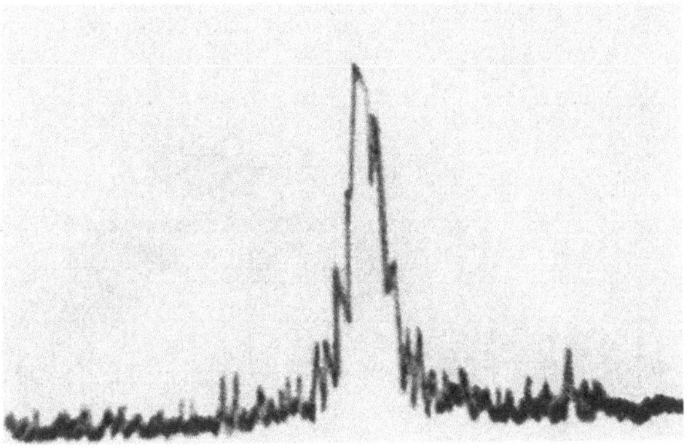

Fig. 1.87 The spectra of the signal at a field $H_{04} = 241$ Oe

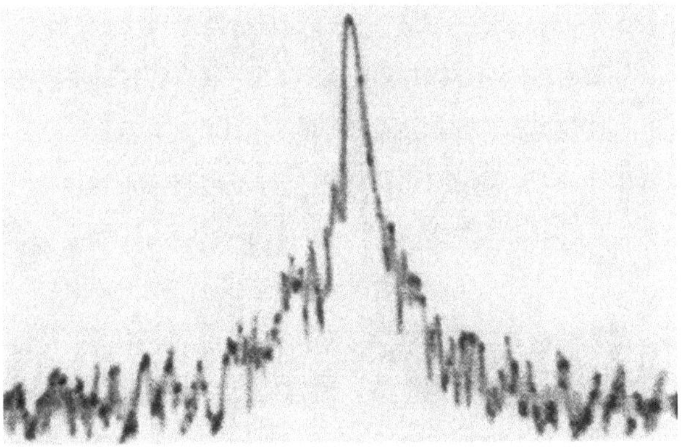

Fig. 1.88 The spectra of the signal at a field $H_{05} = 251$ Oe

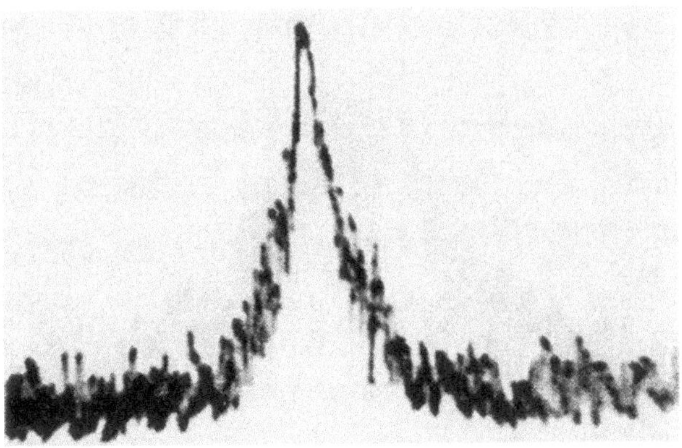

Fig. 1.89 The spectra of the signal at a field $H_{06} = 260$ Oe

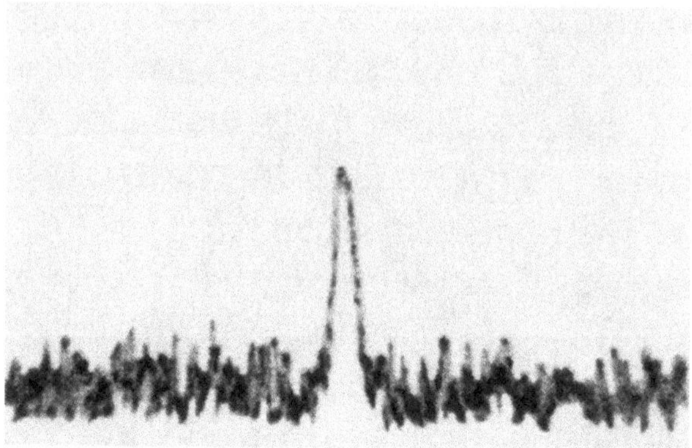

Fig. 1.90 The spectra of pseudonoise signal

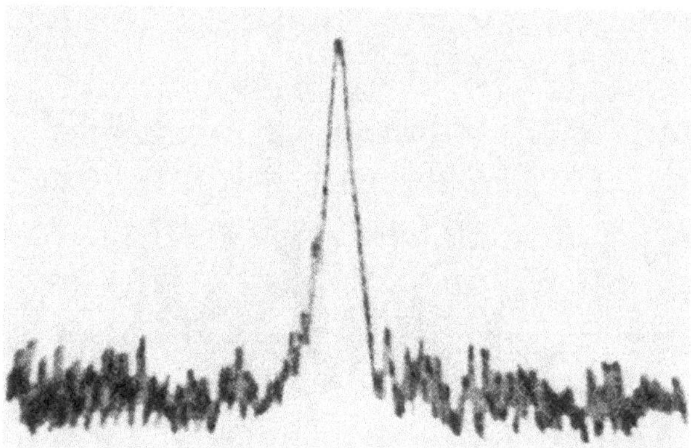

Fig. 1.91 The spectra of pseudonoise signal

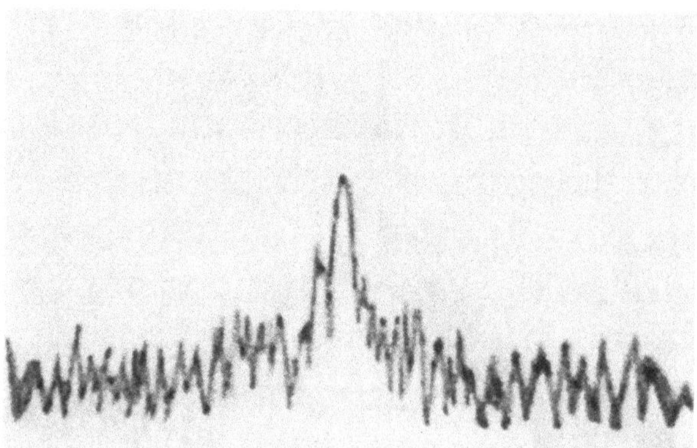

Fig. 1.92 The spectra of pseudonoise signal

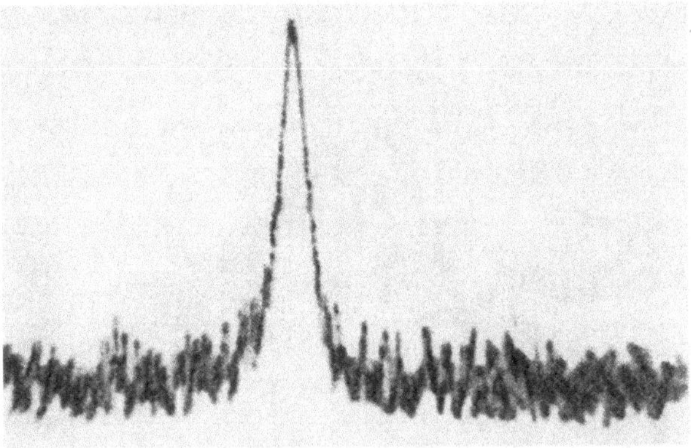

Fig. 1.93 The spectra of pseudonoise signal

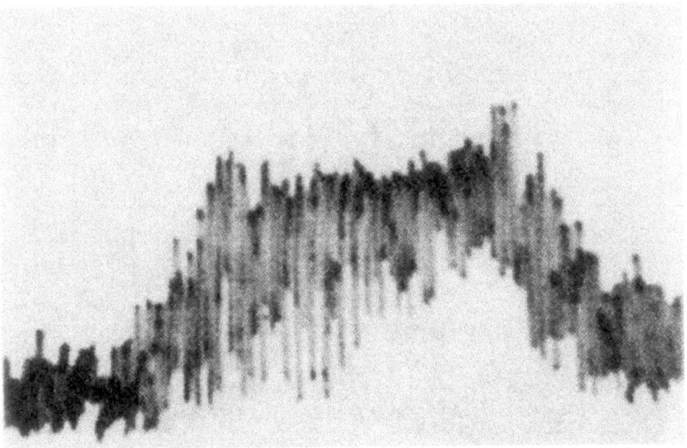

Fig. 1.94 The spectra of pseudonoise signal

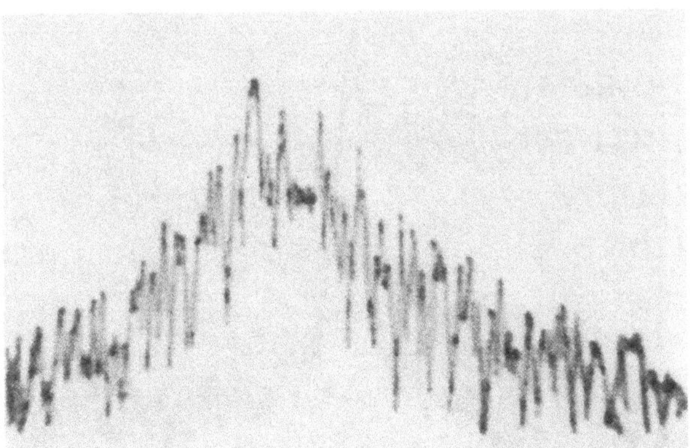

Fig. 1.95 The spectra of pseudonoise signal

Fig. 1.96 The spectra of pseudonoise signal

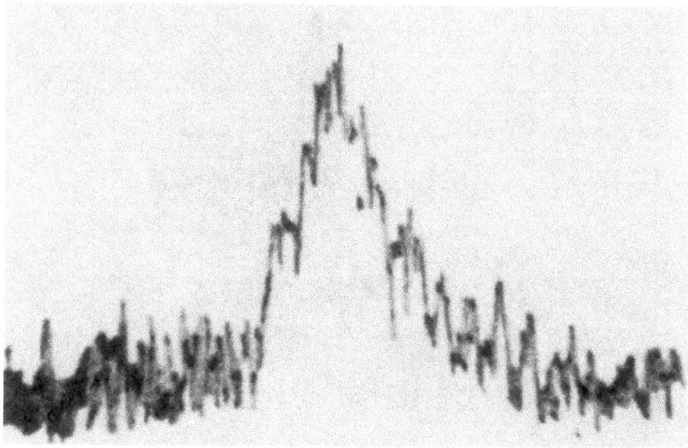

Fig. 1.97 The spectra of pseudonoise signal

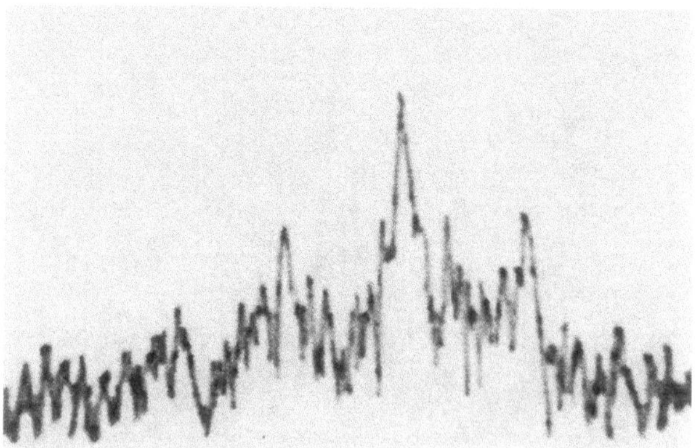

Fig. 1.98 The spectra of pseudonoise signal

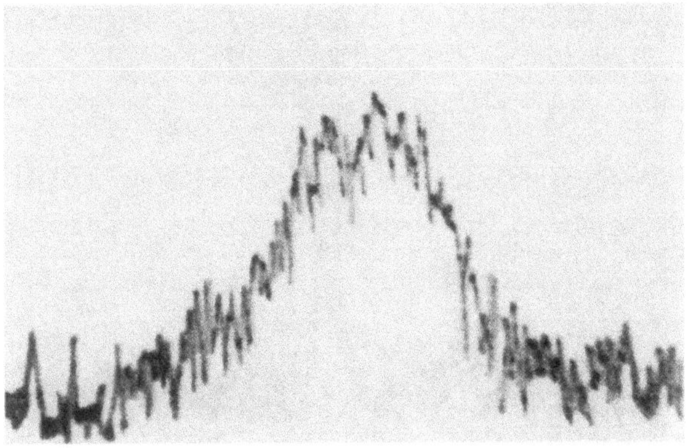

Fig. 1.99 The spectra of pseudonoise signal

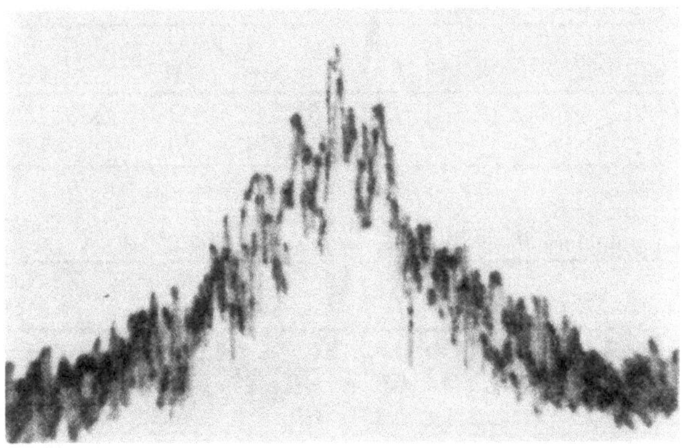

Fig. 1.100 The spectra of pseudonoise signal

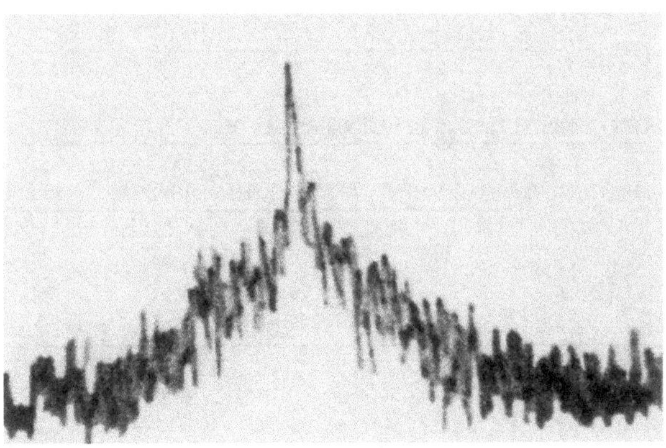

Fig. 1.101 The spectra of pseudonoise signal

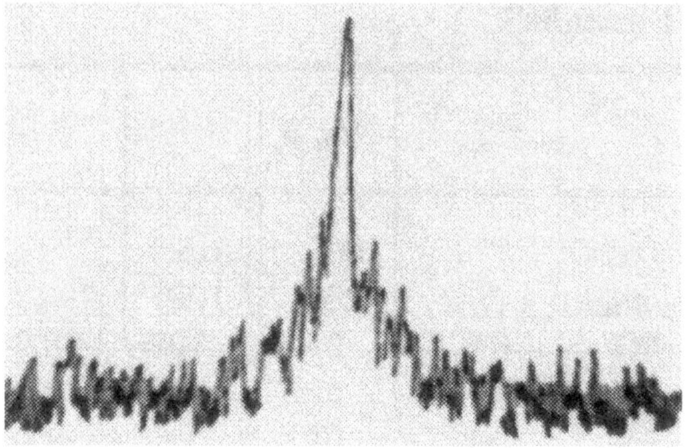

Fig. 1.102 The spectra of pseudonoise signal

Table 1.5 Parameters of signal spectra in various modes

Oscillogram	H_0, Oe	U_c, V	U_e, V	I_c, A	ν_0, MHz	$\Delta\nu_{-3dB}$, kHz	$\Delta\nu_{-60dB}$, kHz	P_{out}, W
Fig. 1.70	0	1.5	1	0.18	652	20	50	0.05
Fig. 1.71	225	1.5	1	0.18	655	30	150	0.05

Table 1.6 Spectra parameters of pseudonoise signals at $P_{out} = 0.5$ W

Oscillogram	H_0, Oe	U_c, V	U_e, V	I_c, A	ν_0, MHz	$\Delta\nu_{-3dB}$, kHz	$\Delta\nu_{-60dB}$, kHz	P_{out}, W
Fig. 1.72	0	9	2.2	0.32	677	30	500	0.5
Fig. 1.73	233	9	2.2	0.32	677	30	500	0.5
Fig. 1.74	235	9	2.2	0.32	676	100	450	0.5
Fig. 1.75	239	9	2.2	0.32	676	150	500	0.5
Fig. 1.76	240	9	2.2	0.32	677	100	600	0.5
Fig. 1.77	241	9	2.2	0.32	677	100	800	0.5
Fig. 1.78	248	21	3.0	0.80	747	200	1,000	0.5
Fig. 1.79	258	9	2.2	0.32	677	50	1,000	0.5

Table 1.7 Parameters of signal spectra in various modes

Oscillogram	H_0, Oe	U_c, V	U_e, V	I_c, A	ν_0, MHz	$\Delta\nu_{-3dB}$, kHz	$\Delta\nu_{-60dB}$, kHz	P_{out}, W
Fig. 1.80	0	15	3	0.73	778.0	15	100	1.5
Fig. 1.81	254.5	15	3	0.73	778.5	100	700	1.5
Fig. 1.82	265.0	15	3	0.73	778.5	50	500	1.5
Fig. 1.83	290.0	15	3	0.73	778.5	20	500	1.5

In the oscillograms (Figs. 1.70 and 1.71), spectra of the signals in an oscillator with heteromagnetic interaction at a low power level $P_{out} = 0.05$ W for $H_0 = 0$ and $H_0 = 225$ Oe are shown. At $H_0 = 225$ Oe widening of the spectral line and

Table 1.8 Parameters of signal spectra in various modes

Oscillogram	H_0, Oe	U_c, V	U_e, V	I_c, A	ν_0, MHz	$\Delta\nu_{-3dB}$, kHz	$\Delta\nu_{-60dB}$, kHz	P_{out}, W
Fig. 1.84	0	21	3	0.8	790	20	130	3.0
Fig. 1.85	100	21	3	0.8	793	25	250	3.0
Fig. 1.86	230	21	3	0.8	791	30	250	3.0
Fig. 1.87	241	21	3	0.8	792	50	200	3.0
Fig. 1.88	251	21	3	0.8	790	30	500	3.0
Fig. 1.88	260	21	3	0.8	790	30	300	3.0

its noisiness (Table 1.5) were observed. In Figs. 1.72–1.79, spectra of pseudonoise signals at a power $P_{out} = 0.5$ W, and in Table 1.6, the corresponding parameters are shown.

In oscillograms 1.77–1.79, the spectra of pseudonoise signals for $H_0 = (233–258)$ Oe at $P_{out} = 0.5$ W are shown. It is obvious that the spectral linewidth of the pseudonoise signal $\Delta\nu_{-3dB}$ changes by 6 times, $\Delta\nu_{-60dB}$ by 2–2.5 times, and the coefficient of the spectrum shape changes from 3.5 up to 20.

In oscillograms 1.80–1.83 and in Table 1.7, the modes and parameters of the spectra of signals at $P_{out} = 1.5$ W for $H_0 = (0–290)$ Oe are given. At change of H_0 uniform widening of the envelope of the spectral line of signal (Figs. 1.80 and 1.81) by ~7 times, passing into the mode of broadband pedestal noisiness (Figs. 1.82 and 1.83) is observed.

In oscillograms 1.84–1.89 and in Table 1.8, the spectra of signals on the output of an oscillator with heteromagnetic interaction are shown, the modes and parameters without magnetic field (Fig. 1.84) and in the presence of a bias field (Figs. 1.85–1.89) at an output power level $P_{out} = 3$ W are given.

At change of the field H_0 from 0 up to 260 Oe a sign-changing frequency deviation within the limits of ±1.5 MHz was observed and a monotonous character of increase of the noisiness level of the spectral lines and pedestal (Fig. 1.89) was noted.

In oscillograms 1.90–1.102, the spectra of various pseudonoise signals at output power levels $P_{out} = 0.05–2.3$ W are shown, and in Table 1.9 the feed modes and parameters of spectra are given.

In oscillograms 1.85, 1.90, 1.91, the spectral lines of signals at output power levels of 3.0, 0.05 and 1.5 W, respectively, at a magnetic field $H_0 = 100$ Oe are shown. Change of P_{out} was carried out due to choice of the feed voltage of a bipolar transistor U_e and U_c that led to preservation of the shape of the spectral line and to change of the frequency within the limits of 60 MHz.

In oscillograms 1.86, 1.92, and 1.93, the spectra of output signals at output power levels of 3.0, 0.05, and 1.5 W, respectively, at a magnetic field $H_0 = 230$ Oe and various feed voltages U_e and U_c are given. At $P_{out} = 1.5$ W and $P_{out} = 3.0$ W the shapes of the spectral lines are kept, and the frequency changes by 14 MHz.

In oscillograms 1.88 and 1.94–1.97, typical spectra of output signals at output power levels of 3.0, 0.5, 1.0, 1.5, and 2.3 W, respectively, for the field $H_0 = 251$ Oe

Table 1.9 The feed modes and the bases parameters of pseudonoise signals for oscillograms, which showed in Figs. 1.90–1.102

Oscillogram	H_0, Oe	U_c, V	U_e, V	I_c, A	ν_0, MHz	$\Delta\nu_{-3dB}$, kHz	$\Delta\nu_{-60dB}$, kHz	P_{out}, W
Fig. 1.90	100	21	3	0.8	735	15	80	0.05
Fig. 1.91	100	21	3	0.8	781	25	200	1.5
Fig. 1.92	230	21	3	0.8	733	30	200	0.05
Fig. 1.93	230	21	3	0.8	777	20	160	1.5
Fig. 1.94	251	21	3	0.8	748	600	800	0.5
Fig. 1.95	251	21	3	0.8	760	50	1,000	1.0
Fig. 1.96	251	21	3	0.8	773	70	800	1.5
Fig. 1.97	251	21	3	0.8	773	50	600	2.3
Fig. 1.98	260	21	3	0.8	747	50	1,600	0.5
Fig. 1.99	260	21	3	0.8	760	150	700	1.0
Fig. 1.100	260	21	3	0.8	775	50	800	1.5
Fig. 1.101	273	21	3	0.8	759	15	700	1.0
Fig. 1.102	273	21	3	0.8	780	15	300	2.0

are given. At increase of the output power level from 0.5 up to 3.0 W a change of the coefficient of the shape of the pseudonoise spectral line practically by an order of magnitude is observed. Envelopes of the spectral lines from nearly rectangular and broadband at $P_{out} = 0.5$ W (Fig. 1.94) to narrow-band ones at $P_{out} = 3$ W (Fig. 1.88) can be realized at change of the spectral linewidth $\Delta\nu_{-3dB}$ practically by 20 times.

In oscillograms 1.89 and 1.98–1.100, the spectra of output signals at power levels of 3.0, 0.5, 1.0, and 1.5 W for a magnetic field $H_0 = 260$ Oe are given. It is obvious that pseudonoise signals with a various width of their spectrum can be generated; the passband $\Delta\nu_{-3dB}$ changes by 5–10 times, but with various widening of the base ($\Delta\nu_{-60dB}$) and uniformity of the PSD. Figure 1.98 shows a pseudonoise signal close to an equidistant spectrum.

In oscillograms 1.101 and 1.102, the spectra of signals with an increased level of their noisiness in the range of Doppler and intermediate tuning out frequencies from the carrier frequency ν_0 down to the level of minus 20 dB from the amplitude level of the carrier frequency at $\Delta\nu_{-3dB} = 15$ kHz and integral output powers $P_{out} = 1$ W and $P_{out} = 2$ W, respectively, are shown. The magnetic field is $H_0 = 273$ Oe.

Let us notice that the above spectra of signals (Figs. 1.70–1.102) are observed at high power levels on an oscillating heteromagnetic transistor at preservation of the technical efficiency of the used CVS (in our case, a transistor). Multifunction interactions were observed on both the reference frequency ν_0 and the harmonic frequencies ν_1 and ν_2.

Our preliminary experiments have confirmed the multifunction character of interactions in the oscillating operating modes of FSS on various types of transistors (bipolar, field ones), at various power levels (low, medium, high), on the harmonic and subharmonic spectral components in the multioctave, superbroadband frequency range. The possibility to control the power and spectral characteristics of output signals by both the bias field, including the modes of small magnetic fields close to self-resonances (FMCR, multidomain, and single-domain modes), and the feed voltages of CVS is shown. From physical reasons, generalized equivalent circuits of bipolar and field powerful magnetotransistors have been designed. On one crystal in the active operating mode of a transistor with FMCR there can be observed: spectrally pure signals, pseudonoise signals with PSD control, and nonuniformity of noise both near the carrier frequency in the range of Doppler tuning out frequencies and at tuning out from it in the range of intermediate tuning out frequencies. Such signals are formed as equidistant frequency spectra with control of the frequency distances between the spectral components and their noisiness level.

A high efficiency of the multifunction interactions in FSS oscillators with the technical efficiency of the basic transistor, reaching 40–55%, has been found.

There were revealed: a complex multiparameter character of the physical interactions in FSS, synchronous control of frequency change of all the spectral components, control of the GFC shape, noise parameters and characteristics of signals on various channels (on the bias field, the level of HF power, feed voltage of the transistor). These circumstances determined the necessity of experimental investigations of the properties of oscillating FSS with various types and parameters of FMCR, their orientations in the external magnetic bias field on medium (tens of mW) and high (hundreds of mW) power levels.

Chapter 2
Properties of Structures with Ferrites of Different Magnetizations

2.1 General Remarks

The complex character of the physical processes, the multifunctionality, and multiparametricity of the interactions in ferrite-transistor structures of various types on low (up to several mW), medium (up to 100 mW), and high (above 1–5 W) power level has defined the necessity of carrying out experiments on FSS for:

- Signals of various kinds (regular, noise-like, noise ones as a frequency synthesizer) on the basic frequency and higher harmonics
- Ferrites of various types, various magnetic parameters, their orientations relative to the semiconductor subsystem and the bias magnetization field $\overline{H}_0$

For definiteness the scheme of an oscillator on a bipolar transistor with a common base has been chosen. FMCR of various ferrite-garnet types (YIG with lowered magnetizations) and with various angles of orientation φ in the bias magnetization field $\overline{H}_0$ were introduced into the area of beam emitter–collector electrodes. The modes of generation of demanded power levels were selected by means of the voltages on emitter U_e and collector U_c of the transistor.

At our experimental research the following factors were varied:

- The brand of FMCR (saturation magnetization $4\pi M_s$, FMR line halfwidth ΔH)
- The angle of FMCR orientation in the external bias magnetization field $\overline{H}_0$; the level of target power
- The kinds of signal spectra (SPS, NS, a signal like a frequency synthesizer as ES FS) on the basic (fundamental) frequency ν_0, on the first ν_1 and second ν_2 harmonics, the width of the spectrum on a -60 dB level, the width of the spectral line at a -3 dB level from the peak value of the signal – $(\Delta\nu_{-3\text{dB}})_{0,1,2}$, and that on the level of the basis or pedestal of the spectral line – $(\Delta\nu_{-60\text{dB}})_{0,1,2}$
- The level of integrated target power P_{out} on three frequency components ν_0, ν_1, ν_2, $P_{\text{out}} = P_{\text{out}}(\nu_0) + P_{\text{out}}(\nu_1) + P_{\text{out}}(\nu_2)$

The following parameters of signal spectra of generating HMT were experimentally investigated on low (mWs) and high (Ws) power levels for spectral components

A.A. Ignatiev and A.V. Lyashenko, *Heteromagnetic Microelectronics: Microsystems of Active Type*, DOI 10.1007/978-1-4419-6002-3_2,

of frequencies on the basic (fundamental) harmonic ν_0 and on the first and second harmonics ν_1 and ν_2 from the magnetic field H_0:

- The change of the central frequencies of signals on the basic and harmonic spectral components $\Delta\nu_{0,1,2}(H_0)$
- The spectral linewidth $(\Delta\nu_{-3\text{dB}})_{0,1,2}(H_0)$ and $\Delta\nu_{-60\text{dB}}(H_0)$
- The integrated target power $P_{\text{out}}(H_0)$
- The kinds of spectra (SPS, NS, ES FS, ES NT FS)

The generating HMT have been made on the basis of a KT962B transistor and FMCR of various brands (KG-8, KG-15, KG-65, KG-140) with various orientations (the angle φ) in the external magnetic field $\overline{H}_0$ at medium and high power levels.

2.2 Structures with Ferrite KG-8

First, consider the properties of HMT on a low power level $P_{\text{out}} = 50\,\text{mW}$ (hereinafter $4\pi M_s = 90\,\text{G}$).

In Fig. 2.1, the dependencies of signal frequency deviations $\Delta\nu_0$, $\Delta\nu_1$, $\Delta\nu_2$ for signals on ν_0, ν_1, ν_2 on the magnetic field H_0 are shown. Figures 2.2 and 2.3 present the dependencies of the spectral characteristics of generated signals $(\Delta\nu_{-3\text{dB}})_{0,1,2}$ and $(\Delta\nu_{-60\text{dB}})_{0,1,2}$, respectively, in the range of the bias field H_0. Figures 2.4 and 2.5 depict similar dependencies in a narrow range of magnetic field H_0.

From Fig. 2.1, it is obvious that at the magnetic field $H_0 \approx 190$ Oe the deviations of frequencies of all the spectral components $\nu_{0,1,2}$ change most strongly.

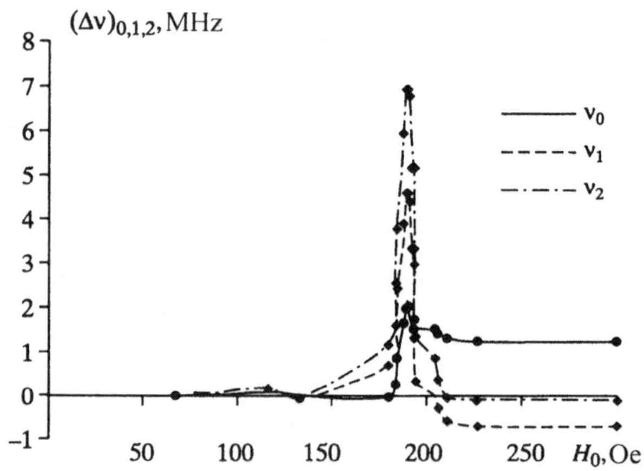

Fig. 2.1 The dependencies of signal frequency deviations $\Delta\nu_0$, $\Delta\nu_1$, and $\Delta\nu_2$ for signals on ν_0, ν_1, and ν_2 on the magnetic field H_0

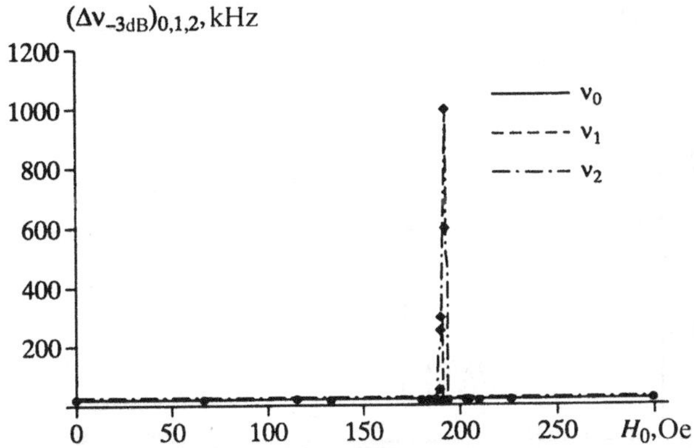

Fig. 2.2 The dependencies of the spectral characteristics of generated signals $(\Delta\nu_{-3dB})_{0,1,2}$ and $(\Delta\nu_{-60dB})_{0,1,2}$

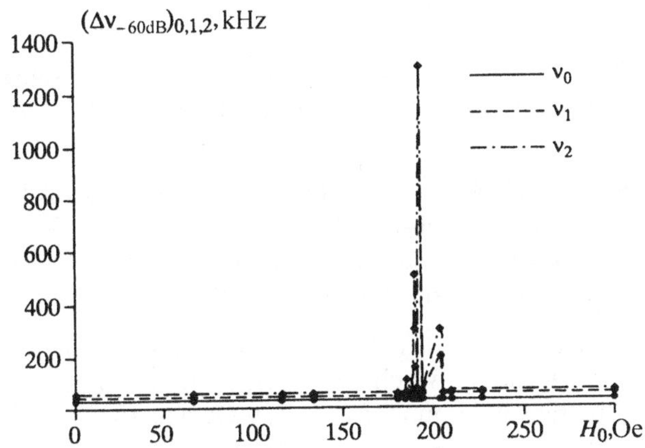

Fig. 2.3 The dependencies of the spectral characteristics of generated signals $(\Delta\nu_{-3dB})_{0,1,2}$ and $(\Delta\nu_{-60dB})_{0,1,2}$

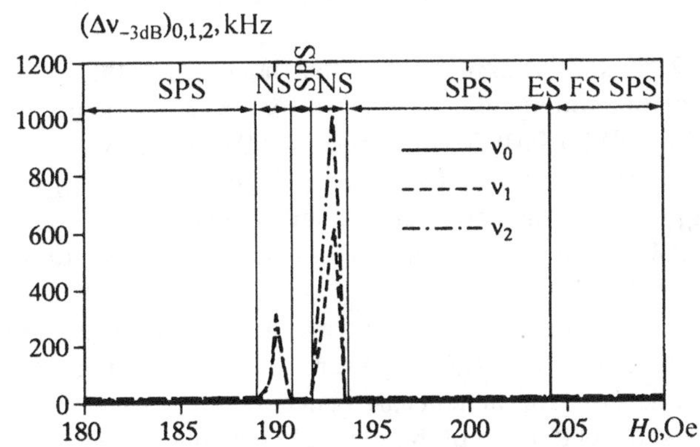

Fig. 2.4 The dependencies of spectral characteristics in a narrow range of magnetic field H_0

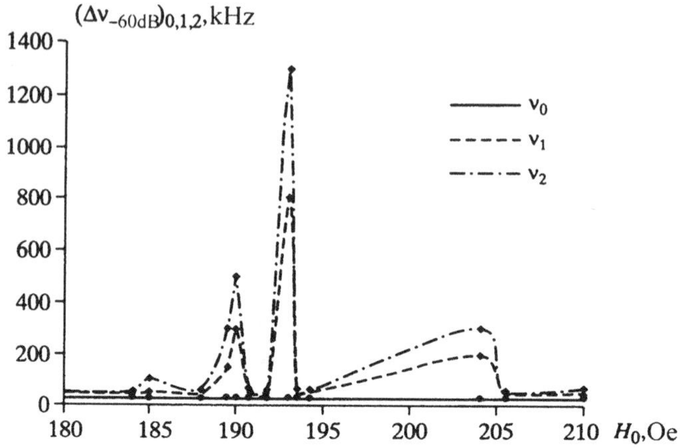

Fig. 2.5 The dependencies of spectral characteristics in a narrow range of magnetic field H_0

In Figs. 2.2–2.5, basic spectral characteristics of generated signals are shown. In Fig. 2.4, the characteristic ranges of changes of the shape of the spectra of generated signals are shown:

- Four areas (modes) in which SPS with the minimal band of the spectral line $(\Delta\nu_{-3dB})_{0,1,2} \approx 20\,\text{kHz}$ are observed
- Two areas of NS in which the spectral linewidth of a signal increases by 15–50
- One area of ES FS (like a frequency synthesizer)

In the autoresonance FMCR mode ($H_0 = 0$) SPS components are generated. Those signals are most broadband for which $(\Delta\nu_{-3dB})_{1,2}$ and $(\Delta\nu_{-60dB})_{1,2}$ are maximal.

At change of the bias field up to $H_0 < 190$ Oe and $H_0 \geq 205$ Oe spectrally pure signals, for which $(\Delta\nu_{-3dB})_{0,1,2} = \text{const}$ and $(\Delta\nu_{-60dB})_{0,1,2} = \text{const}$ were observed.

In Fig. 2.6, the dependence of the integrated target power of a heteromagnetic generator on the field magnitude H_0 is shown. Near $H_0 \approx 190$ Oe the HF power decreases.

In Figs. 2.7–2.12, similar experimental dependencies of the power and spectral characteristics of heteromagnetic generators for the components $\nu_{0,1,2}$ at a high power level (500 mW) for FMCR with $4\pi M_s = 90\,\text{G}$ are shown.

In Fig. 2.7, the dependencies of frequency deviations $(\Delta\nu)_{0,1,2}$ of the spectral components $\nu_{0,1,2}$ on the magnetic field H_0 are presented.

In Figs. 2.8–2.11, changes of the key parameters of the spectra of various kinds of signals are shown.

Figure 2.8 presents the existence ranges of characteristic kinds of signals, namely: two SPS ranges, one NS range, and one ES FS range. In the autoresonance FMCR mode, SPS components are generated.

In Fig. 2.11, the dependencies of the spectral linewidth of generation on a −60 dB level for the frequency components ν_0, ν_1, and ν_2 on the field H_0 are shown.

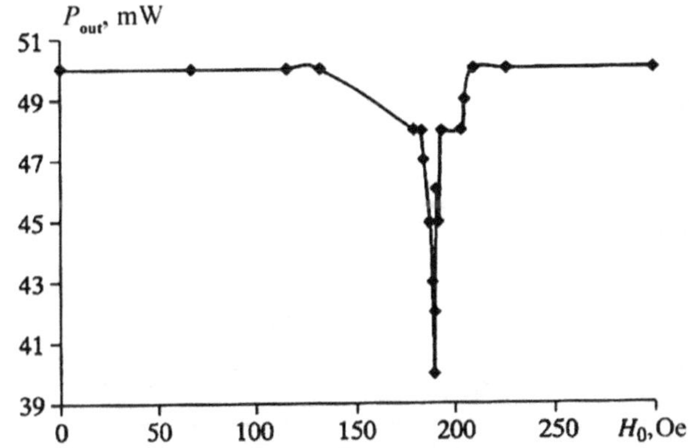

Fig. 2.6 The dependence of the integrated target power of a heteromagnetic generator on the field magnitude H_0

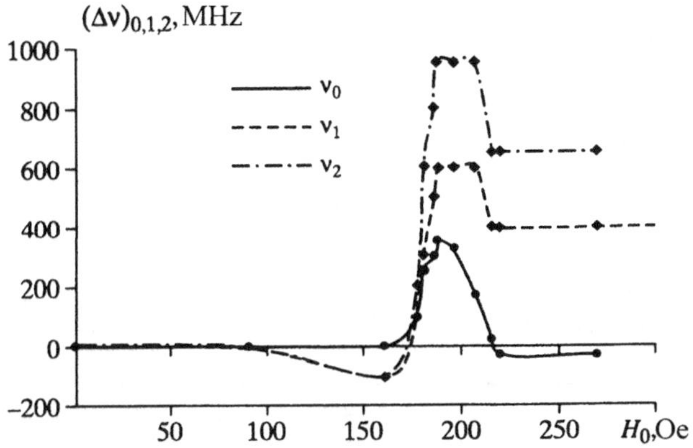

Fig. 2.7 The dependencies of frequency deviations $(\Delta\nu)_{0,1,2}$ of the spectral components $\nu_{0,1,2}$ on the magnetic field H_0

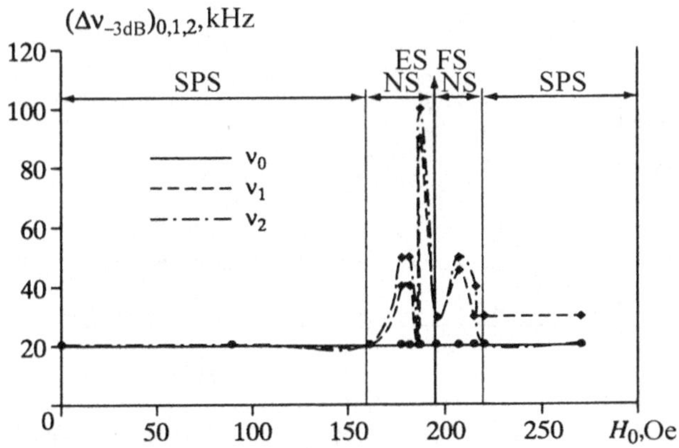

Fig. 2.8 Changes of the key parameters of the spectra of various kinds of signals

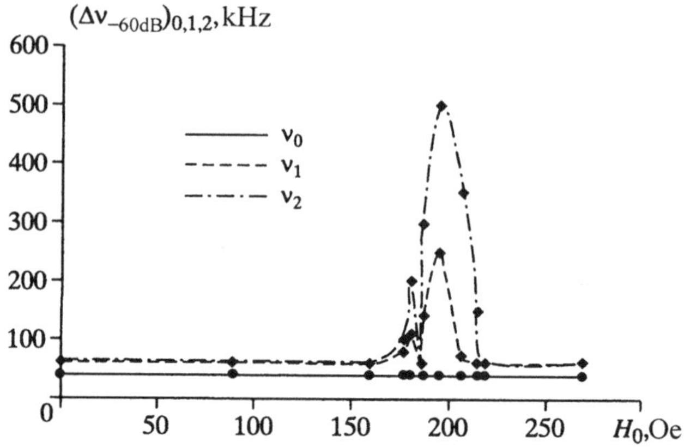

Fig. 2.9 Changes of the key parameters of the spectra of various kinds of signals

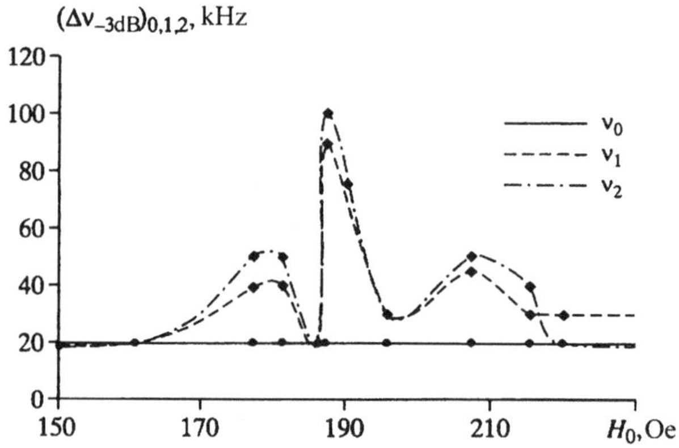

Fig. 2.10 Changes of the key parameters of the spectra of various kinds of signals

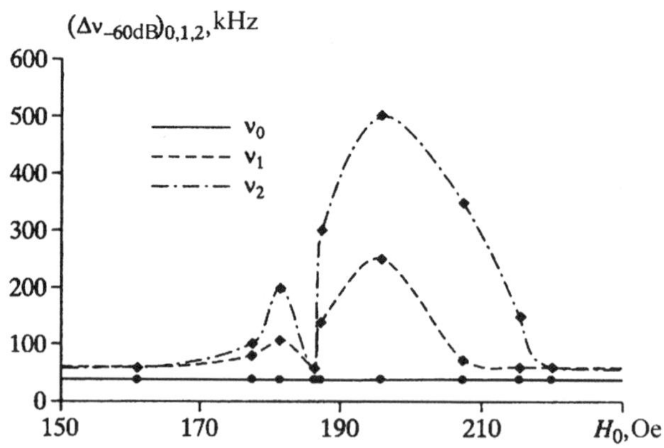

Fig. 2.11 Changes of the key parameters of the spectra of various kinds of signals

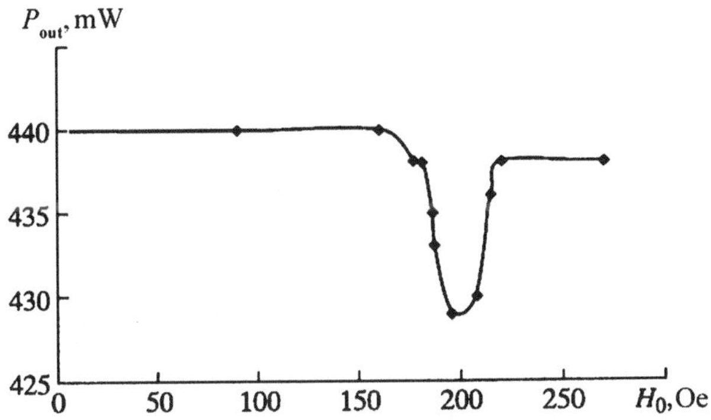

Fig. 2.12 The dependence of the integrated target power P_{out} of the generator on the field H_0

From the data of Figs. 2.8–2.11, it follows that at change of the magnetic field within the limits of $H_0 \leq 160$ Oe and $H_0 \geq 225$ Oe the spectral lines do not change $(\Delta\nu_{-3dB})_{0,1,2} = \text{const}$ and $(\Delta\nu_{-60dB})_{0,1,2} = \text{const}$. The modes NS and ES FS were observed in the range of frequencies, close to the fundamental frequencies of the used structure.

In Fig. 2.12, the dependence of the integrated target power P_{out} of the generator on the field H_0 is presented. At $H_0 = 190$ Oe the target HF power has slightly decreased.

2.2.1 *Angle of Orientation of FMCR $\varphi = 45°$*

In Figs. 2.13–2.18, experimental dependencies of the characteristics of heteromagnetic generators on a low power level (50 mW) are shown for an FMCR with $4\pi M_s = 90$ G.

In Fig. 2.13, dependencies of the frequency deviation $(\Delta\nu)_{0,1,2}$ of the components $\nu_{0,1,2}$ on the magnetic field H_0 are presented. At $H_0 = 193$ Oe the frequency deviation of all the components is maximal.

In Figs. 2.14–2.17, dependencies of the spectral characteristics $(\Delta\nu_{-3dB})_{0,1,2}$, $(\Delta\nu_{-60dB})_{0,1,2}$ of generated signals on the magnetic field H_0 are shown. In Figs. 2.16 and 2.17, the ranges of the most typical kinds of generated signals are shown at change of the magnetic field H_0 within the limits of 0–300 Oe, namely: two SPS ranges, two NS ranges, and two ES FS ranges. In the mode of autoresonance of FMCR, SPS components are generated.

In Fig. 2.18, the dependence of the integrated target power P_{out} of the generator on the bias field H_0 is presented. At $H_0 = 190$ Oe the target HF power of the generator slightly decreases.

In Figs. 2.19–2.22, experimental dependencies of the power and spectral characteristics of HMT for a high power level (500 mW) for an FMCR with $4\pi M_s = 90$ G are shown.

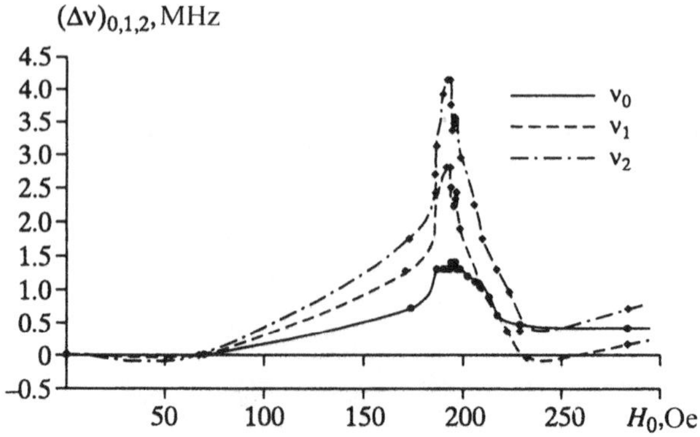

Fig. 2.13 Experimental dependencies of the characteristics of heteromagnetic generators on a low power level (50 mW) are shown for an FMCR with $4\pi M_s = 90\,\mathrm{G}$

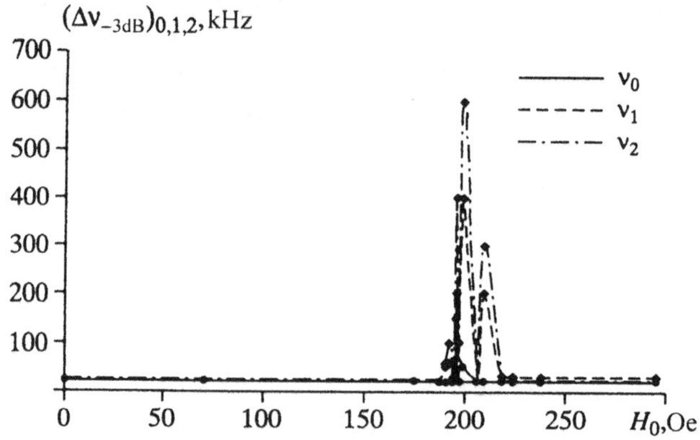

Fig. 2.14 Experimental dependencies of the characteristics of heteromagnetic generators on a low power level (50 mW) are shown for an FMCR with $4\pi M_s = 90\,\mathrm{G}$

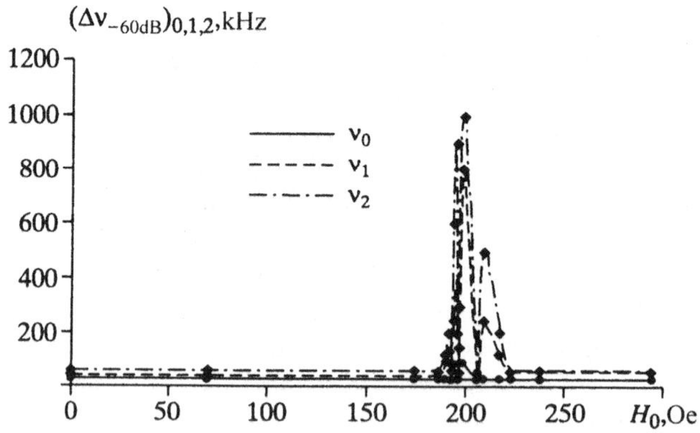

Fig. 2.15 Experimental dependencies of the characteristics of heteromagnetic generators on a low power level (50 mW) are shown for an FMCR with $4\pi M_s = 90\,\mathrm{G}$

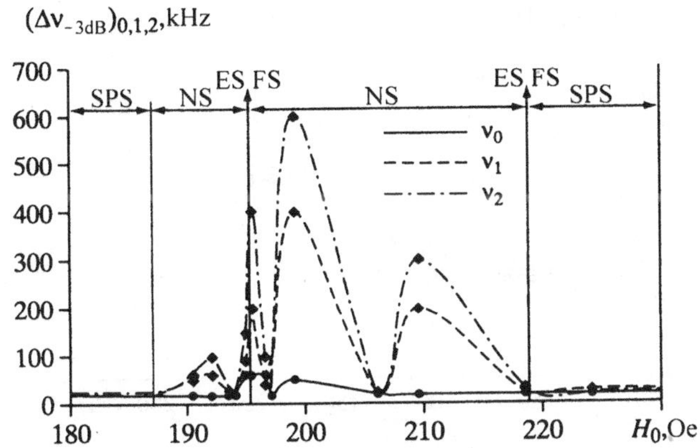

Fig. 2.16 Experimental dependencies of the characteristics of heteromagnetic generators on a low power level (50 mW) are shown for an FMCR with $4\pi M_s = 90\,G$

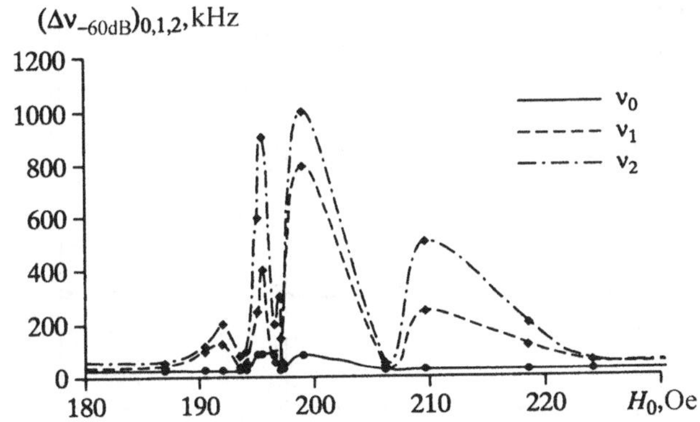

Fig. 2.17 Experimental dependencies of the characteristics of heteromagnetic generators on a low power level (50 mW) are shown for an FMCR with $4\pi M_s = 90\,G$

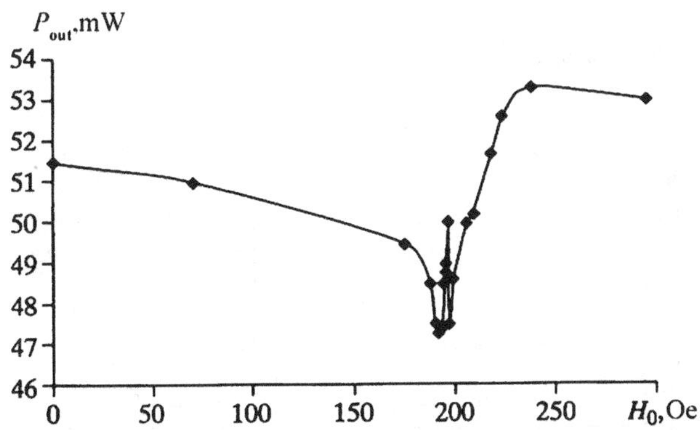

Fig. 2.18 Experimental dependencies of the characteristics of heteromagnetic generators on a low power level (50 mW) are shown for an FMCR with $4\pi M_s = 90\,G$

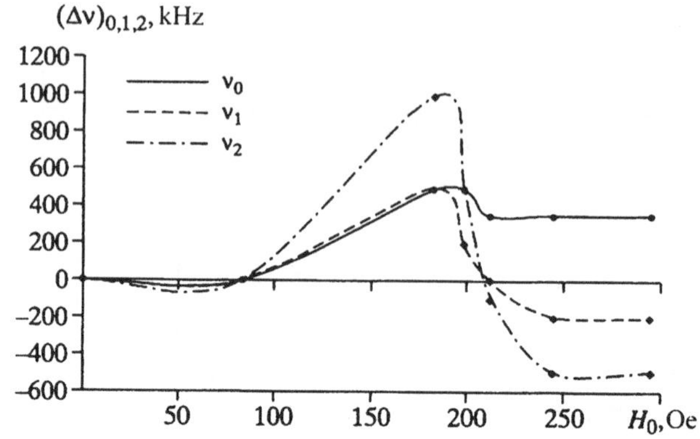

Fig. 2.19 Experimental dependencies of the power and spectral characteristics of HMT for a high power level (500 mW) for an FMCR with $4\pi M_s = 90\,\mathrm{G}$

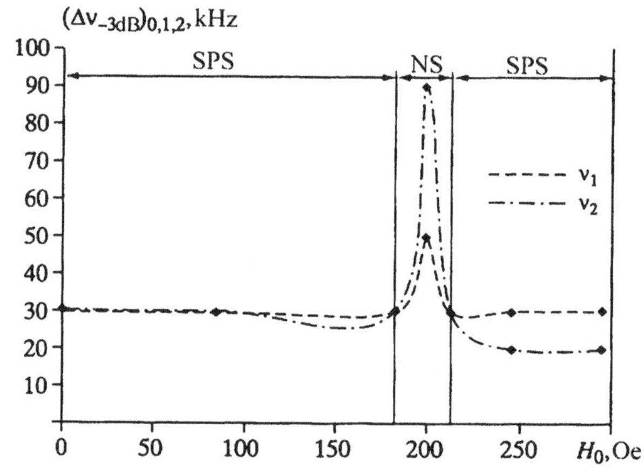

Fig. 2.20 Experimental dependencies of the power and spectral characteristics of HMT for a high power level (500 mW) for an FMCR with $4\pi M_s = 90\,\mathrm{G}$

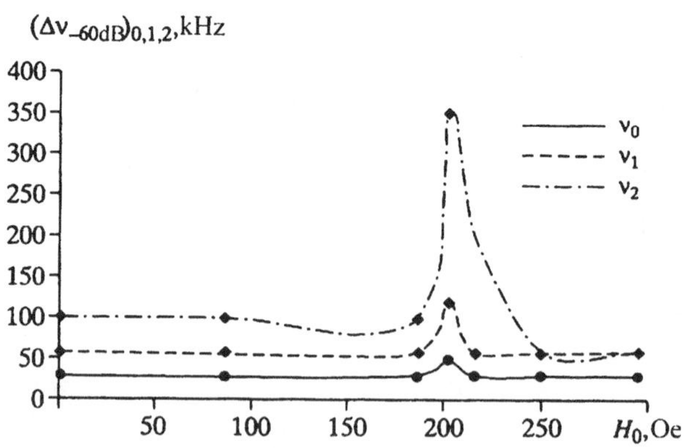

Fig. 2.21 Experimental dependencies of the power and spectral characteristics of HMT for a high power level (500 mW) for an FMCR with $4\pi M_s = 90\,\mathrm{G}$

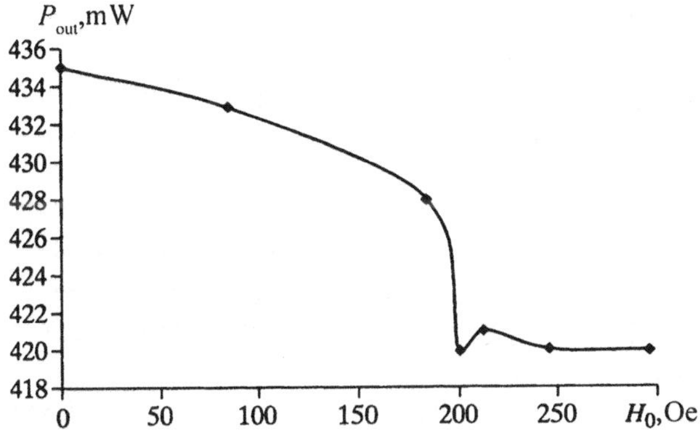

Fig. 2.22 Experimental dependencies of the power and spectral characteristics of HMT for a high power level (500 mW) for an FMCR with $4\pi M_s = 90$ G

In Fig. 2.19, dependencies of the deviations of the central frequencies $(\Delta\nu)_{0,1,2}$ of generated signals on the magnetic field H_0 are presented. The most significant deviation of the central frequencies of signals was observed at a magnetic field $H_0 = 187$ Oe.

In Figs. 2.20–2.21, dependencies of the spectral characteristics $(\Delta\nu_{-3dB})_{0,1,2}$ and $(\Delta\nu_{-60dB})_{0,1,2}$ of various kinds of generated signals for the components $\nu_{0,1,2}$ are shown. At change of the magnetic field H_0 from 0 up to 300 Oe the following kinds of signal spectra s are realized: two SPS areas and an NS area. At magnetic fields $H_0 < 175$ Oe and $H_0 > 220$ Oe the spectral lines are $(\Delta\nu_{-3dB})_{0,1,2} = \text{const}$ and $(\Delta\nu_{-60dB})_{0,1,2} = \text{const}$. In the mode of FMCR autoresonance, SPS components are generated.

In Fig. 2.22, the dependence of the integrated target power P_{out} of the heteromagnetic generator on the bias field H_0 is shown.

2.2.2 Angle of Orientation of FMCR $\varphi = 90°$

In Figs. 2.23–2.28, experimental energetic and spectral characteristics of heteromagnetic generators for the components $\nu_{0,1,2}$ on a low power level (50 mW) for an FMCR with $4\pi M_s = 90$ G are shown.

In Fig. 2.23, dependencies of the maximal deviations of frequencies $(\Delta\nu)_{0,1,2}$ of the spectral components $\nu_{0,1,2}$ on the magnetic field H_0 are presented. At $H_0 = 187$ Oe the deviations of frequencies $\nu_{0,1,2}$ are maximal.

In Figs. 2.24–2.27, dependencies of the spectral characteristics $(\Delta\nu_{-3dB})_{0,1,2}$ and $(\Delta\nu_{-60dB})_{0,1,2}$ for various kinds of generated signals for the components $\nu_{0,1,2}$ are shown. At change of the magnetic field from 0 up to 300 Oe (Fig. 2.26) the following kinds of the spectra of signals are realized, namely: two SPS ranges, two NS ranges,

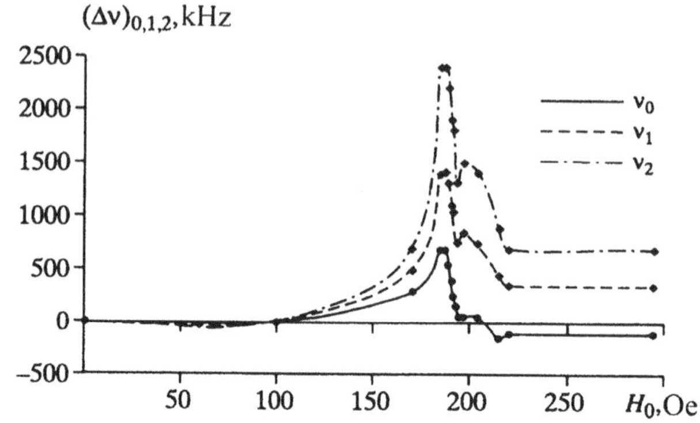

Fig. 2.23 Experimental energetic and spectral characteristics of heteromagnetic generators for the components $\nu_{0,1,2}$ on a low power level (50 mW) for an FMCR with $4\pi M_s = 90\,\mathrm{G}$

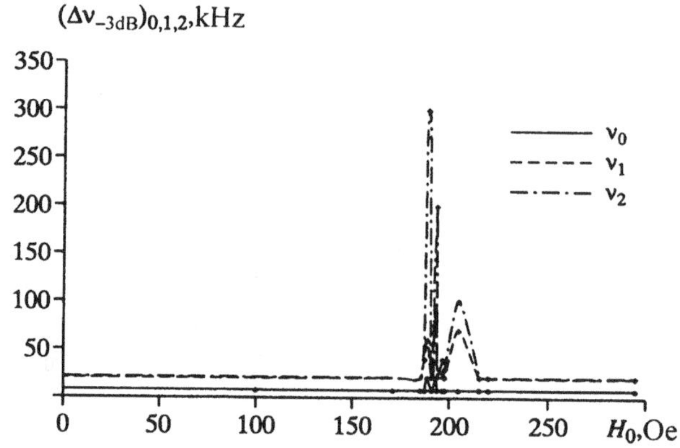

Fig. 2.24 Experimental energetic and spectral characteristics of heteromagnetic generators for the components $\nu_{0,1,2}$ on a low power level (50 mW) for an FMCR with $4\pi M_s = 90\,\mathrm{G}$

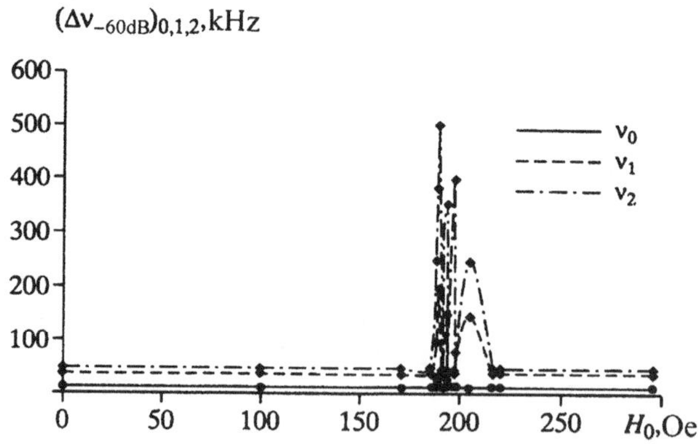

Fig. 2.25 Experimental energetic and spectral characteristics of heteromagnetic generators for the components $\nu_{0,1,2}$ on a low power level (50 mW) for an FMCR with $4\pi M_s = 90\,\mathrm{G}$

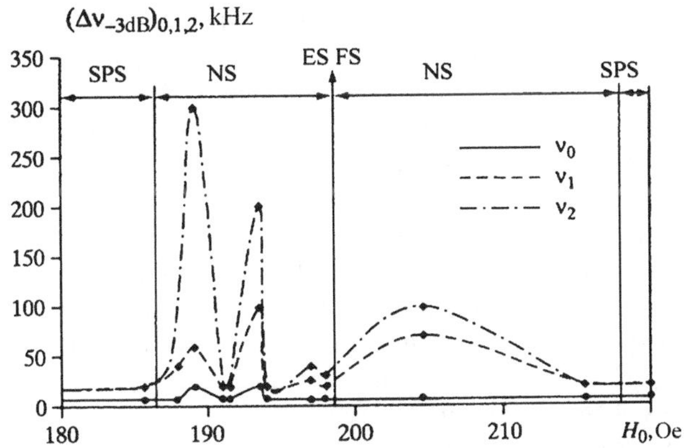

Fig. 2.26 Experimental energetic and spectral characteristics of heteromagnetic generators for the components $\nu_{0,1,2}$ on a low power level (50 mW) for an FMCR with $4\pi M_s = 90\,\mathrm{G}$

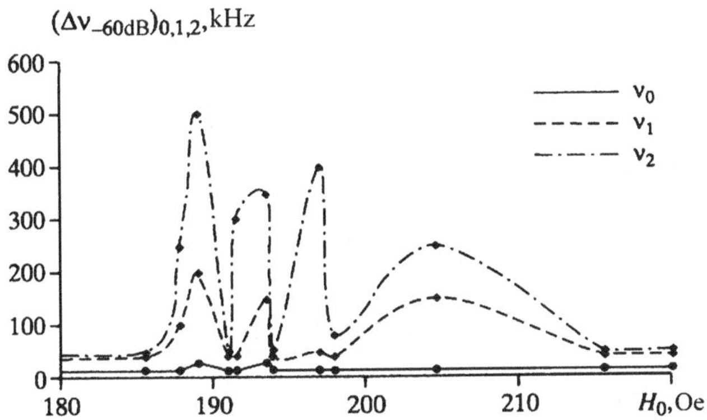

Fig. 2.27 Experimental energetic and spectral characteristics of heteromagnetic generators for the components $\nu_{0,1,2}$ on a low power level (50 mW) for an FMCR with $4\pi M_s = 90\,\mathrm{G}$

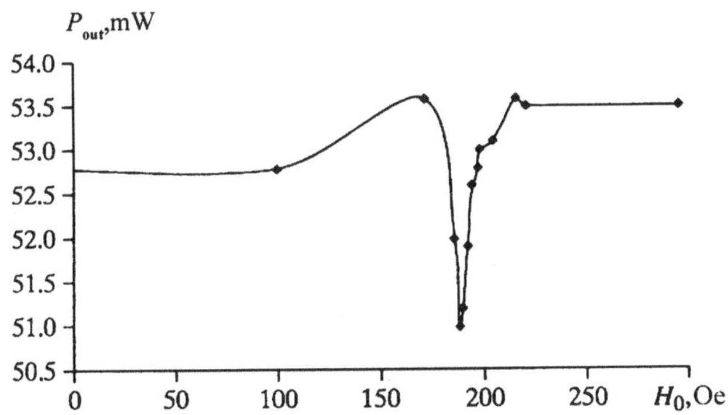

Fig. 2.28 Experimental energetic and spectral characteristics of heteromagnetic generators for the components $\nu_{0,1,2}$ on a low power level (50 mW) for an FMCR with $4\pi M_s = 90\,\mathrm{G}$

and one ES FS range. At magnetic fields $H_0 < 185$ Oe and $H_0 > 215$ Oe the spectral lines are $(\Delta\nu_{-3dB})_{0,1,2} = \text{const}$ and $(\Delta\,\nu_{-60dB})_{0,1,2} = \text{const}$. In the mode of FMCR autoresonance, SPS components are generated.

In Fig. 2.28, the dependence of the integrated target power of the heteromagnetic generator on the magnetic field H_0 is shown. At $H_0 = 190$ Oe the target power decreases.

In Figs. 2.29–2.34, experimental dependencies of the energetic and spectral characteristics of heteromagnetic generators for the components $\nu_{0,1,2}$ at a high power level (0.4 W) for an FMCR with $4\pi M_s = 90$ G are shown.

In Fig. 2.29, dependencies of the maximal deviations of frequencies $(\Delta\nu)_{0,1,2}$ of the spectral components $\nu_{0,1,2}$ on the magnetic field H_0 are presented. At $H_0 = 187$ Oe the deviations of frequencies $\nu_{0,1,2}$ are maximal.

In Figs. 2.30–2.34, dependencies of the spectral characteristics of various kinds of generated signals for the components $\nu_{0,1,2}$ are shown. At change of the magnetic field from 0 up to 300 Oe (Fig. 2.30) the following kinds of signals are consistently realized: two SPS ranges, one NS range, and one ES FS range. At magnetic fields

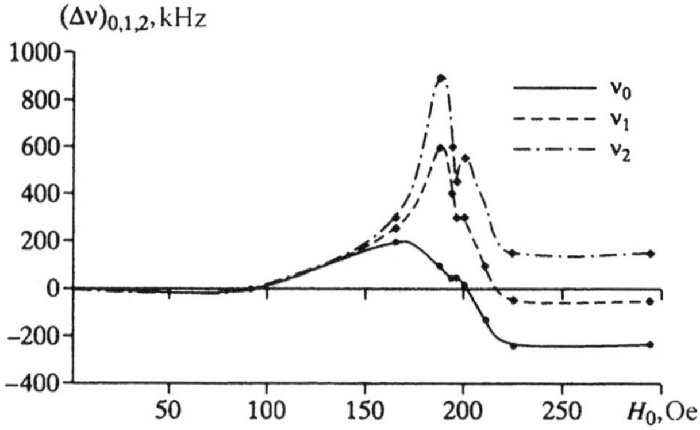

Fig. 2.29 Experimental dependencies of the energetic and spectral characteristics of heteromagnetic generators for the components $\nu_{0,1,2}$ at a high power level (0.4 W) for an FMCR with $4\pi M_s = 90$ G

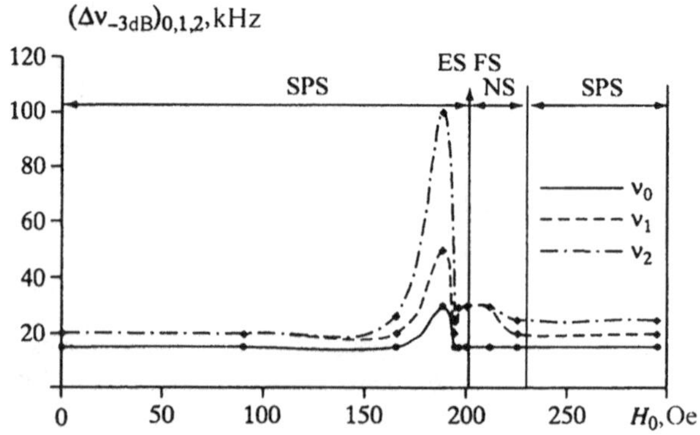

Fig. 2.30 Experimental dependencies of the energetic and spectral characteristics of heteromagnetic generators for the components $\nu_{0,1,2}$ at a high power level (0.4 W) for an FMCR with $4\pi M_s = 90$ G

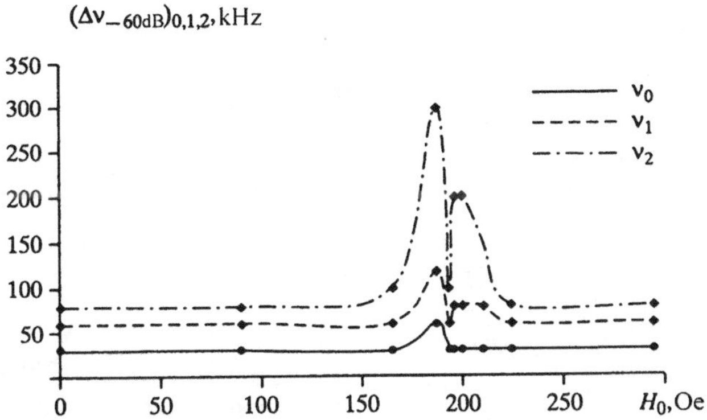

Fig. 2.31 Experimental dependencies of the energetic and spectral characteristics of heteromagnetic generators for the components $\nu_{0,1,2}$ at a high power level (0.4 W) for an FMCR with $4\pi M_s = 90\,\text{G}$

Fig. 2.32 Experimental dependencies of the energetic and spectral characteristics of heteromagnetic generators for the components $\nu_{0,1,2}$ at a high power level (0.4 W) for an FMCR with $4\pi M_s = 90\,\text{G}$

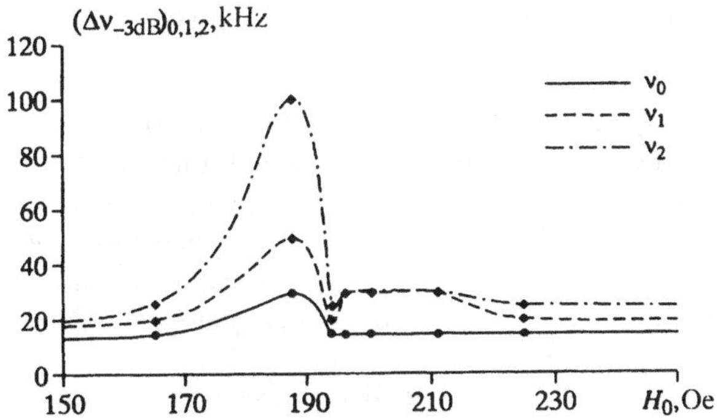

Fig. 2.33 Experimental dependencies of the energetic and spectral characteristics of heteromagnetic generators for the components $\nu_{0,1,2}$ at a high power level (0.4 W) for an FMCR with $4\pi M_s = 90\,\text{G}$

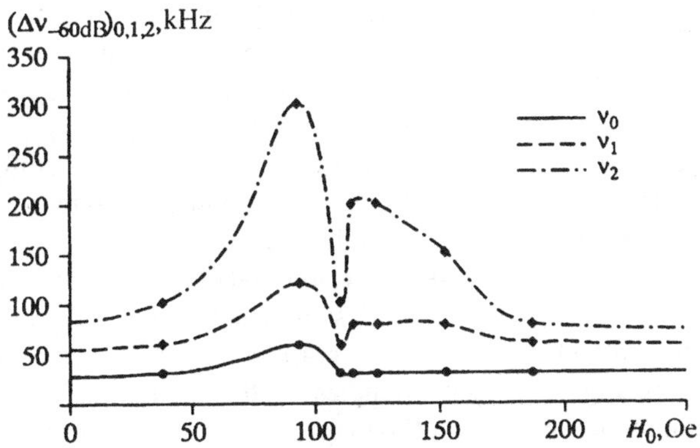

$H_0 < 150$ Oe and $H_0 > 225$ Oe the spectral lines are $(\Delta\nu_{-3dB})_{0,1,2} = \text{const}$ and $(\Delta\nu_{-60dB})_{0,1,2} = \text{const}$. In the mode of FMCR autoresonance, SPS components are generated.

In Fig. 2.34, the dependence of the integrated target power P_{out} of the heteromagnetic generator on the bias field H_0 is shown.

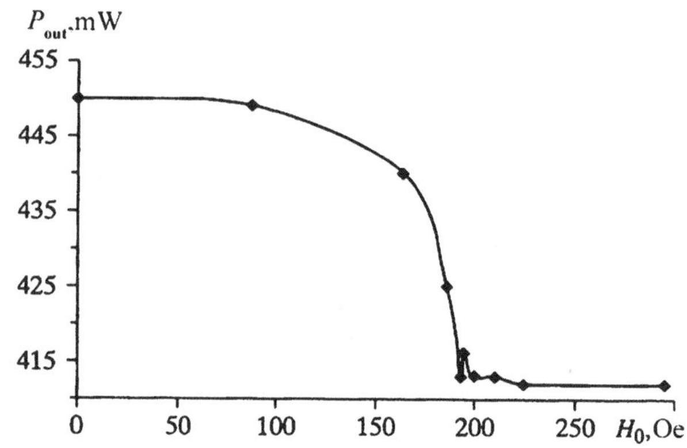

Fig. 2.34 Experimental dependencies of the energetic and spectral characteristics of heteromagnetic generators for the components $\nu_{0,1,2}$ at a high power level (0.4 W) for an FMCR with $4\pi M_s = 90\,\mathrm{G}$

2.3 Structures with Ferrite KG-15

2.3.1 Angle of Orientation of FMCR $\varphi = 0°$

In Figs. 2.35–2.40, experimental characteristics of heteromagnetic structure signals for the spectral components of signals on several frequencies ν_0, ν_1, and ν_2 on a low power level (50 mW) for an FMCR with $4\pi M_s = 190\,\mathrm{G}$ are presented.

In Fig. 2.35, dependencies of the deviations of frequencies $(\Delta\nu)_{0,1,2}$ for the spectral components $\nu_{0,1,2}$ on the magnetic field H_0 are shown. The deviation of frequency is maximal and sign-alternating at changes of the magnetic field from 75 up to 270 Oe.

In Figs. 2.36–2.39, dependencies of some key parameters of the spectral lines of various kinds of signals on several frequencies ν_0, ν_1, and ν_2 are shown. In Fig. 2.38, the areas of existence of some characteristic kinds of signals are shown, namely: three SPS ranges, three NS ranges, and one ES FS range. From the data of Figs. 2.36–2.39 it follows that at change of the magnetic field below of $H_0 \leq 200$ Oe the spectral lines do not change, $(\Delta\nu_{-3\mathrm{dB}})_{0,1,2} = \mathrm{const}$, and $(\Delta\nu_{-60\mathrm{dB}})_{0,1,2} = \mathrm{const}$. In the mode of FMCR autoresonance, SPS components are generated.

In Fig. 2.40, the dependence of the integrated target power P_{out} of the heteromagnetic generator on the bias field H_0 is shown.

In Figs. 2.41–2.46, experimental dependencies of the spectral characteristics of various operating modes of heteromagnetic generators at a high power level (500 mW) for an FMCR with $4\pi M_s = 190\,\mathrm{G}$ are shown.

In Fig. 2.41, dependencies of the deviations of frequencies $(\Delta\nu)_{0,1,2}$ for the components $\nu_{0,1,2}$ on the magnetic field H_0 are presented. The deviation has a sign-alternating character in a magnetic field range from 0 up to 250 Oe.

In Fig. 2.46, the dependence of the integrated target power P_{out} of the heteromagnetic generator on the bias field H_0 is shown.

In Fig. 2.42, ranges of the most typical kinds of the spectra of generated signals are shown at change of the magnetic field, namely: two NS ranges, two SPS ranges,

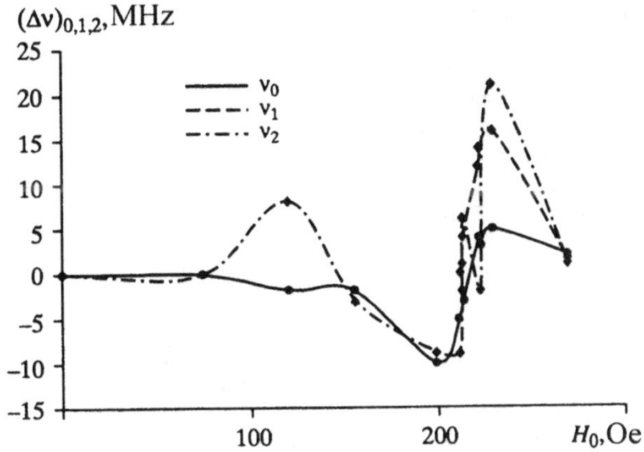

Fig. 2.35 Experimental characteristics of heteromagnetic structure signals for the spectral components of signals on several frequencies ν_0, ν_1, and ν_2 on a low power level (50 mW) for an FMCR with $4\pi M_s = 190\,\mathrm{G}$

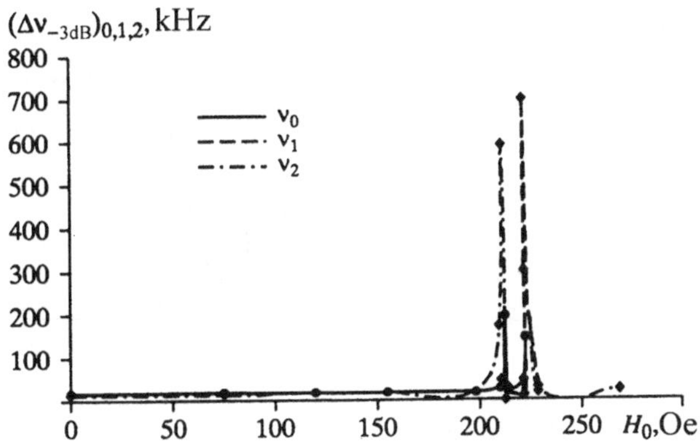

Fig. 2.36 Experimental characteristics of heteromagnetic structure signals for the spectral components of signals on several frequencies ν_0, ν_1, and ν_2 on a low power level (50 mW) for an FMCR with $4\pi M_s = 190\,\mathrm{G}$

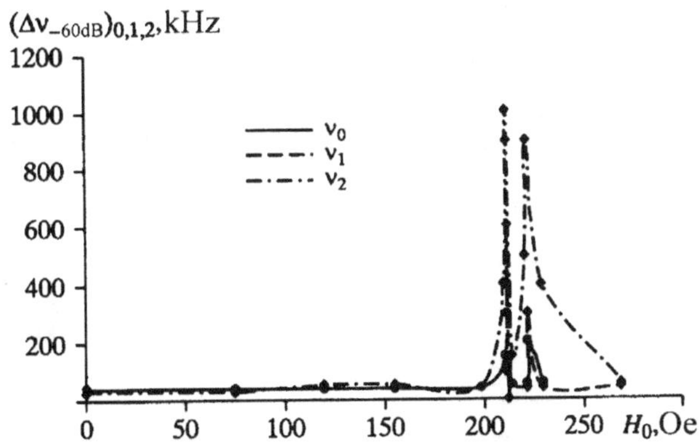

Fig. 2.37 Experimental characteristics of heteromagnetic structure signals for the spectral components of signals on several frequencies ν_0, ν_1, and ν_2 on a low power level (50 mW) for an FMCR with $4\pi M_s = 190\,\mathrm{G}$

and two ES FS ranges. In the mode of FMCR autoresonance we have NS spectral components.

From Figs. 2.42 and 2.43, it is obvious that at change of the magnetic field H_0 from 0 up to 75 Oe the mode of NS generation takes place, and in the autoresonance

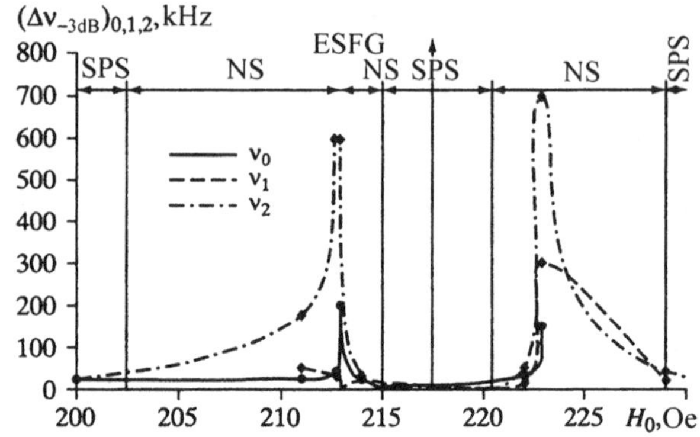

Fig. 2.38 Experimental characteristics of heteromagnetic structure signals for the spectral components of signals on several frequencies ν_0, ν_1, and ν_2 on a low power level (50 mW) for an FMCR with $4\pi M_s = 190\,G$

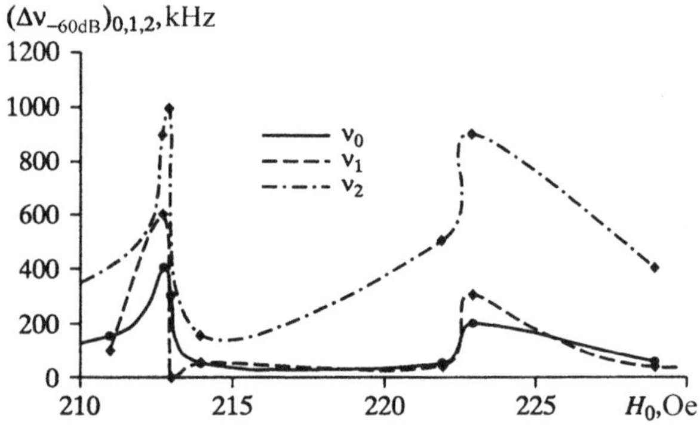

Fig. 2.39 Experimental characteristics of heteromagnetic structure signals for the spectral components of signals on several frequencies ν_0, ν_1, and ν_2 on a low power level (50 mW) for an FMCR with $4\pi M_s = 190\,G$

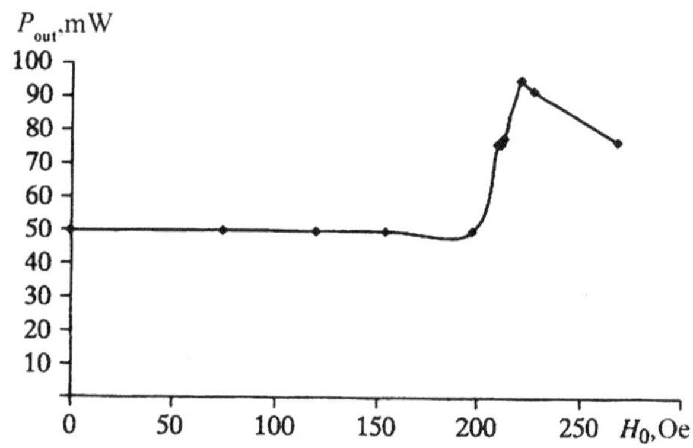

Fig. 2.40 Experimental characteristics of heteromagnetic structure signals for the spectral components of signals on several frequencies ν_0, ν_1, and ν_2 on a low power level (50 mW) for an FMCR with $4\pi M_s = 190\,G$

mode of cubic ferrite (at $H_0 \approx 0$), noise modes of broadband signal generation, for which $(\Delta\nu_{-3dB})_0 < (\Delta\nu_{-3dB})_1 < (\Delta\nu_{-3dB})_2$ and $(\Delta\nu_{-60dB})_0 < (\Delta\nu_{-60dB})_1 < (\Delta\nu_{-60dB})_2$ occur.

At increasing the magnetic field from 0 up to 75 Oe narrowing of all the spectral lines on $\nu_{0,1,2}$ is observed and in a range from 75 up to 200 Oe the spectral lines are $(\Delta\nu_{-3dB})_{0,1,2} = \text{const}$ and $(\Delta\nu_{-60dB})_{0,1,2} = \text{const}$. It speaks for parametrical

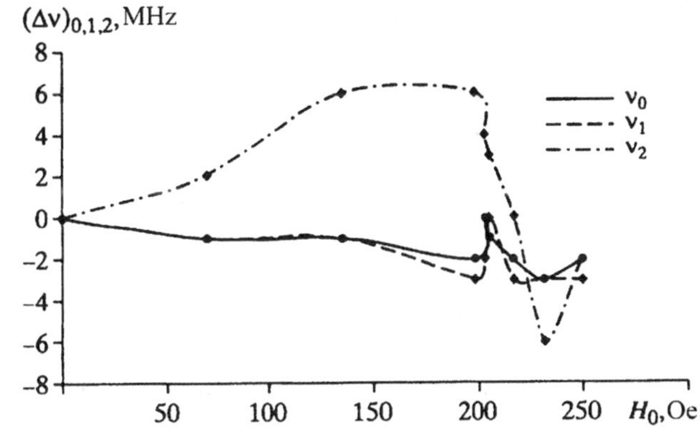

Fig. 2.41 Experimental dependencies of the spectral characteristics of various operating modes of heteromagnetic generators at a high power level (500 mW) for an FMCR with $4\pi M_s = 190\,\mathrm{G}$

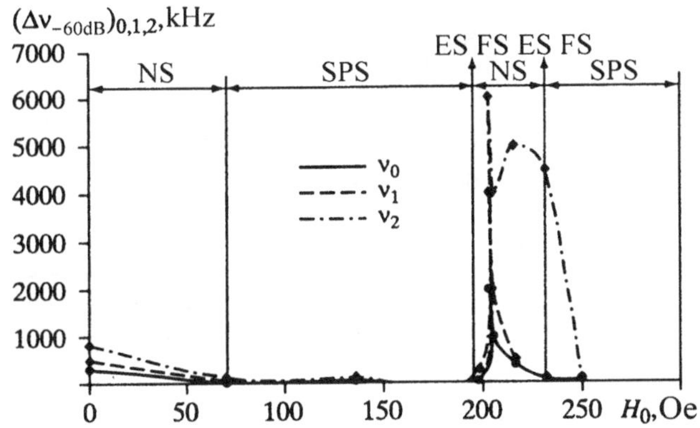

Fig. 2.42 Experimental dependencies of the spectral characteristics of various operating modes of heteromagnetic generators at a high power level (500 mW) for an FMCR with $4\pi M_s = 190\,\mathrm{G}$

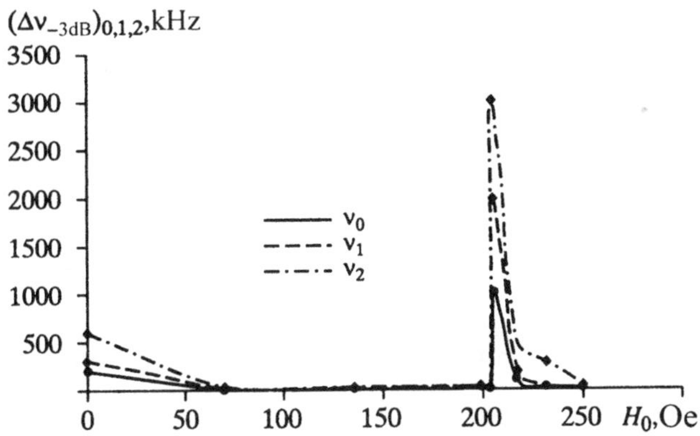

Fig. 2.43 Experimental dependencies of the spectral characteristics of various operating modes of heteromagnetic generators at a high power level (500 mW) for an FMCR with $4\pi M_s = 190\,\mathrm{G}$

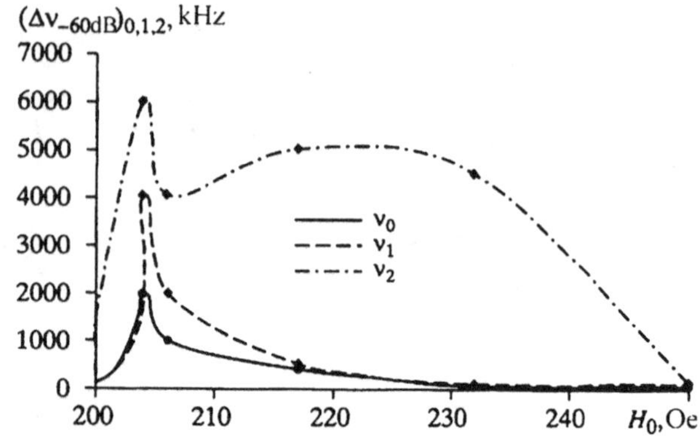

Fig. 2.44 Experimental dependencies of the spectral characteristics of various operating modes of heteromagnetic generators at a high power level (500 mW) for an FMCR with $4\pi M_s = 190$ G

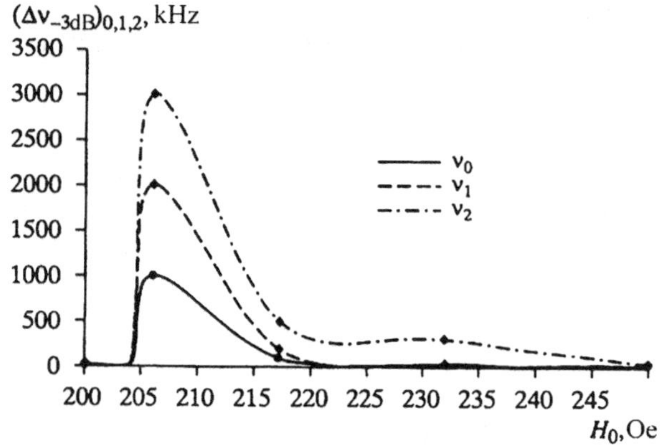

Fig. 2.45 Experimental dependencies of the spectral characteristics of various operating modes of heteromagnetic generators at a high power level (500 mW) for an FMCR with $4\pi M_s = 190$ G

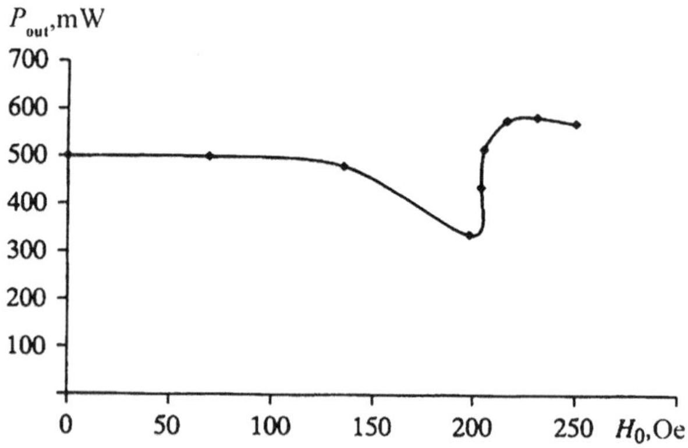

Fig. 2.46 Experimental dependencies of the spectral characteristics of various operating modes of heteromagnetic generators at a high power level (500 mW) for an FMCR with $4\pi M_s = 190$ G

processes of multiplication in the heteromagnetic generator with $\Delta\nu_{n-1} = \text{const}$, $n = 1, 2, 3, \ldots$.

In Fig. 2.46, the dependence of the integrated target power on the magnetic field H_0 is presented. In a range from 150 up to 225 Oe the HF power decreases by 45–50%.

In Figs. 2.47–2.52, experimental dependencies of the spectral characteristics of heteromagnetic generators at a high power level $P_{out} = 1$ W for an FMCR with $4\pi M_s = 190$ G are shown.

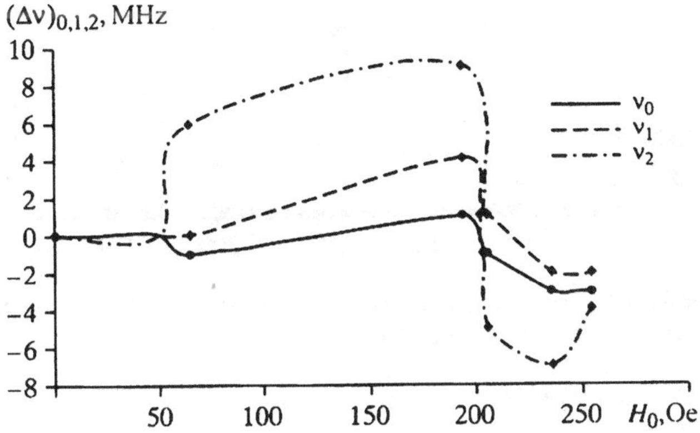

Fig. 2.47 Experimental dependencies of the spectral characteristics of heteromagnetic generators at a high power level $P_{out} = 1$ W for an FMCR with $4\pi M_s = 190$ G

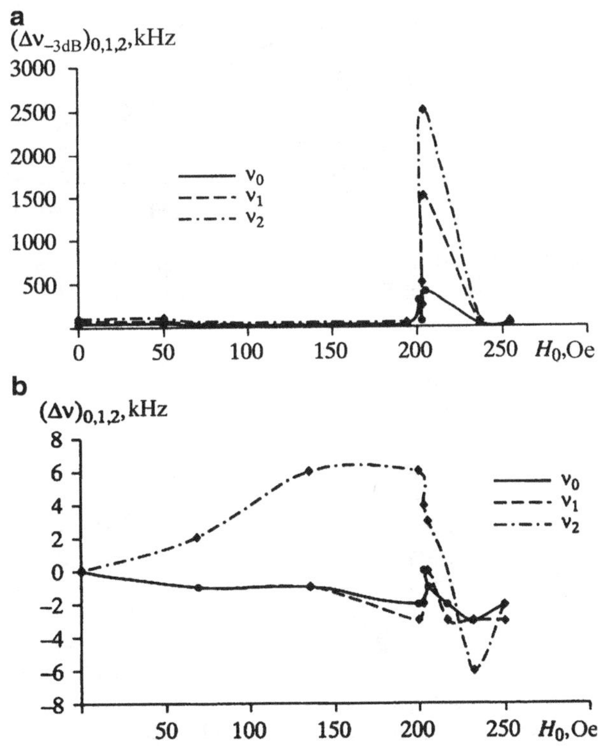

Fig. 2.48 Experimental dependencies of the spectral characteristics of heteromagnetic generators at a high power level $P_{out} = 1$ W for an FMCR with $4\pi M_s = 190$ G

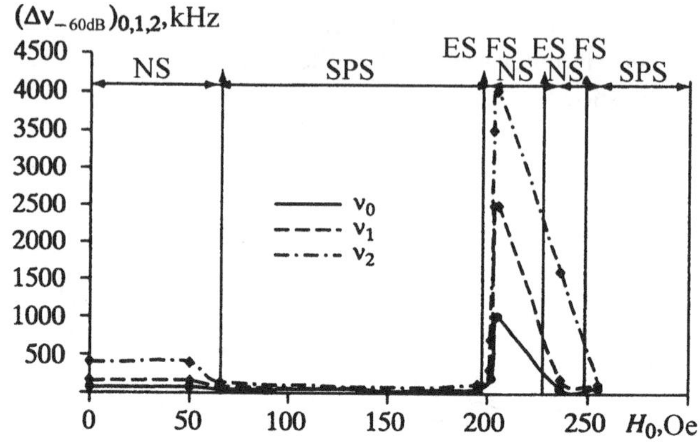

Fig. 2.49 Experimental dependencies of the spectral characteristics of heteromagnetic generators at a high power level $P_{out} = 1$ W for an FMCR with $4\pi M_s = 190$ G

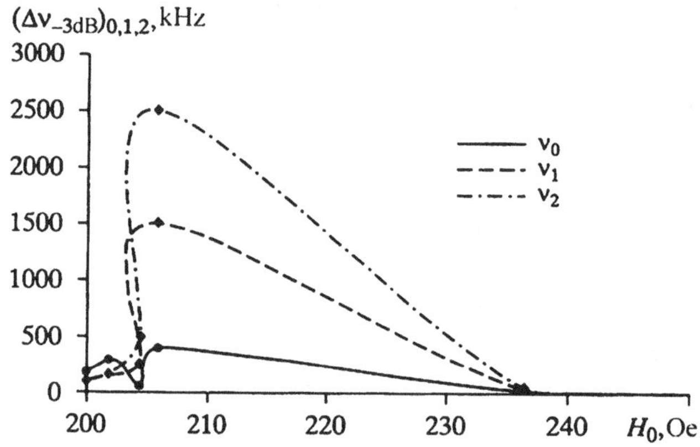

Fig. 2.50 Experimental dependencies of the spectral characteristics of heteromagnetic generators at a high power level $P_{out} = 1$ W for an FMCR with $4\pi M_s = 190$ G

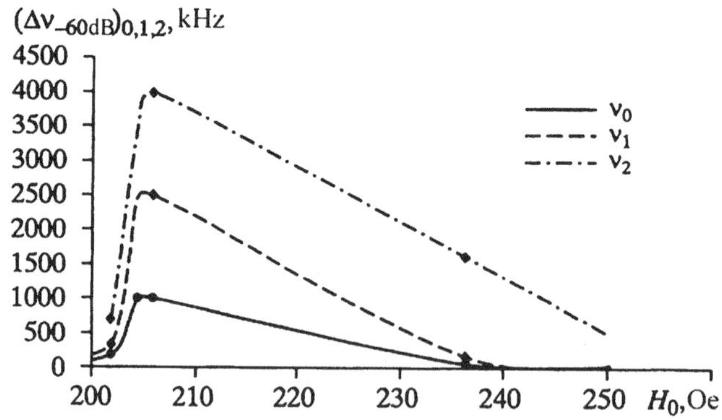

Fig. 2.51 Experimental dependencies of the spectral characteristics of heteromagnetic generators at a high power level $P_{out} = 1$ W for an FMCR with $4\pi M_s = 190$ G

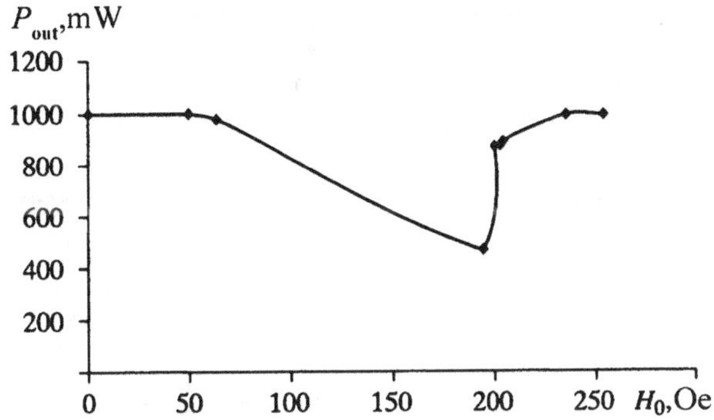

Fig. 2.52 Experimental dependencies of the spectral characteristics of heteromagnetic generators at a high power level $P_{out} = 1$ W for an FMCR with $4\pi M_s = 190$ G

In Fig. 2.47, dependencies of the deviations of frequencies $(\Delta\nu)_{0,1,2}$ for the spectral components $\nu_{0,1,2}$ on the magnetic field H_0 are presented. The deviation has a sign-alternating character in a range from 50 up to 260 Oe.

In Figs. 2.48–2.51, dependencies of the spectral characteristics of generated signals on the magnetic field H_0 are shown.

Figure 2.48 shows the spectral characteristics of signals within a range of 0–250 Oe (a) and from 100 up to 200 Oe (b), where the mode of parametrical multiplication of a signal was observed.

In Fig. 2.49, ranges of the most typical kinds of the spectra of generated signals are shown at changes of the magnetic field, namely: three NS ranges, two SPS ones, and two ES FS ones. In the mode of FMCR autoresonance, NS components are generated.

In Fig. 2.52, the dependence of the integrated target power P_{out} of the heteromagnetic generator on the bias field H_0 is shown.

2.3.2 Angle of Orientation of FMCR $\varphi = 45°$

In Fig. 2.53–2.58, experimental dependencies of the characteristics of heteromagnetic generators on a low power level (50 mW) are shown for an FMCR with $4\pi M_s = 190$ G.

In Fig. 2.53, dependencies of the deviations of frequencies $(\Delta\nu)_{0,1,2}$ for the spectral components $\nu_{0,1,2}$ on the magnetic field H_0 are presented. The deviation of frequency has a sign-alternating character in a range from 50 up to 260 Oe.

In Fig. 2.56, ranges of characteristic kinds of the spectra of generated signals are shown at changes of the magnetic field, namely: two SPS ranges, three NS ranges, and two ES FS ones. From Figs. 2.54–2.56, it is obvious that in the range from 50 up to 200 Oe the modes of parametrical signal multiplication take place. In the mode of FMCR autoresonance, SPS components are generated.

In Fig. 2.58, the dependence of the integrated target power P_{out} of the heteromagnetic generator on the bias field H_0 is shown.

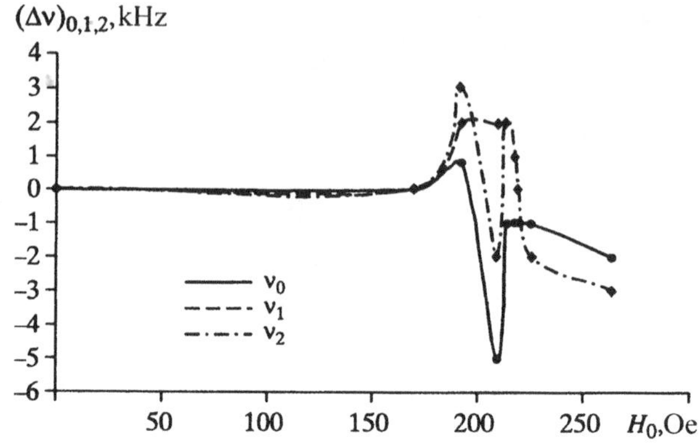

Fig. 2.53 Experimental dependencies of the characteristics of heteromagnetic generators on a low power level (50 mW) are shown for an FMCR with $4\pi M_s = 190\,\mathrm{G}$

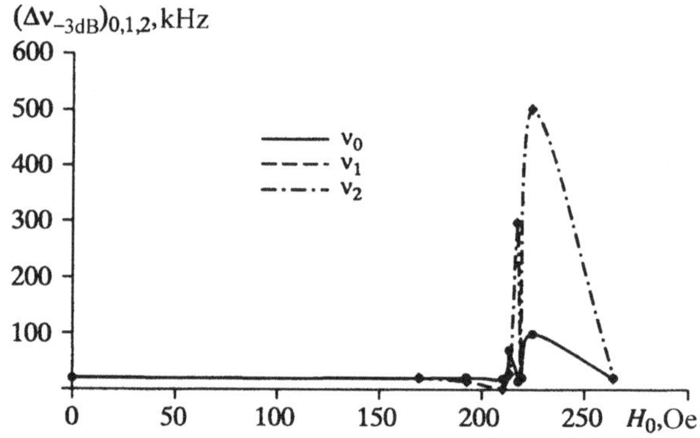

Fig. 2.54 Experimental dependencies of the characteristics of heteromagnetic generators on a low power level (50 mW) are shown for an FMCR with $4\pi M_s = 190\,\mathrm{G}$

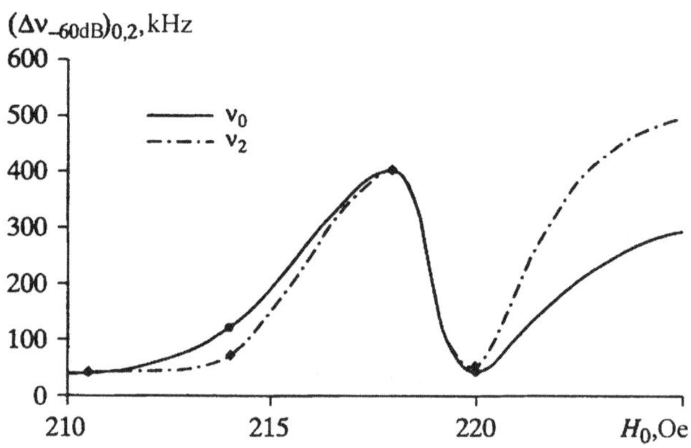

Fig. 2.55 Experimental dependencies of the characteristics of heteromagnetic generators on a low power level (50 mW) are shown for an FMCR with $4\pi M_s = 190\,\mathrm{G}$

In Figs. 2.59–2.62, experimental dependencies of the characteristics of the heteromagnetic generators at a high level of power (500 mW) for an FMCR with $4\pi M_s = 190\,\mathrm{G}$ are shown.

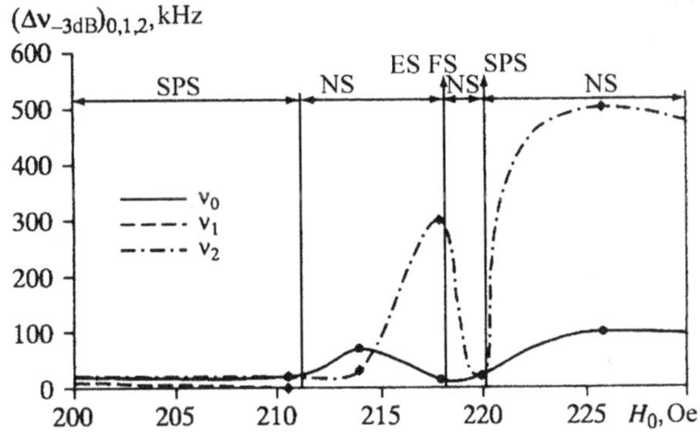

Fig. 2.56 Experimental dependencies of the characteristics of heteromagnetic generators on a low power level (50 mW) are shown for an FMCR with $4\pi M_s = 190\,\mathrm{G}$

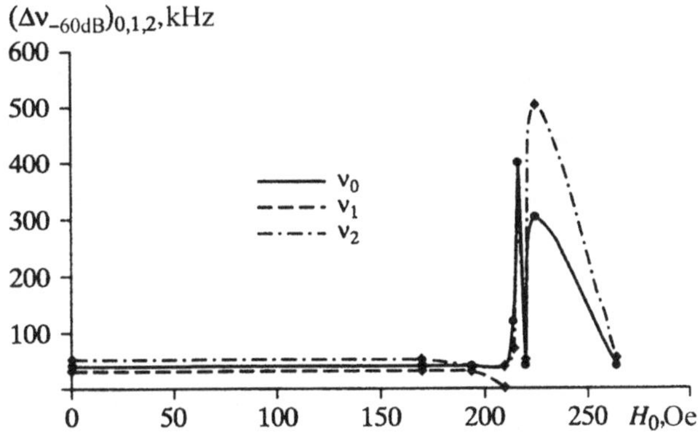

Fig. 2.57 Experimental dependencies of the characteristics of heteromagnetic generators on a low power level (50 mW) are shown for an FMCR with $4\pi M_s = 190\,\mathrm{G}$

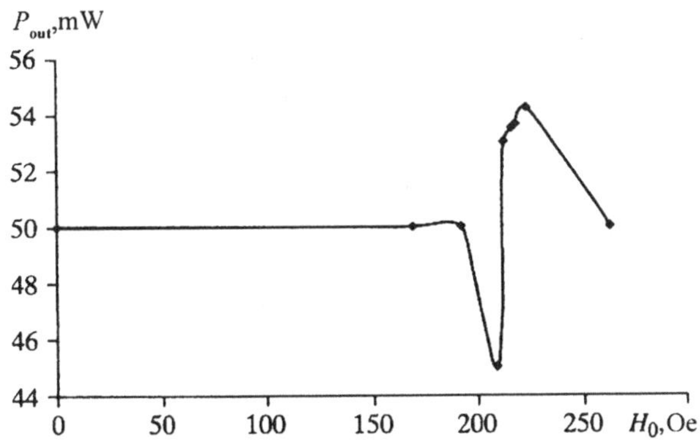

Fig. 2.58 Experimental dependencies of the characteristics of heteromagnetic generators on a low power level (50 mW) are shown for an FMCR with $4\pi M_s = 190\,\mathrm{G}$

In Fig. 2.59, dependencies of the deviations of frequencies $(\Delta\nu)_{0,1,2}$ for the spectral components $\nu_{0,1,2}$ on the magnetic field H_0 are presented. The deviation of frequency has a sign-alternating character from 180 up to 260 Oe.

In Fig. 2.60, ranges of the most typical kinds of the spectra of generated signals are shown at changes of the magnetic field, namely: one SPS range, one

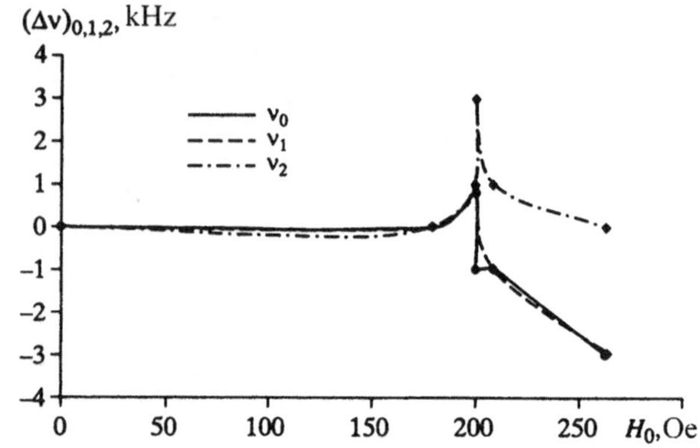

Fig. 2.59 Experimental dependencies of the characteristics of the heteromagnetic generators at a high level of power (500 mW) for an FMCR with $4\pi M_s = 190\,\text{G}$

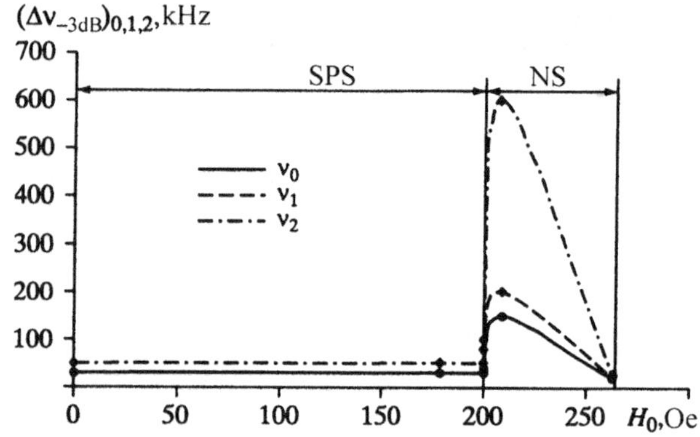

Fig. 2.60 Experimental dependencies of the characteristics of the heteromagnetic generators at a high level of power (500 mW) for an FMCR with $4\pi M_s = 190\,\text{G}$

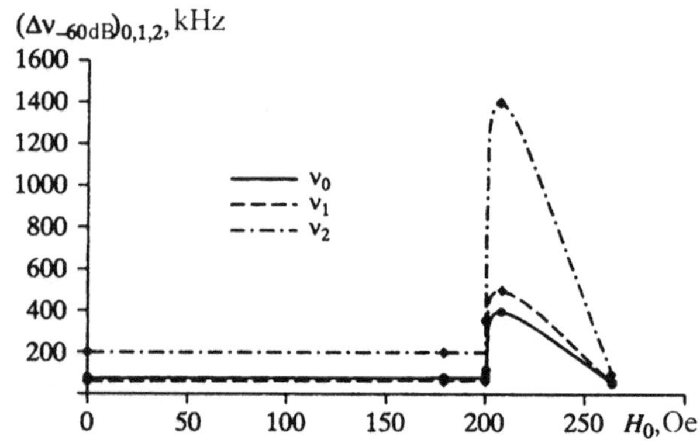

Fig. 2.61 Experimental dependencies of the characteristics of the heteromagnetic generators at a high level of power (500 mW) for an FMCR with $4\pi M_s = 190\,\text{G}$

NS range. In a range from 25 up to 200 Oe, the process of parametrical signal multiplication takes place. In the mode of FMCR autoresonance, SPS components are generated.

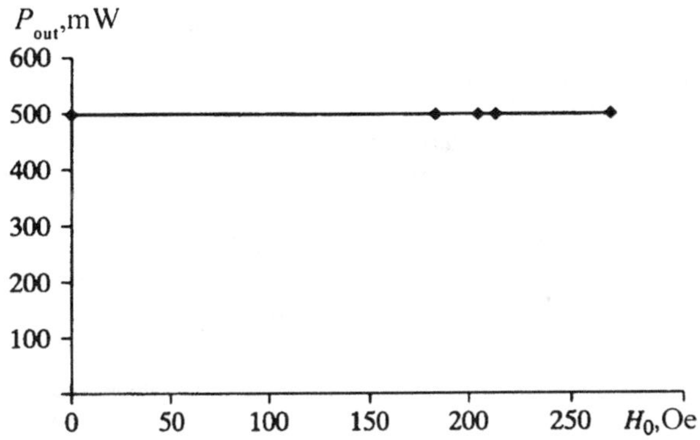

Fig. 2.62 Experimental dependencies of the characteristics of the heteromagnetic generators at a high level of power (500 mW) for an FMCR with $4\pi M_s = 190\,\text{G}$

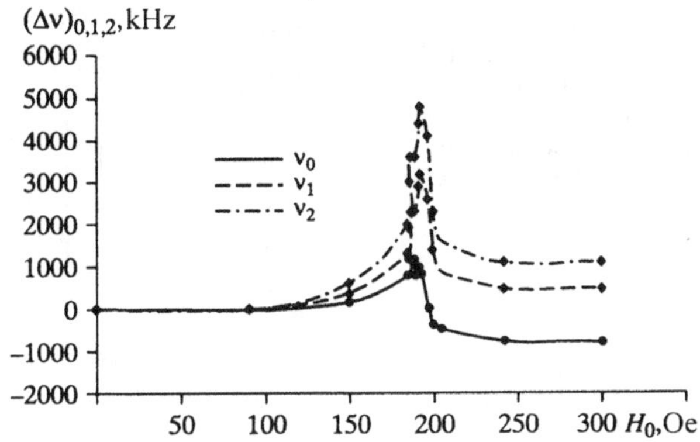

Fig. 2.63 Experimental dependencies of the parameters of the spectral characteristics of heteromagnetic generators on a low power level (50 mW) are shown for an FMCR with $4\pi M_s = 190\,\text{G}$

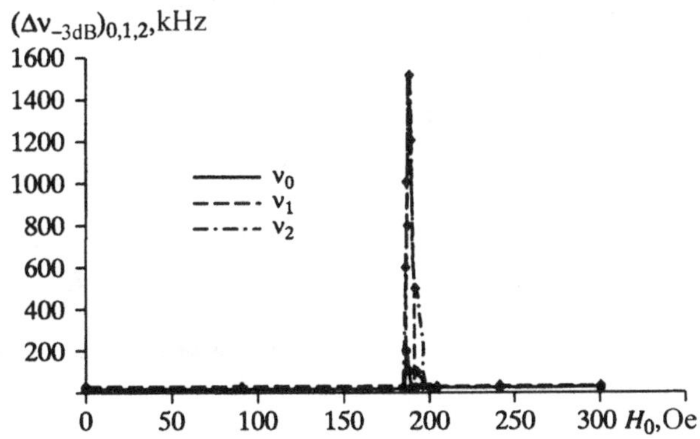

Fig. 2.64 Experimental dependencies of the parameters of the spectral characteristics of heteromagnetic generators on a low power level (50 mW) are shown for an FMCR with $4\pi M_s = 190\,\text{G}$

2.3.3 Angle of Orientation of FMCR $\varphi = 90°$

In Figs. 2.63–2.68, experimental dependencies of the parameters of the spectral characteristics of heteromagnetic generators on a low power level (50 mW) are shown for an FMCR with $4\pi M_s = 190\,\text{G}$.

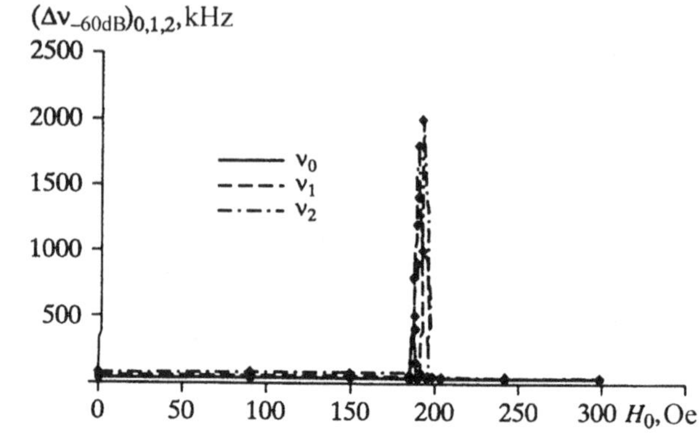

Fig. 2.65 Experimental dependencies of the parameters of the spectral characteristics of heteromagnetic generators on a low power level (50 mW) are shown for an FMCR with $4\pi M_s = 190\,\mathrm{G}$

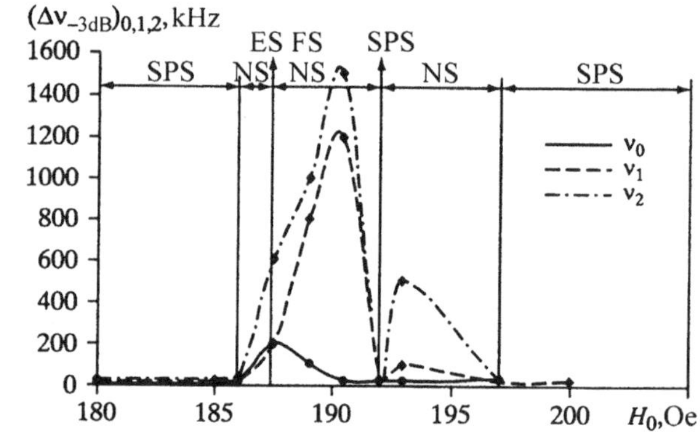

Fig. 2.66 Experimental dependencies of the parameters of the spectral characteristics of heteromagnetic generators on a low power level (50 mW) are shown for an FMCR with $4\pi M_s = 190\,\mathrm{G}$

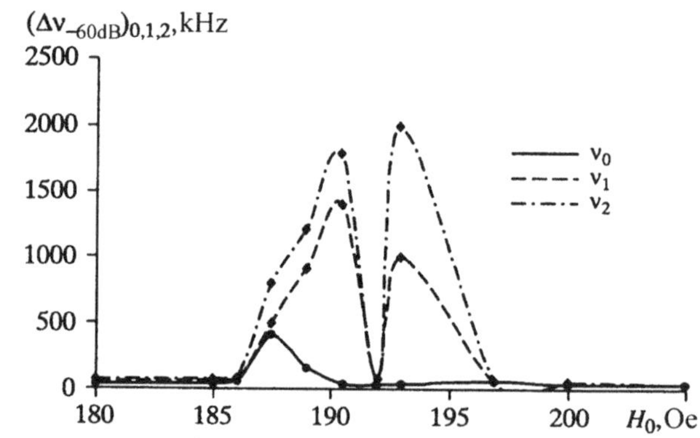

Fig. 2.67 Experimental dependencies of the parameters of the spectral characteristics of heteromagnetic generators on a low power level (50 mW) are shown for an FMCR with $4\pi M_s = 190\,\mathrm{G}$

In Fig. 2.63, dependencies of the deviations of frequencies $(\Delta\nu)_{0,1,2}$ for the components $\nu_{0,1,2}$ on the magnetic field H_0 are presented. The deviation has a sign-alternating character from 100 up to 300 Oe.

In Figs. 2.64–2.67, dependencies of the spectral characteristics of generated signals on the magnetic field H_0 are shown. In Fig. 2.66, ranges of the most typical kinds of the spectra of generated signals are shown at changes of the magnetic field, namely: three SPS ranges, three NS ranges, and one ES FS range.

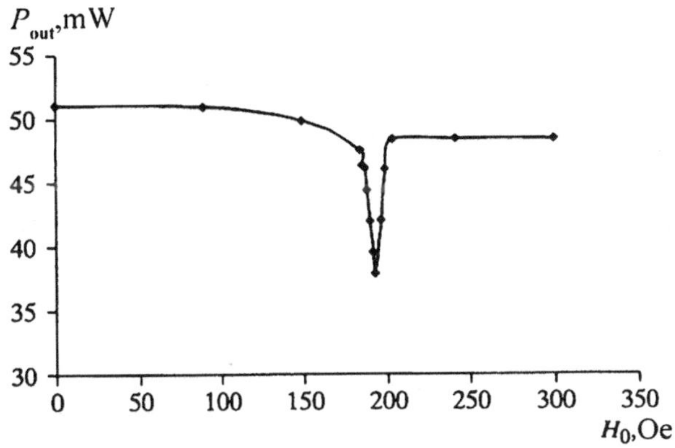

Fig. 2.68 Experimental dependencies of the parameters of the spectral characteristics of heteromagnetic generators on a low power level (50 mW) are shown for an FMCR with $4\pi M_s = 190\,\mathrm{G}$

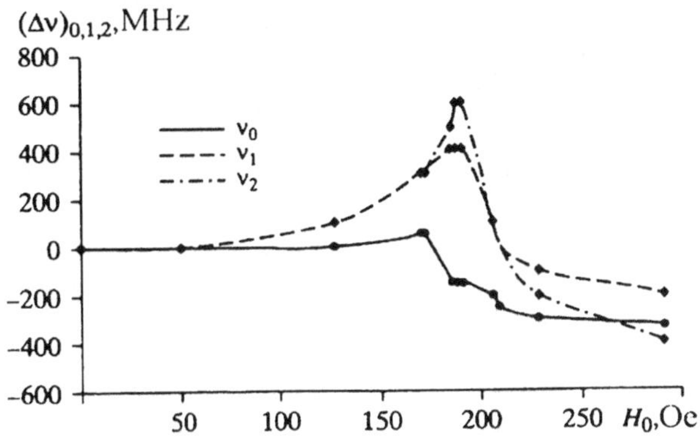

Fig. 2.69 Experimental dependencies of the parameters of the spectral characteristics of the heteromagnetic generator at a high level of power (500 mW) for an FMCR with $4\pi M_s = 190\,\mathrm{G}$

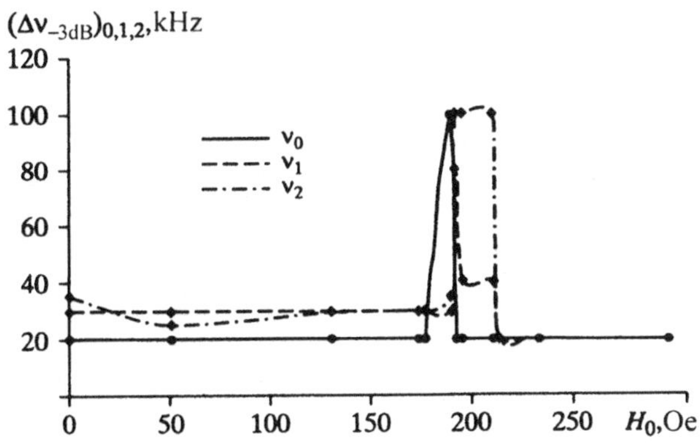

Fig. 2.70 Experimental dependencies of the parameters of the spectral characteristics of the heteromagnetic generator at a high level of power (500 mW) for an FMCR with $4\pi M_s = 190\,\mathrm{G}$

From Figs. 2.63–2.67, one can see that in the range from 100 up to 185 Oe parametrical multiplication of a signal takes place. In the mode of FMCR autoresonance, SPS components are generated.

In Figs. 2.69–2.73, experimental dependencies of the parameters of the spectral characteristics of the heteromagnetic generator at a high level of power (500 mW) for an FMCR with $4\pi M_s = 190\,\mathrm{G}$ are shown.

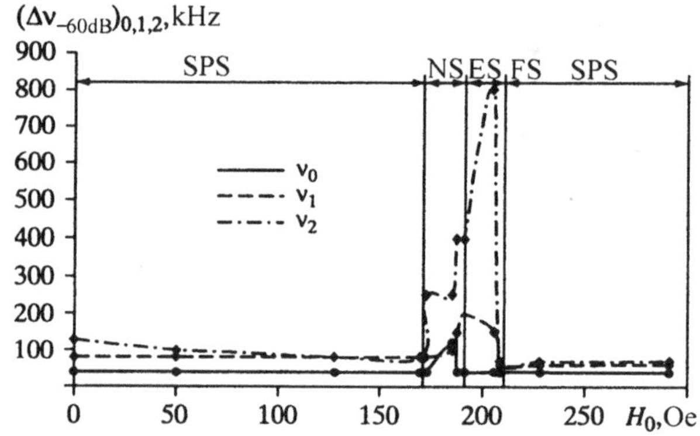

Fig. 2.71 Experimental dependencies of the parameters of the spectral characteristics of the heteromagnetic generator at a high level of power (500 mW) for an FMCR with $4\pi M_s = 190\,\mathrm{G}$

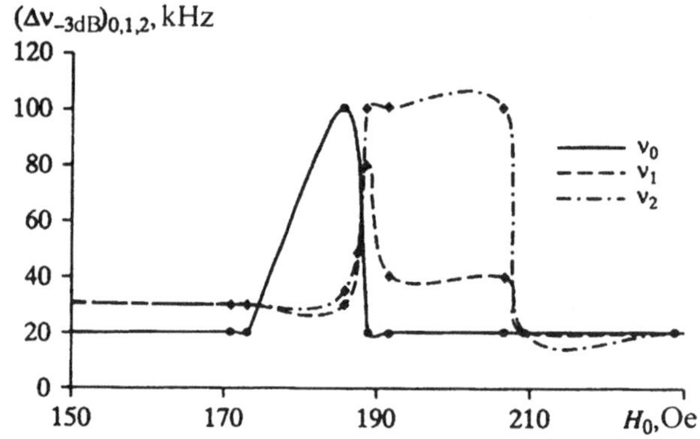

Fig. 2.72 Experimental dependencies of the parameters of the spectral characteristics of the heteromagnetic generator at a high level of power (500 mW) for an FMCR with $4\pi M_s = 190\,\mathrm{G}$

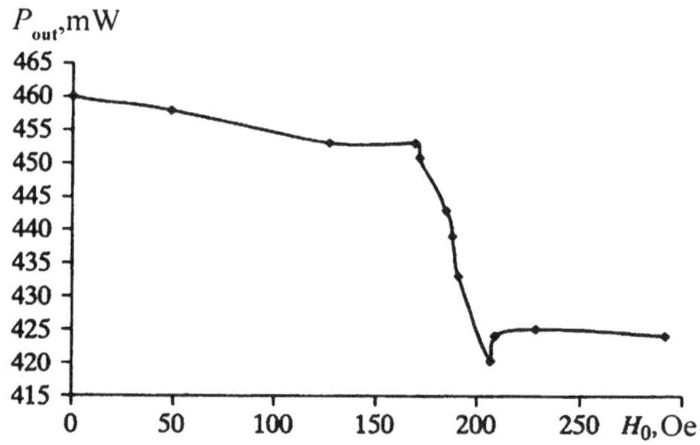

Fig. 2.73 Experimental dependencies of the parameters of the spectral characteristics of the heteromagnetic generator at a high level of power (500 mW) for an FMCR with $4\pi M_s = 190\,\mathrm{G}$

In Fig. 2.69, dependencies of the deviations of frequencies $(\Delta\nu)_{0,1,2}$ for the spectral components $\nu_{0,1,2}$ on the magnetic field H_0 are presented. The deviation is sign-alternating from 50 up to 290 Oe.

In Figs. 2.70–2.72, dependencies of the key parameters of the spectral characteristics of generated signals on the magnetic field H_0 are shown. In Fig. 2.71, ranges of the most typical kinds of generated signals are shown at changes of the magnetic field, namely: two SPS ranges, one NS range, and one ES FS range. Some modes close to parametrical signal multiplication (Figs. 2.70–2.72) in the range of magnetic field from 50 up to 170 Oe and from 220 up to 290 Oe are noted. In the mode of FMCR autoresonance, SPS components are generated.

In Fig. 2.73, the dependence of the integrated target power P_{out} for a heteromagnetic generator on the bias field H_0 is presented.

2.4 Structures with Ferrite KG-50

The results of our researches of the energetic and spectral characteristics of heteromagnetic structures on the basis of the transistor KT962B and the ferrite KG-50 with various orientations in an external magnetic field H_0 at an angle φ at low and high levels of power are presented in Figs. 2.74–2.105.

2.4.1 Angle of Orientation of FMCR $\varphi = 0°$

In Figs. 2.74–2.77, experimental characteristics of some key parameters of the spectra of signals in heteromagnetic structures for the components of signals on several frequencies $\nu_{0,1,2}$ on a low power level (50 mW) for an FMCR with $4\pi M_s = 620$ G are presented.

In Fig. 2.74, dependencies of the deviations of frequencies $(\Delta\nu)_{0,1,2}$ for the spectral components $\nu_{0,1,2}$ on the magnetic field H_0 are shown. The deviation is maximal in a narrow range of the magnetic field (180–200 Oe).

In Figs. 2.75 and 2.76, dependencies of the key parameters of the spectral lines of various kinds of signals are shown. In Fig. 2.75, the existence ranges of some characteristic kinds of signals are shown, namely: two SPS ranges and one NS range. From the data of Figs. 2.75 and 2.76 it is obvious that from 80 up to 140 Oe and from 220 up to 820 Oe processes of parametrical signal multiplication take place. In the mode of FMCR autoresonance, SPS components are generated.

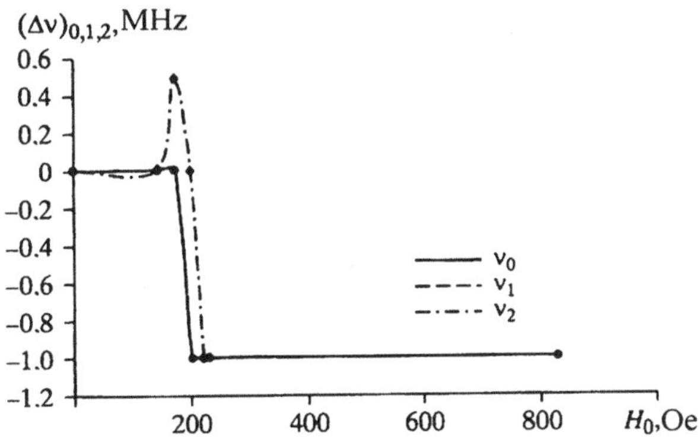

Fig. 2.74 Experimental characteristics of some key parameters of the spectra of signals in heteromagnetic structures for the components of signals on several frequencies $\nu_{0,1,2}$ on a low power level (50 mW) for an FMCR with $4\pi M_s = 620$ G

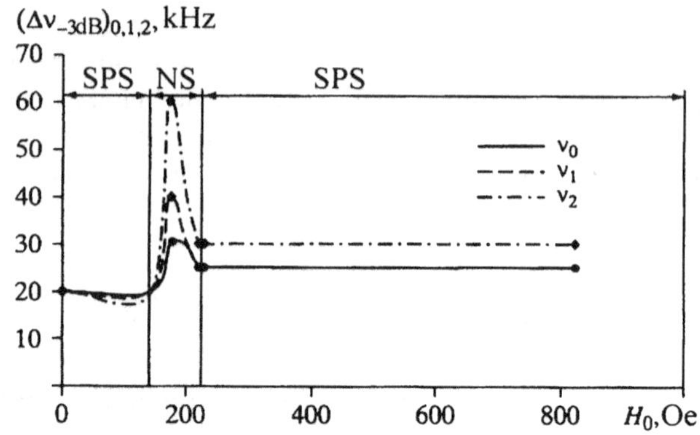

Fig. 2.75 Experimental characteristics of some key parameters of the spectra of signals in heteromagnetic structures for the components of signals on several frequencies $\nu_{0,1,2}$ on a low power level (50 mW) for an FMCR with $4\pi M_s = 620\,\text{G}$

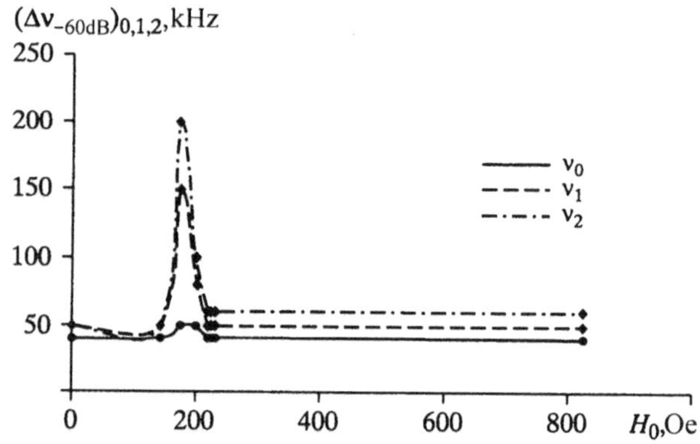

Fig. 2.76 Experimental characteristics of some key parameters of the spectra of signals in heteromagnetic structures for the components of signals on several frequencies $\nu_{0,1,2}$ on a low power level (50 mW) for an FMCR with $4\pi M_s = 620\,\text{G}$

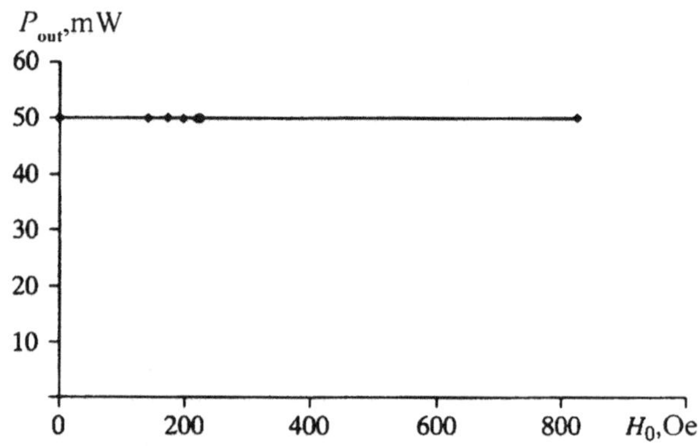

Fig. 2.77 Experimental characteristics of some key parameters of the spectra of signals in heteromagnetic structures for the components of signals on several frequencies $\nu_{0,1,2}$ on a low power level (50 mW) for an FMCR with $4\pi M_s = 620\,\text{G}$

In Fig. 2.77, the dependence of the integrated target power P_{out} of the heteromagnetic generator on the bias field H_0 is presented.

In Figs. 2.78–2.83, experimental dependencies of the characteristics of heteromagnetic generators at a high level of power (500 mW) for an FMCR with $4\pi M_s = 620\,\text{G}$ are presented. In Fig. 2.78, dependencies of the deviations of frequencies $(\Delta\nu)_{0,1,2}$ for the spectral components $\nu_{0,1,2}$ on the magnetic field H_0 are shown. The deviation is maximal from 0 up to 260 Oe.

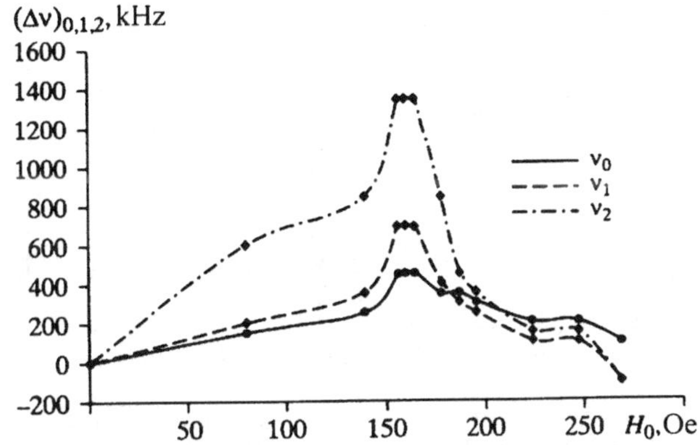

Fig. 2.78 Experimental dependencies of the characteristics of heteromagnetic generators at a high level of power (500 mW) for an FMCR with $4\pi M_s = 620\,\mathrm{G}$

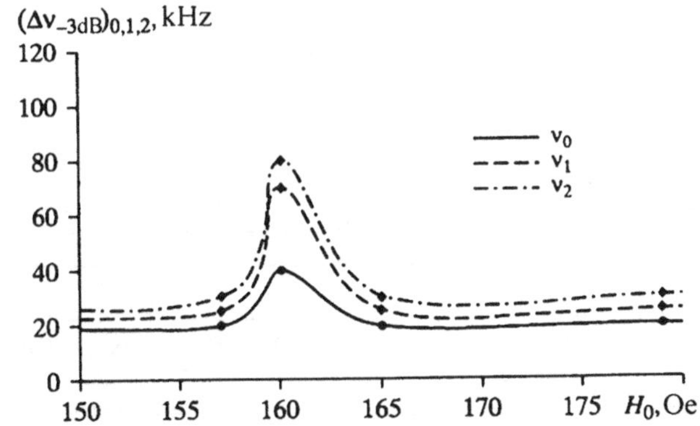

Fig. 2.79 Experimental dependencies of the characteristics of heteromagnetic generators at a high level of power (500 mW) for an FMCR with $4\pi M_s = 620\,\mathrm{G}$

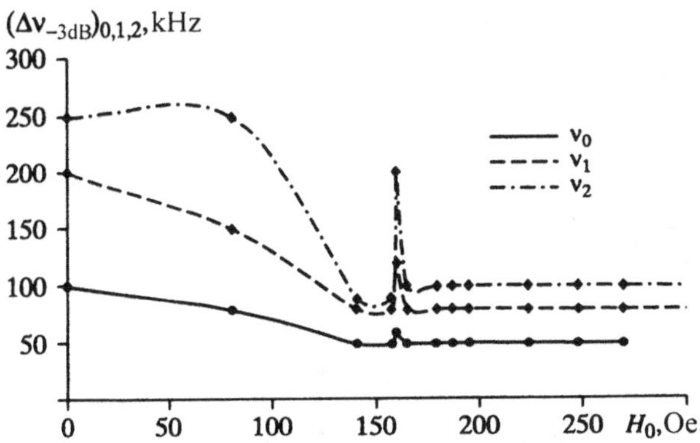

Fig. 2.80 Experimental dependencies of the characteristics of heteromagnetic generators at a high level of power (500 mW) for an FMCR with $4\pi M_s = 620\,\mathrm{G}$

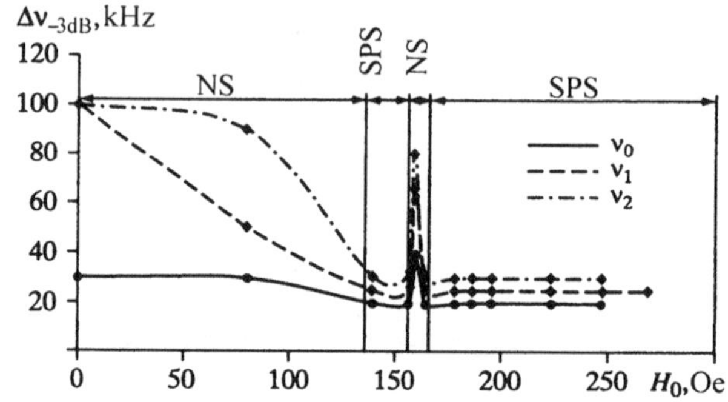

Fig. 2.81 Experimental dependencies of the characteristics of heteromagnetic generators at a high level of power (500 mW) for an FMCR with $4\pi M_s = 620\,\mathrm{G}$

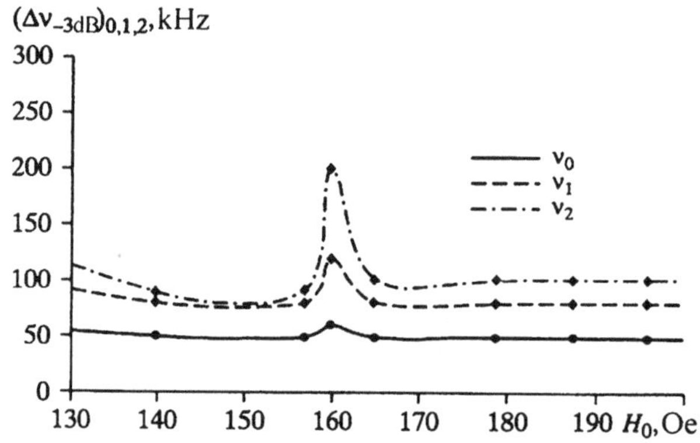

Fig. 2.82 Experimental dependencies of the characteristics of heteromagnetic generators at a high level of power (500 mW) for an FMCR with $4\pi M_s = 620\,\mathrm{G}$

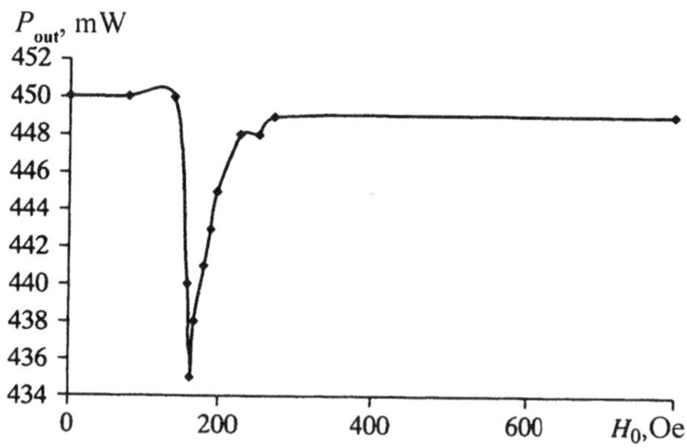

Fig. 2.83 Experimental dependencies of the characteristics of heteromagnetic generators at a high level of power (500 mW) for an FMCR with $4\pi M_s = 620\,\mathrm{G}$

In Figs. 2.79–2.82, dependencies of the spectral characteristics of generated signals on the magnetic field H_0 are shown. In Fig. 2.79, ranges of the most typical kinds of generated signals are shown at changes of the magnetic field, namely: two NS ranges and two SPS ones. Modes close to parametrical multiplication are noted for magnetic fields $H_0 = 150$ Oe and $H_0 > 170$ Oe (Figs. 2.80–2.82). At $H_0 = 156$ Oe in the mode of FMCR autoresonance, NS components are generated.

In Fig. 2.80, the dependence of the spectral linewidth of generation on a level -60 dB for the frequency components ν_0, ν_1, and ν_2 on the field H_0 is shown.

In Fig. 2.83, dependencies of the integrated target power P_{out} on the magnetic field H_0 are presented. At $H_0 = 156$ Oe insignificant reduction of the target power takes place.

2.4.2 Angle of Orientation of FMCR $\varphi = 45°$

In Figs. 2.84–2.87, experimental characteristics of heteromagnetic structures for spectral components of signals on several frequencies $\nu_{0,1,2}$ on a low level of power (50 mW) for an FMCR with $4\pi M_s = 620$ G are presented.

In Fig. 2.84, dependencies of the deviations of frequencies $(\Delta\nu)_{0,1,2}$ for the spectral components $\nu_{0,1,2}$ on the magnetic field H_0 are presented. The deviation of frequency is maximal from 190 up to 300 Oe.

In Figs. 2.85 and 2.86, dependencies of the key parameters of the spectral lines of various kinds of signals are shown. In Fig. 2.85, the ranges of existence of some characteristic kinds of signals are shown, namely: two SPS ranges and one NS range. In the mode of FMCR autoresonance, SPS components are generated.

In Fig. 2.87, the dependence of the integrated target power P_{out} on the magnetic field H_0 is presented.

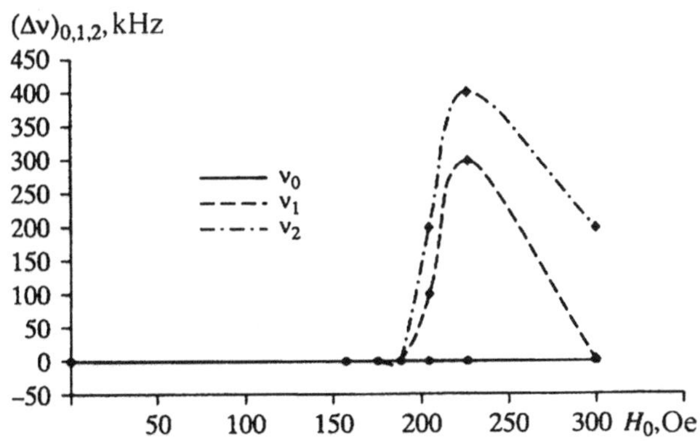

Fig. 2.84 Experimental characteristics of heteromagnetic structures for spectral components of signals on several frequencies $\nu_{0,1,2}$ on a low level of power (50 mW) for an FMCR with $4\pi M_s = 620$ G

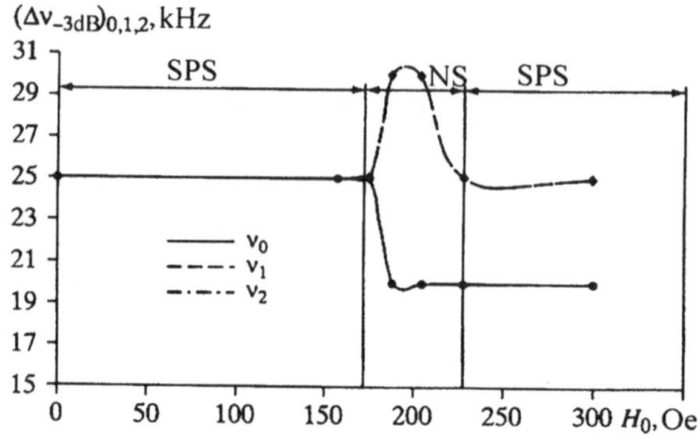

Fig. 2.85 Experimental characteristics of heteromagnetic structures for spectral components of signals on several frequencies $\nu_{0,1,2}$ on a low level of power (50 mW) for an FMCR with $4\pi M_s =$ 620 G

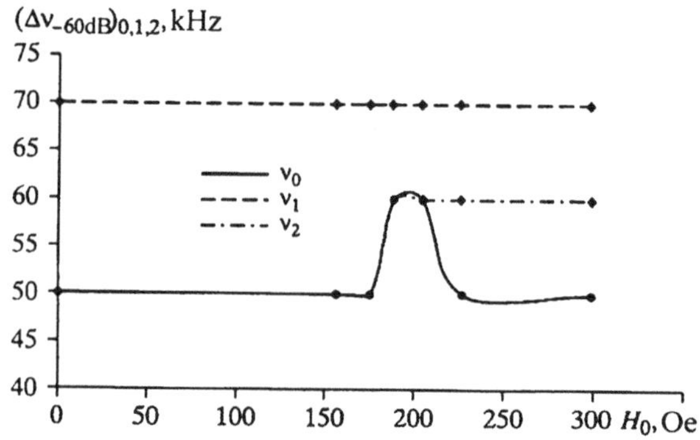

Fig. 2.86 Experimental characteristics of heteromagnetic structures for spectral components of signals on several frequencies $\nu_{0,1,2}$ on a low level of power (50 mW) for an FMCR with $4\pi M_s =$ 620 G

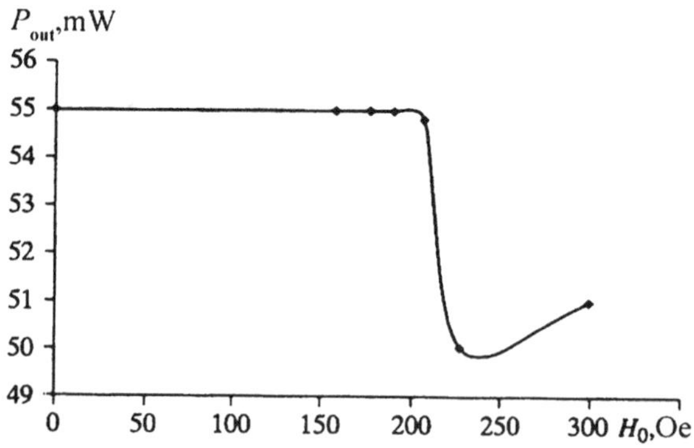

Fig. 2.87 Experimental characteristics of heteromagnetic structures for spectral components of signals on several frequencies $\nu_{0,1,2}$ on a low level of power (50 mW) for an FMCR with $4\pi M_s =$ 620 G

In Figs. 2.88–2.93, experimental characteristics of heteromagnetic structures at a high level of power (500 mW) for an FMCR with $4\pi M_s = 620$ G are presented.

In Fig. 2.88, dependencies of the deviations of frequencies $(\Delta\nu)_{0,1,2}$ for the spectral components $\nu_{0,1,2}$ on the magnetic field H_0 are shown. The deviation is maximal from 0 up to 325 Oe.

In Figs. 2.88–2.92, dependencies of the key parameters of the spectral lines of various kinds of signals are shown. In Fig. 2.89, the ranges of existence of some characteristic kinds of spectra of signals are shown, namely: two NS ranges and two SPS ones. From Figs. 2.90–2.92, it is obvious that there are narrow ranges of the magnetic field within which the processes come nearer to parametrical multiplication. In the mode of FMCR autoresonance, NS components are generated.

In Fig. 2.93, the dependence of the integrated target power P_{out} on the magnetic field H_0 is shown.

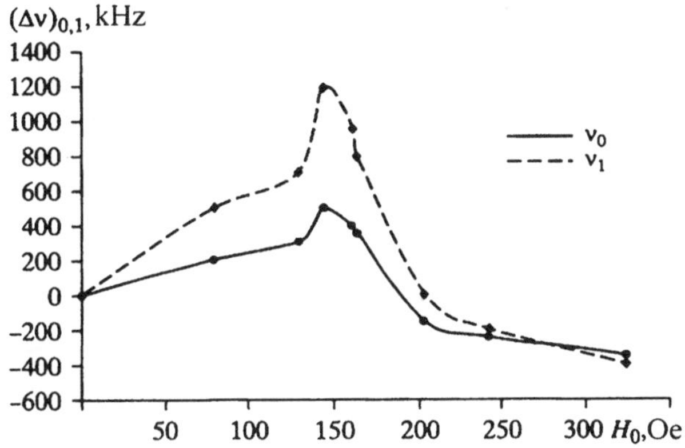

Fig. 2.88 Experimental characteristics of heteromagnetic structures at a high level of power (500 mW) for an FMCR with $4\pi M_s = 620$ G

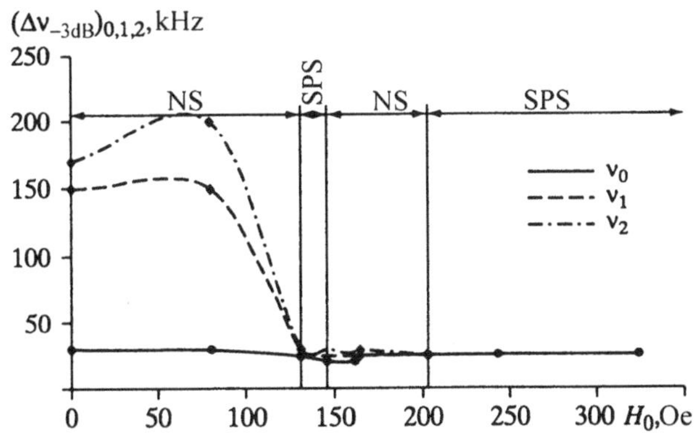

Fig. 2.89 Experimental characteristics of heteromagnetic structures at a high level of power (500 mW) for an FMCR with $4\pi M_s = 620$ G

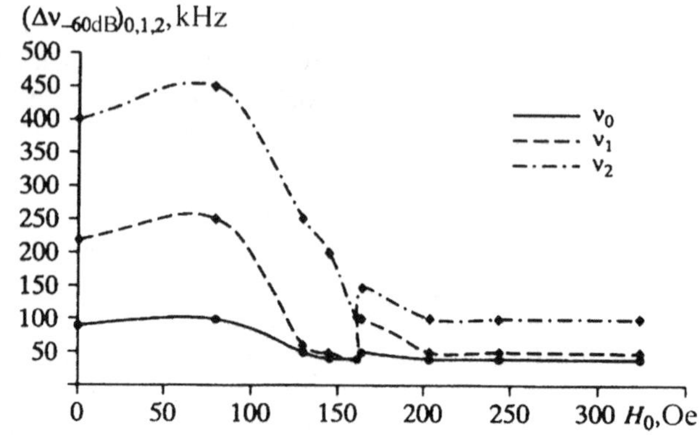

Fig. 2.90 Experimental characteristics of heteromagnetic structures at a high level of power (500 mW) for an FMCR with $4\pi M_s = 620\,\text{G}$

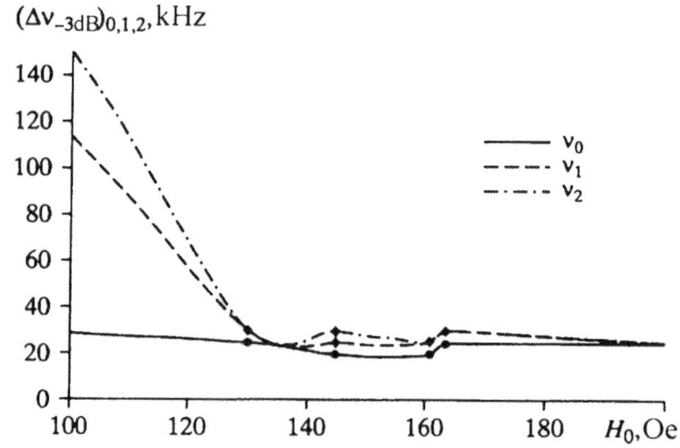

Fig. 2.91 Experimental characteristics of heteromagnetic structures at a high level of power (500 mW) for an FMCR with $4\pi M_s = 620\,\text{G}$

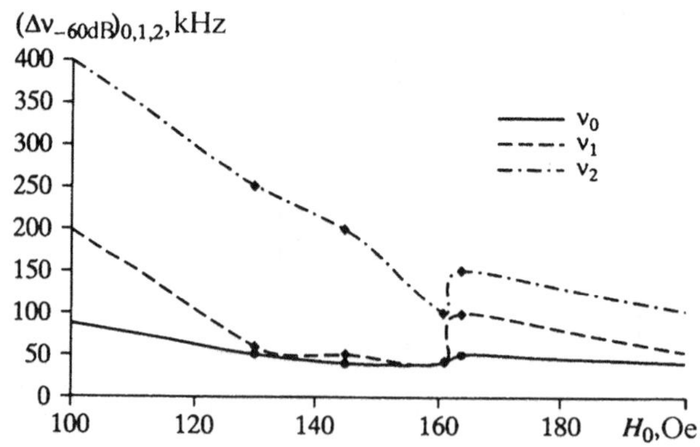

Fig. 2.92 Experimental characteristics of heteromagnetic structures at a high level of power (500 mW) for an FMCR with $4\pi M_s = 620\,\text{G}$

2.4.3 Angle of Orientation of FMCR $\varphi = 90°$

In Figs. 2.94–2.97, experimental characteristics of heteromagnetic generators on a low level of power (50 mW) for an FMCR with $4\pi M_s = 620$ G are presented.

In Fig. 2.94, dependencies of the deviations of frequencies $(\Delta\nu)_{0,1,2}$ for the spectral components $\nu_{0,1,2}$ on the magnetic field H_0 are shown.

In Figs. 2.95 and 2.96, dependencies of the spectral lines $(\Delta\nu_{-3dB})_{0,1,2}$ and $(\Delta\nu_{-60dB})_{0,1,2}$ of generated signals are shown. It is seen that in a significant range of magnetic field H_0 the processes have parametrical character or are close to it.

In Fig. 2.96, the ranges of existence of some characteristic kinds of signals are shown, namely: two SPS ranges and one NS range.

In Fig. 2.97, the dependence of the integrated target power P_{out} on the bias field H_0 is presented.

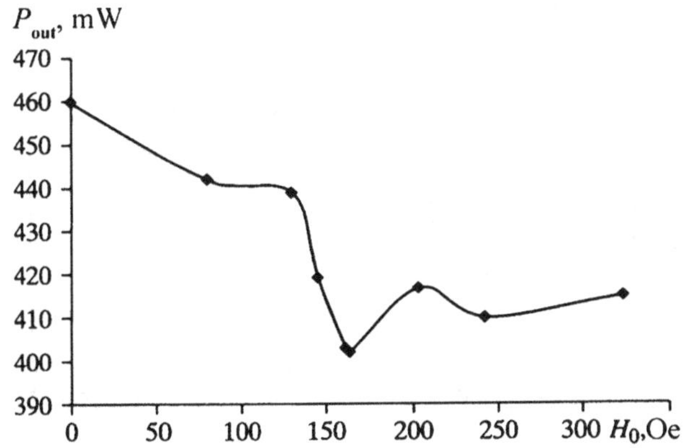

Fig. 2.93 Experimental characteristics of heteromagnetic structures at a high level of power (500 mW) for an FMCR with $4\pi M_s = 620$ G

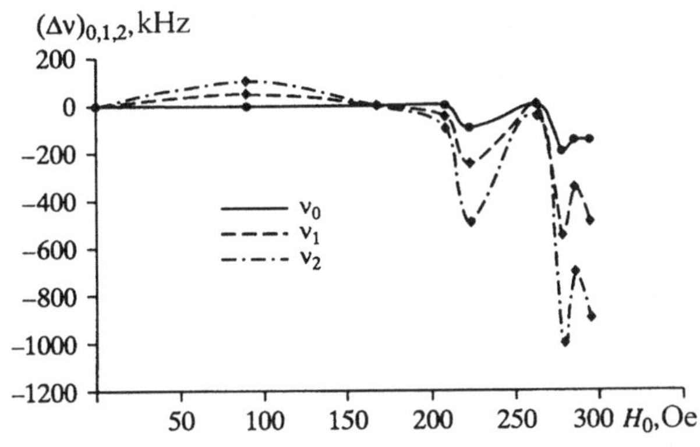

Fig. 2.94 Experimental characteristics of heteromagnetic generators on a low level of power (50 mW) for an FMCR with $4\pi M_s = 620$ G

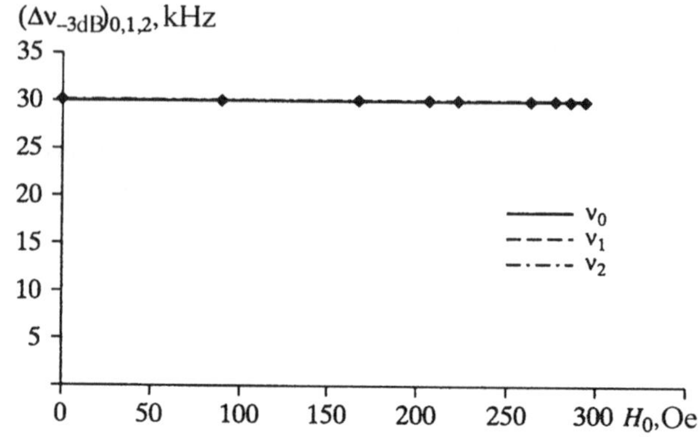

Fig. 2.95 Experimental characteristics of heteromagnetic generators on a low level of power (50 mW) for an FMCR with $4\pi M_s = 620\,G$

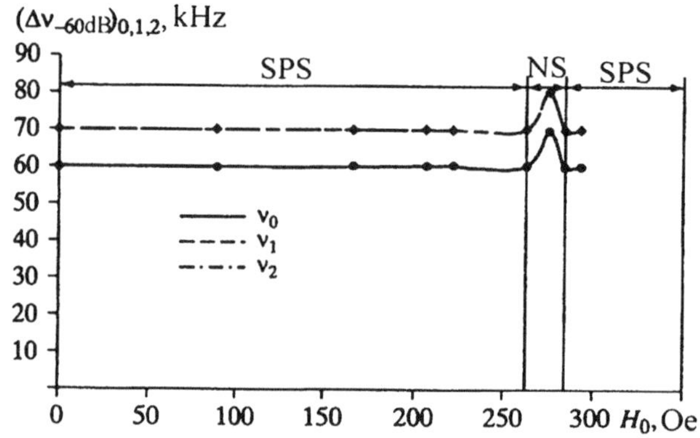

Fig. 2.96 Experimental characteristics of heteromagnetic generators on a low level of power (50 mW) for an FMCR with $4\pi M_s = 620\,G$

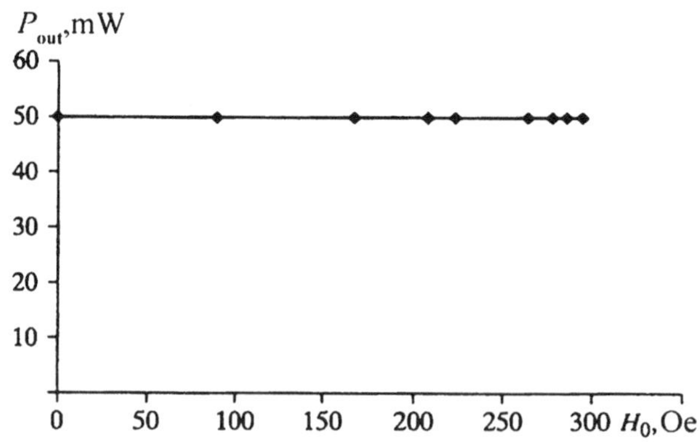

Fig. 2.97 Experimental characteristics of heteromagnetic generators on a low level of power (50 mW) for an FMCR with $4\pi M_s = 620\,G$

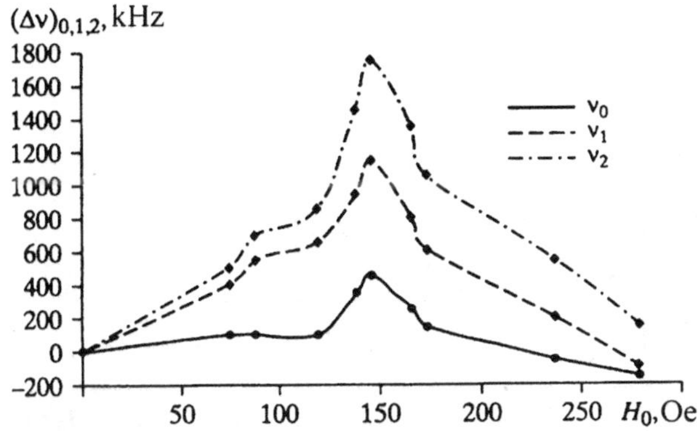

Fig. 2.98 The key parameters of the spectral experimental characteristics of heteromagnetic structures at a high level of power (500 mW)

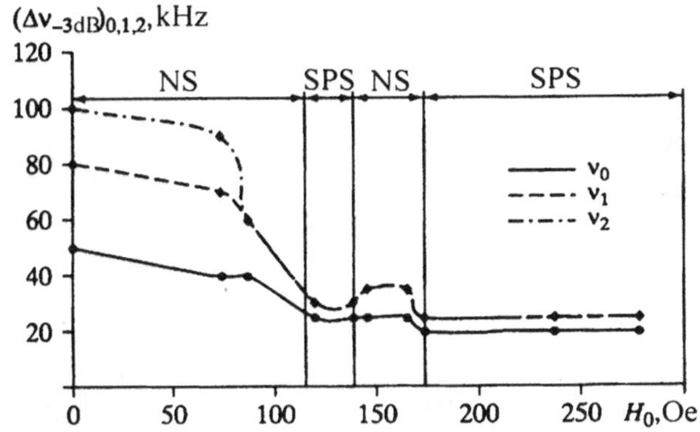

Fig. 2.99 The key parameters of the spectral experimental characteristics of heteromagnetic structures at a high level of power (500 mW)

In Figs. 2.98–2.101, the key parameters of the spectral experimental characteristics of heteromagnetic structures at a high level of power (500 mW) are presented.

In Fig. 2.99, the ranges of existence of some characteristic kinds of the spectra of signals are shown, namely: two SPS ranges and two NS ones. In the mode of FMCR autoresonance, NS components are generated.

In Fig. 2.101, the dependence of the integrated target power P_{out} of the heteromagnetic generator on the bias field H_0 is presented.

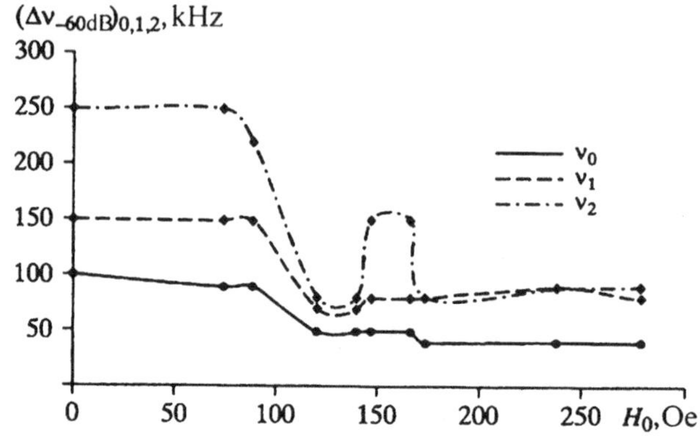

Fig. 2.100 The key parameters of the spectral experimental characteristics of heteromagnetic structures at a high level of power (500 mW)

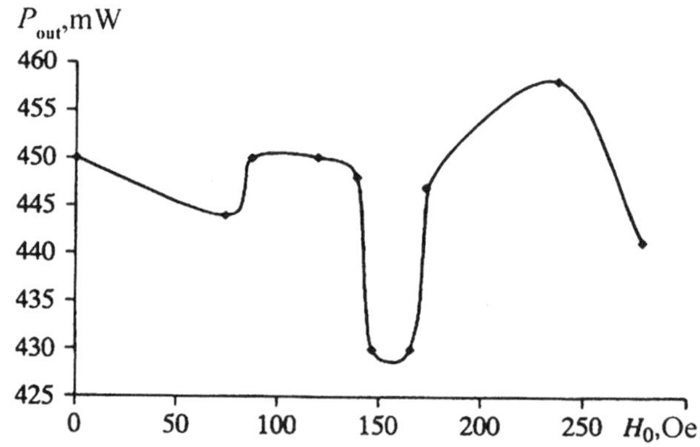

Fig. 2.101 The key parameters of the spectral experimental characteristics of heteromagnetic structures at a high level of power (500 mW)

2.5 Structures with Ferrites KG-65 and KG-140

2.5.1 *Angle of Orientation of FMCR $\varphi = 90°$*

In Figs. 2.102–2.105, experimental characteristics of the key parameters and kinds of spectra of signals in heteromagnetic structures for the ferrite KG-140 at a high level of power (500 mW) for an FMCR with $4\pi M_s = 1{,}750$ G are presented.

In Fig. 2.102, the dependence of the central spectral components ν_0, ν_1, and ν_2 on the field H_0 is given.

In Figs. 2.103 and 2.104, dependencies of the spectral lines of generated signals on the magnetic field magnitude are shown. In Fig. 2.103, the ranges of existence of some characteristic kinds of signals are presented, namely: SPS, NS, and SPS.

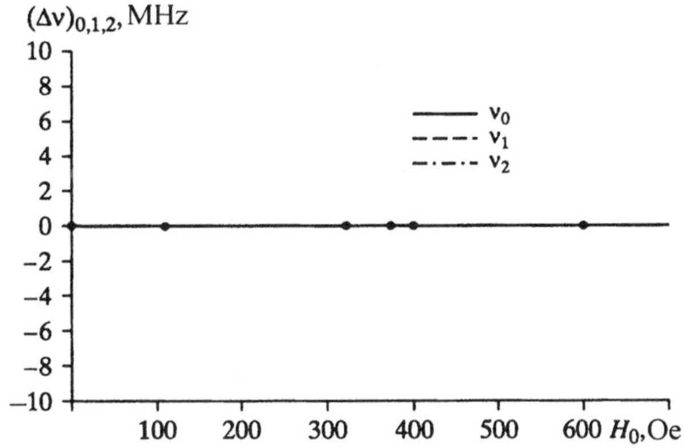

Fig. 2.102 Experimental characteristics of the key parameters and kinds of spectra of signals in heteromagnetic structures for the ferrite KG-140 at a high level of power (500 mW) for an FMCR with $4\pi M_s = 1{,}750\,\mathrm{G}$

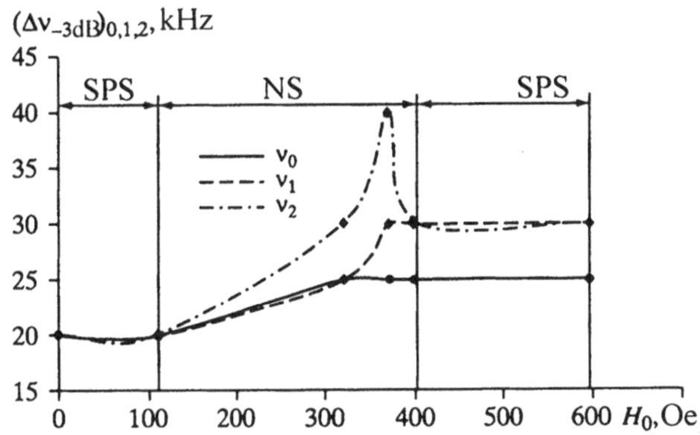

Fig. 2.103 Experimental characteristics of the key parameters and kinds of spectra of signals in heteromagnetic structures for the ferrite KG-140 at a high level of power (500 mW) for an FMCR with $4\pi M_s = 1{,}750\,\mathrm{G}$

In Fig. 2.105, the dependence of the integrated target power P_{out} of the heteromagnetic generator on the bias field H_0 is presented.

The generalized experimental dependencies presented above allow one:

- To resolve some characteristic kinds of signals and to analyze the dynamics of their change in a wide band of frequencies above the higher frequency of the used transistor structure
- To derive functional dependencies necessary for processing the parameters of HMT in the UHF and HHF ranges with the ferrite working in an unsaturated nonlinear mode

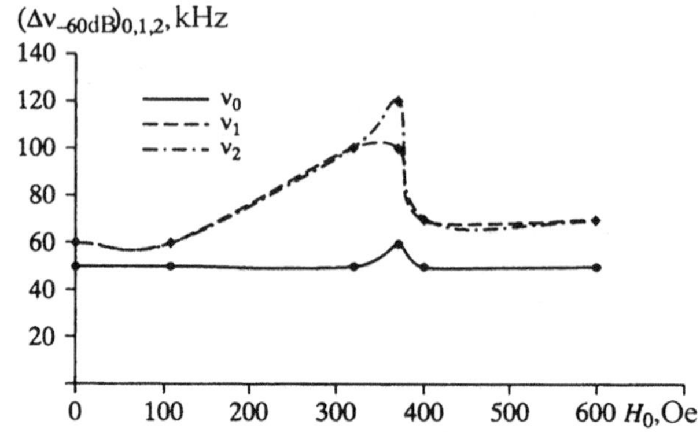

Fig. 2.104 Experimental characteristics of the key parameters and kinds of spectra of signals in heteromagnetic structures for the ferrite KG-140 at a high level of power (500 mW) for an FMCR with $4\pi M_s = 1{,}750\,\mathrm{G}$

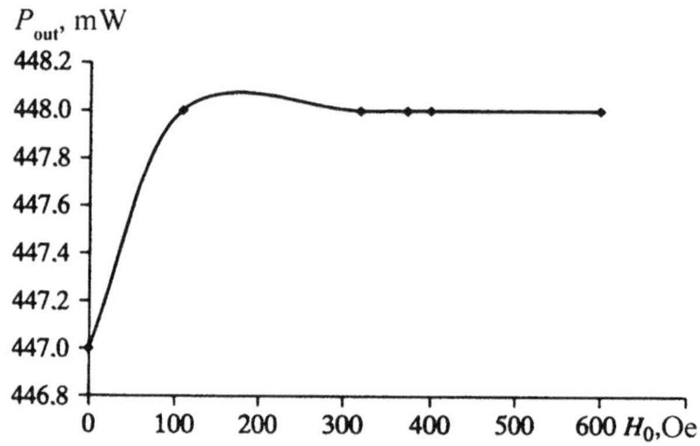

Fig. 2.105 Experimental characteristics of the key parameters and kinds of spectra of signals in heteromagnetic structures for the ferrite KG-140 at a high level of power (500 mW) for an FMCR with $4\pi M_s = 1{,}750\,\mathrm{G}$

- To derive functional dependencies for processing the parameters of the equivalent circuits of heteromagnetic transistor and heteromagnetic diode structures in the UHF and HHF ranges
- To give recommendations for the choice of proper parameters of generating and mixing heteromagnetic structures with an expanded dynamic range of HF power, management limits of the key parameters of signals of various types: spectrally clean ones, noise-type, noises, evenly spaced grid frequencies, and various modes of noisy spectral lines as well
- To give recommendations for requirements to the operating parameters on both the semiconductor and ferrite subsystems

2.6 Generalization of Experimental Data

In Table 2.1, data for FMCR of various brands and saturation magnetizations (YIG resonators), the angles of their orientations in an external magnetic field H_0, spectra types of generated signals, and the order of their following are shown for a changing bias field H_0: from $H_0 \approx 0$ (an autoresonance mode) up to 300 Oe as for the middle (50 mW), and high (500 mW) levels of power.

Table 2.1 Data for FMCR of various brands and saturation magnetizations (YIG resonators), the angles of their orientations in an external magnetic field H_0, spectra types of generated signals, and the order of their following are shown for a changing bias field H_0: from $H_0 \approx 0$ (an autoresonance mode) up to 300 Oe as for the middle (50 mW), and high (500 mW) levels of power

		$P_{out} = 50$ mW		$P_{out} = 500$ mW	
FMCR, ferrite brand: $4\pi M_s$, G	Angle of orientation in external magnetic field $\overline{H}_0$, $\varphi(°)$	Kinds of spectra of signals on one structure at increasing H_0	Spectrum of signal at autoresonance FMCR ($H_0 \approx 0$ Oe)	Kinds of spectra of signals on one structure at increasing H_0	Spectrum of signal at autoresonance FMCR ($H_0 \approx 0$ Oe)
KG-8, 90	0	SPS, NS, ES FS, NS, SPS	SPS	SPS, NS, ES FS, NS, SPS	SPS
	45	SPS, NS, ES FS, NS, ES FS, SPS	SPS	SPS, NS, SPS	SPS
	90	SPS, NS, ES FS, NS, SPS	SPS	SPS, ES FS, NS, SPS	SPS
KG-15, 190	0	SPS, NS, ES FS, NS, SPS, NS, SPS	SPS	NS, SPS, ES FS, NS, ES FS, NS, ES FS	NS
	45	SPS, NS, ES FS, NS, SPS, NS	SPS	SPS, NS	SPS
	90	SPS, NS, ES FS, NS, SPS, NS, SPS	SPS	SPS, NS, ES FS, SPS	SPS
KG-50, 620	0	SPS, NS, SPS	SPS	NS, SPS, NS, SPS	NS
	45	SPS, NS, SPS	SPS	NS, SPS, NS, SPS	NS
	90	SPS, NS, SPS	SPS	NS, SPS, NS, SPS	NS
KG-140, 1750	0	–	–	SPS, NS, SPS	SPS

From the resulted data some laws are seen at formation of various kinds of spectra of signals, and on the middle level of power ($P_{out} \approx 50\,\text{mW}$) the kinds of spectra of signals and the order of their following for various FMCR are kept practically independent on the angle of orientation of the FMCR in the field $\overline{H}_0$. Naturally, this conclusion does not cover the quantitative characteristics and parameters of the formed kinds of the spectra of signals. At a high level of power ($P_{out} \approx 500\,\text{mW}$) no deduce is possible to infer, though, as a whole, the prevalence of noise modes, including generation of NS spectra is appreciable at autoresonance (at $H_0 \approx 0$).

Chapter 3
Control Over Energy and Spectral Characteristics

3.1 Control Over Characteristics of Spectral-Pure Signals

For generation modes of various types of SPS below are given experimental dependencies of the maximal frequency deviation $(\Delta\nu_{max})^{SPS}_{0,1,2}$, spectral linewidths $(\nu_{-3dB})^{SPS}_{0,1,2}$ on the orientation angle φ for FMCR in a magnetotransistor for the alloyed ferrite with a cubic structure KG-8, KG-15, KG-50 with saturation magnetizations $4\pi M_s = 90$, 190, and 620 G, respectively, on low ($P_{out} = 50$ mW) and high ($P_{out} = 500$ mW) levels of integral power.

3.1.1 Structures with Various Orientations in a Magnetic Field

3.1.1.1 FMCR from KG-8

From Fig. 3.1, it follows that the frequency deviation $(\Delta\nu_{max})^{SPS}_{0,1,2}$ has its greatest value at $\varphi = 0°$ for all the spectral components ($n = 1, 2, 3$) on the medium (50 mW) power level.

At increase of the power level by an order of magnitude, up to 500 mW (Fig. 3.2), the frequency deviation $(\Delta\nu_{max})^{SPS}_{0,1,2}$ is maximal at $\varphi \cong 45°$, $(\Delta\nu_{max})^{SPS}_{1} \cong (\Delta\nu_{max})^{SPS}_{0}$, and for the harmonics with $n = 2 - (\Delta\nu_{max})^{SPS}_{2} \cong 2(\Delta\nu_{max})^{SPS}_{0,1}$.

The frequency deviation $(\Delta\nu_{max})^{SPS}_{0,1,2}$ on a low power level ($P_{out} \cong 50$ mW) is a weak function of φ (Fig. 3.3).

The shape of spectral lines depends on the angle of orientation φ in a complex manner for various harmonics of signals for generation powers $P_{out} = 50$ mW and $P_{out} = 500$ mW.

From Figs. 3.4 and 3.5, it is obvious that at the power level $P_{out} = 50$ mW the shape of the spectral line of a signal on the basic frequency ν_0 is practically independent of φ. The most significant is the broadening of the base of the spectral line of a signal $(\Delta\nu_{-60dB})^{SPS}_{2}$ at $\varphi \cong 0°$. When $\varphi \geq 45°$ the width of the basis of spectral lines is the same $(\Delta\nu_{-60dB})^{SPS}_{1} \cong (\Delta\nu_{-60dB})^{SPS}_{2}$, $(\Delta\nu_{-3dB})^{SPS}_{1} > (\Delta\nu_{-3dB})^{SPS}_{0,2}$ and when $\varphi > 45°$ both broadening of $(\Delta\nu_{-3dB})^{SPS}_{2}$ and narrowing of $(\nu_{-3dB})^{SPS}_{1}$

A.A. Ignatiev and A.V. Lyashenko, *Heteromagnetic Microelectronics: Microsystems of Active Type*, DOI 10.1007/978-1-4419-6002-3_3,

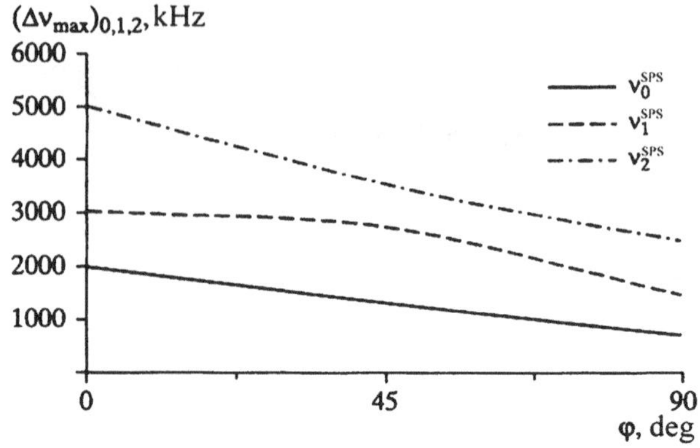

Fig. 3.1 The frequency deviation $(\Delta\nu_{\max})_{0,1,2}^{\mathrm{SPS}}$ has its greatest value at $\varphi = 0°$ for all the spectral components ($n = 1, 2, 3$) on the medium (50 mW) power level

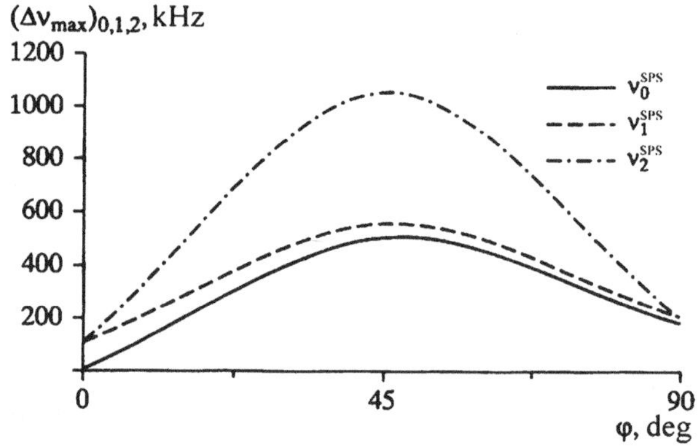

Fig. 3.2 The frequency deviation $(\Delta\nu_{\max})_{0,1,2}^{\mathrm{SPS}}$ is maximal at $\varphi \cong 45°$, $(\Delta\nu_{\max})_1^{\mathrm{SPS}} \cong (\Delta\nu_{\max})_0^{\mathrm{SPS}}$, and for the harmonics with $n = 2 - (\Delta\nu_{\max})_2^{\mathrm{SPS}} \cong 2(\Delta\nu_{\max})_{0,1}^{\mathrm{SPS}}$

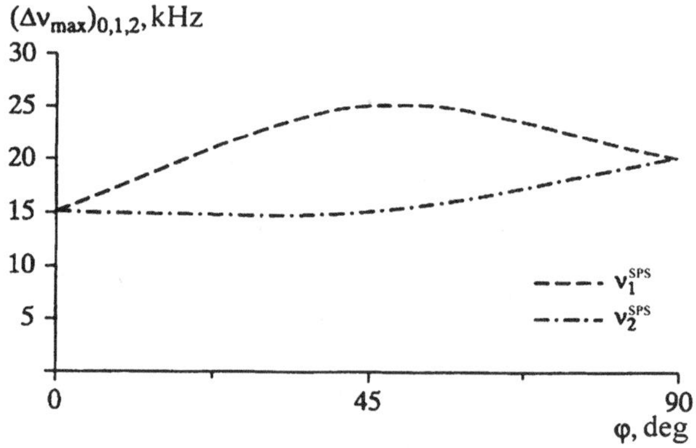

Fig. 3.3 The frequency deviation $(\Delta\nu_{\max})_{0,1,2}^{\mathrm{SPS}}$ on a low power level ($P_{\mathrm{out}} \cong 50\,\mathrm{mW}$) is a weak function of φ

take place, which speaks for complex processes in FSS and various opportunities of management of the shapes of spectral lines in generating modes.

At high power levels ($P \cong 500\,\mathrm{mW}$), the shape of the spectral lines of harmonic components (Figs. 3.6 and 3.7) is a complex function on the orientation angle of ferrite φ. At $\varphi \cong 0°$, the value $(\nu_{-3\,\mathrm{dB}})_0^{\mathrm{SPS}} \cong (\nu_{-3\,\mathrm{dB}})_2^{\mathrm{SPS}}$, $(\nu_{-3\mathrm{dB}})_1^{\mathrm{SPS}} > (\nu_{-3\mathrm{dB}})_{0,2}^{\mathrm{SPS}}$

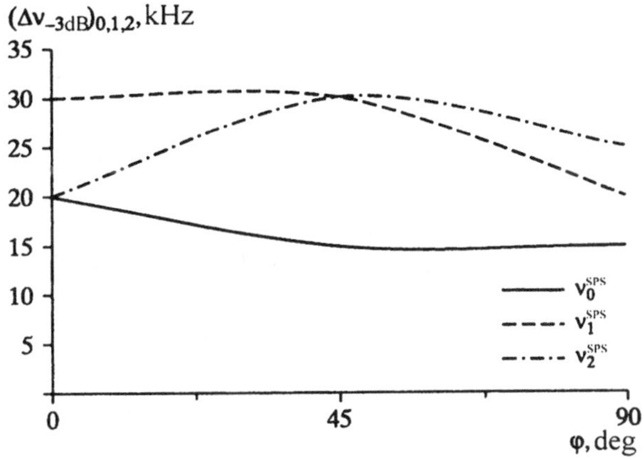

Fig. 3.4 At the power level $P_{out} = 50\,\mathrm{mW}$ the shape of the spectral line of a signal on the basic frequency ν_0 is practically independent of φ

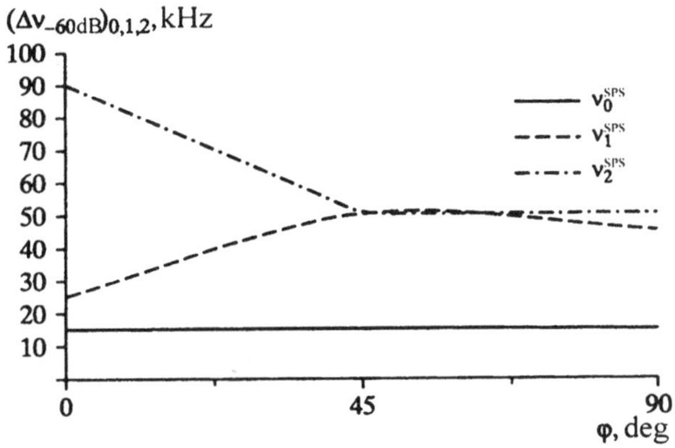

Fig. 3.5 At the power level $P_{out} = 50\,\mathrm{mW}$ the shape of the spectral line of a signal on the basic frequency ν_0 is practically independent of φ

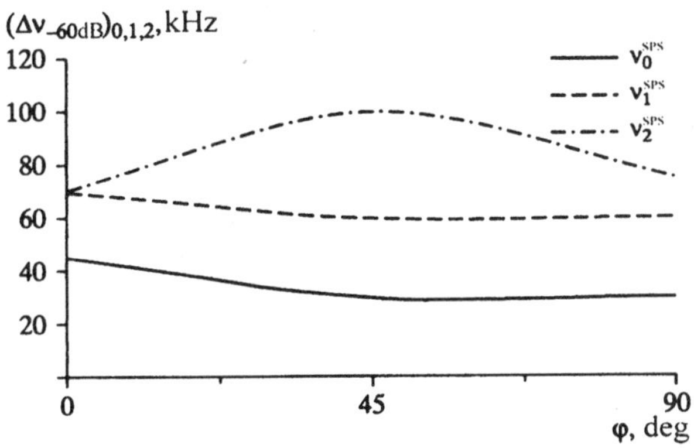

Fig. 3.6 The shape of the spectral lines of harmonic components is a complex function on the orientation angle of ferrite φ

but $(\nu_{-60\mathrm{dB}})_1^{\mathrm{SPS}} \cong (\nu_{-60\mathrm{dB}})_2^{\mathrm{SPS}}$. At $\varphi \cong 45°$, $(\nu_{-3\mathrm{dB}})_1^{\mathrm{SPS}} \cong (\nu_{-3\mathrm{dB}})_2^{\mathrm{SPS}}$ holds, but $(\nu_{-60\mathrm{dB}})_0^{\mathrm{SPS}} > (\nu_{-60\mathrm{dB}})_1^{\mathrm{SPS}} > (\nu_{-60\mathrm{dB}})_2^{\mathrm{SPS}}$ and for a signal on the frequency ν_2 the spectral line broadening is maximal. At increase in φ from 45° up to 90°, synchronous narrowing of the spectral lines of signals on the frequencies $\nu_{1,2}^{\mathrm{SPS}}$ was observed.

3.1.1.2 FMCR of KG-15

From Fig. 3.7, one can see that the frequency deviation $(\Delta\nu_{max})^{SPS}_{0,1,2}$ is a complex function of the orientation angle of ferrite φ for the spectral components on the frequencies ν^{SPS}_0, ν^{SPS}_1, ν^{SPS}_2 on the medium ($P \cong 50\,\mathrm{mW}$) power level. At $\varphi = 0°$ the deviation of frequency has its greatest value for a signal on the frequency ν^{SPS}_2, and $(\Delta\nu_{max})^{SPS}_2 > (\Delta\nu_{max})^{SPS}_0 > (\Delta\nu_{max})^{SPS}_1$. At increase in φ up to 90° an almost monotonous reduction of the values of s of frequency deviation $(\Delta\nu_{max})^{SPS}_{0,1,2}$ is observed.

At increase in the output power level up to 500 mW (Fig. 3.8) at $\varphi = 0°$ the values are $(\Delta\nu_{max})^{SPS}_0 < (\Delta\nu_{max})^{SPS}_1 > (\Delta\nu_{max})^{SPS}_2$, and with growth of the number of harmonic the deviation of frequency increases. At increase in φ up to 45°, modes with $(\Delta\nu_{max})^{SPS}_{0,1,2} \cong$ const are observed. At change of φ from 45° to 90° a reduction of the values of $(\Delta\nu_{max})^{SPS}_{0,1,2}$ takes place.

The shape of the spectral lines for low-power signals (Figs. 3.9 and 3.10) on $\nu^{SPS}_{0,1,2}$ is a complex function of φ. At $\varphi = 0°$, the spectral lines have their values

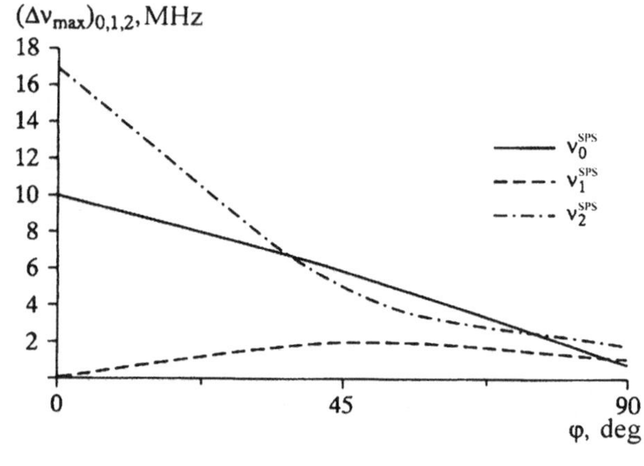

Fig. 3.7 The frequency deviation $(\Delta\nu_{max})^{SPS}_{0,1,2}$ is a complex function of the orientation angle of ferrite φ for the spectral components on the frequencies ν^{SPS}_0, ν^{SPS}_1, ν^{SPS}_2 on the medium ($P \cong 50\,\mathrm{mW}$) power level

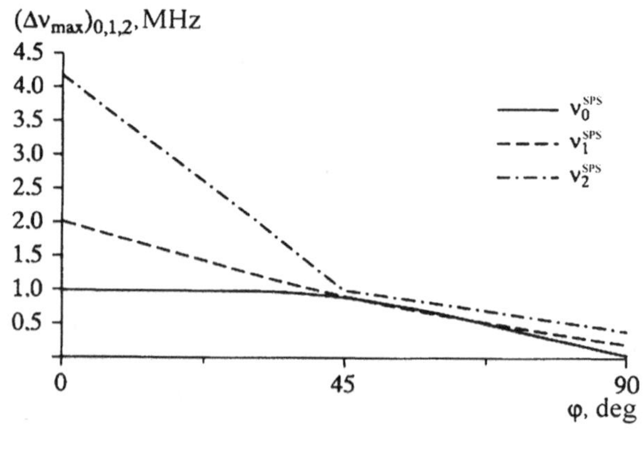

Fig. 3.8 At increase in the output power level up to 500 mW at $\varphi = 0°$ the values are $(\Delta\nu_{max})^{SPS}_0 < (\Delta\nu_{max})^{SPS}_1 > (\Delta\nu_{max})^{SPS}_2$, and with growth of the number of harmonic the deviation of frequency increases

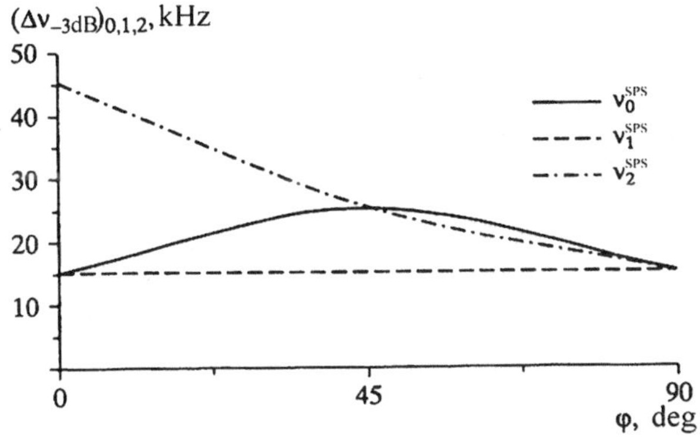

Fig. 3.9 The shape of the spectral lines for low-power signals on $\nu_{0,1,2}^{SPS}$ is a complex function of φ

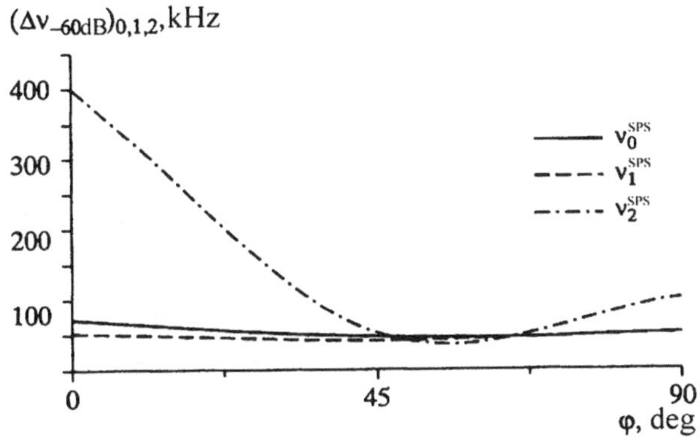

Fig. 3.10 The shape of the spectral lines for low-power signals on $\nu_{0,1,2}^{SPS}$ is a complex function of φ

$(\Delta\nu_{-3dB})_0^{SPS} \cong (\Delta\nu_{-3dB})_1^{SPS} < (\Delta\nu_{-3dB})_2^{SPS}$ and $(\Delta\nu_{-60dB})_0^{SPS} \cong (\Delta\nu_{-60dB})_1^{SPS} < (\Delta\nu_{-60dB})_2^{SPS}$. With an increase of φ up to 45° the line on the basic frequency $(\Delta\nu_{-3dB})_0^{SPS}$ widens, the width of the line for $(\Delta\nu_{-3dB})_1^{SPS}$ does not vary, and the width of the line $(\Delta\nu_{-3dB})_2^{SPS}$ decreases. At $\varphi \cong 45°$, the values are $(\Delta\nu_{-3dB})_0^{SPS} \cong (\Delta\nu_{-3dB})_2^{SPS}$ and $(\nu_{-3dB})_{0,2}^{SPS} > (\Delta\nu_{-3dB})_1^{SPS}$, and $(\Delta\nu_{-60dB})_{0,1,2}^{SPS}$ are minimal. At an increase of φ from 45° up to 90° a minor alteration of the shape of all the components $\nu_{0,1,2}^{SPS}$ is observed.

The shape of the spectral lines of signals for high power (Figs. 3.11 and 3.12) as a function of φ varies uniformly, and the broadening of lines is most considerable at $\varphi \cong 45°$.

3.1.1.3 FMCR of KG-50

At the medium power level ($P_{out} \cong 50$ mW) deviation of frequency $(\Delta\nu_{max})_{0,1,2}^{SPS}$ starts to appear from $\varphi > 45°$, and $(\Delta\nu_{max})_0^{SPS} < (\Delta\nu_{max})_1^{SPS} < (\Delta\nu_{max})_2^{SPS}$ (Fig. 3.13).

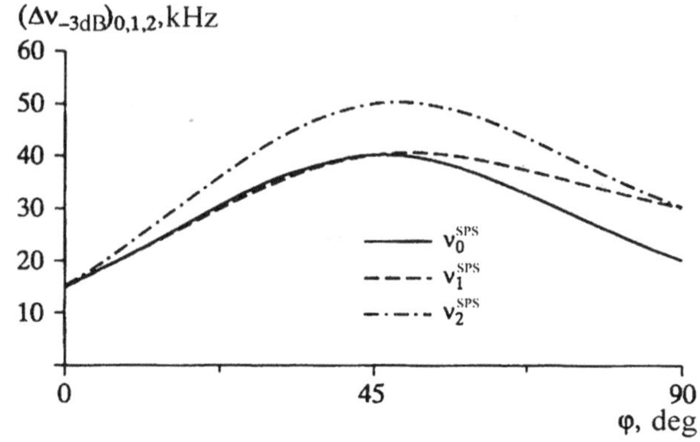

Fig. 3.11 The shape of the spectral lines of signals for high power as a function of φ varies uniformly, and the broadening of lines is most considerable at $\varphi \cong 45°$

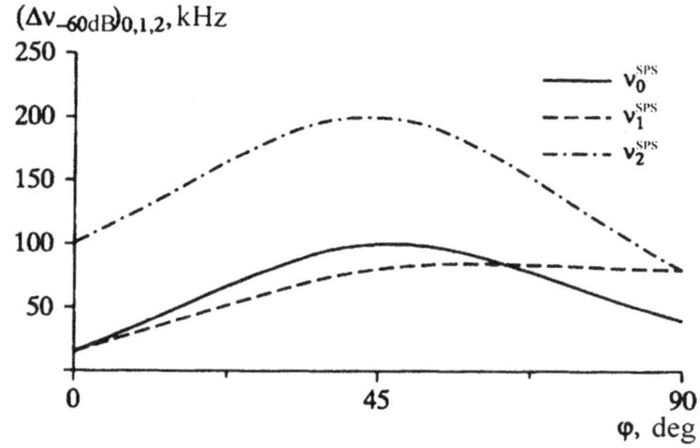

Fig. 3.12 The shape of the spectral lines of signals for high power as a function of φ varies uniformly, and the broadening of lines is most considerable at $\varphi \cong 45°$

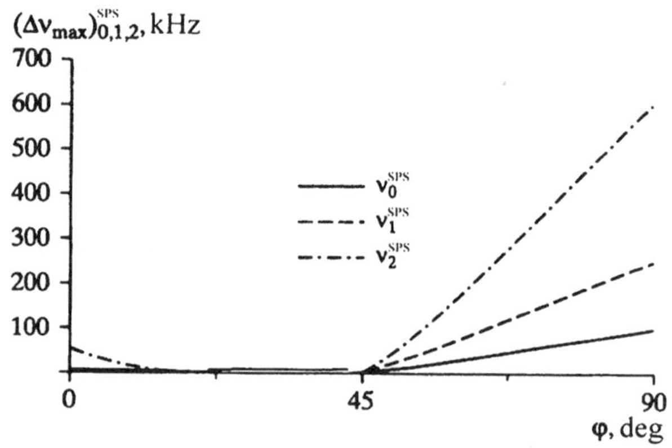

Fig. 3.13 At the medium power level ($P_{out} \cong 50\,\text{mW}$) deviation of frequency $(\Delta\nu_{max})^{SPS}_{0,1,2}$ starts to appear from $\varphi > 45°$, and $(\Delta\nu_{max})^{SPS}_{0} < (\Delta\nu_{max})^{SPS}_{1} < (\Delta\nu_{max})^{SPS}_{2}$

At a high power level ($P_{out} = 500\,\text{mW}$) the deviation of frequency $(\Delta\nu_{max})^{SPS}_{0,1,2}$ is maximal at $\varphi \cong 45°$ and increases with the number of harmonic (Fig. 3.14).

The shape of the spectral lines on the medium power level is a complex function of φ (Figs. 3.15 and 3.16). For the basic frequency ν_0^{SPS} increasing in φ from $0°$ to $90°$ leads to monotonous broadening of the spectral line $(\Delta\nu_{-3dB})_0^{SPS}$ and $(\Delta\nu_{-60dB})_0^{SPS}$. Within the limits of $\varphi = (0° \div 45°)$ the values $(\Delta\nu_{-3dB})_1^{SPS}$ and

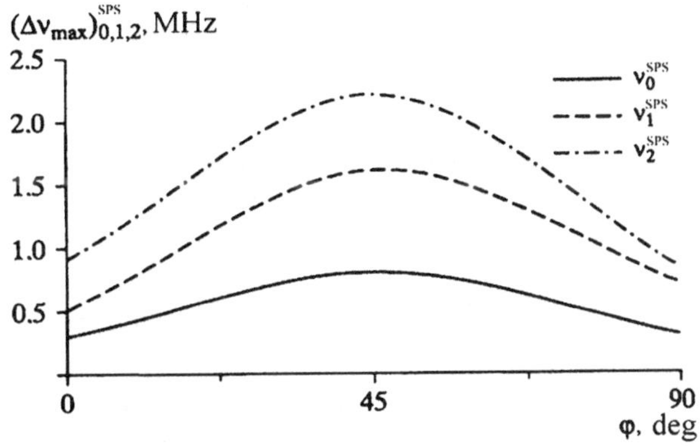

Fig. 3.14 At a high power level ($P_{out} = 500\,\text{mW}$) the deviation of frequency $(\Delta\nu_{max})^{SPS}_{0,1,2}$ is maximal at $\varphi \cong 45°$ and increases with the number of harmonic

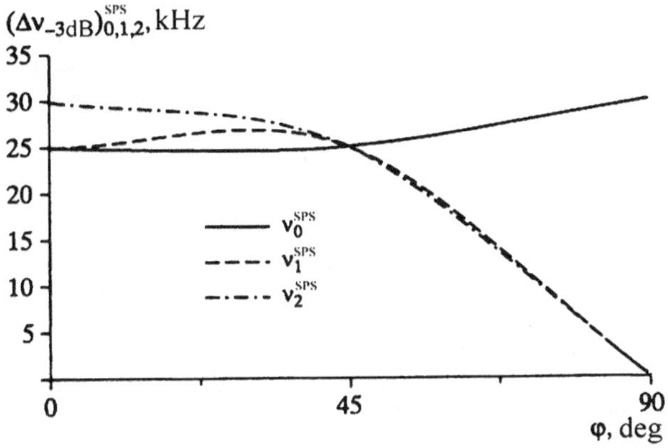

Fig. 3.15 The shape of the spectral lines on the medium power level is a complex function of φ

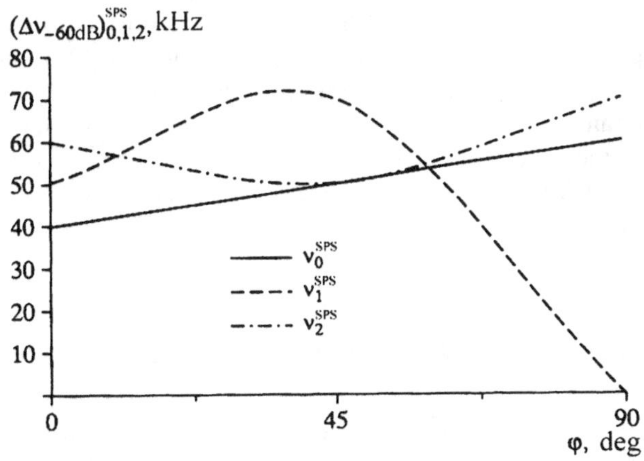

Fig. 3.16 The shape of the spectral lines on the medium power level is a complex function of φ

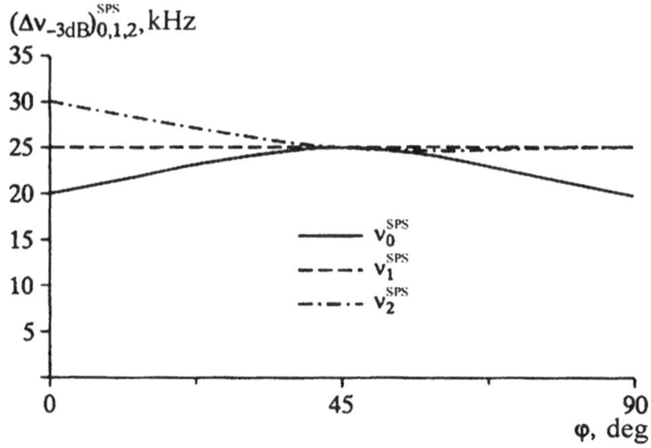

Fig. 3.17 At a high power level (P_{out} = 500 mW, Fig. 3.17) for the spectral component on the basic frequency ν_0^{SPS} at an increase in φ an insignificant broadening of the spectral line $(\Delta\nu_{-3dB})_0^{SPS}$ is observed, which is maximal at $\varphi \cong 45°$, and at change of φ within (45° ÷ 90°) this spectral line is narrowed down to its initial value at $\varphi = 0°$

$(\Delta\nu_{-60dB})_1^{SPS}$ slightly increase, but $(\Delta\nu_{-3dB})_2^{SPS}$ and $(\Delta\nu_{-60dB})_2^{SPS}$ decrease, and the slope is $(\Delta\nu_{-60dB})_1^{SPS}/\Delta\varphi > 0$, and $(\Delta\nu_{-60dB})_2^{SPS}/\Delta\varphi < 0$. At $\varphi \cong 45°$, the mode with $(\Delta\nu_{-3dB})_{0,1,2}^{SPS} \cong$ const takes place, and $(\Delta\nu_{-60dB})_0^{SPS} \cong (\Delta\nu_{-60dB})_2^{SPS}$. At $\varphi \cong 45°$, the pedestals of the spectral lines are $(\Delta\nu_{-60dB})_{0,1,2}^{SPS} \cong$ const. Since angles $\varphi > 45°$, $(\Delta\nu_{-3dB})_1^{SPS} \cong (\Delta\nu_{-3dB})_2^{SPS}$, and the slope is $(\Delta\nu_{-3dB})_{1,2}^{SPS}/\Delta\varphi < 0$, and $(\Delta\nu_{-3dB})_0^{SPS}/\Delta\varphi > 0$. For $\varphi > 45° (\Delta\nu_{-60dB})_{0,2}^{SPS}/\Delta\varphi > 0$ and $(\Delta\nu_{-60dB})_0^{SPS}/\Delta\varphi < (\Delta\nu_{-60dB})_2^{SPS}/\Delta\varphi$, but $(\Delta\nu_{-60dB})_1^{SPS}/\Delta\varphi < 0$.

At a high power level (P_{out} = 500 mW, Fig. 3.17) for the spectral component on the basic frequency ν_0^{SPS} at an increase in φ an insignificant broadening of the spectral line $(\Delta\nu_{-3dB})_0^{SPS}$ is observed, which is maximal at $\varphi \cong 45°$, and at change of φ within (45° ÷ 90°) this spectral line is narrowed down to its initial value at $\varphi = 0°$. The pedestal for the spectral line of frequencies ν_0^{SPS} at increase in φ is slightly narrowed (Fig. 3.18). The spectral linewidth $(\Delta\nu_{-3dB})_1^{SPS}$ on the frequency ν_1^{SPS} does not change for $\varphi = (0° \div 90°)$, but the pedestal has its minimal width at $\varphi \cong 45°$. For the spectral component on the frequency ν_2^{SPS}, with increase in φ some reduction of $(\Delta\nu_{-3dB})_2^{SPS}$ was observed at a constant width of the pedestal of the spectrum $(\Delta\nu_{-60dB})_2^{SPS} \cong$ const.

3.1.2 Structures with Ferrites of Various Magnetization

Below our experimental dependencies of the parameters $[(\Delta\nu_{max})_{0,1,2}^{SPS}$, $(\Delta\nu_{-3dB})_{0,1,2}^{SPS}$ and $(\Delta\nu_{-60\,dB})_{0,1,2}^{SPS}]$ of the spectral lines of HMT, generated SPS on the saturation magnetizations ($4\pi M_s$) of YIG-cubic ferrite for angles $\varphi = 0°$, 45°, 90° in a field bias $\overline{H}_0$ are resulted at both medium (50 mW) and high (500 mW) power levels.

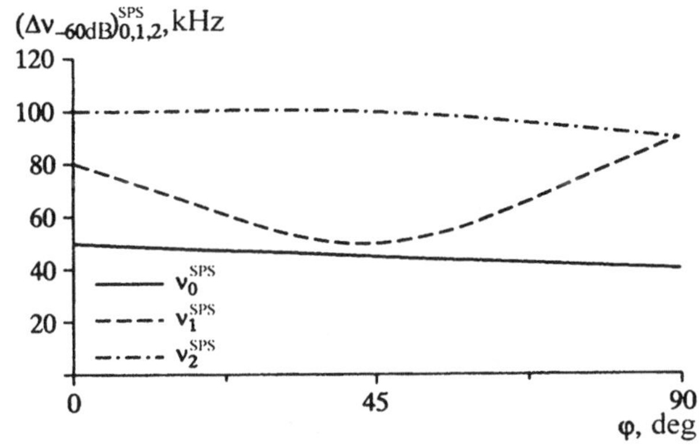

Fig. 3.18 The pedestal for the spectral line of frequencies ν_0^{SPS} at increase in φ is slightly narrowed

Fig. 3.19 On the medium power average level (50 mW) for the ferrite KG-15 with $4\pi M_s \cong 190$ G the value of the maximal deviation of frequency is $(\Delta\nu_{max})_0^{SPS} < (\Delta\nu_{max})_2^{SPS}$, but $(\Delta\nu_{max})_{0,2}^{SPS} > (\Delta\nu_{max})_1^{SPS}$

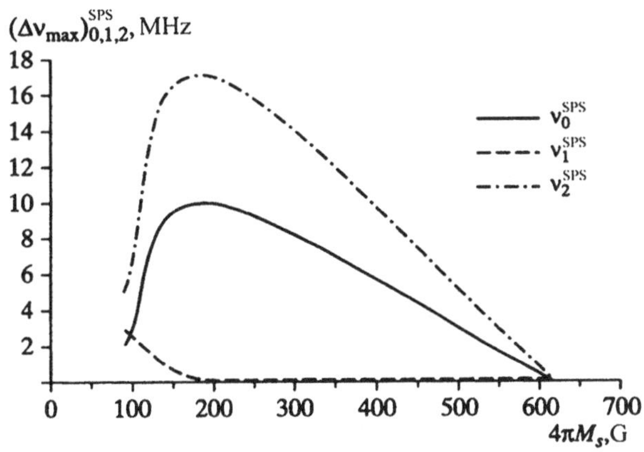

3.1.2.1 FMCR Orientation Angle $\varphi = 0°$

From the data in Fig. 3.19, it is obvious that on the medium power average level (50 mW) for the ferrite KG-15 with $4\pi M_s \cong 190$ G the value of the maximal deviation of frequency is $(\Delta\nu_{max})_0^{SPS} < (\Delta\nu_{max})_2^{SPS}$, but $(\Delta\nu_{max})_{0,2}^{SPS} > (\Delta\nu_{max})_1^{SPS}$.

At the high power level (500 mW) with growth of the number of harmonic n the maximal deviation of the spectral components of signals $(\Delta\nu_{max})_{n-1}^{SPS}$ $(n = 1, 2, 3, \ldots)$ monotonously increases, and at $4\pi M_s \cong 190$ G for FMCR made of KG-15 $(\Delta\nu_{max})_0^{SPS} < (\Delta\nu_{max})_1^{SPS} < (\Delta\nu_{max})_2^{SPS}$ and the deviation reaches its maximal values (Fig. 3.20).

The shape of the spectral line depends on the value of ferrite magnetization $4\pi M_s$ as follows (Figs. 3.21–3.24).

On the medium power level (Figs.3.21 and 3.22) for FMCR with its saturation magnetization $4\pi M_s \cong 90$ G (KG-8) the value is $(\nu_{-3dB})_{0,1,2}^{SPS} \cong$ const, $(\nu_{-60dB})_0^{SPS} \cong (\nu_{-60dB})_1^{SPS}$, but $(\nu_{-60dB})_2^{SPS} > (\nu_{-60dB})_{0,1}^{SPS}$. The narrowest spectral

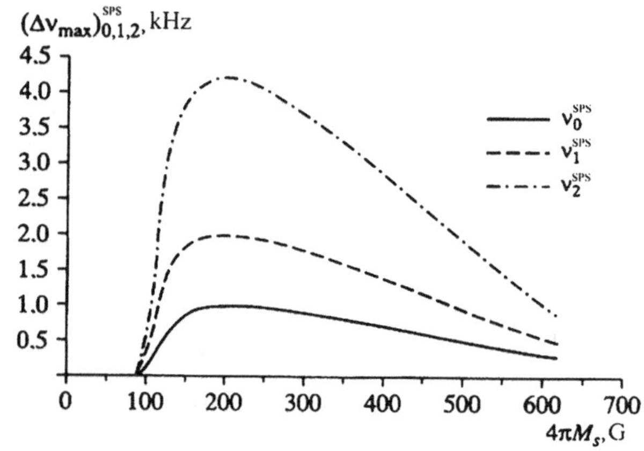

Fig. 3.20 At the high power level (500 mW) with growth of the number of harmonic n the maximal deviation of the spectral components of signals $(\Delta\nu_{\max})^{\text{SPS}}_{n-1}$ $(n = 1, 2, 3, \ldots)$ monotonously increases, and at $4\pi M_s \cong 190$ G for FMCR made of KG-15 $(\Delta\nu_{\max})^{\text{SPS}}_0 < (\Delta\nu_{\max})^{\text{SPS}}_1 < (\Delta\nu_{\max})^{\text{SPS}}_2$ and the deviation reaches its maximal values

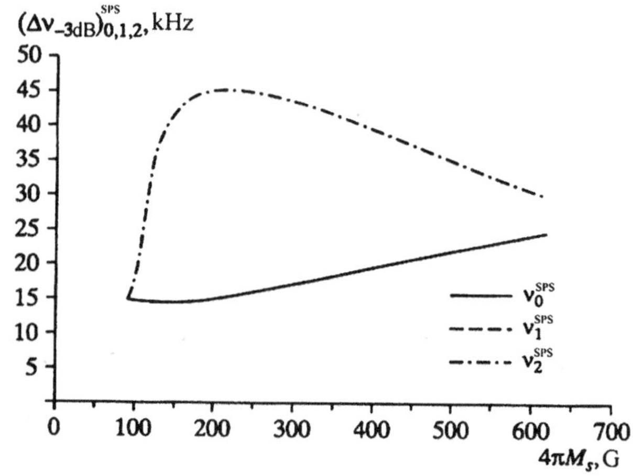

Fig. 3.21 The shape of the spectral line depends on the value of ferrite magnetization $4\pi M_s$ as follows

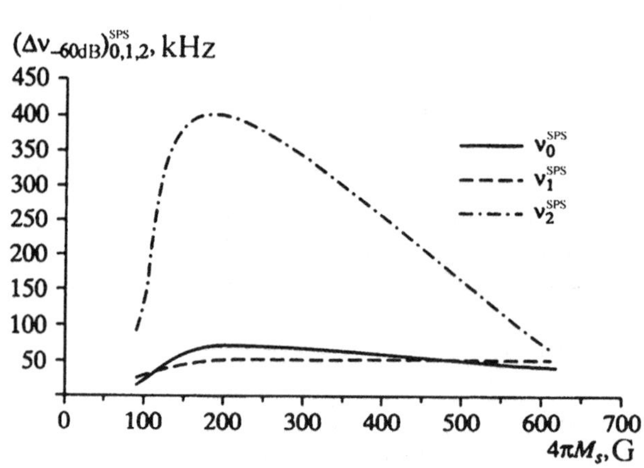

Fig. 3.22 The shape of the spectral line depends on the value of ferrite magnetization $4\pi M_s$ as follows

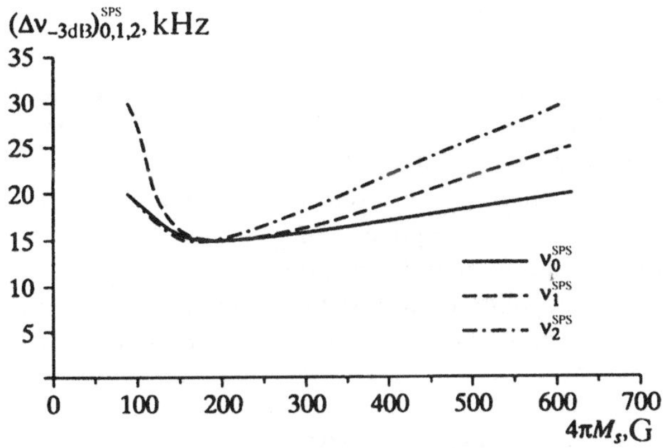

Fig. 3.23 The shape of the spectral line depends on the value of ferrite magnetization $4\pi M_s$ as follows

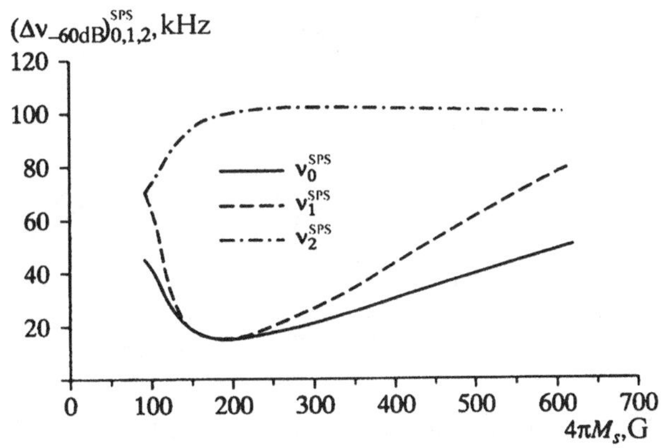

Fig. 3.24 The shape of the spectral line depends on the value of ferrite magnetization $4\pi M_s$ as follows

lines were observed at the magnetization $4\pi M_s \cong 190$ G (KG-15) for the spectral components of signals with $n = 0, 1$ $(\nu_{-3dB})_{0,1}^{SPS} < (\nu_{-3dB})_2^{SPS}$, and $(\nu_{-60dB})_0^{SPS} \cong (\nu_{-60dB})_1^{SPS}$ and $(\nu_{-60dB})_2^{SPS} >> (\nu_{-60dB})_{0,1}^{SPS}$. At increase in FMCR's magnetization, broadening of the lines at a level of -3 dB of the spectral components of signals with $n = 0, 1$ and reduction of the spectral linewidth with $n = 2$ were observed, and at the level of the spectral line pedestal $(\nu_{-60dB})_{0,1,2}^{SPS} \cong$ const and has its minimal value for magnetizations $4\pi M_s \cong 600$ G.

At the high power level (500 mW) at $4\pi M_s \cong 190$ G the narrowest spectral lines $(\nu_{-3dB})_{0,1,2}^{SPS} = \min$ and $(\Delta\nu_{-60dB})_{0,1,2}^{SPS} = \min$ were observed, but $(\Delta\nu_{-60dB})_2^{SPS} >> (\Delta\nu_{-60dB})_{0,1}^{SPS}$. At reduction of the FMCR magnetization (transfer to KG-8) and its increase (transfer to KG-50) the spectral components broadened to some measure, and for signals with the frequency ν_2 the value is $(\Delta\nu_{-60dB})_2^{SPS} \cong$ const and did not depend on the saturation magnetization of ferrite.

3.1.2.2 FMCR Orientation Angle $\varphi = 45°$

From the data in Fig. 3.25, it is obvious that the deviation of frequency on the medium power level (50 mW) is maximal for FMCR with its saturation magnetizations $4\pi M_s \cong 190$ G (KG-15).

At the high power level (500 mW) the deviation of frequency varies differently for the spectral components of generated signals on the frequencies $\nu_{0,1,2}$ depending on the saturation magnetization of ferrite (Fig. 3.26). At changes of the magnetization $4\pi M_s$ within (90–190) G the laws of deviation changes for the signals ν_0^{SPS} also ν_1^{SPS} are close, but practically there is no frequency change for ν_2^{SPS}. At changes of

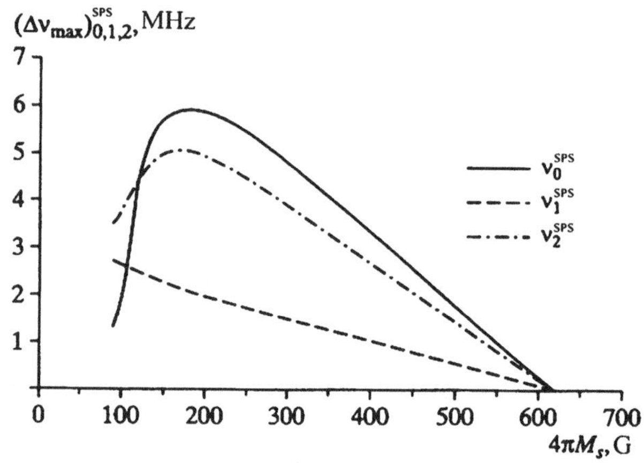

Fig. 3.25 The deviation of frequency on the medium power level (50 mW) is maximal for FMCR with its saturation magnetizations $4\pi M_s \cong 190$ G (KG-15)

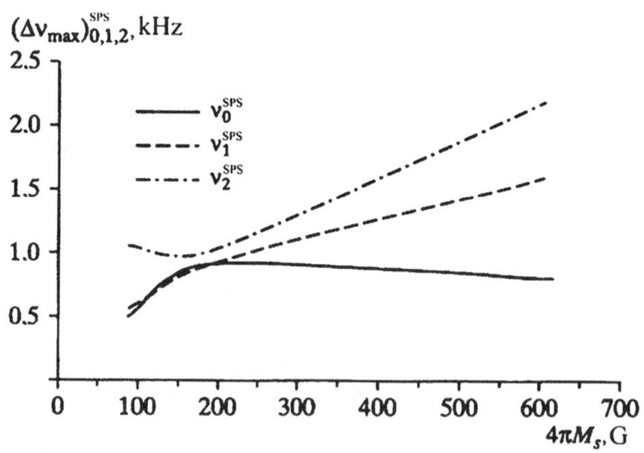

Fig. 3.26 At the high power level (500 mW) the deviation of frequency varies differently for the spectral components of generated signals on the frequencies $\nu_{0,1,2}$ depending on the saturation magnetization of ferrite

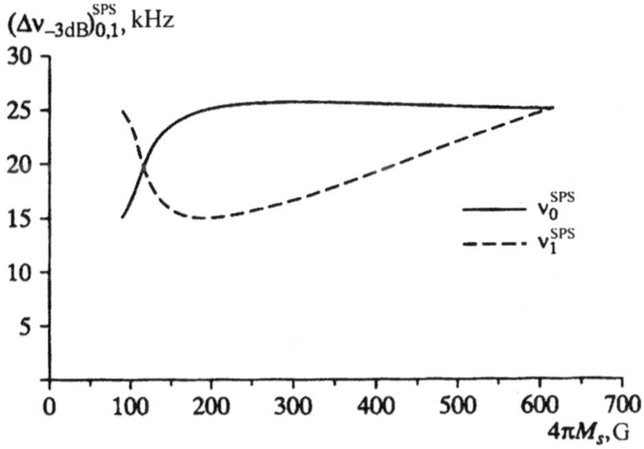

Fig. 3.27 The features of formation of the spectral lines of signals on the medium power levels depending on the saturation magnetization of ferrite $4\pi M_s$

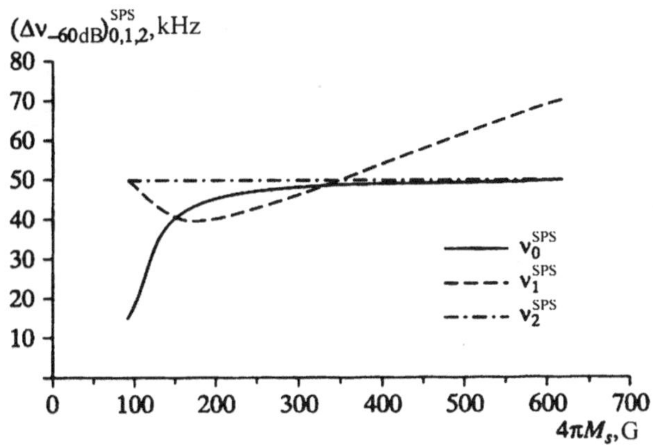

Fig. 3.28 The features of formation of the spectral lines of signals on the medium power levels depending on the saturation magnetization of ferrite $4\pi M_s$

$4\pi M_s$ within (190–600) G a weak falling dependence of $(\Delta\nu_{max})_0^{SPS}$ takes place, but for the maximum harmonic components with $n = 1, 2$ the deviation of frequency increases at increase in the FMCR magnetization.

Consider the features of formation of the spectral lines of signals on the medium (Figs. 3.27 and 3.28) and high (Figs. 3.29 and 3.30) power levels depending on the saturation magnetization of ferrite $4\pi M_s$.

On the medium (Figs. 2.27 and 3.28) power level (50 mW) the narrowest spectral line is registered for the FMCR with its magnetization $4\pi M_s \cong 90$ G (KG-8). With increase in the magnetization up to $4\pi M_s \cong 190$ G (KG-15) the spectral line on basic frequency ν_0 insignificantly broadens, but on the harmonic with $n = 1$ $(\Delta\nu_{-3dB})_1^{SPS} < (\Delta\nu_{-3dB})_0^{SPS}$, and the broadening of the spectral lines on the basis

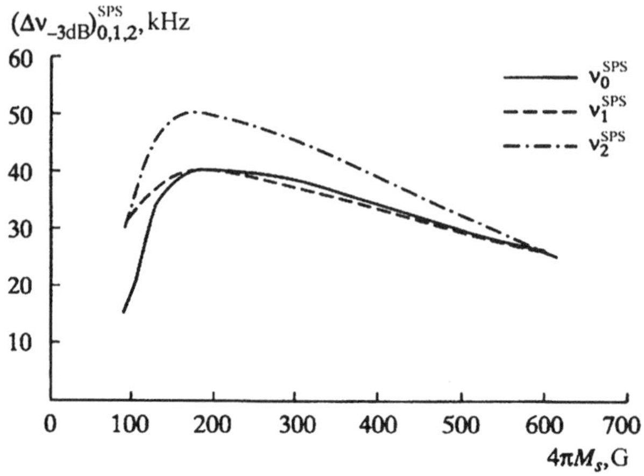

Fig. 3.29 The features of formation of the spectral lines of signals on the high power levels depending on the saturation magnetization of ferrite $4\pi M_s$

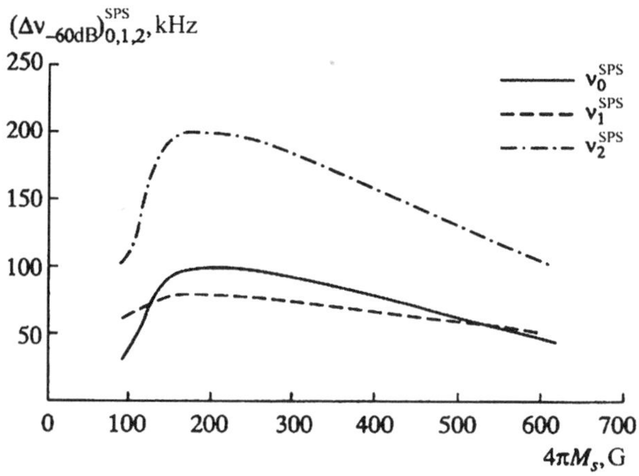

Fig. 3.30 The features of formation of the spectral lines of signals on the high power levels depending on the saturation magnetization of ferrite $4\pi M_s$

for the spectral lines of signals on the basic and harmonic components varies slightly $(\Delta\nu_{-60\text{dB}})^{\text{SPS}}_{0,1,2} \cong$ const. At increase of the saturation magnetization of ferrite up to $4\pi M_s = 600$ G (KG-50) for the spectral lines $(\Delta\nu_{-3\text{dB}})^{\text{SPS}}_0 \cong (\Delta\nu_{-3\text{dB}})^{\text{SPS}}_1 \cong (\Delta\nu_{-3\text{dB}})^{\text{SPS}}_2$, $(\Delta\nu_{-60\text{dB}})^{\text{SPS}}_0 \cong (\Delta\nu_{-60\text{dB}})^{\text{SPS}}_2$, but $(\Delta\nu_{-60\text{dB}})^{\text{SPS}}_1 > (\Delta\nu_{-60\text{dB}})^{\text{SPS}}_{0,2}$.

On the high (Figs. 3.29 and 3.30) power level (500 mW) the narrowest spectral lines were observed for the FMCR with its magnetization $4\pi M_s \cong 90$ G (KG-8). For the FMCR with $4\pi M_s \cong 190$ G, the greatest broadening of the spectral lines takes place for the components on the frequencies $\nu_{0,1,2}$, and $(\Delta\nu_{-3\text{dB}})^{\text{SPS}}_0 < (\Delta\nu_{-3\text{dB}})^{\text{SPS}}_1$, but $(\Delta\nu_{-60\text{dB}})^{\text{SPS}}_1 < (\Delta\nu_{-60\text{dB}})^{\text{SPS}}_0 << (\Delta\nu_{-60\text{dB}})^{\text{SPS}}_2$.

At increase in FMCR's saturation magnetization up to $4\pi M_s \cong 600$ G (KG-50), the spectral linewidth at a level of −3 and −60 dB decreases, and $(\Delta\nu_{-3dB})_0^{SPS} \cong (\Delta\nu_{-3dB})_1^{SPS}$ and $(\Delta\nu_{-60dB})_0^{SPS} \cong (\Delta\nu_{-60dB})_2^{SPS}$.

3.1.2.3 FMCR Orientation Angle $\varphi = 90°$

At the medium power level (50 mW) the deviation of frequency has its maximal value for the FMCR with its saturation magnetization $4\pi M_s = 90$ G (KG-8), and $(\Delta\nu_{max})_0^{SPS} < (\Delta\nu_{max})_1^{SPS} < (\Delta\nu_{max})_2^{SPS}$ (Fig. 3.31). With increase in FMCR's saturation magnetization, the deviation of frequency for the basic ν_0^{SPS} and higher harmonic spectral components ν_1^{SPS}, ν_2^{SPS} goes down.

At the high power level (500 mW) with growth of FMCR's saturation magnetization, the deviation of frequency almost linearly increases and $(\Delta\nu_{max})_0^{SPS} < (\Delta\nu_{max})_1^{SPS} < (\Delta\nu_{max})_2^{SPS}$ (Fig. 3.32).

The spectral lines of generated signals in this case depend on the value of ferrite magnetization $4\pi M_s$ in a complex manner. On a low power level (Figs. 3.33 and 3.34) at the magnetization $4\pi M_s = 90$ G (KG-8) the slope of $(\Delta\nu_{-3dB})_{0,2}^{SPS}$ is various. The most narrow-band spectral lines on the level −60 dB were observed at the magnetization $4\pi M_s = 90$ G (KG-8), and $(\Delta\nu_{-60dB})_0^{SPS} < (\Delta\nu_{-60dB})_1^{SPS} << (\Delta\nu_{-3dB})_2^{SPS}$.

At the magnetization $4\pi M_s = 190$ G the values are $(\Delta\nu_{-3dB})_0^{SPS} \cong (\Delta\nu_{-3dB})_2^{SPS}$ and $(\Delta\nu_{-60dB})_0^{SPS} \cong (\Delta\nu_{-60dB})_1^{SPS}$.

At a high power level (Figs. 3.35 and 3.36) the narrowest spectral lines of signals were observed in the structures with FMCR having their magnetizations $4\pi M_s = 90$ G (KG-8), at the magnetization $4\pi M_s = 190$ G (KG-15) the values

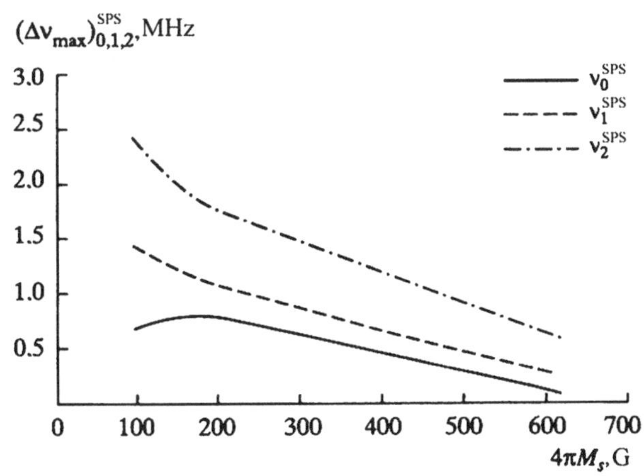

Fig. 3.31 At the medium power level (50 mW) the deviation of frequency has its maximal value for the FMCR with its saturation magnetization $4\pi M_s = 90$ G (KG-8), and $(\Delta\nu_{max})_0^{SPS} < (\Delta\nu_{max})_1^{SPS} < (\Delta\nu_{max})_2^{SPS}$

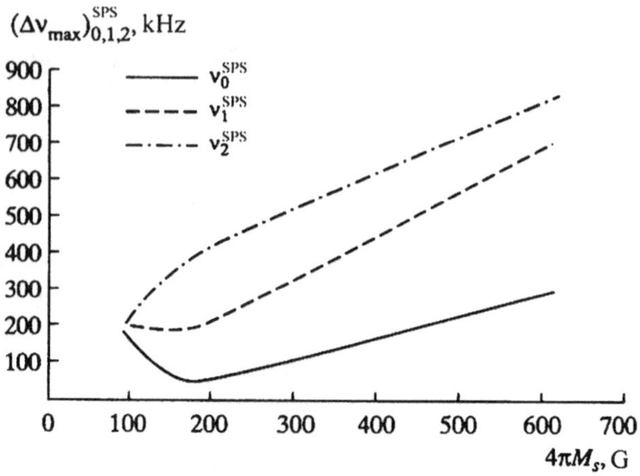

Fig. 3.32 At the high power level (500 mW) with growth of FMCR's saturation magnetization, the deviation of frequency almost linearly increases and $(\Delta\nu_{\max})_0^{SPS} < (\Delta\nu_{\max})_1^{SPS} < (\Delta\nu_{\max})_2^{SPS}$

Fig. 3.33 On a low power level at the magnetization $4\pi M_s = 90\,G$ (KG-8) the slope of $(\Delta\nu_{-3dB})_{0,2}^{SPS}$ is various

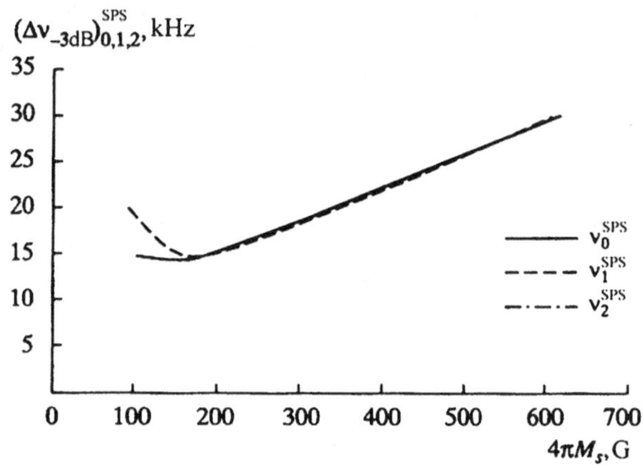

Fig. 3.34 On a low power level at the magnetization $4\pi M_s = 90\,G$ (KG-8) the slope of $(\Delta\nu_{-3dB})_{0,2}^{SPS}$ is various

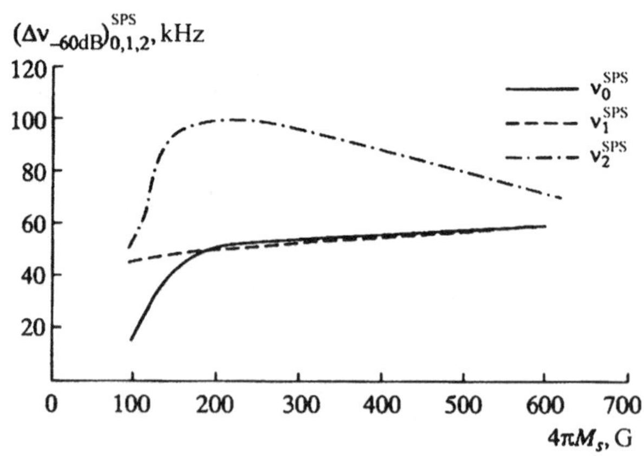

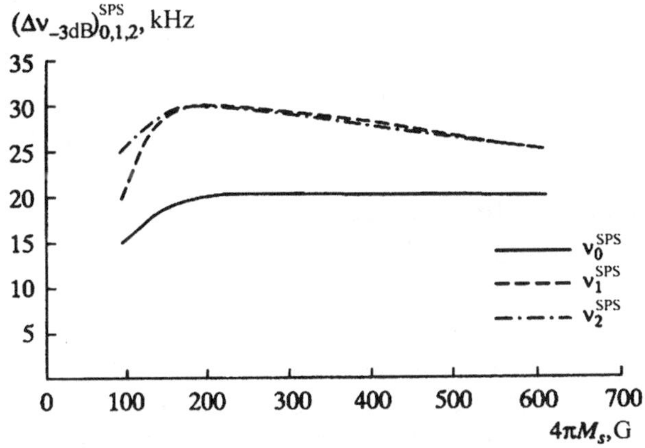

Fig. 3.35 At a high power level the narrowest spectral lines of signals were observed in the structures with FMCR having their magnetizations $4\pi M_s = 90\,\text{G}$ (KG-8), at the magnetization $4\pi M_s = 190\,\text{G}$ (KG-15) the values were $(\Delta\nu_{-3dB})^{SPS}_1 = (\Delta\nu_{-3dB})^{SPS}_2$ and $(\Delta\nu_{-60dB})^{SPS}_1 = (\Delta\nu_{-60dB})^{SPS}_2$

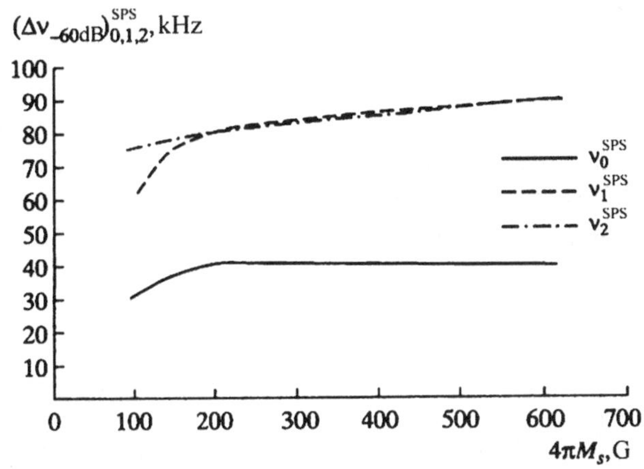

Fig. 3.36 At a high power level the narrowest spectral lines of signals were observed in the structures with FMCR having their magnetizations $4\pi M_s = 90\,\text{G}$ (KG-8), at the magnetization $4\pi M_s = 190\,\text{G}$ (KG-15) the values were $(\Delta\nu_{-3dB})^{SPS}_1 = (\Delta\nu_{-3dB})^{SPS}_2$ and $(\Delta\nu_{-60dB})^{SPS}_1 = (\Delta\nu_{-60dB})^{SPS}_2$

were $(\Delta\nu_{-3dB})^{SPS}_1 = (\Delta\nu_{-3dB})^{SPS}_2$ and $(\Delta\nu_{-60dB})^{SPS}_1 = (\Delta\nu_{-60dB})^{SPS}_2$. At increase in FMCR's magnetization the spectral linewidth changed slightly.

The resulted experimental dependencies allow estimations of the expected parameters at generation of spectrally pure signals by heteromagnetic structures, namely:

- The deviation of frequencies of the basic $(\Delta\nu)^{SPS}_0$ and higher harmonic components $(\Delta\nu)^{SPS}_{n-1} (n = 1, 2)$

- The shape of spectral lines by the values $(\Delta\nu_{-3\mathrm{dB}})_{n-1}^{\mathrm{SPS}}$ and $(\Delta\nu_{-60\mathrm{dB}})_{n-1}^{\mathrm{SPS}}$ for $n = 1, 2, \ldots$
- The level of integral output power P_{out}.

3.2 Control Over Characteristics of Pseudonoise and Noise Signals

Experimental dependencies of the key parameters of PS and NS spectra in generating heteromagnetic structures on a medium (50 mW) and high (500 mW) power levels on the FMCR orientation angle φ in an external magnetic field H_0 for alloyed ferrites with a cubic structure with saturation magnetizations $4\pi M_s = 90$, 190, and 620 G for the spectral components ν_0^{NS}, ν_1^{NS}, ν_2^{NS} are considered.

3.2.1 Structures with Various Orientations in a Magnetic Field

3.2.1.1 FMCR of KG-8

On the medium power level (50 mW) the maximal deviation of NS frequency for various spectral components was observed at the FMCR orientation $\varphi = 0°$ (Fig. 3.37). With increase of φ from 0° to 90° an almost linear decrease in the dependencies $(\Delta\nu_{\max})_{1,2}^{\mathrm{NS}}$ takes place, and $(\Delta\nu_{\max})_0^{\mathrm{NS}}$ varies slightly.

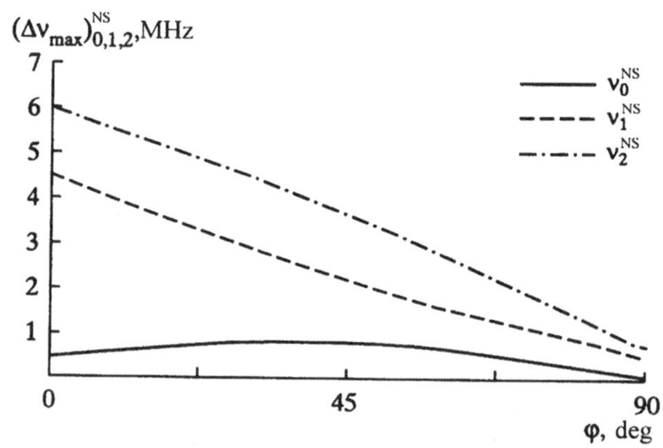

Fig. 3.37 On the medium power level (50 mW) the maximal deviation of NS frequency for various spectral components was observed at the FMCR orientation $\varphi = 0°$

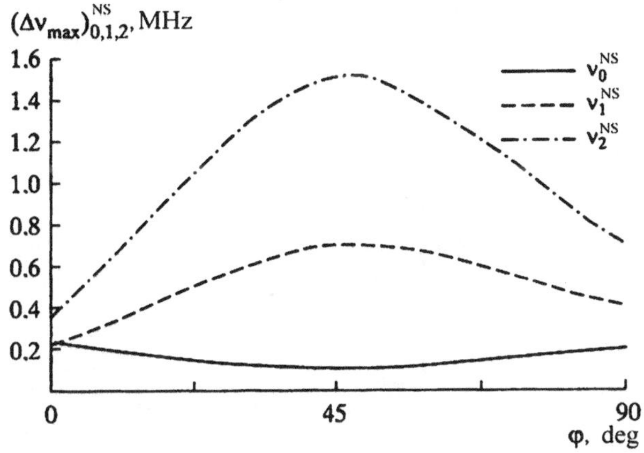

Fig. 3.38 At the high power level (500 mW) the maximal deviation of NS frequency for the components on the frequencies $\nu_{1,2}^{NS}$ takes place at $\varphi = 45°$

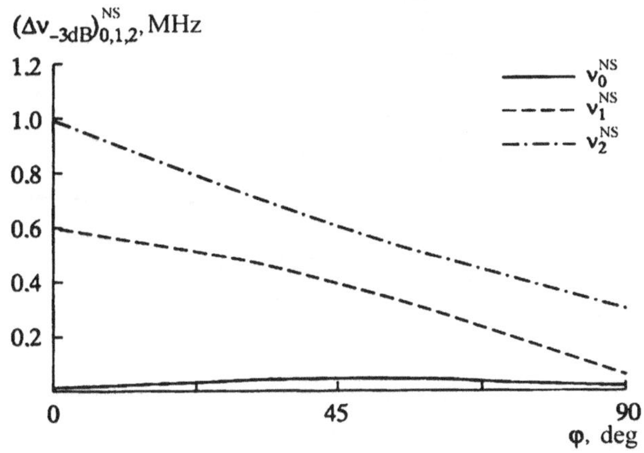

Fig. 3.39 The NS envelope on the low power level is most broadband at $\varphi = 0°$ for the frequency components $\nu_{1,2}^{NS}$

At the high power level (500 mW) the maximal deviation of NS frequency for the components on the frequencies $\nu_{1,2}^{NS}$ takes place at $\varphi = 45°$ (Fig. 3.38).

The NS envelope on the low power level (Fig. 3.39 and 3.40) is most broadband at $\varphi = 0°$ for the frequency components $\nu_{1,2}^{NS}$. At increase in φ up to 90° the band of NS frequencies decreases.

At the high power level (Figs. 3.41 and 3.42) the NS spectral linewidth for the frequency components of $\nu_{0,1,2}^{NS}$ changes rather slightly.

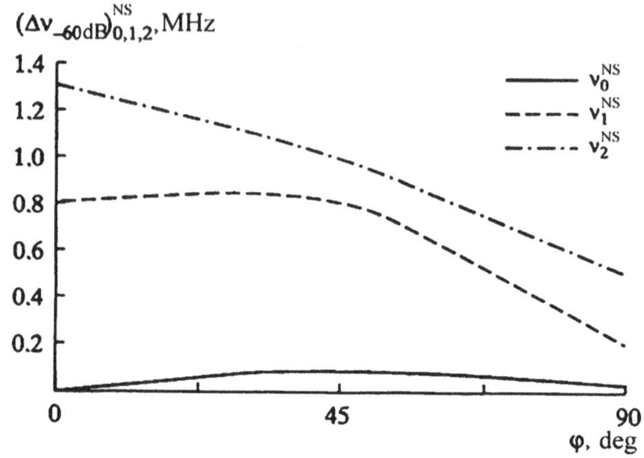

Fig. 3.40 The NS envelope on the low power level is most broadband at $\varphi = 0°$ for the frequency components $\nu^{NS}_{1,2}$

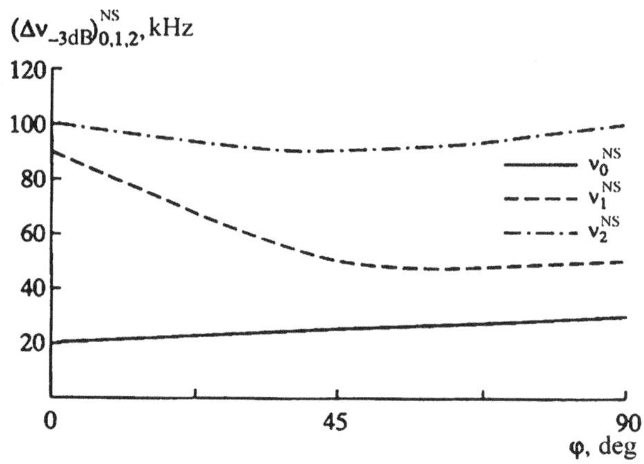

Fig. 3.41 At the high power level the NS spectral linewidth for the frequency components of $\nu^{NS}_{0,1,2}$ changes rather slightly

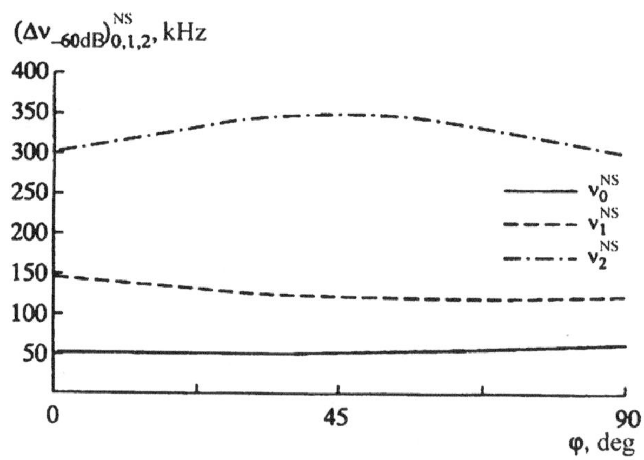

Fig. 3.42 At the high power level the NS spectral linewidth for the frequency components of $\nu^{NS}_{0,1,2}$ changes rather slightly

3.2.1.2 FMCR of KG-15

The deviation of NS frequency on the medium power level (50 mW) is most significant for signals on the frequencies $\nu_{1,2}^{NS}$ at change of the ferrite orientation angle φ within 0–45° (Fig. 3.43).

At the high power level (500 mW) the most broadband spectral NS components were observed for the frequencies ν_2^{NS}, and with increase in φ a reduction of the range of the NS central frequency deviation (Fig. 3.44) took place.

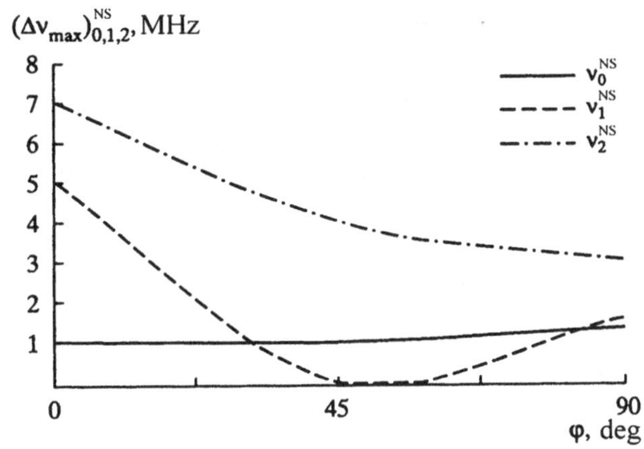

Fig. 3.43 The deviation of NS frequency on the medium power level (50 mW) is most significant for signals on the frequencies $\nu_{1,2}^{NS}$ at change of the ferrite orientation angle φ within 0–45°

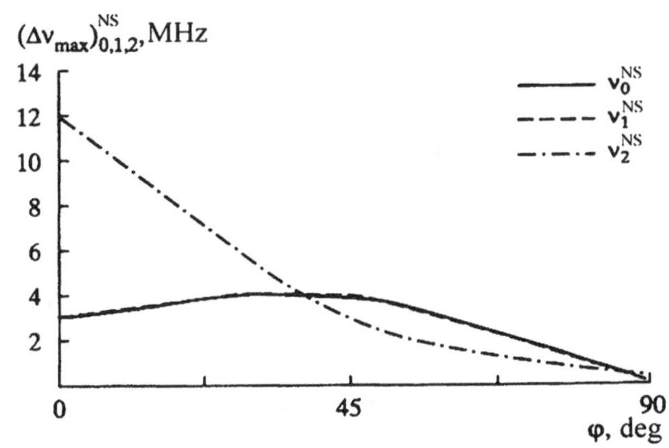

Fig. 3.44 At the high power level (500 mW) the most broadband spectral NS components were observed for the frequencies ν_2^{NS}, and with increase in φ a reduction of the range of the NS central frequency deviation took place

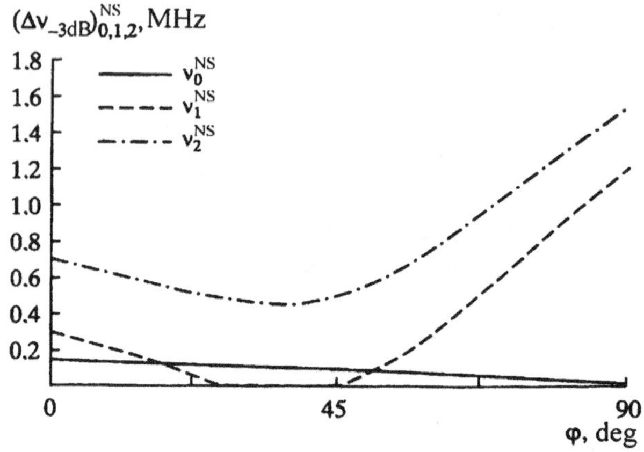

Fig. 3.45 The NS envelope on the medium power level for the spectral components on the frequency ν_0^{NS} decreases at a change of φ from 0° to 90°

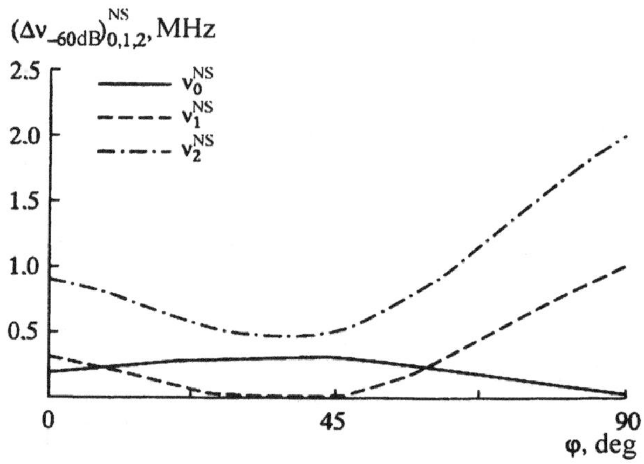

Fig. 3.46 The NS envelope on the medium power level for the spectral components on the frequency ν_0^{NS} at $\varphi = 45°$ a weak increase of $(\Delta\nu_{-60\,dB})_0^{NS}$ takes place

The NS envelope on the medium power level (Figs. 3.45 and 3.46) for the spectral components on the frequency ν_0^{NS} decreases at a change of φ from 0° to 90° (Fig. 3.45), but at $\varphi = 45°$ a weak increase of $(\Delta\nu_{-60dB})_0^{NS}$ takes place (Fig. 3.46). The most broadband NS modes take place for the frequency components $\nu_{1,2}^{NS}$ at FMCR's orientation angle $\varphi = 90°$.

At the high power level the NS envelopes for the frequencies $\nu_{0,1,2}^{NS}$ are most broadband at $\varphi = 0°$ and with increase in φ they decrease for $(\Delta\nu_{-3dB})_{0,1,2}^{NS}$ and $(\Delta\nu_{-60dB})_{0,1,2}^{NS}$ (Figs. 3.47 and 3.48).

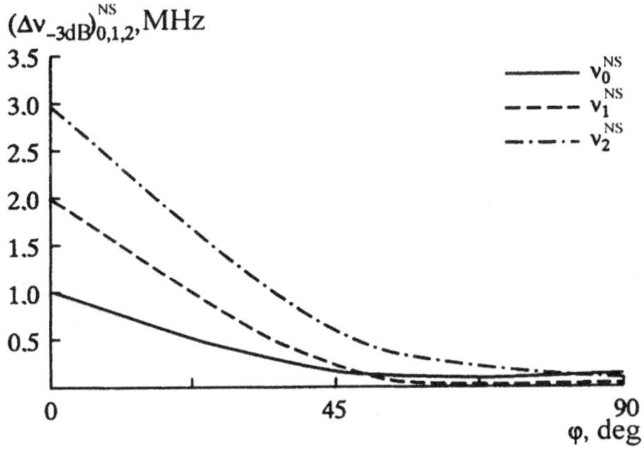

Fig. 3.47 At the high power level the NS envelopes for the frequencies $\nu^{NS}_{0,1,2}$ are most broadband at $\varphi = 0°$ and with increase in φ they decrease for $(\Delta\nu_{-3\,dB})^{NS}_{0,1,2}$ and $(\Delta\nu_{-60\,dB})^{NS}_{0,1,2}$

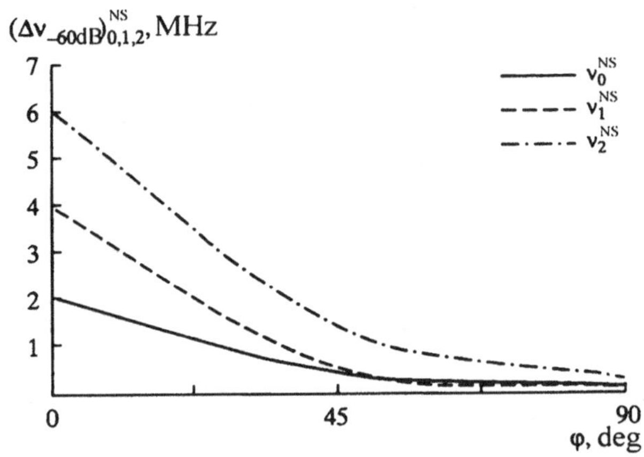

Fig. 3.48 At the high power level the NS envelopes for the frequencies $\nu^{NS}_{0,1,2}$ are most broadband at $\varphi = 0°$ and with increase in φ they decrease for $(\Delta\nu_{-3\,dB})^{NS}_{0,1,2}$ and $(\Delta\nu_{-60\,dB})^{NS}_{0,1,2}$

3.2.1.3 FMCR of KG-50

On the medium power level (50 mW) the deviation of the spectral components of the frequencies $\nu^{NS}_{0,1,2}$ NS is maximal at $\varphi = 0°$, and at $\varphi = 45°$ the deviation is minimal (Fig. 3.49).

At the high power level (500 mW) the deviation of the spectral components of the NS frequencies $\nu^{NS}_{0,1,2}$ is maximal for all the components at $\varphi = 45°$ (Fig. 3.50).

The NS spectral line envelopes at the medium power level (Figs. 3.51 and 3.52) are most broadband for the central frequencies $\nu_{0,1,2}$ at $\varphi = 0°$.

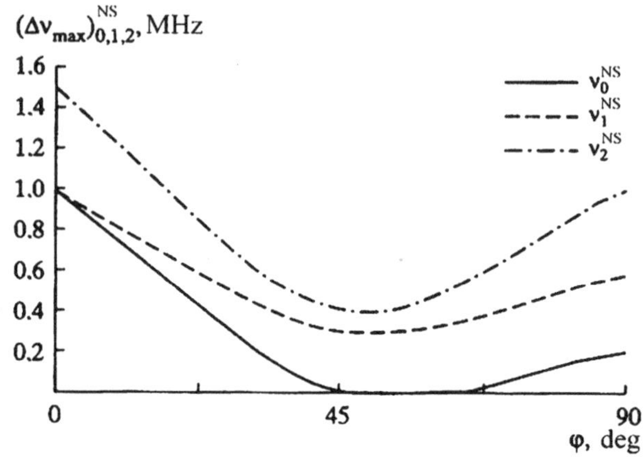

Fig. 3.49 On the medium power level (50 mW) the deviation of the spectral components of the frequencies $\nu_{0,1,2}^{NS}$ NS is maximal at $\varphi = 0°$, and at $\varphi = 45°$ the deviation is minimal

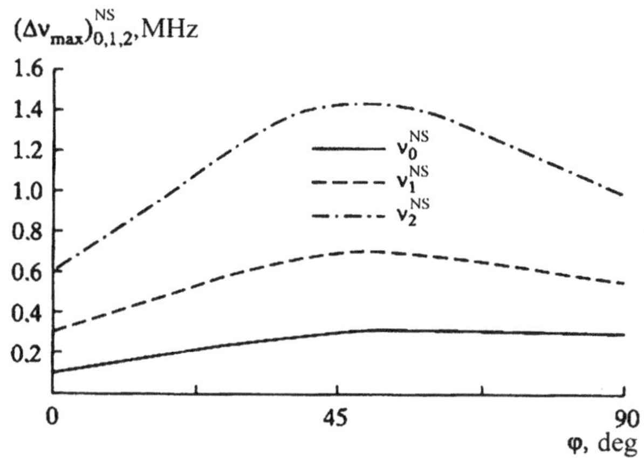

Fig. 3.50 At the high power level (500 mW) the deviation of the spectral components of the NS frequencies $\nu_{0,1,2}^{NS}$ is maximal for all the components at $\varphi = 45°$

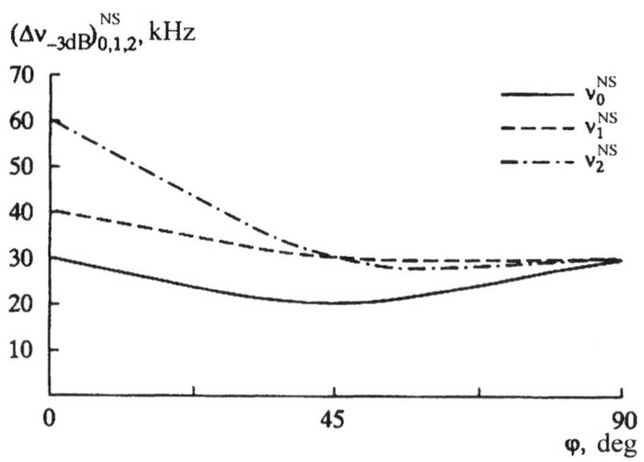

Fig. 3.51 The NS spectral line envelopes at the medium power level are most broadband for the central frequencies $\nu_{0,1,2}$ at $\varphi = 0°$

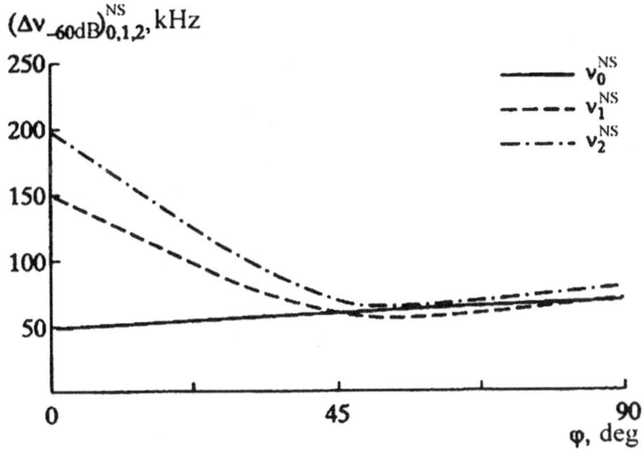

Fig. 3.52 The NS spectral line envelopes at the medium power level are most broadband for the central frequencies $\nu_{0,1,2}$ at $\varphi = 0°$

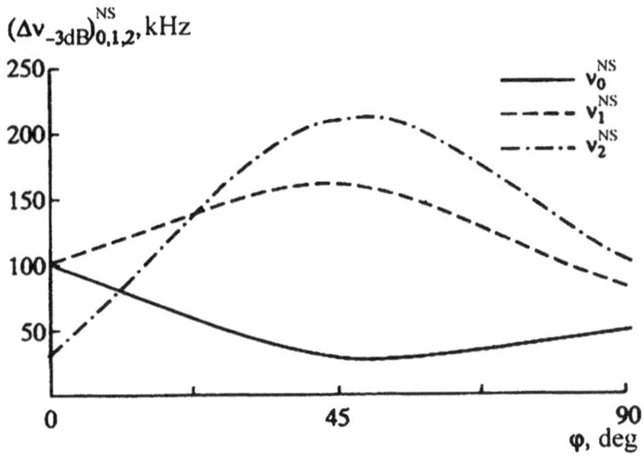

Fig. 3.53 At a high power level (500 mW) the NS spectral line envelopes for the components with the frequencies $\nu_{1,2}^{NS}$ are most broadband at $\varphi \cong 45°$

At increase in φ up to 90° a reduction of the spectral linewidth of the NS components $\nu_{0,1,2}^{NS}$ is observed and at $\varphi \cong 90° (\Delta\nu_{-3dB})_{0,1,2}^{NS} = \text{const}$ and $(\Delta\nu_{-60dB})_{0,1,2}^{NS} = \text{const}$.

At a high power level (500 mW) the NS spectral line envelopes for the components with the frequencies $\nu_{1,2}^{NS}$ are most broadband at $\varphi \cong 45°$ (Figs. 3.53 and 3.54).

3.2.2 Structures with Ferrites of Various Magnetization

3.2.2.1 FMCR Orientation Angle $\varphi = 0°$

On both medium (50 mW) and high (500 mW) power levels the deviation of the NS central frequency is maximal for the structures with FMCR having its orientation angle $\varphi \cong 0°$ and saturation magnetization $4\pi M_s = 190\,\text{G}$ (Figs. 3.55 and 3.56, respectively).

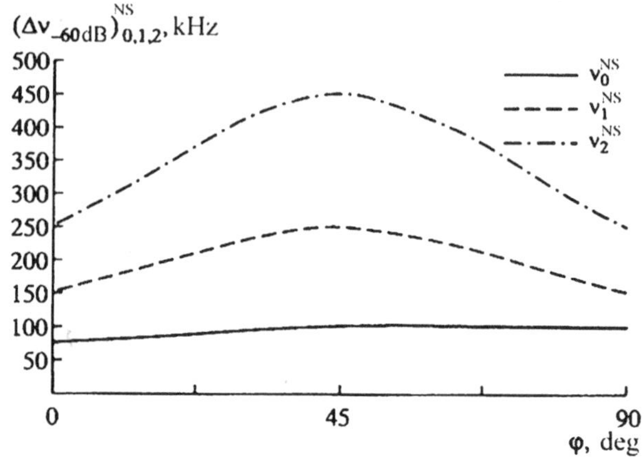

Fig. 3.54 At a high power level (500 mW) the NS spectral line envelopes for the components with the frequencies $\nu_{1,2}^{NS}$ are most broadband at $\varphi \cong 45°$

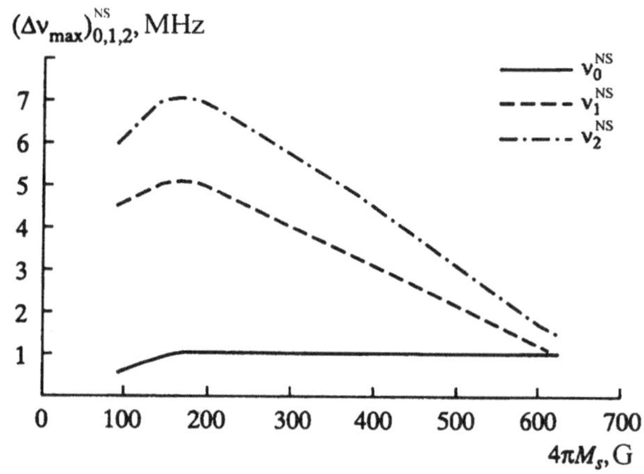

Fig. 3.55 On both medium (50 mW) and high (500 mW) power levels the deviation of the NS central frequency is maximal for the structures with FMCR having its orientation angle $\varphi \cong 0°$ and saturation magnetization $4\pi M_s = 190\,\mathrm{G}$

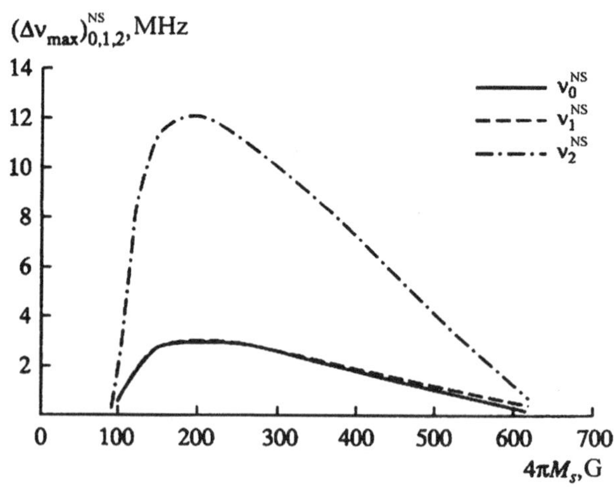

Fig. 3.56 On both medium (50 mW) and high (500 mW) power levels the deviation of the NS central frequency is maximal for the structures with FMCR having its orientation angle $\varphi \cong 0°$ and saturation magnetization $4\pi M_s = 190\,\mathrm{G}$

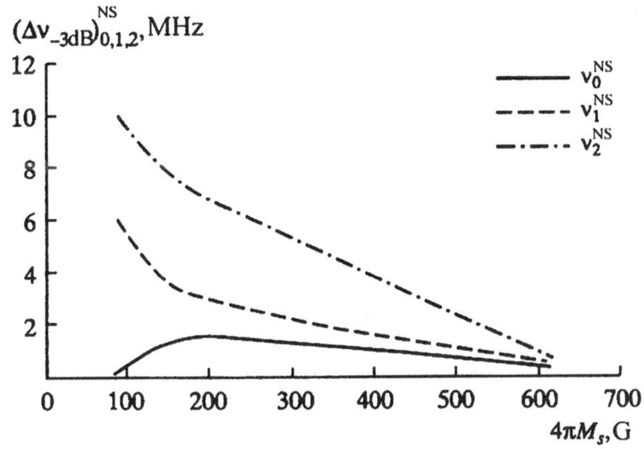

Fig. 3.57 The NS spectral line envelopes on the medium power level are most broadband for the components with the central frequencies $\nu^{NS}_{1,2}$

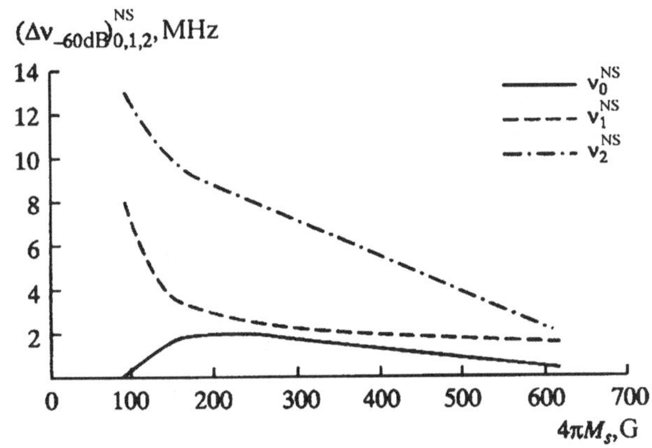

Fig. 3.58 The NS spectral line envelopes on the medium power level are most broadband for the components with the central frequencies $\nu^{NS}_{1,2}$

The NS spectral line envelopes on the medium power level (Figs. 3.57 and 3.58) are most broadband for the components with the central frequencies $\nu^{NS}_{1,2}$. At increase in FMCR's saturation magnetization, more narrow-band modes and a tendency to formation of signals with $(\Delta\nu_{-3\text{dB}})^{NS}_{0,1,2} \cong$ const and $(\Delta\nu_{-60\text{dB}})^{NS}_{0,1,2} \cong$ const take place.

At a high power level (Figs. 3.59 and 3.60) the most NS broadband modes are for the FMCR with its magnetization $4\pi M_s \cong 190$ G (KG-15). For the FMCRs with their magnetizations $4\pi M_s \cong 90$ G and $4\pi M_s \cong 600$ G, modes with $(\Delta\nu_{-3\text{dB}})^{NS}_{0,1,2} \cong$ const and $(\Delta\nu_{-60\text{dB}})^{NS}_{0,1,2} \cong$ const take place.

3.2.2.2 FMCR Orientation Angle $\varphi = 45°$

On the medium power level (50 mW) the maximal deviation of the central frequency of the NS spectral components $\nu^{NS}_{0,1,2}$ is observed at FMCR's saturation magnetization $4\pi M_s = 90$ G, with an increase in magnetization the deviation of frequency decreases (Fig. 3.61).

Fig. 3.59 At a high power level the most NS broadband modes are for the FMCR with its magnetization $4\pi M_s \cong 190$ G (KG-15)

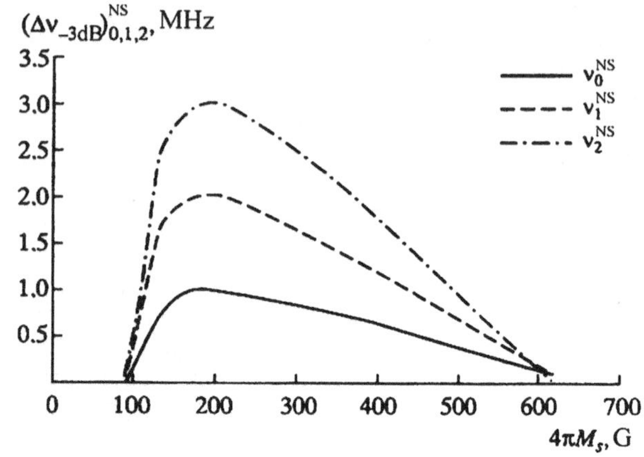

Fig. 3.60 At a high power level the most NS broadband modes are for the FMCR with its magnetization $4\pi M_s \cong 190$ G (KG-15)

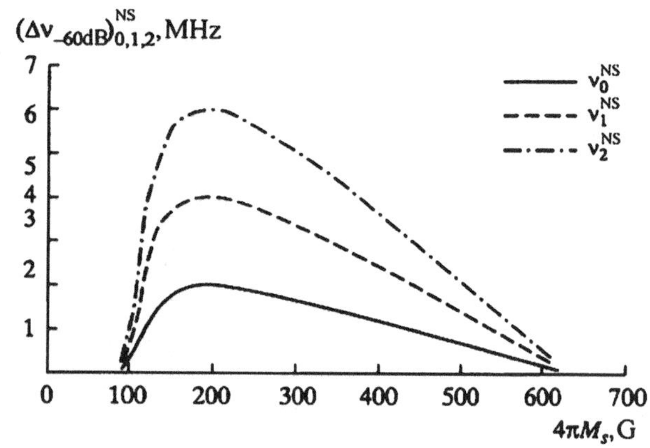

Fig. 3.61 On the medium power level (50 mW) the maximal deviation of the central frequency of the NS spectral components $\nu_{0,1,2}^{NS}$ is observed at FMCR's saturation magnetization $4\pi M_s = 90$ G, with an increase in magnetization the deviation of frequency decreases

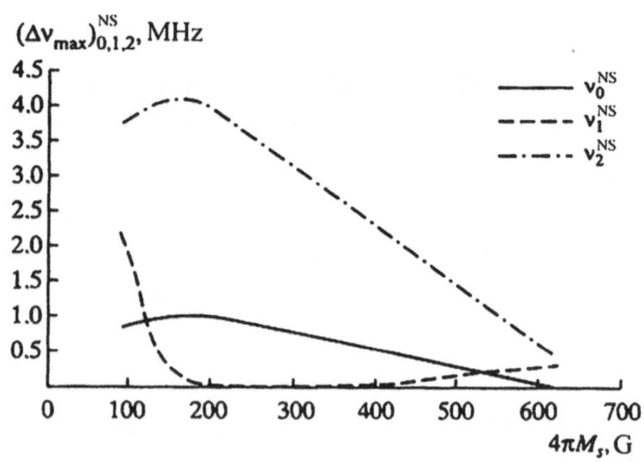

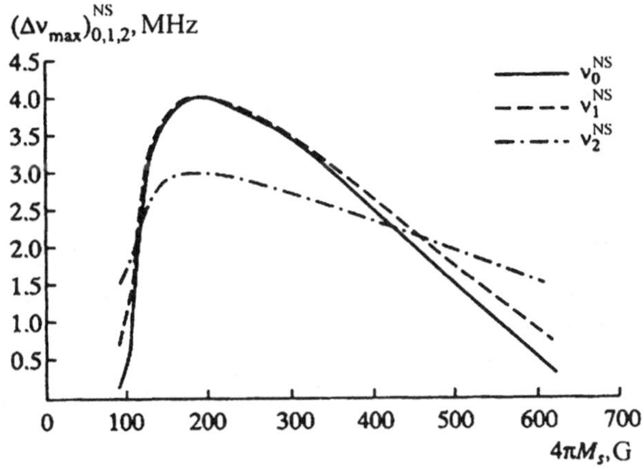

Fig. 3.62 At a high power level (500 mW) the maximal frequency deviation of the NS spectral components was observed at FMCR's magnetization $4\pi M_s \cong 190$ G

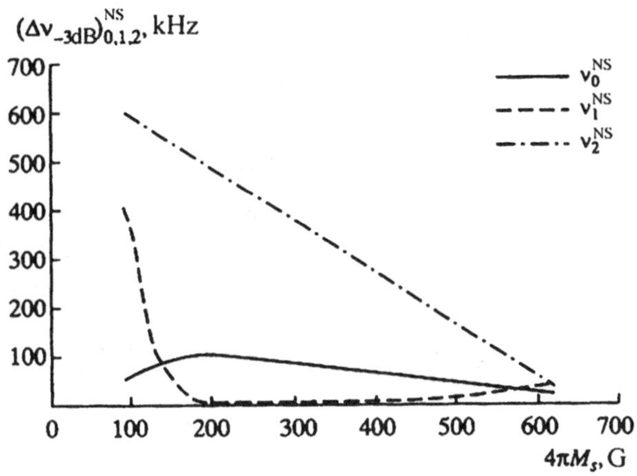

Fig. 3.63 Judging by the NS spectrum envelope on the medium power level the most broadband mode was observed at FMCR's magnetization $4\pi M_s = 90$ G. With an increase in the FMCR magnetization the NS spectrum envelope is narrowed for all the components and $(\Delta\nu_{-3dB})^{NS}_{0,1,2} = \text{const}$, $(\Delta\nu_{-60dB})^{NS}_{0,1,2} = \text{const}$

At a high power level (500 mW) the maximal frequency deviation of the NS spectral components was observed at FMCR's magnetization $4\pi M_s \cong 190$ G (Fig. 3.62).

Judging by the NS spectrum envelope on the medium power level the most broadband mode was observed at FMCR's magnetization $4\pi M_s = 90$ G. With an increase in the FMCR magnetization the NS spectrum envelope is narrowed for all the components and $(\Delta\nu_{-3dB})^{NS}_{0,1,2} = \text{const}$, $(\Delta\nu_{-60dB})^{NS}_{0,1,2} = \text{const}$ (Figs. 3.63 and 3.64).

At a high power level for all the spectral components on the frequency $\nu^{NS}_{0,1,2}$ the NS spectrum envelope is most broadband at FMCR's magnetization $4\pi M_s = 190$ G (KG-15) (Figs. 3.65 and 3.66).

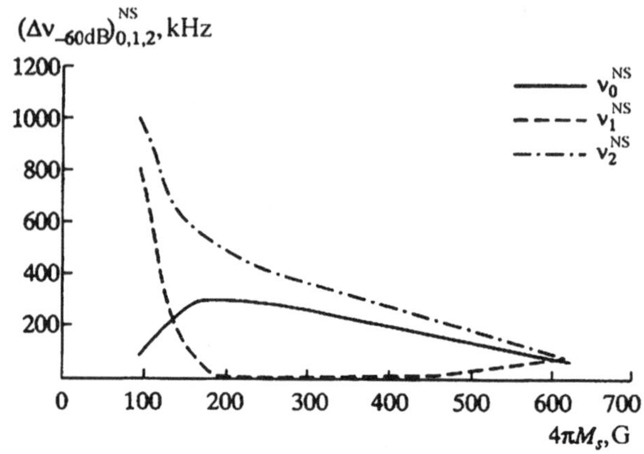

Fig. 3.64 Judging by the NS spectrum envelope on the medium power level the most broadband mode was observed at FMCR's magnetization $4\pi M_s = 90\,\mathrm{G}$. With an increase in the FMCR magnetization the NS spectrum envelope is narrowed for all the components and $(\Delta\nu_{-3dB})^{NS}_{0,1,2} = \mathrm{const}$, $(\Delta\nu_{-60dB})^{NS}_{0,1,2} = \mathrm{const}$

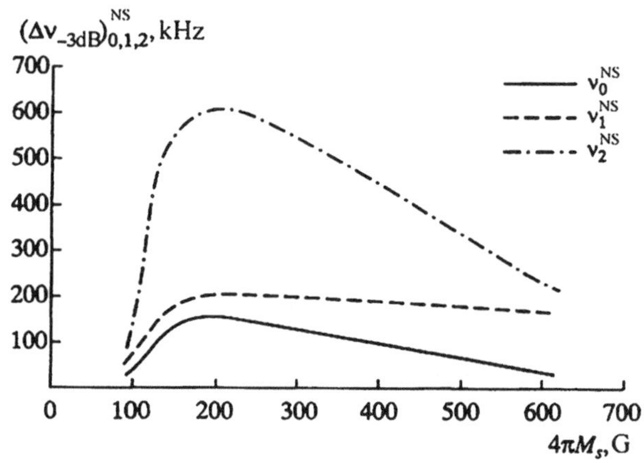

Fig. 3.65 At a high power level for all the spectral components on the frequency $\nu^{NS}_{0,1,2}$ the NS spectrum envelope is most broadband at FMCR's magnetization $4\pi M_s = 190\,\mathrm{G}$ (KG-15)

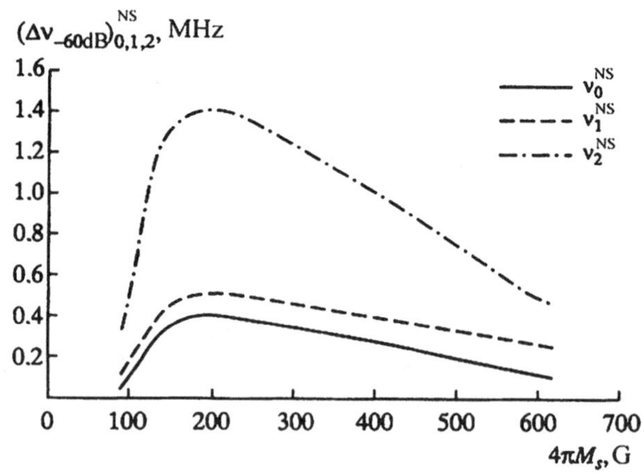

Fig. 3.66 At a high power level for all the spectral components on the frequency $\nu^{NS}_{0,1,2}$ the NS spectrum envelope is most broadband at FMCR's magnetization $4\pi M_s = 190\,\mathrm{G}$ (KG-15)

3.2.2.3 FMCR Orientation Angle $\varphi = 90°$

On the medium power level (50 mW) the deviation of the central NS frequency is maximal for all the spectral components at FMCR's magnetization $4\pi M_s = 190$ G (Fig. 3.67).

At a high power level (500 mW) the deviation of the central frequency of the spectral components on the NS frequency $\nu_{0,1,2}^{NS}$ is maximal at FMCR's magnetization $4\pi M_s = 600$ G (KG-50) and at $4\pi M_s = 90$ G (KG-8) (Fig. 3.68).

The NS spectrum envelopes on the medium power level (Figs. 3.69 and 3.70) are most broadband for the spectral components $\nu_{1,2}$ at the FMCR saturation magnetization of $4\pi M_s = 190$ G (KG-15).

At a high power level the NS spectrum envelopes (Figs. 3.71 and 3.72) most strongly change at a level −3 dB at variations of FMCR's magnetization $4\pi M_s$. At $4\pi M_s = 90\,\mathrm{G}(\Delta\nu_{-3\mathrm{dB}})_0^{NS} < (\Delta\nu_{-3\mathrm{dB}})_1^{NS} << (\Delta\nu_{-3\mathrm{dB}})_2^{NS}$ and $(\Delta\nu_{-60\mathrm{dB}})_0^{NS} < (\Delta\nu_{-60\mathrm{dB}})_1^{NS} < (\Delta\nu_{-60\mathrm{dB}})_2^{NS}$.

At FMCR's magnetization $4\pi M_s = 190$ G the NS envelope for the basic frequency νM_0 has the strongest broadening, but $(\Delta\nu_{-3\mathrm{dB}})_0^{NS} > (\Delta\nu_{-3\mathrm{dB}})_{1,2}^{NS}$.

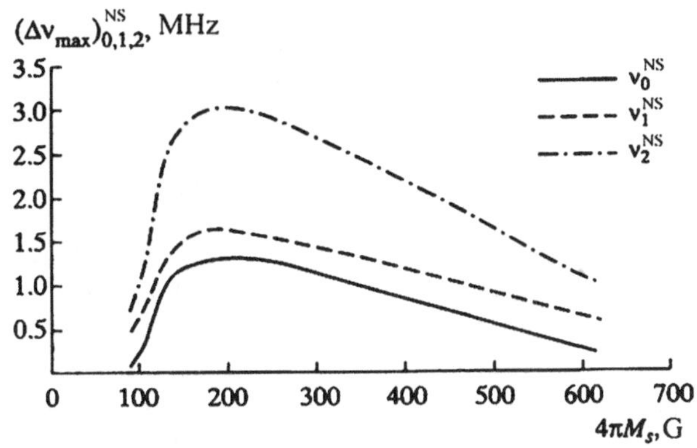

Fig. 3.67 On the medium power level (50 mW) the deviation of the central NS frequency is maximal for all the spectral components at FMCR's magnetization $4\pi M_s = 190$ G

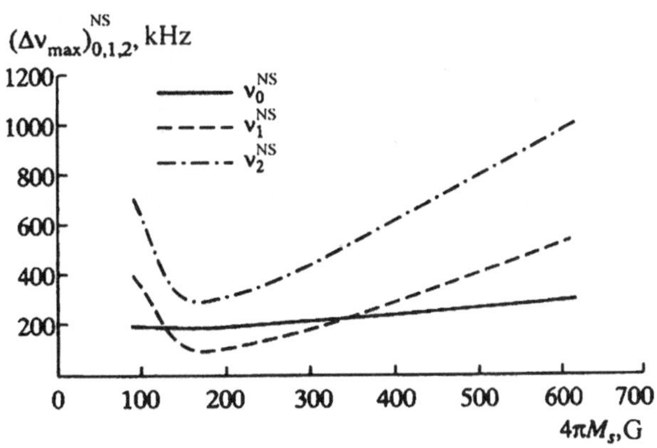

Fig. 3.68 At a high power level (500 mW) the deviation of the central frequency of the spectral components on the NS frequency $\nu_{0,1,2}^{NS}$ is maximal at FMCR's magnetization $4\pi M_s = 600$ G (KG-50) and at $4\pi M_s = 90$ G (KG-8)

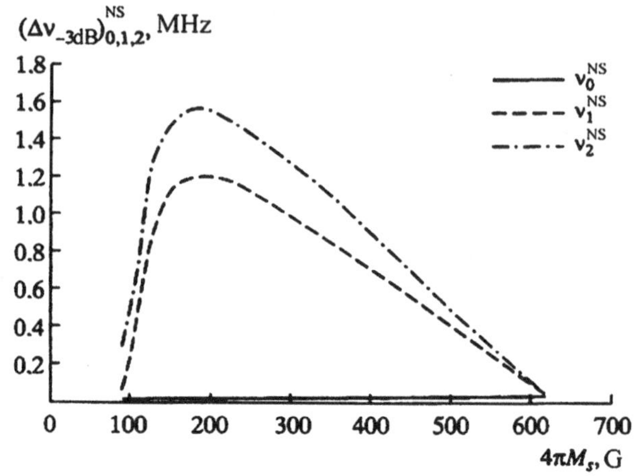

Fig. 3.69 The NS spectrum envelopes on the medium power level are most broadband for the spectral components $\nu_{1,2}$ at the FMCR saturation magnetization of $4\pi M_s = 190\,\text{G}$ (KG-15)

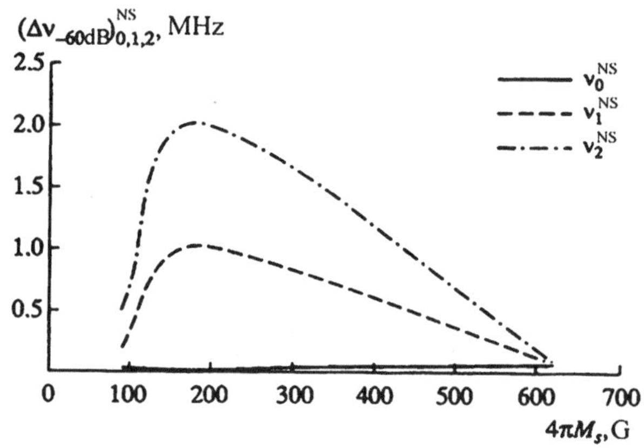

Fig. 3.70 The NS spectrum envelopes on the medium power level are most broadband for the spectral components $\nu_{1,2}$ at the FMCR saturation magnetization of $4\pi M_s = 190\,\text{G}$ (KG-15)

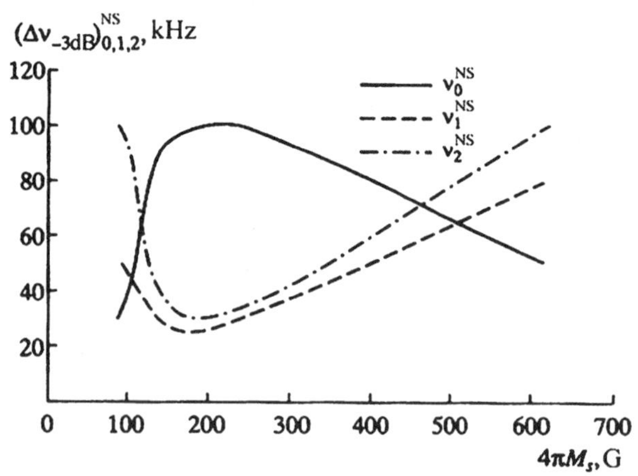

Fig. 3.71 At a high power level the NS spectrum envelopes most strongly change at a level $-3\,\text{dB}$ at variations of FMCR's magnetization $4\pi M_s$. At $4\pi M_s = 90\,\text{G}(\Delta\nu_{-3\text{dB}})_0^{\text{NS}} < (\Delta\nu_{-3\text{dB}})_1^{\text{NS}} << (\Delta\nu_{-3\text{dB}})_2^{\text{NS}}$ and $(\Delta\nu_{-60\text{dB}})_0^{\text{NS}} < (\Delta\nu_{-60\text{dB}})_1^{\text{NS}} < (\Delta\nu_{-60\text{dB}})_2^{\text{NS}}$

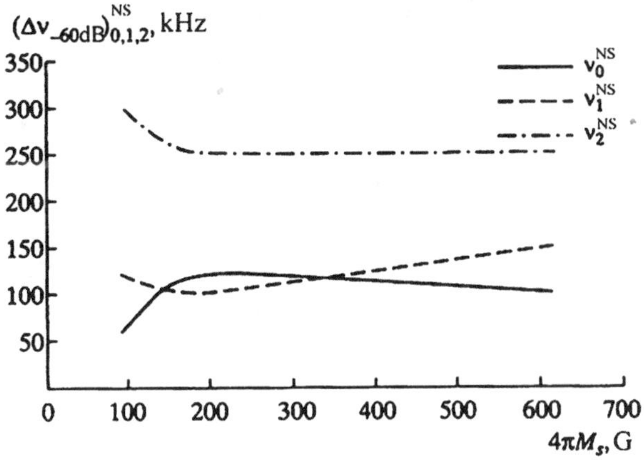

Fig. 3.72 At a high power level the NS spectrum envelopes most strongly change at a level −3 dB at variations of FMCR's magnetization $4\pi M_s$. At $4\pi M_s = 90\,\text{G}(\Delta\nu_{-3dB})_0^{NS} < (\Delta\nu_{-3dB})_1^{NS} << (\Delta\nu_{-3dB})_2^{NS}$ and $(\Delta\nu_{-60dB})_0^{NS} < (\Delta\nu_{-60dB})_1^{NS} < (\Delta\nu_{-60dB})_2^{NS}$

For the magnetization $4\pi M_s = 600\,\text{G}(\Delta\nu_{-3dB})_1^{NS} < (\Delta\nu_{-3dB})_2^{NS} > (\Delta\nu_{-3dB})_0^{NS}$.

The basis of the NS spectra envelope at a level −60 dB weakly depends on FMCR's magnetization $4\pi M_s$ within the limits of 90–600 G, and $(\Delta\nu_{-60dB})_0^{NS} \cong (\Delta\nu_{-60dB})_1^{NS}$ and $(\Delta\nu_{-60dB})_{0,1}^{NS} < (\Delta\nu_{-60dB})_2^{NS}$.

The resulted dependencies allow one to estimate the key parameters of generated noise and noise-like signals, namely:

- The deviation of the central frequencies of the basic and higher harmonic components $(\Delta\nu_{max})_{0,1,2}^{NS}$
- The shape of the spectral line envelope by the parameters $(\Delta\nu_{-3dB})_{0,1,2}^{NS}$ and $(\Delta\nu_{-60dB})_{0,1,2}^{NS}$
- The level of integral output power P_{out}.

3.3 Control Over Characteristics of Evenly Spaced Grids of Signal Frequencies

The results of our researches of management of the characteristics of evenly spaced frequency spectra (ES FS) for FMCR on the basis of various types of ferrites and their orientations in an external magnetic field H_0 are considered at medium and high power levels.

3.3.1 *Structures with Various Orientations in a Magnetic Field*

3.3.1.1 FMCR of KG-8

In Figs. 3.73 and 3.74, the characteristics of the ES FS spectra envelopes of heteromagnetic structures are resulted at the medium power level for the FMCR on the

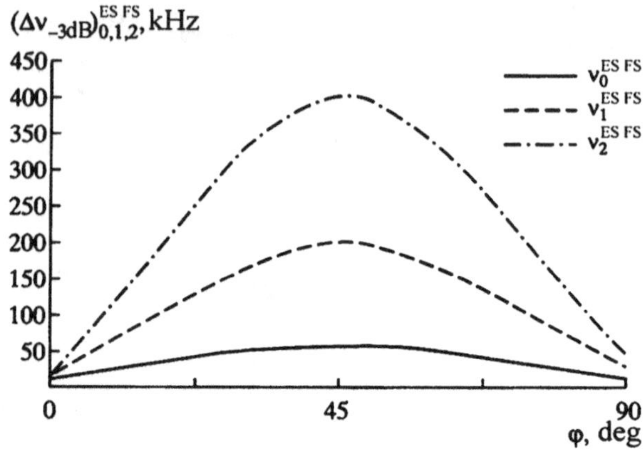

Fig. 3.73 The characteristics of the ES FS spectra envelopes of heteromagnetic structures are resulted at the medium power level for the FMCR on the basis of KG-8 ferrite on the orientation angle φ

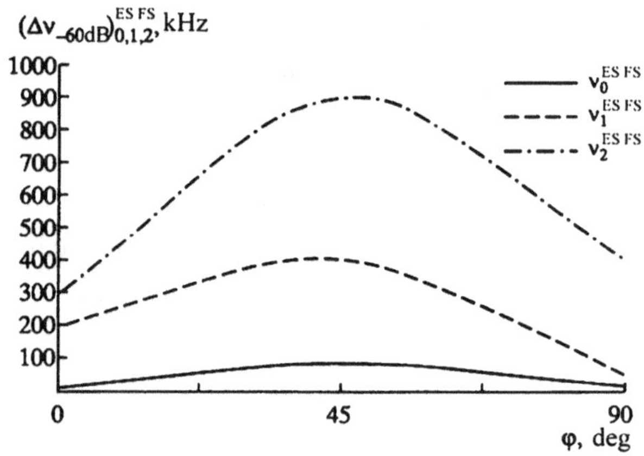

Fig. 3.74 The characteristics of the ES FS spectra envelopes of heteromagnetic structures are resulted at the medium power level for the FMCR on the basis of KG-8 ferrite on the orientation angle φ

basis of KG-8 ferrite on the orientation angle φ. It is obvious that the most broadband modes are observed at FMCR's orientation of $\varphi \cong 45°$, and as the number of harmonic increases, the expansion of the ES FS spectrum increases as well. At $\varphi \cong 0°, 90°$, narrow-band modes with $(\Delta\nu_{-3\text{dB}})^{\text{ES FS}}_{0,1,2} \cong$ const take place.

At a high power level (Figs. 3.75 and 3.76) the most broadband ES FS modes were observed at $\varphi = 0°$ and $90°$.

3.3.1.2 FMCR of KG-15

In Fig. 3.77, the dependencies of the maximal deviations of the central frequencies $(\Delta\nu_{\max})^{\text{ES FS}}_{0,1,2}$ of the ES FS envelopes for $\nu^{\text{ES FS}}_{0,1,2}$ depending on φ are shown. At $\varphi = 90°$ the deviation of frequency is maximal.

The equidistant frequencies spectrum envelopes on the medium power level (Figs. 3.78 and 3.79) for signals in the field of central frequencies $\nu^{\text{ES FS}}_{0,1,2}$ are most broadband at $\varphi = 0°$ and $90°$ for $\nu^{\text{ES FS}}_{1,2}$. At a level of -3 dB the ES FS spectra

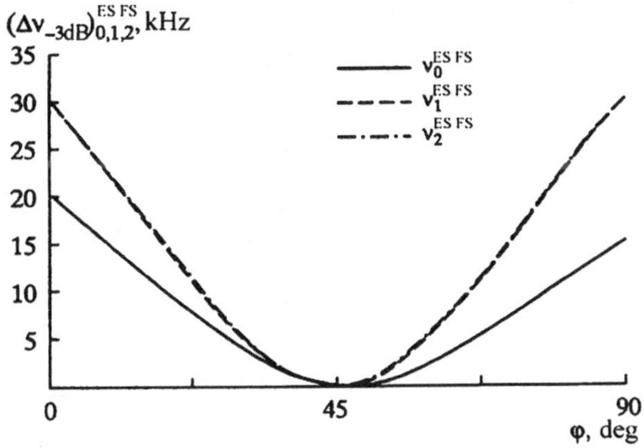

Fig. 3.75 At a high power level the most broadband ES FS modes were observed at $\varphi = 0°$ and $90°$

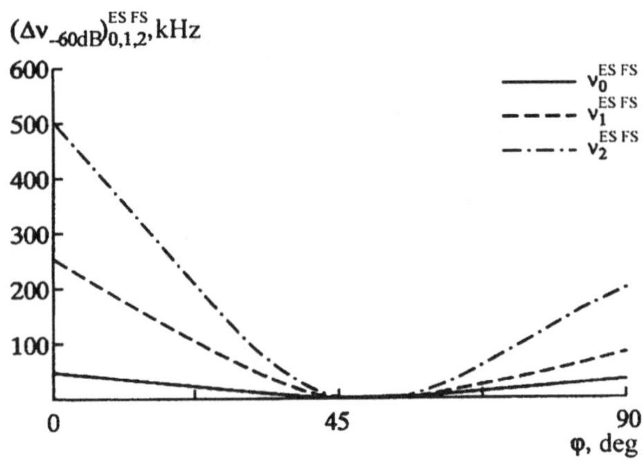

Fig. 3.76 At a high power level the most broadband ES FS modes were observed at $\varphi = 0°$ and $90°$

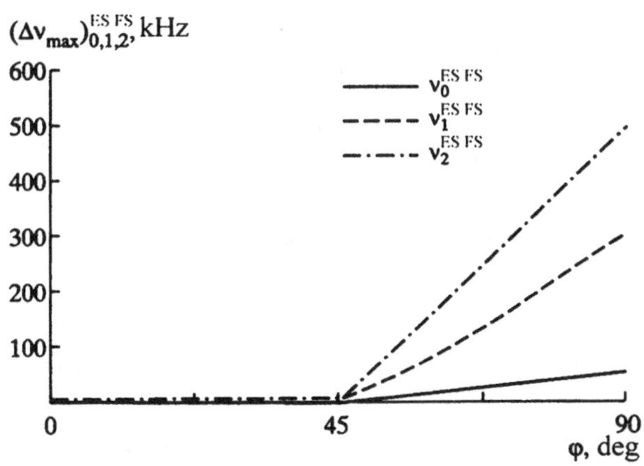

Fig. 3.77 The dependencies of the maximal deviations of the central frequencies $(\Delta\nu_{max})^{ES\,FS}_{0,1,2}$ of the ES FS envelopes for $\nu^{ES\,FS}_{0,1,2}$ depending on φ are shown

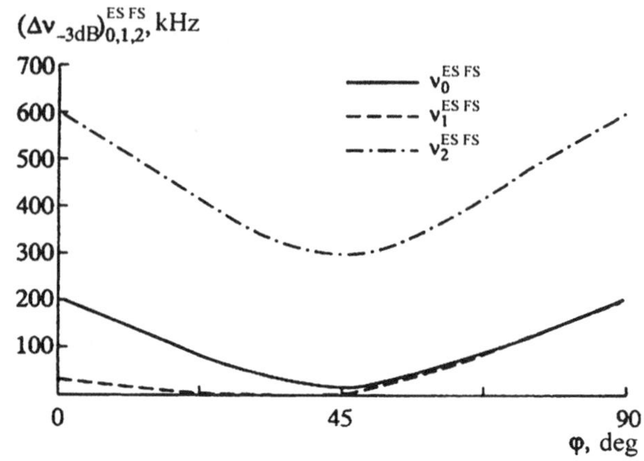

Fig. 3.78 The equidistant frequencies spectrum envelopes on the medium power level for signals in the field of central frequencies $\nu_{0,1,2}^{\mathrm{ES\,FS}}$ are most broadband at $\varphi = 0°$ and 90° for $\nu_{1,2}^{\mathrm{ES\,FS}}$

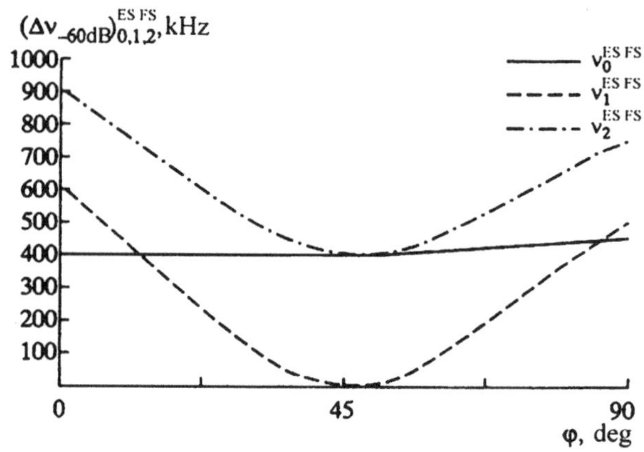

Fig. 3.79 The equidistant frequencies spectrum envelopes on the medium power level for signals in the field of central frequencies $\nu_{0,1,2}^{\mathrm{ES\,FS}}$ are most broadband at $\varphi = 0°$ and 90° for $\nu_{1,2}^{\mathrm{ES\,FS}}$

for the central frequencies $\nu_0^{\mathrm{ES\,FS}}$ at $\varphi = 45°$ have a minimum $(\Delta\nu_{-3\mathrm{dB}})_0^{\mathrm{ES\,FS}}$, but $(\Delta\nu_{-60\mathrm{dB}})_0^{\mathrm{ES\,FS}}$ depends weakly on FMCR's orientation in the heteromagnetic structure.

At a high power level (Figs. 3.80 and 3.81) the maximal broadening of the ES FS spectrum envelopes at a level −3 dB was observed at $\varphi \cong 0°$ for the signal with $(\Delta\nu_{-3\mathrm{dB}})_2^{\mathrm{ES\,FS}}$ and $(\Delta\nu_{-60\mathrm{dB}})_2^{\mathrm{ES\,FS}}$. At $\varphi = 90°$ takes place enlargement spectral components for frequencies $\nu_{1,2}^{\mathrm{ES\,FS}}$.

3.3.2 Structures with Ferrites of Various Magnetization

3.3.2.1 FMCR Orientation Angle $\varphi = 0°$

The deviation of the central frequencies of the ES FS spectral components for $\nu_{0,1,2}^{\mathrm{ES\,FS}}$ as a function of FMCR's saturation magnetization at change of $4\pi M_s = (200\text{–}600)$ G increases with growth of the number of harmonic (Fig. 3.82).

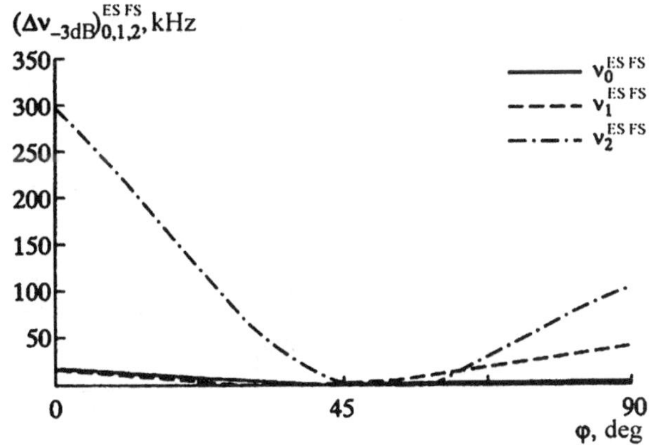

Fig. 3.80 At a high power level the maximal broadening of the ES FS spectrum envelopes at a level −3 dB was observed at $\varphi \cong 0°$ for the signal with $(\Delta\nu_{-3dB})_2^{ES\ FS}$ and $(\Delta\nu_{-60dB})_2^{ES\ FS}$

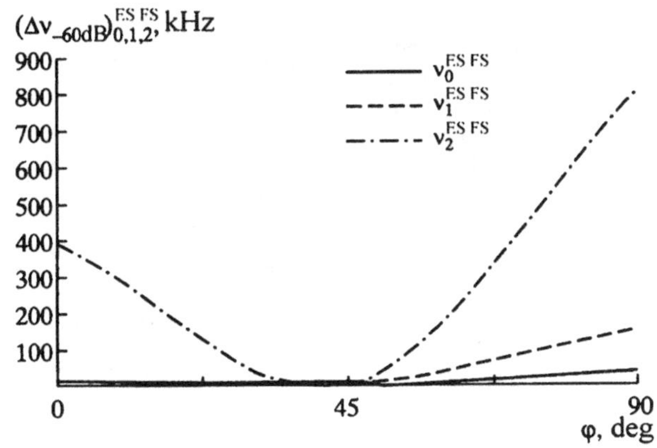

Fig. 3.81 At a high power level the maximal broadening of the ES FS spectrum envelopes at a level −3 dB was observed at $\varphi \cong 0°$ for the signal with $(\Delta\nu_{-3dB})_2^{ES\ FS}$ and $(\Delta\nu_{-60dB})_2^{ES\ FS}$

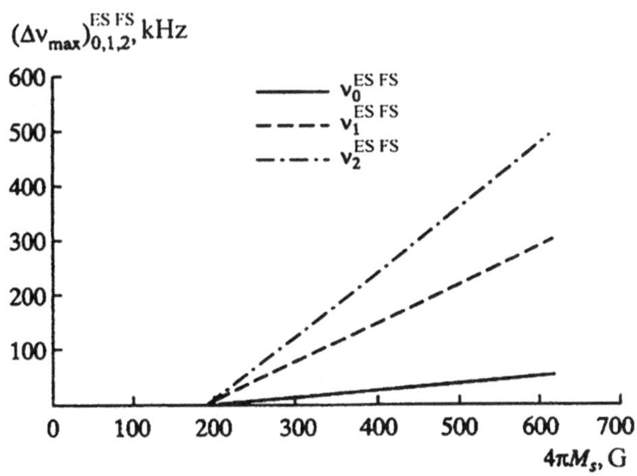

Fig. 3.82 The deviation of the central frequencies of the ES FS spectral components for $\nu_{0,1,2}^{ES\ FS}$ as a function of FMCR's saturation magnetization at change of $4\pi M_s = (200–600)$ G increases with growth of the number of harmonic

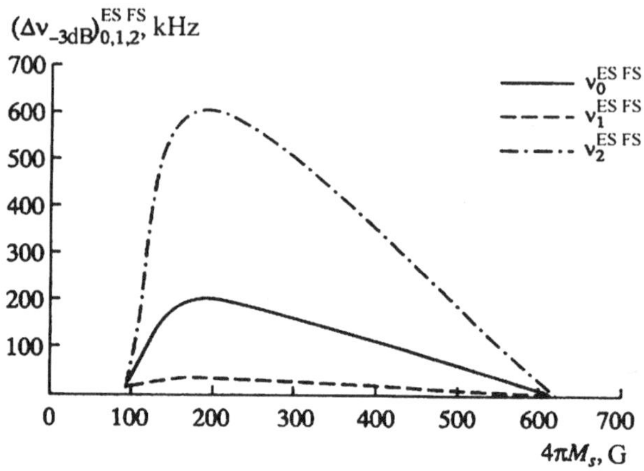

Fig. 3.83 The dependencies of the equidistant frequency spectrum envelopes on the medium level generated power are resulted

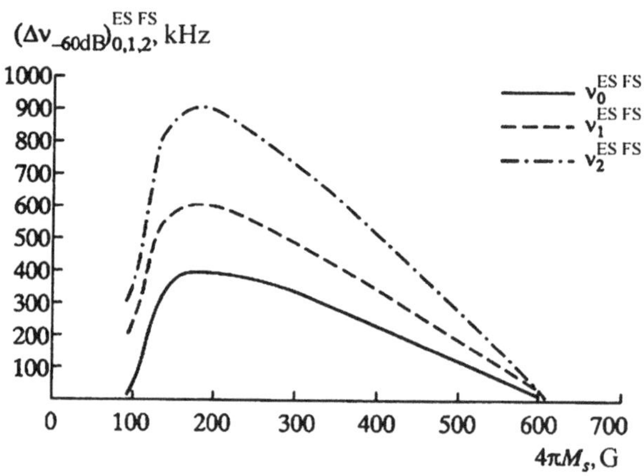

Fig. 3.84 The dependencies of the equidistant frequency spectrum envelopes on the medium level generated power are resulted

In Figs. 3.83 and 3.84, the dependencies of the equidistant frequency spectrum envelopes on the medium level generated power are presented. The greatest broadening of the ES FS spectra was observed at FMCR's magnetization $4\pi M_s \cong 190$ G (KG-15).

At a high power level 0.25 W (Figs. 3.85 and 3.86) the broadening of the ES FS spectra envelopes for the FMCR with a magnetization $4\pi M_s \cong 90$ G was maximal for $(\Delta\nu_{-3dB})_2^{ES\,FS}$, and the parameters $(\Delta\nu_{-3dB})_{0,1}^{ES\,FS}$ almost did not change. At $4\pi M_s \cong 190$ G the greatest broadening of the line $(\Delta\nu_{-3dB})_2^{ES\,FS}$ took place. For the basis of the ES FS envelope at FMCR's saturation magnetization FMCR $4\pi M_s \cong 90$ G, $(\Delta\nu_{-60dB})_0^{ES\,FS} < (\Delta\nu_{-60dB})_1^{ES\,FS} < (\Delta\nu_{-60dB})_2^{ES\,FS}$ takes place.

In Figs. 3.87 and 3.88, the parameters $(\Delta\nu_{-3dB})_{0,1,2}^{ES\,FS}$ and $(\Delta\nu_{-60dB})_{0,1,2}^{ES\,FS}$ from FMCR's magnetization $4\pi M_s$ at a high power level.

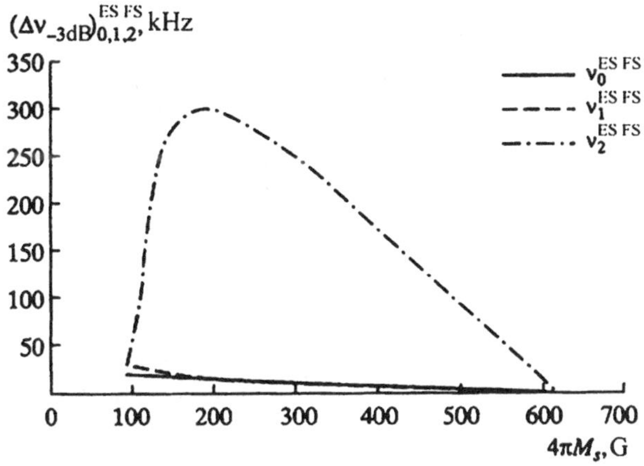

Fig. 3.85 At a high power level the broadening of the ES FS spectra envelopes for the FMCR with a magnetization $4\pi M_s \cong 90$ G was maximal for $(\Delta\nu_{-3dB})_2^{ES\ FS}$, and the parameters $(\Delta\nu_{-3dB})_{0,1}^{ES\ FS}$ almost did not change

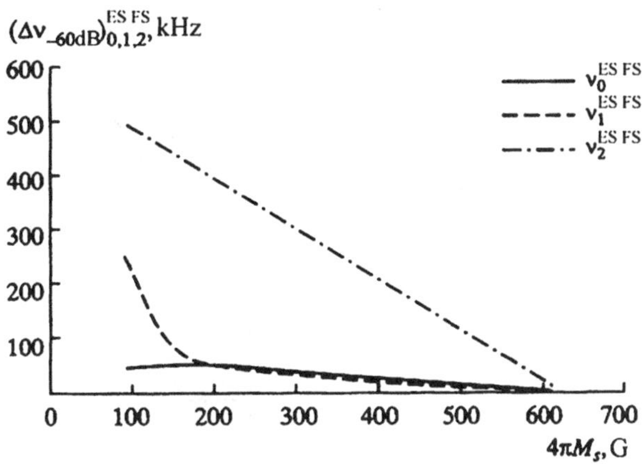

Fig. 3.86 At a high power level the broadening of the ES FS spectra envelopes for the FMCR with a magnetization $4\pi M_s \cong 90$ G was maximal for $(\Delta\nu_{-3dB})_2^{ES\ FS}$, and the parameters $(\Delta\nu_{-3dB})_{0,1}^{ES\ FS}$ almost did not change

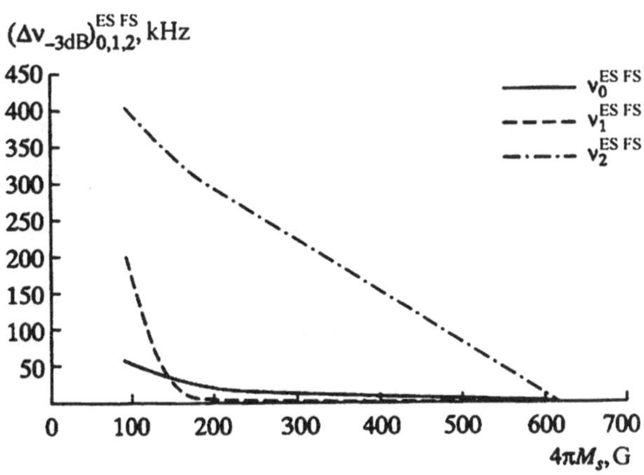

Fig. 3.87 The parameters $(\Delta\nu_{-3dB})_{0,1,2}^{ES\ FS}$ and $(\Delta\nu_{-60dB})_{0,1,2}^{ES\ FS}$ from FMCR's magnetization $4\pi M_s$ at power 0.5 W

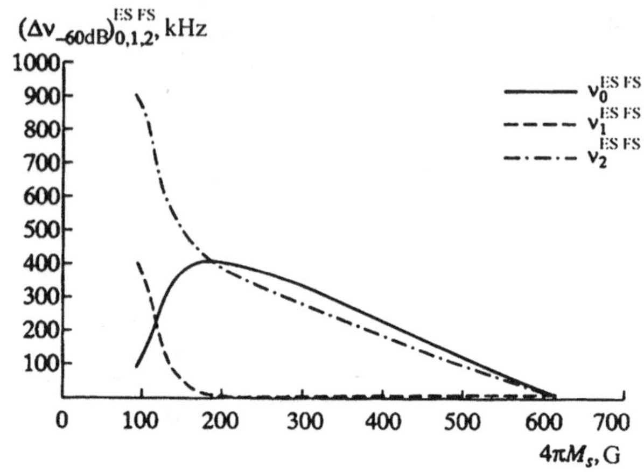

Fig. 3.88 The parameters $(\Delta\nu_{-3dB})_{0,1,2}^{ES\,FS}$ and $(\Delta\nu_{-60dB})_{0,1,2}^{ES\,FS}$ from FMCR's magnetization $4\pi M_s$ at power 0.5 W

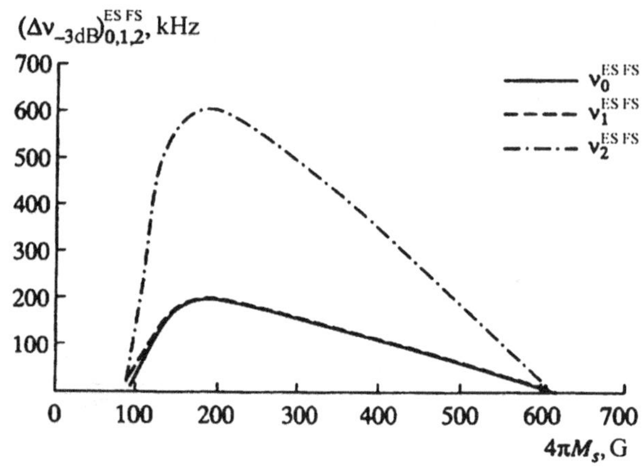

Fig. 3.89 On the medium power level the ES FS spectrum envelopes are maximally widened for the FMCR with its magnetization $4\pi M_s = 190$ G for all the spectral components of generated signals $\nu_{0,1,2}^{ES\,FS}$

3.3.2.2 FMCR Orientation Angle $\varphi = 90°$

On the medium power level the ES FS spectrum envelopes (Figs. 3.89 and 3.90) are maximally widened for the FMCR with its magnetization $4\pi M_s = 190$ G for all the spectral components of generated signals $\nu_{0,1,2}^{ES\,FS}$.

At a high power level the ES FS spectrum envelope ES FS (Figs. 3.91 and 3.92) is also maximally widened for the FMCR with its magnetization $4\pi M_s = 190$ G.

The resulted experimental dependencies allow one to estimate the key parameters of the generated equidistant frequency spectrum, namely:

- The deviation of the central frequencies of the components $(\Delta\nu_{max})_{0,1,2}^{ES\,FS}$
- The spectrum width and the shape of the envelope of the spectral components $(\Delta\nu_{-3dB})_{0,1,2}^{ES\,FS}$., $(\Delta\nu_{-60dB})_{0,1,2}^{ES\,FS}$
- The central output power P_{out}.

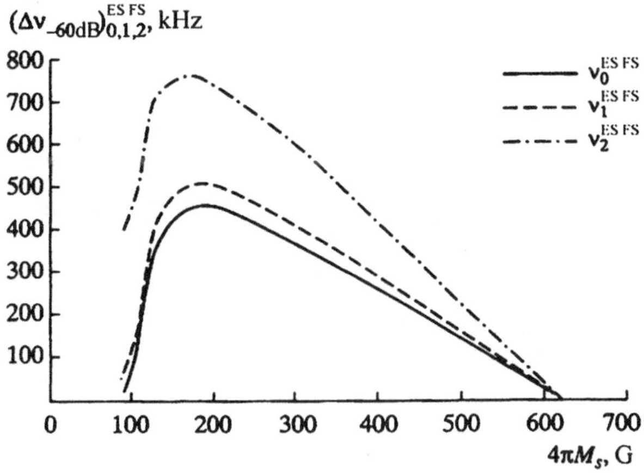

Fig. 3.90 On the medium power level the ES FS spectrum envelopes are maximally widened for the FMCR with its magnetization $4\pi M_s = 190\,\text{G}$ for all the spectral components of generated signals $\nu_{0,1,2}^{ES\,FS}$

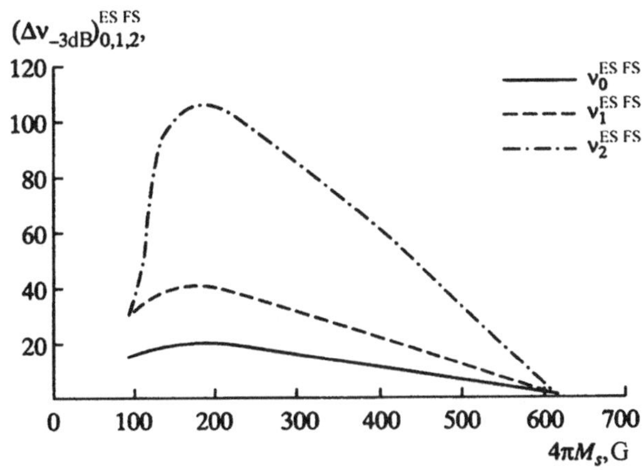

Fig. 3.91 At a high power level the ES FS spectrum envelope ES FS is also maximally widened for the FMCR with its magnetization $4\pi M_s = 190\,\text{G}$

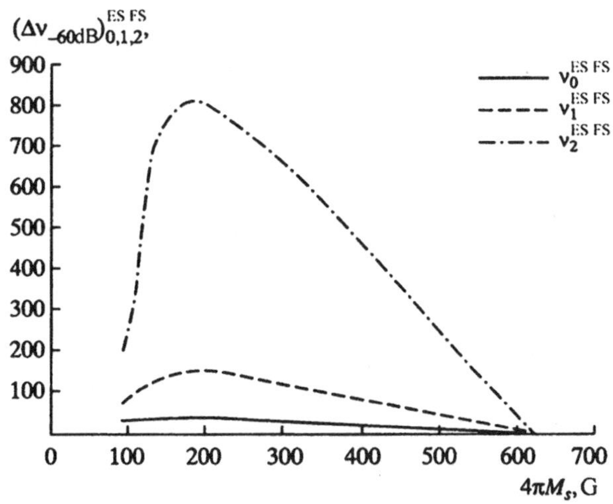

Fig. 3.92 At a high power level the ES FS spectrum envelope ES FS is also maximally widened for the FMCR with its magnetization $4\pi M_s = 190\,\text{G}$

Chapter 4
Generalization Control Characteristics in Generative Structures

In connection with the complex character of the investigated physical mechanisms in generating HMT at a certain stage of the development of this lead the results of our investigations of spectral characteristics for various angles φ of FMCR orientations in an external magnetic field H_0 and various magnetizations satiations at middle and high power levels have been generalized.

Below no specific kinds of spectra are discussed, but generalized dependences of the following quantities on the FMCR orientation angles φ and saturation magnetizations $4\pi M_s$ are resulted: the maximum deviations of signals $(\Delta\nu_{max}(\varphi))_{0,1,2}$, parameters of spectral lines $(\Delta\nu_{-3dB}(\varphi))_{0,1,2}$ and $(\Delta\nu_{-60dB}(\varphi))_{0,1,2}$, the maximum deviations $(\Delta\nu_{max}(4\pi M_s))_{0,1,2}$, parameters of spectral lines $(\Delta\nu_{-3dB}(4\pi M_s))_{0,1,2}$ and $(\Delta\nu_{-60dB}(4\pi M_s))_{0,1,2}$.

These data are useful at classification of the physical mechanisms and give estimations of achievable parameters. In the previous chapter the similar material for various spectra kinds of signals generated in HMT, namely, SPS, NS, ES FS has been systematized. Though each of these kinds of spectra can also be investigated and classified in even more details – by the quality of signals (noise levels, the range of their tuning out from the carrier frequency, the reorganization steepness of spectra parameters, including noise ones, at various values of the operating parameters – pressure, magnetic field, the level of power generated), by the extremely achievable parameters of spectra and their optimization in specific modes, researches on various HMT types (*e.g.*, bipolar and field ones) with the purpose to optimize the ways and designs can also be carried out.

4.1 Structure Characteristics with Various Orientations

4.1.1 Structures with KG-8 *FMCR*

In Fig. 4.1 dependences of the maximum frequency deviations of signals for the spectral components $\nu_{0,1,2}$ on the FMCR orientation angle φ in a magnetic field $\overline{H}_0$ at a medium (50 mW) power capacity are resulted. It is visible that $(\Delta\nu_{max})_{0,1,2}$ for $\varphi = 0°$.

A.A. Ignatiev and A.V. Lyashenko, *Heteromagnetic Microelectronics: Microsystems of Active Type*, DOI 10.1007/978-1-4419-6002-3_4,

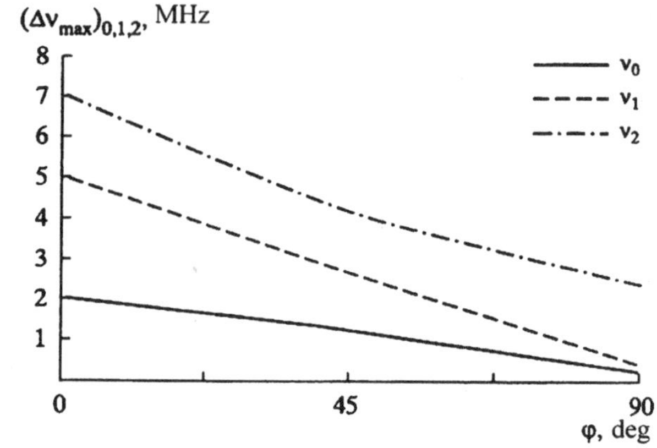

Fig. 4.1 The dependences of the maximum frequency deviations of signals $(\Delta\nu_{max})_{0,1,2}$ on the FMCR orientation angle φ in a magnetic field at power level 50 mW

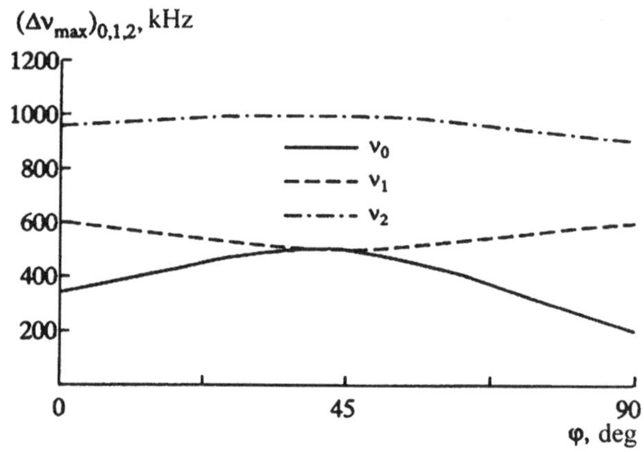

Fig. 4.2 The dependences of the maximum frequency deviations of signals $(\Delta\nu_{max})_{0,1,2}$ on the FMCR orientation angle φ in a magnetic field at power level 500 mW

On the high (500 mW) power level the deviation of signals is maximum at $\varphi = 0°$ and 90° (Fig. 4.2).

Changes of the parameters of spectra of generated signals at the medium power level as functions of the FMCR orientation angle φ produce the dependences shown in Figs. 4.3 and 4.4. The maximum widening of the spectral lines was observed at $\varphi = 0°$.

At the high power level (Figs. 4.5 and 4.6) the parameters $(\Delta\nu_{-3dB})_{0,1,2}$ are weak functions of the FMCR orientation angle φ, but widening of the spectra for $(\Delta\nu_{-60dB})_{0,1,2}$ is maximum for the FMCR orientation $\varphi = 0°$.

4.1.2 Structures with KG-*15 FMCR*

In Figs. 4.7 and 4.8 dependences of $(\Delta\nu_{max})_{0,1,2}$ (the maximum deviations of signal frequencies at the medium and high power levels), respectively, are shown. At $\varphi = 0°$ the deviations of frequencies have the maximum values.

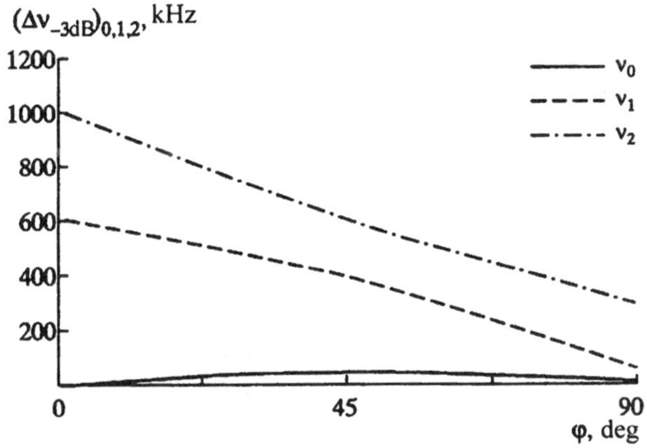

Fig. 4.3 The dependences of the spectral lines width $(\Delta\nu_{-3\mathrm{dB}})_{0,1,2}$ on the FMCR orientation angle φ in a magnetic field at power level 50 mW

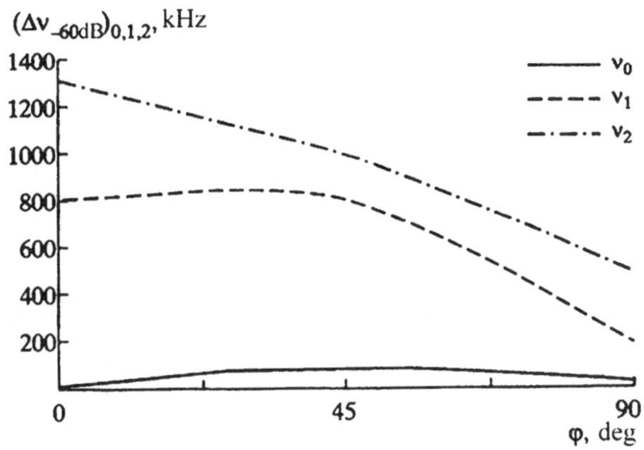

Fig. 4.4 The dependences of the base width of the spectral lines $(\Delta\nu_{-60\mathrm{dB}})_{0,1,2}$ on the FMCR orientation angle φ in a magnetic field at power level 50 mW

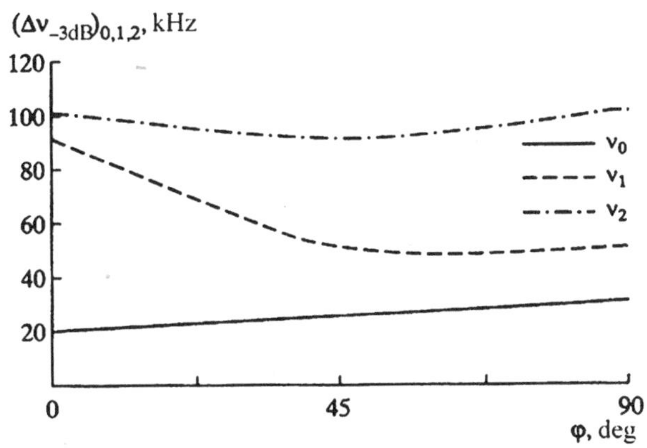

Fig. 4.5 The dependences of the spectral lines width $(\Delta\nu_{-3\mathrm{dB}})_{0,1,2}$ on the FMCR orientation angle φ in a magnetic field at power level 500 mW

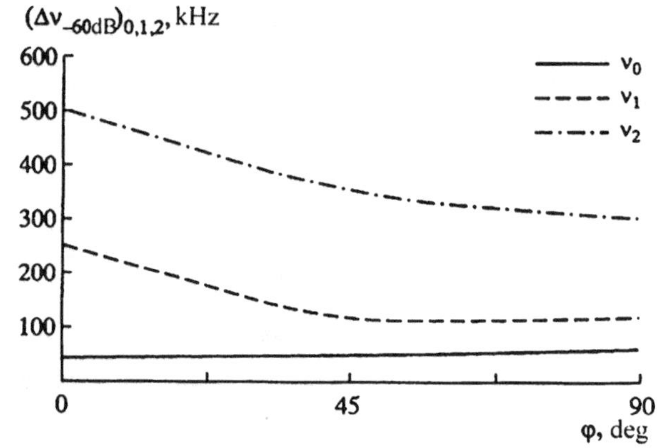

Fig. 4.6 The dependences of the base width of the spectral lines $(\Delta\nu_{-60dB})_{0,1,2}$ on the FMCR orientation angle φ in a magnetic field at power level 500 mW

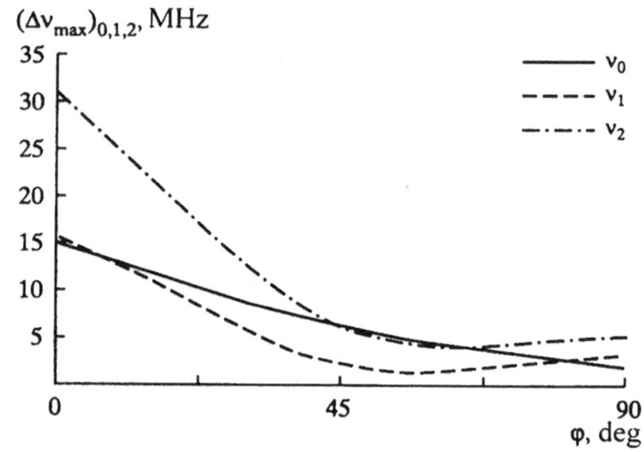

Fig. 4.7 The dependences of the maximum frequency deviations of signals $(\Delta\nu_{max})_{0,1,2}$ on the FMCR orientation angle φ in a magnetic field at power level 50 mW

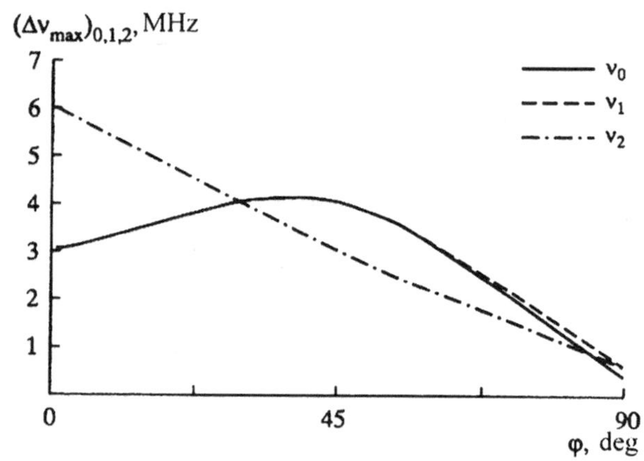

Fig. 4.8 The dependences of the maximum frequency deviations of signals $(\Delta\nu_{max})_{0,1,2}$ on the FMCR orientation angle φ in a magnetic field at power level 500 mW

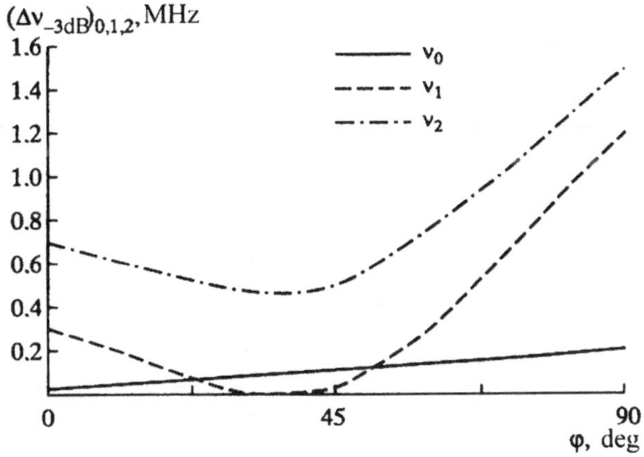

Fig. 4.9 The dependences of the spectral lines width $(\Delta\nu_{-3dB})_{0,1,2}$ on the FMCR orientation angle φ in a magnetic field at power level 50 mW

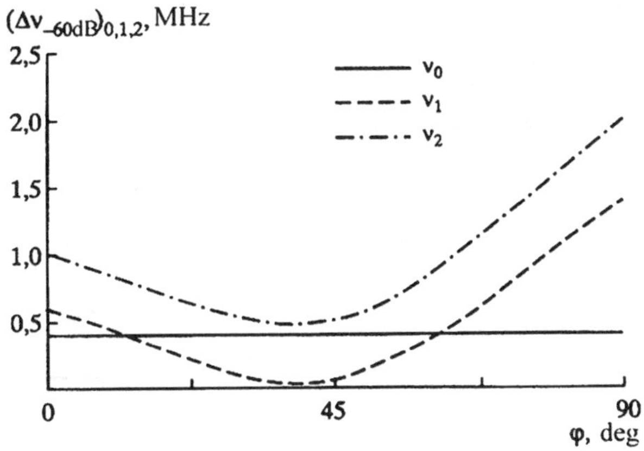

Fig. 4.10 The dependences of the base width of the spectral lines $(\Delta\nu_{-60dB})_{0,1,2}$ on the FMCR orientation angle φ in a magnetic field at power level 50 mW

In Figs. 4.9 and 4.10 the dependences illustrating the change of the parameters of spectral lines of signals on frequencies $\nu_{0,1,2}$ on the FMCR orientation angle φ at the medium power level are resulted. At $\varphi = 0°$ the spectral lines $(\Delta\nu_{-3dB})_{0,1,2}$ and $(\Delta\nu_{-60dB})_{0,1,2}$ broaden most essentially.

At the high power level (Figs. 4.11 and 4.12) at $\varphi = 0°$ the spectral lines $(\Delta\nu_{-3dB})_{0,1,2}$ and $(\Delta\nu_{-60dB})_{0,1,2}$ broaden maximally.

4.1.3 Structures with **KG-50** *FMCR*

In Figs. 4.13 and 4.14 dependences of the maximum deviations of frequencies for components $\nu_{0,1,2}$ on the FMCR orientation angle φ are resulted for the medium and high power levels, respectively. At the medium power level (Fig. 4.13) the signal frequency deviations are maximum at $\varphi = 45°$ and a little bit lower at $\varphi = 90°$.

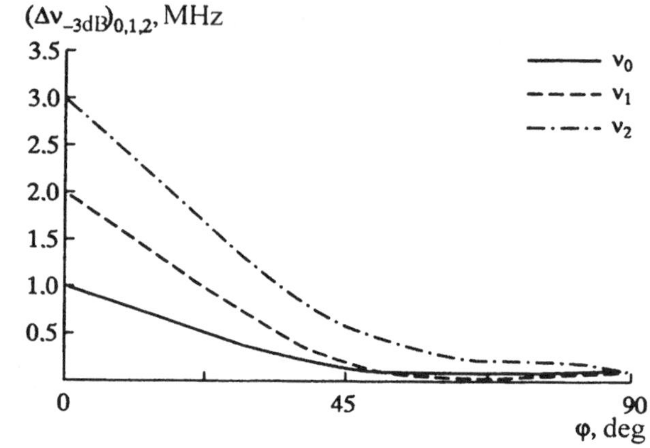

Fig. 4.11 The dependences of the spectral lines width $(\Delta\nu_{-3\mathrm{dB}})_{0,1,2}$ on the FMCR orientation angle φ in a magnetic field at power level 500 mW

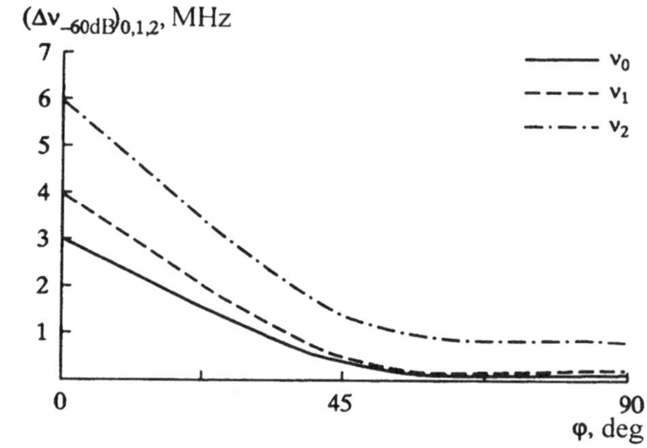

Fig. 4.12 The dependences of the base width of the spectral lines $(\Delta\nu_{-60\mathrm{dB}})_{0,1,2}$ on the FMCR orientation angle φ in a magnetic field at power level 500 mW

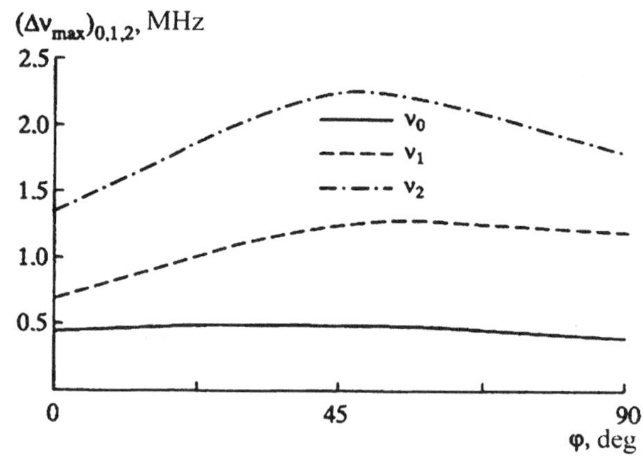

Fig. 4.13 The dependences of the maximum frequency deviations of signals $(\Delta\nu_{\max})_{0,1,2}$ on the FMCR orientation angle φ in a magnetic field at power level 50 mW

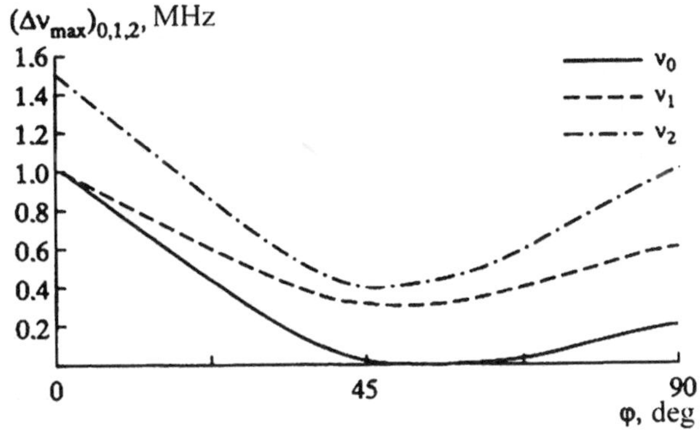

Fig. 4.14 The dependences of the maximum frequency deviations of signals $(\Delta\nu_{max})_{0,1,2}$ on the FMCR orientation angle φ in a magnetic field at power level 500 W

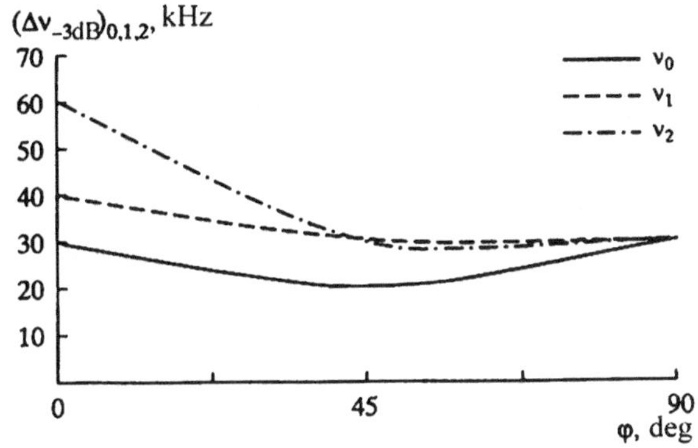

Fig. 4.15 The dependences of the spectral lines width $(\Delta\nu_{-3dB})_{0,1,2}$ on the FMCR orientation angle φ in a magnetic field at power level 50 W

At the high power level (Fig. 4.14) the deviations of frequencies are maximum at the ferrite orientation $\varphi = 0°$.

The spectral characteristics of signals at the medium power level (Figs. 4.15 and 4.16) most essentially change at the FMCR orientation $\varphi = 0°$.

At the high power level (Figs. 4.17 and 4.18) the spectral characteristics of generated signals most essentially change at the FMCR orientation $\varphi = 45°$.

4.2 Structure Characteristics with Various Magnetizations

As follows from the data resulted above, the target parameters of generated signals are strongly influenced by the angle of FMCR orientation φ in a magnetic field H_0 (the polarization of an HF magnetic field, excitation of magnetization oscillations in the FMCR), saturation magnetization $4\pi M_s$, the level of HF power of oscillation excitation in the ferrite.

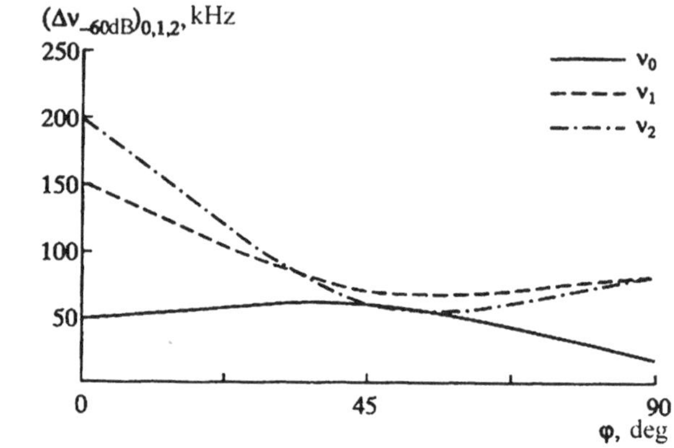

Fig. 4.16 The dependences of the base width of the spectral lines $(\Delta\nu_{-60\text{dB}})_{0,1,2}$ on the FMCR orientation angle φ in a magnetic field at power level 50 W

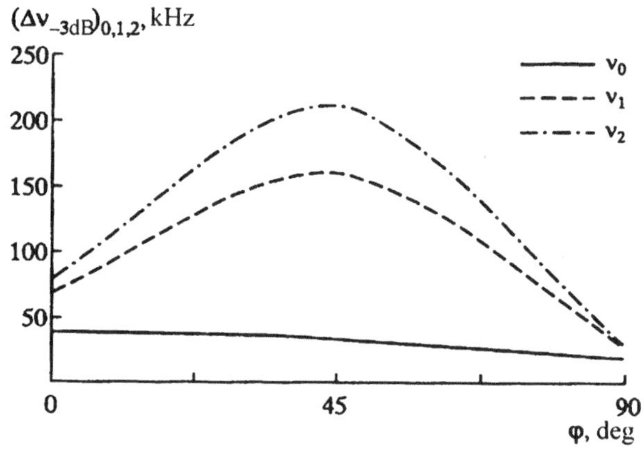

Fig. 4.17 The dependences of the spectral lines width $(\Delta\nu_{-3\text{dB}})_{0,1,2}$ on the FMCR orientation angle φ in a magnetic field at power level 500 mW

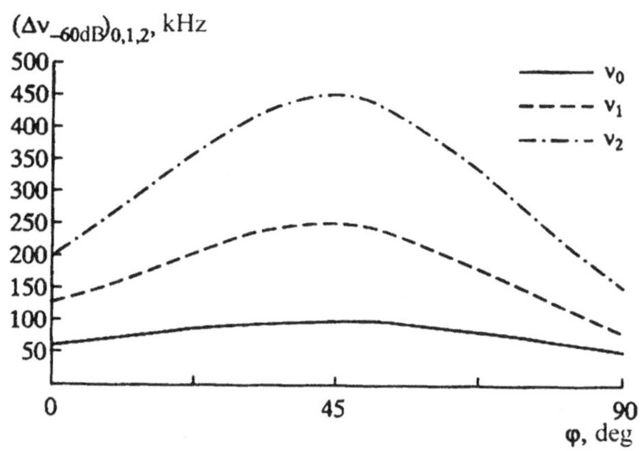

Fig. 4.18 The dependences of the base width of the spectral lines $(\Delta\nu_{-60\text{dB}})_{0,1,2}$ on the FMCR orientation angle φ in a magnetic field at power level 500 mW

4.2.1 FMCR Orientation Angle $\varphi = 0°$

In Figs. 4.19 and 4.20 dependences of the maximum deviations of signal frequencies $\nu_{0,1,2}$ on the FMCR saturation magnetization at the medium and high power levels, respectively, are resulted. At the FMCR magnetization $4\pi M_s = 190$ G deviation is maximum.

In Figs. 4.21 and 4.22 dependences of the spectral characteristics of generated signals on the FMCR magnetization $4\pi M_s$ at the medium power level are shown.

The most essential control over the parameters of spectral lines was observed at $H_0 = 90$ G. At transition to high power levels (Figs. 4.23 and 4.24) the maximum control over the spectral characteristics takes place at the FMCR magnetization $4\pi M_s = 190$ G.

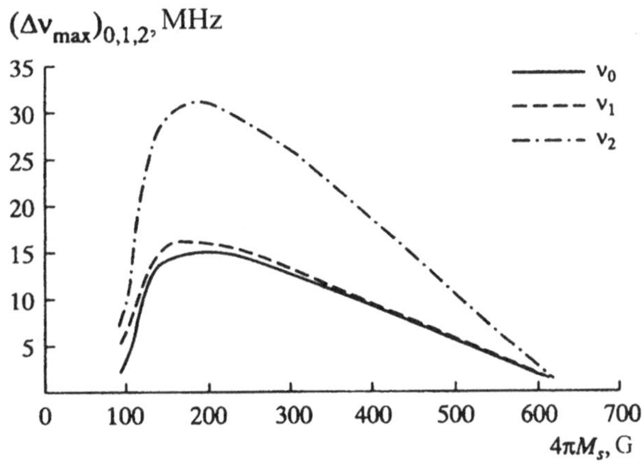

Fig. 4.19 The dependences of the maximum deviations of signal frequencies $(\Delta\nu_{max})_{0,1,2}$ on the FMCR saturation magnetization at the orientation angle 0° in a magnetic field and at power level 50 mW

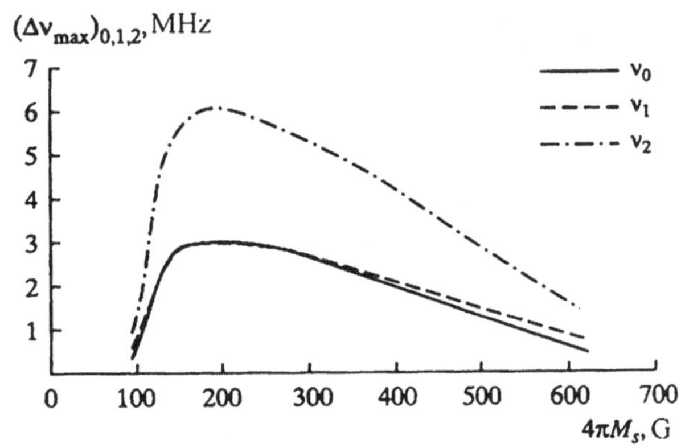

Fig. 4.20 The dependences of the maximum deviations of signal frequencies $(\Delta\nu_{max})_{0,1,2}$ on the FMCR saturation magnetization at the orientation angle 0° in a magnetic field and at power level 500 mW

Fig. 4.21 The dependences of the spectral lines width $(\Delta\nu_{-3dB})_{0,1,2}$ on the FMCR saturation magnetization at the orientation angle 0° in a magnetic field and at power level 50 mW

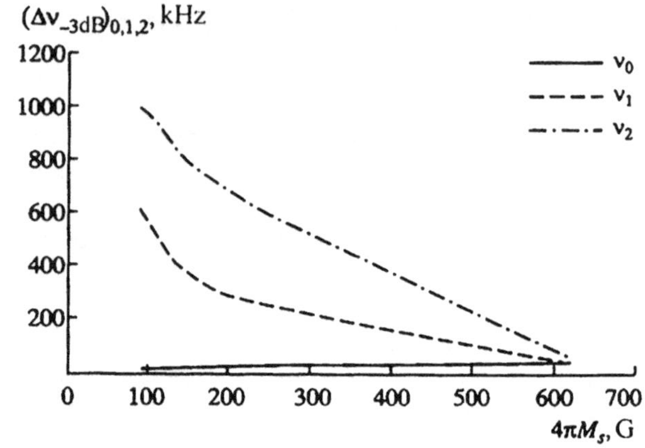

Fig. 4.22 The dependences of the base width of the spectral lines $(\Delta\nu_{-60dB})_{0,1,2}$ on the FMCR saturation magnetization at the orientation angle 0° in a magnetic field and at power level 500 mW

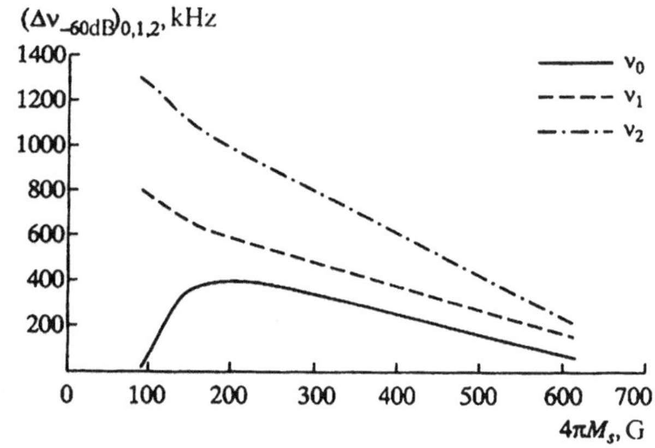

Fig. 4.23 The control over the spectral line width $(\Delta\nu_{-3dB})_{0,1,2}$ for the various FMCR saturation magnetization at the orientation angle 0° in a magnetic field and at power level 500 mW

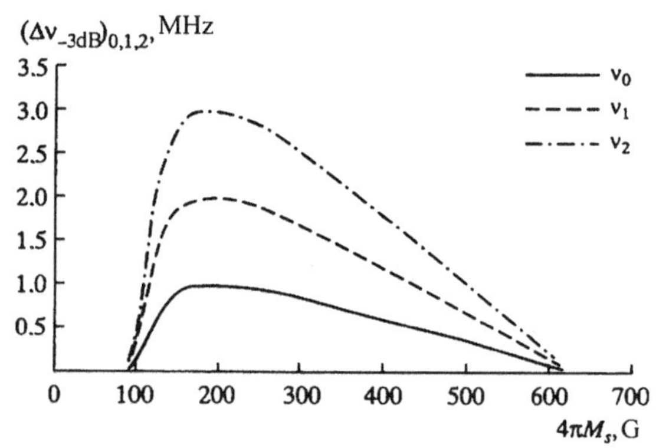

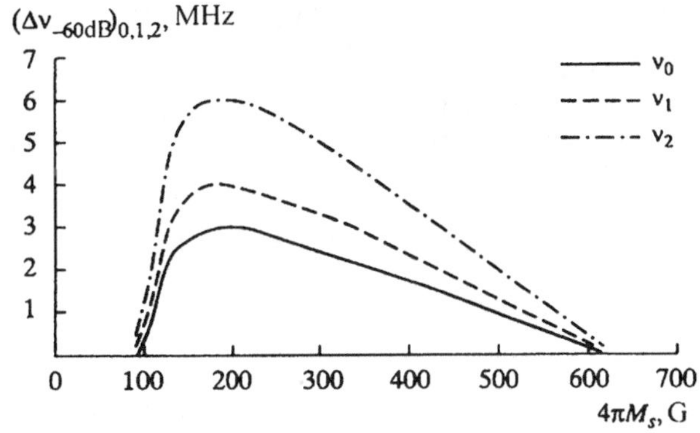

Fig. 4.24 The control over base width of the spectral line $(\Delta\nu_{-60dB})_{0,1,2}$ for the various FMCR saturation magnetization at the orientation angle 0° in a magnetic field and at power level 500 mW

4.2.2 FMCR Orientation Angle $\varphi = 45°$

In Figs. 4.25 and 4.26 dependences of the maximum deviations of frequencies $\nu_{0,1,2}$ of generated signals on the FMCR saturation magnetization at high and medium power levels are shown. At the magnetization $4\pi M_s = 190$ G the deviation of signals is maximum.

In Figs. 4.27 and 4.28 the dependences showing control over the spectral characteristics of signals at the medium power level are resulted. At the FMCR magnetization $4\pi M_s = 90$ G control over the characteristics is as much as possible.

At a high power level (Figs. 4.29 and 4.30) the maximum limits of control over spectral characteristics are noted at the FMCR magnetization $4\pi M_s = 190$ G.

4.2.3 FMCR Orientation Angle $\varphi = 90°$

In Figs. 4.31 and 4.32 dependences of the maximum deviations of frequencies $\nu_{0,1,2}$ of generated signals on the FMCR saturation magnetization at medium and high power levels are shown. At a medium power level (Fig. 4.31) the deviations of frequencies are maximum at the FMCR magnetization $4\pi M_s = 190$ G, while at a high power level at $4\pi M_s = 600$ G (Fig. 4.32).

At medium (Figs. 4.33 and 4.34) and high (Figs. 4.35 and 4.36) power levels control over spectral characteristics was carried out within the maximum limits at the FMCR magnetization $4\pi M_s = 190$ G.

The resulted generalized dependences, together with the data presented in chapters 2, 3, allow estimations of the expected parameters in various generation modes of heteromagnetic structures at medium (50 mW) and high (0.5 W) power levels for:

- Spectrally pure signals
- Evenly spaced frequency spectrum
- Noise-type signals

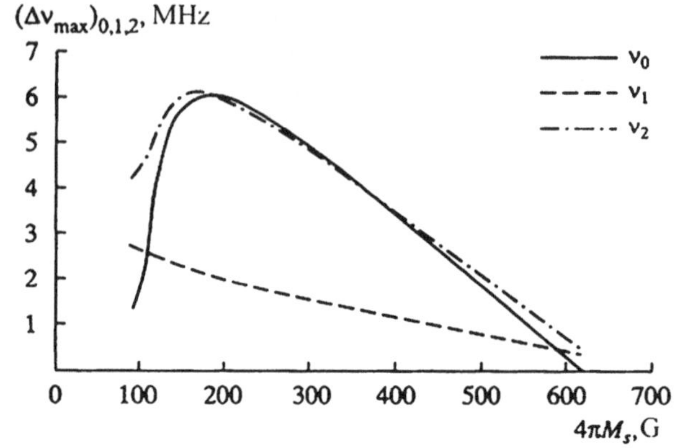

Fig. 4.25 The dependences of the maximum deviations of signal frequencies $(\Delta\nu_{max})_{0,1,2}$ on the FMCR saturation magnetization at the orientation angle 45° in a magnetic field and at power level 50 mW

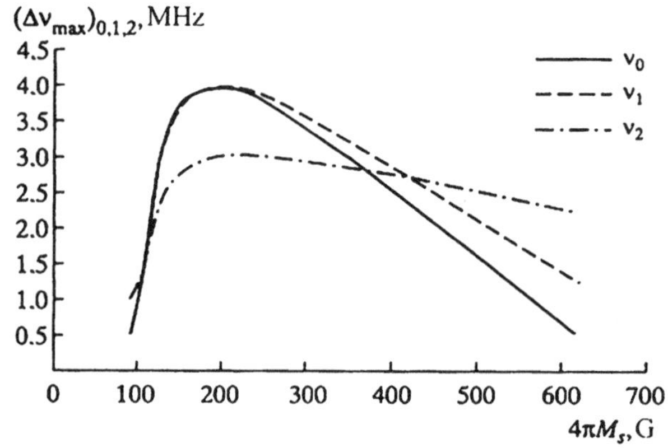

Fig. 4.26 The dependences of the maximum deviations of signal frequencies $(\Delta\nu_{max})_{0,1,2}$ on the FMCR saturation magnetization at the orientation angle 45° in a magnetic field and at power level 500 mW

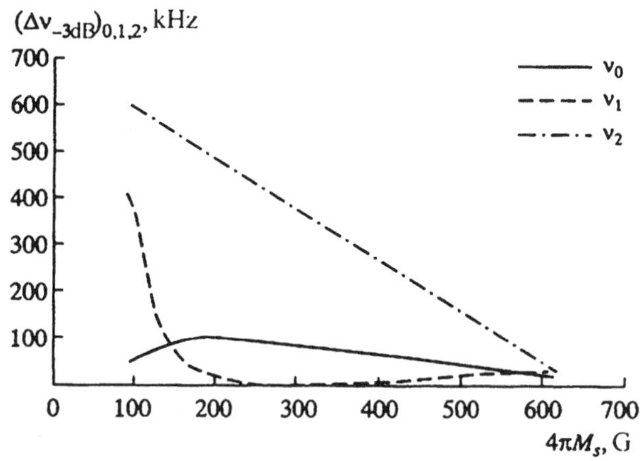

Fig. 4.27 The dependences of the spectral lines width $(\Delta\nu_{-3dB})_{0,1,2}$ on the FMCR saturation magnetization at the orientation angle 45° in a magnetic field and at power level 50 mW

Fig. 4.28 The dependences of the base width of the spectral lines $(\Delta\nu_{-60dB})_{0,1,2}$ on the FMCR saturation magnetization at the orientation angle 45° in a magnetic field and at power level 50 mW

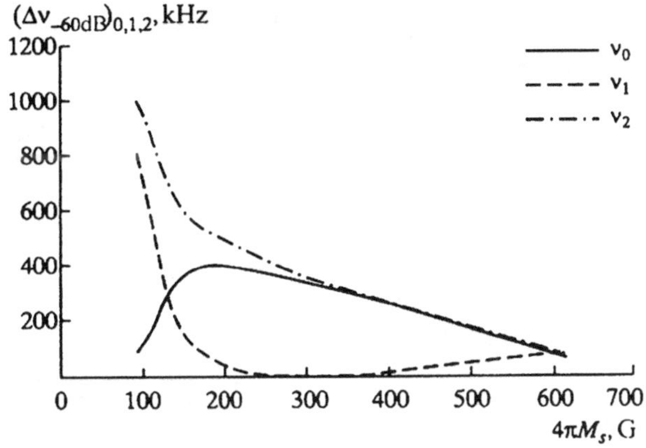

Fig. 4.29 The control over the spectral line width $(\Delta\nu_{-3dB})_{0,1,2}$ for the various FMCR saturation magnetization at the orientation angle 45° in a magnetic field and at power level 500 mW

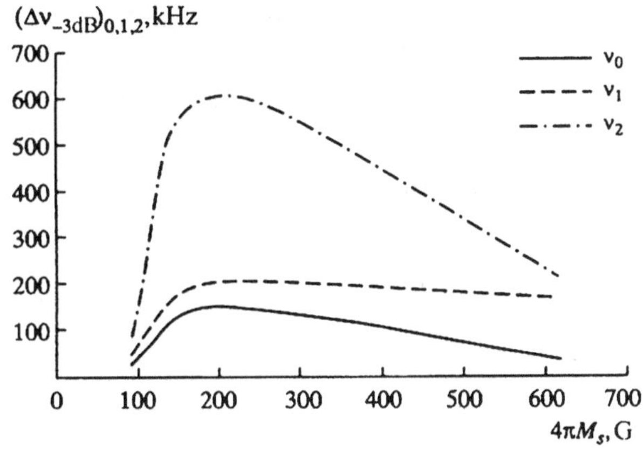

Fig. 4.30 The control over base width of the spectral line $(\Delta\nu_{-60dB})_{0,1,2}$ for the various FMCR saturation magnetization at the orientation angle 45° in a magnetic field and at power level 500 mW

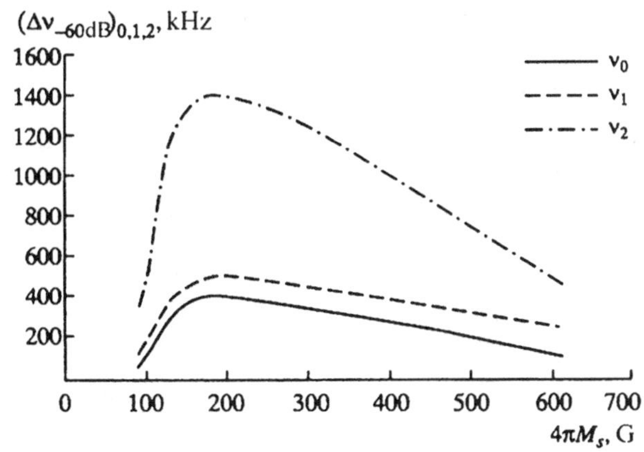

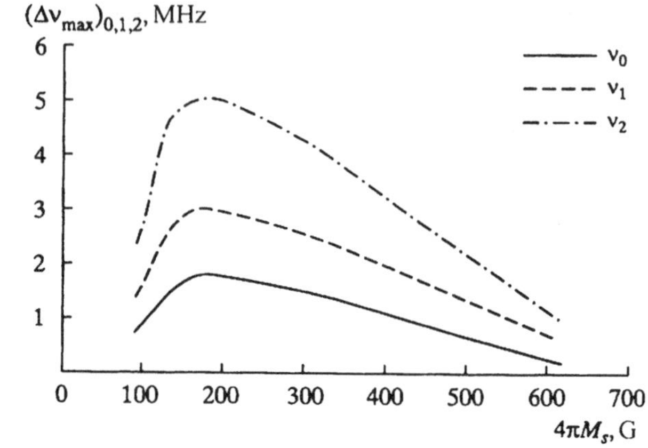

Fig. 4.31 The dependences of the maximum deviations of signal frequencies $(\Delta\nu_{max})_{0,1,2}$ on the FMCR saturation magnetization at the orientation angle 90° in a magnetic field and at power level 50 mW

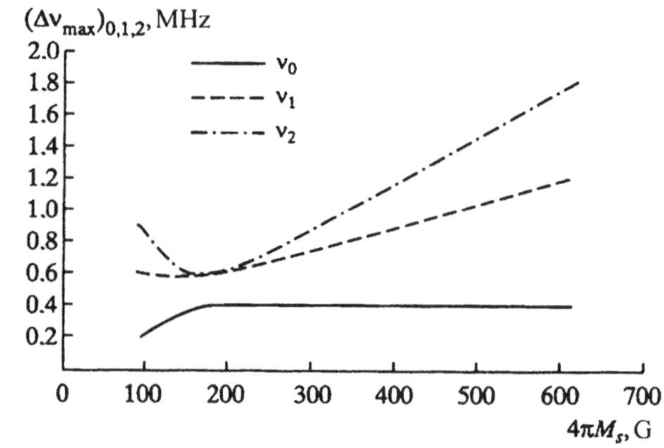

Fig. 4.32 The dependences of the maximum deviations of signal frequencies $(\Delta\nu_{max})_{0,1,2}$ on the FMCR saturation magnetization at the orientation angle 90° in a magnetic field and at power level 500 mW

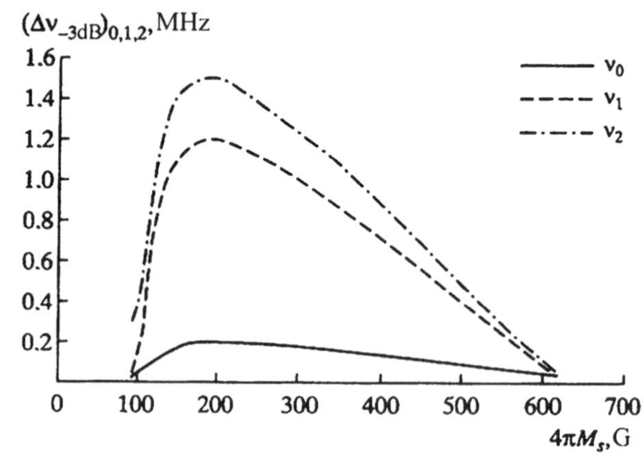

Fig. 4.33 The dependences of the spectral lines width $(\Delta\nu_{-3dB})_{0,1,2}$ on the FMCR saturation magnetization at the orientation angle 90° in a magnetic field and at power level 50 mW

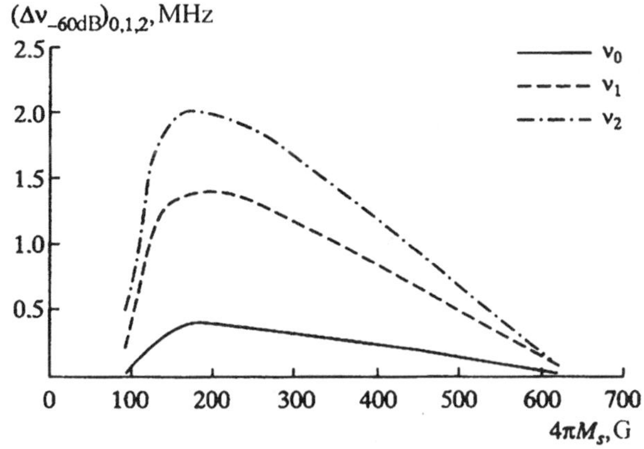

Fig. 4.34 The dependences of the base width of the spectral lines $(\Delta\nu_{-60dB})_{0,1,2}$ on the FMCR saturation magnetization at the orientation angle 90° in a magnetic field and at power level 500 mW

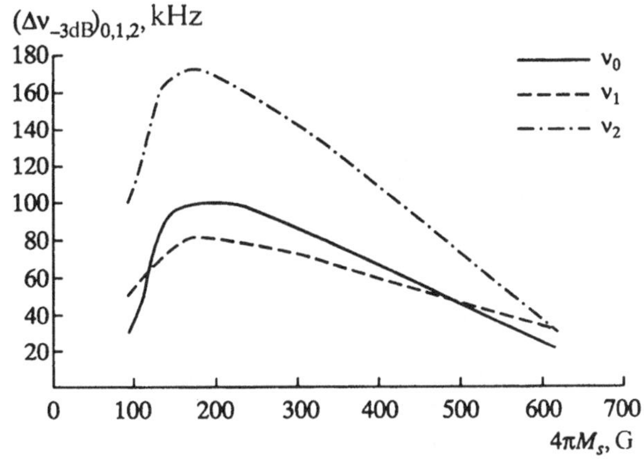

Fig. 4.35 The control over the spectral line width $(\Delta\nu_{-3dB})_{0,1,2}$ for the various FMCR saturation magnetization at the orientation angle 90° in a magnetic field and at power level 500 mW

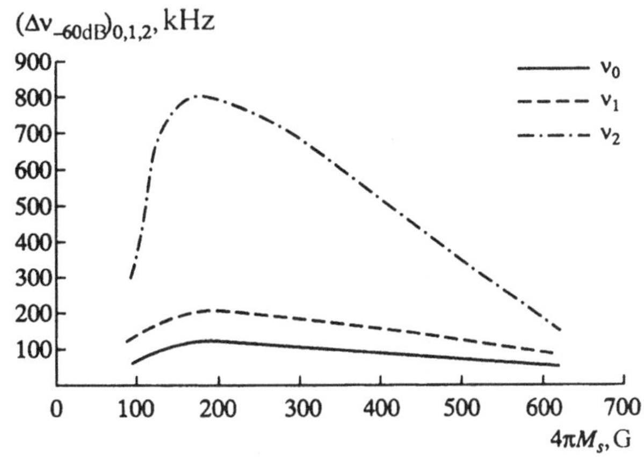

Fig. 4.36 The control over base width of the spectral line $(\Delta\nu_{-60dB})_{0,1,2}$ for the various FMCR saturation magnetization at the orientation angle 90° in a magnetic field and at power level 500 mW

Table 4.1 Mechanisms of heteromagnetic interactions in FTS-generating modes

No.	Kind of signal and mode	Characteristic dependences	Mechanisms of interactions in heteromagnetic structure	Basic elements providing interaction in structure
1	2	3	4	5
1.	Spectrally pure signal of parametric interaction:	(a). The width of a spectral line and its shape do not change for ν_0^{SPS}, ν_{n-1}^{SPS} $(n = 1, 2, 3, \ldots)$ and ν_{-m+1}^{SPS}, $(m = 1, 2, 3, \ldots)$	Parametric processes of multiplication for harmonious components $(\nu_{n-1}, n = 1, 2, 3, \ldots)$ and parametric processes of division for subharmonic components $(\nu_{-m+1}, m = 1, 2, 3, \ldots)$	One-connected system in a circuit with a positive feedback with an oscillatory contour in FMCR domain structure on one of its own frequencies $(\nu_{p1}, \nu_{p2}, \nu_{t1}, \nu_{t2})$ matching with own frequency of fluctuations in the transistor ν_0
	– on basic frequency $\nu_0, n = 1$	$(\Delta\nu_{-3dB})_0^{SPS} = (\Delta\nu_{-3dB})_1^{SPS} =$ $(\Delta\nu_{-3dB})_2^{SPS} = \ldots =$ $(\Delta\nu_{-3dB})_{n-1}^{SPS} = \ldots =$ $(\Delta\nu_{-3dB})_{-1}^{SPS} =$ $(\Delta\nu_{-3dB})_{-2}^{SPS} = \ldots =$ $(\Delta\nu_{-3dB})_{-m+1}^{SPS}$		
	– on harmonious components $\nu_{n-1} = n \cdot \nu_0, n = 1, 2, 3, \ldots$	$(\Delta\nu_{-60dB})_0^{SPS} = (\Delta\nu_{-60dB})_1^{SPS} =$ $(\Delta\nu_{-60dB})_2^{SPS} = \ldots =$ $(\Delta\nu_{-60dB})_{n-1}^{SPS} = (\Delta\nu_{-60dB})_{-1}^{SPS} =$ $(\Delta\nu_{-60dB})_{-2}^{SPS} = \ldots =$ $(\Delta\nu_{-60dB})_{-m+1}^{SPS}$		

(continued)

Table 4.1 (continued)

No.	Kind of signal and mode	Characteristic dependences	Mechanisms of interactions in heteromagnetic structure	Basic elements providing interaction in structure
1	2	3	4	5
	– on subharmonic components $\nu_{m+1} = \nu_0/m, m = 1, 2, 3, \ldots$	(b). The amplitude of spectral components does not change $P(\nu_0) = P(\nu_1) = \ldots = P(\nu_{n-1}) = P(\nu_{-1}) = \ldots = P(\nu_{-m+1}) = \text{const}$, where $n = 1, 2, 3, \ldots$ and $m = 1, 2, 3 \ldots$		
		(c). All spectral components synchronously reconstruct by frequency with a constant steepness at change of magnetic field H_0		
2.	Noisiness of pedestal of spectrally pure signal of parametric interaction near the bearing in the field of Doppler $\pm F_{DOP}$ and intermediate $\pm F_{int}$ frequencies of tuning out	(a). The width of a spectral line and its shape do not change for $\nu_0, \nu_{n-1}, \nu_{-m+1}$	Combination of parametric processes of multiplication and division of signal of the basic frequency ν_1 of spectral comp onents in the field of harmonic ν_{n-1} and subharmonic ν_{-m+1} frequencies with processes of unstable parametric modulation with oscillation frequencies of interdomain borders $\nu_d \cong F_{DOP}$ or $\nu_d \cong F_{int}$ (double parametric processes of multiplication, division and parametrical unstable modulation)	Two-connected system with parametric interactions in a circuit with positive feedback with oscillatory contours in domain FMCR structure on one of own frequencies ($\nu_{p1}, \nu_{p2}, \nu_{t1}, \nu_{t2}$), matching with own oscillation frequency in transistor ν_0 and parametric unstable noise interactions on oscillation frequency of interdomain borders ν_d

(continued)

Table 4.1 (continued)

No.	Kind of signal and mode	Characteristic dependences	Mechanisms of interactions in heteromagnetic structure	Basic elements providing interaction in structure
1	2	3	4	5
		(b). Noisiness of pedestals of all spectral components within the limits of constant tuning out from the bearing one in Doppler $\pm F_{DOP}$ and intermediate $\pm F_{int}$ frequency ranges of tuning out (c). The signal/noise ratio on constant tuning out F_{DOP}, F_{int} from the bearing one(s) does not change with increase in the number of harmonic n or subharmonics m (d). Amplitudes of spectral components do not change $P(\nu_0) = P(\nu_1) = \ldots = P(\nu_{n-1}) = P(\nu_{-1}) = \ldots = P(\nu_{-m+1}) = \text{const}$, where $n = 1, 2, 3, \ldots$ and $m = 1, 2, 3 \ldots$ (e). All spectral components synchronously are reconstructed on frequency with a constant steepness at change of magnetic field H_0		

(continued)

Table 4.1 (continued)

No.	Kind of signal and mode	Characteristic dependences	Mechanisms of interactions in heteromagnetic structure	Basic elements providing interaction in structure
1	2	3	4	5
3.	Modes of multiplication of signal of the basic frequency ν_1 and formation of a spectrum of harmonic components on frequencies $\nu_{n-1}, n = 1, 2, 3, \ldots$	(a). The width of the spectral line with the harmonic number increases $(\Delta\nu_{-3dB})_0 < (\Delta\nu_{-3dB})_1 < \ldots < (\Delta\nu_{-3dB})_{n-1}, n = 1, 2, 3, \ldots)$ and $(\Delta\nu_{-60dB})_0 < (\Delta\nu_{-60dB})_1 < \ldots < (\Delta\nu_{-60dB})_{-m+1}, m = 1, 2, 3, \ldots)$ (b). The amplitude of the harmonic components decreases $P(\nu_0) > P(\nu_1) > \ldots > P(\nu_{n-1}), n = 1, 2, 3, \ldots$ (c). All the spectral components synchronously reconstruct by frequency at change of the magnetic field H_0. The steepness of the frequency deviation $\Delta\nu_{-n+1}$ increases with the number of harmonic n	Nonlinear parametric processes of signal multiplication of basic frequency ν_0	One-coherent system in a circuit with positive feedback with oscillatory contour in FMCR domain structure on one of own frequencies $(\nu_{p1}, \nu_{p2}, \nu_{t1}, \nu_{t2})$ matchingwith own frequency of oscillations in transistor ν_0.

(continued)

Table 4.1 (continued)

No.	Kind of signal and mode	Characteristic dependences	Mechanisms of interactions in heteromagnetic structure	Basic elements providing interaction in structure
1	2	3	4	5
4.	Mode of noisiness of harmonic components of spectral lines on frequencies ν_{n-1}, $n = 1, 2, 3, \ldots$	(a). The width of the spectral line with the harmonic number increases, and $(\Delta\nu_{-3\text{dB}})_{n-1} = n(\Delta\nu_{-3\text{dB}})_0$ $(\Delta\nu_{-60\text{dB}})_{n-1} = n(\Delta\nu_{-60\text{dB}})_0$, $n = 1, 2, 3, \ldots$ (b). The amplitude of harmonic components decreases $P(\nu_0) > P(\nu_1) > \ldots > P(\nu_{n-1})$, $n = 1, 2, 3, \ldots$	Multiplication of signal of the basic frequency ν_0 at frequency modulation of borders of oscillations of domains or nonlinear resonance on one of frequencies of domain structure in the band of own frequencies of the transistor generator	Two-connected system in circuit with positive feedback with oscillatory contour in FMCR domain structure on one of own frequencies ($\nu_{p1}, \nu_{p2}, \nu_{t1}, \nu_{t2}$), matching with own frequency of oscillations in transistor ν_0 and frequency of unstable oscillations of interdomain borders ν_{d} Processes of nonlinear resonance on one of own frequencies of the FMCR domain structure, providing additional (or basic) channel of unstable oscillations on frequencies close to oscillations of interdomain borders ν_{d}

(continued)

Table 4.1 (continued)

No.	Kind of signal and mode	Characteristic dependences	Mechanisms of interactions in heteromagnetic structure	Basic elements providing interaction in structure
1	2	3	4	5
		(c). The signal/noise ratio on constant tuning out from the bearing frequency decreases with increase in the harmonic number n		
		(d). The steepness of frequency deviation of spectral components ν_{n-1} increases with increase in the harmonic number n		
5.	Evenly spaced frequency spectrum of spectrally pure signals in modes of parametric multiplication, division of signal of basic frequency	(a). Spectrally pure components of evenly spaced frequency spectrum in frequency ranges of harmonics and subharmonics have the identical shape and parameters	Parametric processes of multiplication and division of signal of basic frequency ν_0 at parametric frequency modulation by frequency of oscillations of interdomain borders in ferrite	Two-connected system with parametric interactions in circuit with positive feedback with oscillatory contours in FMCR domain structure on one of own frequencies ($\nu_{p1}, \nu_{p2}, \nu_{t1}, \nu_{t2}$), matching with own frequency of oscillations in transistor ν_0 and parametric frequency modulation on frequency of oscillations of domain borders ν_d
		(b). The amplitude of spectral components in a group of evenly spaced frequency grid decreases from one component to another by $-10\,dB$		

(continued)

Table 4.1 (continued)

No.	Kind of signal and mode	Characteristic dependences	Mechanisms of interactions in heteromagnetic structure	Basic elements providing interaction in structure
1	2	3	4	5
		(c). All the components of evenly spaced frequency spectrum synchronously reconstruct by frequency with a constant steepness at change of magnetic field H_0 (d). Synchronous discrete control over evenly spaced (frequency distances) spectral groups at change of field bias H_0 and the HF power level in the system		
6.	Evenly spaced frequency spectrum with a various noise level on the frequency spectrum envelope	(a). The width of spectral lines does not change for ν_0, ν_{n-1} and ν_{-m+1}, $n = 1, 2, 3, \ldots$, $m = 1, 2, 3, \ldots$	Parametric processes of multiplication and division of signal on basic frequency ν_1 at parametric unstable frequency modulation with frequency of oscillations of interdomain borders in ferrite ν_d	Two-connected system with parametric interactions in circuit with positive feedback with oscillatory contours in FMCR domain structure on one of own frequencies (ν_{p1}, ν_{p2}, ν_{t1}, ν_{t2}), matching with own frequency of oscillations in transistor ν_0 and parametric frequency modulation on frequency of oscillations of domain borders ν_d

(continued)

Table 4.1 (continued)

No.	Kind of signal and mode	Characteristic dependences	Mechanisms of interactions in heteromagnetic structure	Basic elements providing interaction in structure
1	2	3	4	5
		(b). Synchronous noisiness of pedestals in the field of each spectral component of frequency grid with control over noisiness level and frequency tuning out		
		(c). Synchronous reorganization by frequency of all groups of evenly spaced frequency spectrum with a constant steepness at change of magnetic field H_0		
7.	White noise in a multioctave frequency range	(a). Uniform spectral density of noise power in a multioctave frequency range	Mode of collapse of magnetic behaviors of ferromagnetic subsystem of FMCR at certain amount of magnetic field H_0	Multicoheret FMCR system in mode of collapse of magnetic properties at intensive noise (thermal) modulation of own frequencies of domains
			Off-orientation of domains participating in interaction with transistor's HF magnetic fields. Thermal intensive modulation of own frequencies of domains	

4.3 Physical Mechanisms of Heteromagnetic Interactions

The mechanisms of generator multipurpose heteromagnetic interactions in FTS in an expanded dynamic power range have been experimentally investigated.

Main attention is given to studying typical laws at formation of spectra of various kinds of signals, namely, SPS, NS, ES FS on the basic and harmonious frequency components $\nu_{0,1,2}$–$(\Delta\nu_{-3\text{dB}})_{0,1,2}$ and $(\Delta\nu_{-60\text{dB}})_{0,1,2}$, basic characteristics of control over their parameters (the deviation of frequencies $(\Delta\nu_{\max})_{0,1,2}$, the width and shape of spectral lines $(\Delta\nu_{-3\text{dB}})_{0,1,2}$ and $(\Delta\nu_{-60\text{dB}})_{0,1,2}$ at changes of the values of FMCR magnetization and its orientation in a magnetic field.

In Table 4.1 the basic mechanisms of heteromagnetic interactions in the investigated structures are resulted.

Part II
Process Modeling in Heteromagnetic Structures

Techniques and results of our modeling of the parameters of equivalent circuits of HMG of low and high power levels are considered with one- and multiplanimetric models for generation modes of spectrally pure, harmonic, and subharmonic signal components, signals as evenly spaced frequency spectra and noise-type signals as examples.

Chapter 5
Heteromagnetic Oscillator

Single-Circuit Models

5.1 Equivalent Circuit of a High-Power Bipolar Transistor

The equivalent circuit of Gummel–Poon's model of the bipolar UHF transistor is presented in Fig. 5.1, where the basic notation [39–41] is retained.

As opposed to the classical models by Ebers–Moll and Linvill, this model allows one to consider:

- Reduction of the transfer factor and the boundary frequency at high values of the collector current.
- Final target conductivity of the transistor and its dependence on the current of the base.
- Voltage dependence of the barrier transition capacities.
- Influence of parasitic capacities and the conductions of the base and collector on the transmission factors.

This model provides a sufficiently high accuracy of the description of both static and dynamic characteristics of bipolar UHF transistors in the mode of big signals. A generalized charge of the transistor is used in it, which allows the current transferred from the emitter to the collector through p–n junction voltages to be expressed.

The basic equation for the collector current is

$$I_{cc} = \text{const}\frac{\exp(qV_{eb}/kT) - \exp(qV_{cb}/kT)}{Q_b}, \tag{5.1}$$

where V_{eb} is the emitter–base junction voltage, V_{cb} the collector–base junction voltage, Q_b the full charge accumulated in the base of the transistor, q the electron charge, k Boltzmann's constant, and T is the absolute transistor temperature.

Equation (5.1) is derived in [41] by exact integration of the equations of charge transfer from the emitter to collector with effects of carrier recombination in the transistor neglected. At preset voltages V_{eb} and V_{cb}, the exponents in (5.1) can be calculated exactly. The integral charge in the base is a function of power sources. The collector current I_c of the transistor (Fig. 5.2) consists from three components, namely,

$$I_c = I_{cc} - I_{bc} + I_A. \tag{5.2}$$

A.A. Ignatiev and A.V. Lyashenko, *Heteromagnetic Microelectronics: Microsystems of Active Type*, DOI 10.1007/978-1-4419-6002-3_5,

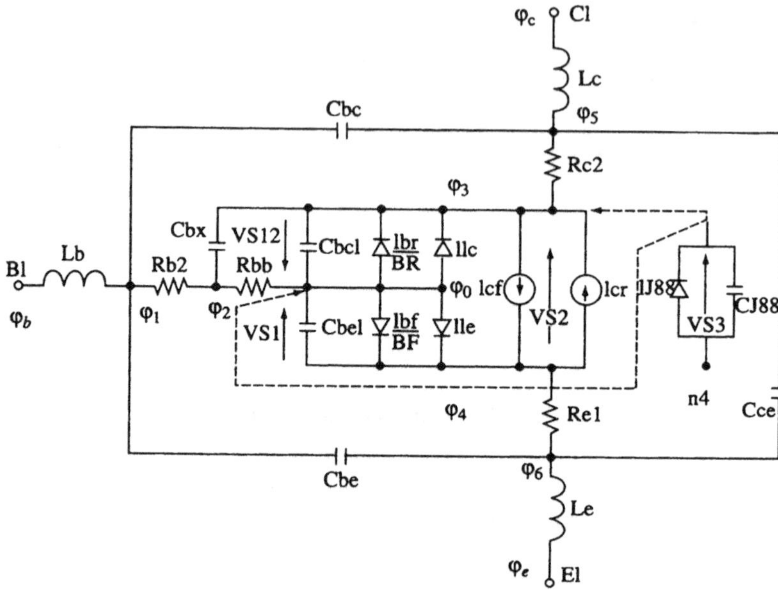

Fig. 5.1 The equivalent circuit of Gummel–Poon's model of the bipolar UHF transistor

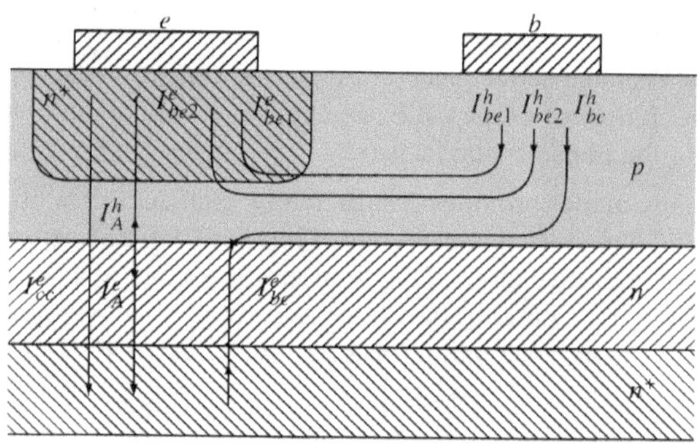

Fig. 5.2 The collector current I_c of the transistor

The total charge in the base of the transistor is

$$Q_b = Q_{b0} + Q_e + Q_c + B \cdot \tau_f I_f + \tau_r I_r, \tag{5.3}$$

where Q_{b0} is the balanced full charge in the base, Q_e the capacity of the emitter junction, Q_c the capacity of the collector junction, $B \cdot \tau_f I_f$ the direct diffusion capacity of the accumulated charge, I_f the direct current of the emitter–base junction; $\tau_r I_r$ the return diffusion capacity of the accumulated charge; I_r the reverse current of the collector–base junction; τ_f the direct, and τ_r the return time of transfer; the parameter B describes the effect of dislodgment of the base. It monotonously increases with increasing $|I_e|$ and decreases with increasing $|V_{eb}|$. The direct current of the emitter–base junction is

$$I_f = \text{const}\frac{(\exp(V_{eb}/kT) - 1)}{Q_b}. \tag{5.4}$$

The return current of the collector–base junction is

$$I_r = \text{const}\frac{(\exp(qV_{cb}/kT) - 1)}{Q_b}. \tag{5.5}$$

The total recombination rate h in the given model is introduced phenomenologically and consists of three components:

$$h = I_{be1} + I_{be2} + I_{bc}, \tag{5.6}$$

where I_{be1} is the current of the holes from the base area to inject into the emitter one; I_{be2} the current of the holes from the base area to recombine in the emitter junction; and I_{bc} is the current of the holes from the base area to recombine in the collector junction.

The value of the current of holes is

$$I_{be1} = \text{const}(\exp(qV_{cb}/kT) - 1). \tag{5.7}$$

The dependence I_{be2} has the following form

$$I_{be2} = \text{const}(\exp(qV_{eb}/n_e kT) - 1), \tag{5.8}$$

where n_e is a constant, $1 < n_e < 2$.

The dependence I_{bc} has the form

$$I_{bc} = \text{const}(\exp(qI_{cb}/n_c kT) - 1). \tag{5.9}$$

The holes generated in the field of the emitter–base junction due to the avalanche process pass in the collector and base areas with the recombination rate

$$h = I_b + I_A, \tag{5.10}$$

the current on the transistor base being

$$I_b = I_{b1} + I_{be2} + I_{bc} - I_A. \tag{5.11}$$

The equivalent parameters of Gummel–Poon's model are resulted in Table 5.1. As typical, the parameters used as the first approximation of the transistor's model are chosen. Some parameters insignificant at the given investigation phase are put equal to zero.

Table 5.1 The equivalent parameters of Gummel–Poon's model

Designation	Parameter	Unit	Typical value
IS	The cutoff current	A	1×10^{-16}
ISE	Saturation base–emitter leakage current	A	0.0
ISC	Saturation base–collector leakage current	A	0.0
BF	Ideal direct transfer factor by current	–	100
BR	Ideal return transfer factor by current	–	1.0
NF	Current emission factor	–	1.0
NE	Base–emitter emission leakage factor	–	1.5
NR	Return emission factor by current	–	1.0
NC	Base–collector emission factor	–	2.0
VA	Initial direct displacement voltage	B	0.0
VB	Initial return displacement voltage	B	0.0
RBM	Minimum base resistance	OM	0.0
RB	Maximal base resistance in the common base circuit	OM	0.0
IRB	Current at which the base resistance decreases by 2	A	0.0
TF	Ideal direct transfer time	C	0.0
TR	Ideal return transfer time	C	0.0
VJE	Initial base–emitter potential	B	0.75
VJC	Initial base–collector potential	B	0.75
CJC	Initial base–collector capacity	F	0.0
XCJC	Base–collector capacity fraction (factor) connected to the internal base resistance	F	1.0
ISS	Saturation leakage current from the substrate	A	0.0
NS	Substrate emission factor	–	1.0
CJS	Initial base–collector capacity	F	0.0
VJS	Initial substrate potential	B	0.75
KF	Flickering noise factor	–	0.0
AF	Flickering noise exponent	–	1.0
$V_t = k\mathrm{TJ}/q$	Temperature voltage	J/C	To be calculated
k	Boltzmann's constant	J/C	1.38062×10^{-23}
q	Electron charge	C	$1.6021892 \times 10^{-19}$
TJ	Absolute temperature of the transistor structure	K	293.15

For modeling of p–n junctions four diodes connected pairwise-parallel are used. To consider the influence of p–n junctions on each other, the equivalent circuit includes two current generators between the emitter and the collector (one models the direct current from the emitter through the base to the collector, the other one does the return current from the collector through the base to the emitter). To take account of the diffusion and threshold capacities of p–n junctions, capacities are put parallel to the diodes, and to provide for the volume resistance of the crystal the corresponding resistances are introduced to the transistor's terminals. These parameters describe the internal model of the transistor. The external model incorporates resistances, inductances, and capacities of the terminals due to the beam-boiled electrodes, surface effects on the transistor crystal, influence of the box.

Base width modulation is considered by introduction of a dependence of generator currents on the p–n junction voltage and diode currents. The main advantage of

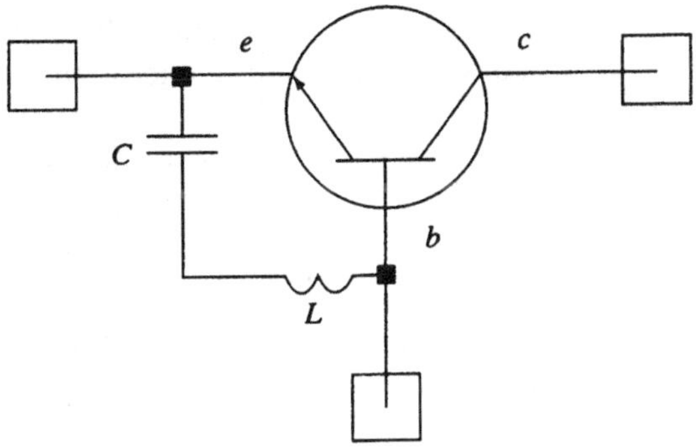

Fig. 5.3 A built-in matching LC-chain in the base–emitter circuit

this model over various low-level signal or high-frequency two-port network models is that it precisely enough reflects the physical processes in the powerful transistor and is independent of the way the transistor placed in the circuit. Besides, the equations of the model do not depend on the operating mode (cutoff, active mode, saturation). It is especially important for development of powerful bipolar HMTs in multifunctional modes.

To increase the input resistance, KT962A transistors have a built-in matching *LC*-chain in the base–emitter circuit (Fig. 5.3). Its parameters are taken from the transistor's specification.

In view of the parasitic resistances of diffusion areas Gummel–Poon's model contains 26 independent parameters falling into groups by the way of their experimental determination. For determination of the parameters of the model it is necessary to have some characteristics of the transistor in both static and dynamic modes. By analyzing static input and output characteristics it is possible to estimate the following most important parameters of the model: IS, ISE, ISC, BF, BR, NF, NE, NR, NC, VA, VB.

The most significant currents in the model are the currents of generators I_{cf} and I_{cr} which, in turn, depend on the currents through the p–n junctions from the base to the emitter and collector. The base current I_{b} in a static mode develops the currents of four diodes: I_{bf}/BF, I_{br}/BF, I_{le}, I_{lc}. The diodes with the currents I_{bf}/BF and I_{br}/BF are principal as the values of I_{bf} and I_{br} control the currents of generators I_{cf} and I_{cr}:

$$I_{\mathrm{cf}} = \frac{I_{\mathrm{bf}}}{K_{\mathrm{qb}}},$$

$$I_{\mathrm{cr}} = \frac{I_{\mathrm{br}}}{K_{\mathrm{qb}}}, \tag{5.12}$$

where K_{qb} is a factor considering base width modulation and depending on the diode voltages and currents I_{bf} and I_{br}.

The diodes with currents I_{le} and I_{lc} are additional ones. They set a fraction of the current from the base through the p–n junction so that it does not influence the current of the corresponding generator. It is necessary under high voltages, when the current transferred from the emitter to the collector exceeds its experimentally obtained value.

The current through the p–n junction is approximated by the dependence

$$I = I_0 \left[\exp\left(\frac{U}{m \times \varphi_t} \right) - 1 \right], \tag{5.13}$$

where I_0 is the saturation current of the p–n junction at its inclusion in the opposite direction, U the p–n junction voltage, $\varphi_t = kT/q$ the thermal potential ($\varphi_t \approx 0.026\,\mathrm{V}$), m is a constant to be chosen for agreement with experiment.

The currents of diodes I_{bf}, I_{br}, I_{le}, and I_{lc} are possible to set for a given working point by the following parameters: $I_{bf} = I_{bf}$ (IS, NF), $I_{br} = I_{br}$ (ISE, NF), $I_{le} = I_{le}$ (ISC, NC). If the second parameters (NF, NE, NC) are too small, the current through the corresponding diode will grow more sharply at increase of the p–n junction voltage, *i.e.*, the second parameter mainly influences the VAC steepness (at a constant voltage on the diode). In our case, it is not always true because at change of the parameter the voltage on the diode more often changes as well. Thus for achievement of the desirable result, it is necessary to change at once several parameters, not one.

Except for these characteristics, the volume resistance of the base, emitter, and collector (they strongly influence the voltage distribution on p–n junctions, and, accordingly, on the currents of diodes and generators) are important.

At analysis of the work of the transistor in static modes (input and target characteristics, VAC p–n junctions, etc.) it is possible to determine the parameters of all the diodes and generators.

The dynamic characteristics in the model are determined by the capacities of p–n junctions and their dependences on the electric mode, and the influence of parasitic capacities between the transistor's terminals and the case.

As no direct measurement of the equivalent parameters of the used model was possible, their determination was carried out by optimization computation and joining with experimental results. At such an approach the initial information are the results of measurements carried out by accepted techniques on the standard measuring equipment. On the basis of the obtained information by means of our developed programs of optimization such values of the parameters of the model which provide the best agreement between measurement results and modeling results were determined. The research was made on various transistor structures. The error function was defined by

$$\xi = \sum \left[\frac{(V_{bi} - V_{bmi})}{V_{bmi}} \right]^2 + \left[\frac{(I_{ci} - I_{cmi})}{I_{cmi}} \right]^2, \tag{5.14}$$

where V_{cmi} and I_{cmi} are the measured values of the collector voltage and current, and V_{bi} and I_{bi} are the calculated values of the collector voltage and current.

By means of a specially developed program the total error was minimized and it is possible to determine the parameters IS, ISE , ISC, ISS, NS, NF, NE, NR, NC, R_{b2}, R_{c2}, R_{e1}, R_{bb}, BF, BR, VA, VB, ICR, ICF, RB, RBM.

As the number of the unknown parameters is great and the degree of the influence of these parameters on the total error is various, the parameters should be divided into three groups, namely:

1. IS, ISE, ISC, ISS, NS, NF, NE, NR, NC
2. $R_{b2}, R_{c2}, R_{e1}, R_{bb}$
3. BF, BR, VA, VB, ICR, ICF, RB, RBM

At the first step those parameters of the model were determined, which strongly influence the input and output characteristics of the transistor. At the second step, the parameters over all the groups were refined.

A similar approach for a more simplified model of the transistor with charge control was used in [41]. For calculation of the static characteristics the set of nonlinear algebraic equations derived from (5.8 to 5.25) under the condition of equality to zero of all the derivatives with respect to time was solved.

The starting point close enough to the modeled curve was by an iterative method. In the space of parameters, an area to be separated into intervals was allocated. The interval-containing points close to the sought solution was found. Further the found area was separated into intervals again, and the area containing points closer to the solution was sought. This cycle was repeated until the interval found at the current step reduced to the preset value. The method is unsuitable for finding all points on the curve because of its poor convergence, and the errors arising in curve condensation areas in the parameter space or the big curvature of these curves. Therefore, the algorithm described above was used for finding an initial approximation only. Further, an algorithm of movement along the solution curve with the use of Newton's method and the forecast of new points from those found at the previous step was used.

The dynamic parameters of a transistor can be taken from the reference data of either the transistor of the given type or its foreign analog.

5.2 Modeling of Static Characteristics of a Powerful Bipolar Transistor

As an example, the results of our modeling of the KT962A transistors are presented. The static characteristics of this transistor were investigated as shown in Fig. 5.4.

The currents of emitter I_e and collector I_c were measured by ammeters, the voltages V_{eb}, V_{cb} by voltmeters. Resistor R was joined in the collector circuit for restriction of the current through the p–n junctions, its resistance to be selected for each specific measurement ($2\,\Omega$–$2\,k\Omega$).

At measurements of the input and target characteristics, it is necessary to maintain constant values of the voltage V_{cb}, and current I_e, respectively. The results of our measurements of the input and output characteristics are presented in Tables 5.2 and 5.3, respectively. The plots of the dependences obtained from (5.15) to (5.28) are shown in Figs. 5.5 and 5.6.

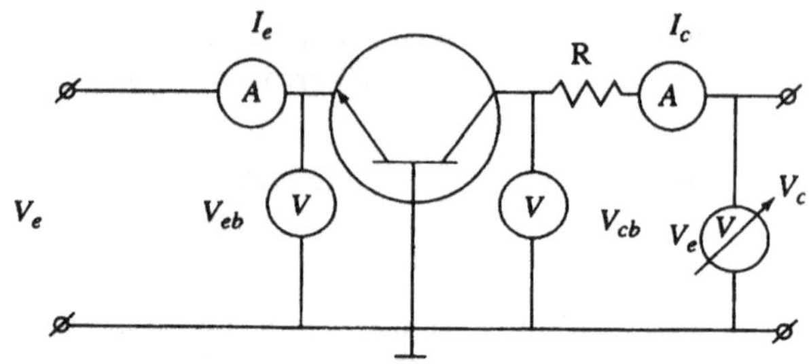

Fig. 5.4 The circuit for investigation of characteristics of the bipolar transistor

Table 5.2 The results of measurements of the input characteristics

$V_{bc} = 0$ V			$V_{bc} = 5$ V			$V_{bc} = 10$ V		
I_e, mA	V_{be}, V	I_c, mA	I_e, mA	V_{be}, V	I_c, mA	I_e, mA	V_{be}, V	I_c, mA
0	0.378	0	0	0.363	0	0	0.339	0
0	0.407	0	0	0.457	0	0	0.424	0
0	0.493	0	2.0	0.588	0	2.4	0.611	0
2.4	0.64	0	6.4	0.672	0	4.7	0.661	0
4.7	0.701	0	27.0	0.715	0	28.0	0.716	0
20.0	0.748	0	36.0	0.722	30.0	90.0	0.781	60.0
41.0	0.776	0	100.0	0.751	90.0	–	–	–
100.0	0.812	0	–	–	–	–	–	–

The parameters of the model were determined by the LSQ method (the technique described above). The value of the total error by (5.14) was $\xi = 0.003$, that corresponded to the average deviations of currents and voltages from their measured values no more than 5.4%.

5.3 Basic Model Equations

The equations of the model of a powerful UHF transistor used as the basic element for a generator with heteromagnetic interactions are resulted below.

$$I_b = \frac{I_{bf}}{BF} + I_{le} + \frac{I_{br}}{BR} - I_{lc}, \tag{5.15}$$

$$I_c = \frac{I_{ef}}{K_{qb}} - \frac{I_{br}}{K_{qb}} - \frac{I_{br}}{BR} - I_{lc}, \tag{5.16}$$

$$I_{bf} = IS\left[\left(\frac{\exp(V_{s1})}{NE \times V_t}\right) - 1\right], \tag{5.17}$$

$$I_{le} = ISE\left[\left(\frac{\exp(V_{s1})}{NE \times V_t}\right) - 1\right], \tag{5.18}$$

$$I_{br} = IS\left[\left(\frac{\exp(V_{s12})}{NR \times V_t}\right) - 1\right], \tag{5.19}$$

Table 5.3 The results of measurements of the output characteristics

$I_e = 0\,\text{mA}$			$I_e = 0.5\,\text{mA}$			$I_e = 1\,\text{mA}$		
V_{be}	I_c	V_{bc}	V_{be}	I_c	V_{bc}	V_{be}	I_c	V_{bc}
−0.003	0	0.003	0.642	0.11	−0.61	0.663	0.11	−0.64
−0.17	0.09	2.55	0.624	0.45	0.13	0.652	0.59	−0.61
−0.304	0.15	4.93	0.616	0.47	3.07	0.642	0.90	0.61
−0.43	0.22	7.72	0.609	0.48	6.11	0.638	0.91	3.41
−0.532	0.27	10.28	0.601	0.5	9.02	0.635	0.92	6.4
−0.634	0.32	13.15	0.592	0.51	11.94	0.633	0.93	9.41
−0.717	0.36	15.65	0.581	0.52	14.80	0.63	0.94	12.31
−0.797	0.41	18.19	0.567	0.54	17.40	0.627	0.95	15.35
−0.875	0.45	20.90	0.541	0.56	20.4	0.625	0.96	18.00
−0.951	0.51	23.60	0.494	0.59	23.10	0.622	0.97	20.80
−1.018	0.55	26.00	0.435	0.64	25.50	0.619	0.98	23.50
$I_e = 5\,\text{mA}$			$I_e = 10\,\text{mA}$			$I_e = 500\,\text{mA}$		
0.714	0.12	−0.69	0.736	0.12	−0.71	0.929	0.72	−0.86
0.712	0.61	−0.69	0.735	0.59	−0.71	0.929	3.30	−0.86
0.709	1.15	−0.68	0.733	1.15	−0.71	0.929	6.30	−0.86
0.706	1.70	−0.68	0.732	1.70	−0.71	0.929	9.10	−0.85
0.703	2.10	−0.67	0.730	2.20	−0.71	1.233	51.00	0.50
0.699	2.70	−0.66	0.729	2.70	−0.70	1.234	120.0	3.16
0.695	3.10	−0.65	0.727	3.20	−0.7	1.235	190.0	6.67
0.691	3.60	−0.64	0.726	3.60	−0.7	1.237	250.0	10.22
0.688	4.10	−0.62	0.724	4.10	−0.69	1.239	300.0	14.17
0.685	4.40	0.13	0.722	4.60	−0.69	1.242	410.0	21.11
0.684	4.50	2.71	0.721	5.20	−0.69	1.244	480.0	26.30
0.682	4.50	9.50	0.705	8.50	−0.63	–	–	–
0.680	4.50	15.00	0.702	9.10	2.02	–	–	–
0.679	4.55	19.30	0.702	9.10	4.84	–	–	–
0.678	4.60	24.60	0.701	9.20	7.83	–	–	–
–	–	–	0.700	9.20	10.69	–	–	–
–	–	–	0.699	9.20	13.70	–	–	–
–	–	–	0.698	9.20	16.46	–	–	–
–	–	–	0.697	9.20	19.43	–	–	–

$$I_{lc} = \text{ISC}\left[\left(\frac{\exp(V_{s12})}{\text{NC} \times V_t}\right) - 1\right], \tag{5.20}$$

$$I_{cf} = \frac{I_{bf}}{K_{qb}}, \tag{5.21}$$

$$I_{cr} = \frac{I_{br}}{K_{qb}}, \tag{5.22}$$

$$I_{jss} = I_{ss}\left[\left(\frac{\exp(V_{s3})}{\text{NS} \times V_t}\right) - 1\right], \tag{5.23}$$

$$K_{qb} = \frac{K_{q1}}{2}(1 + \sqrt{1 + 4K_{q2}}), \tag{5.24}$$

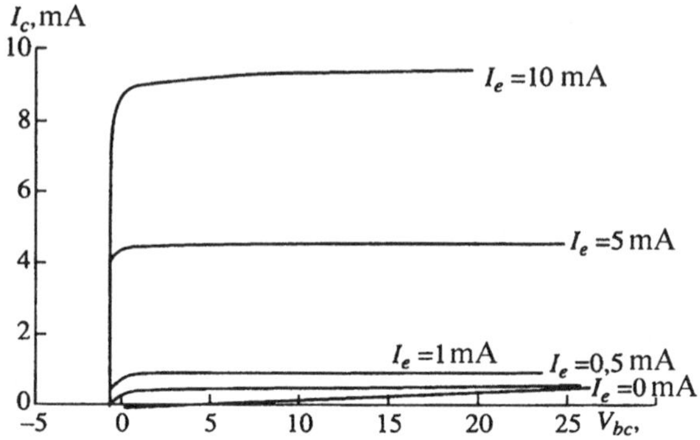

Fig. 5.5 The plots of the dependences obtained from (5.15) to (5.28)

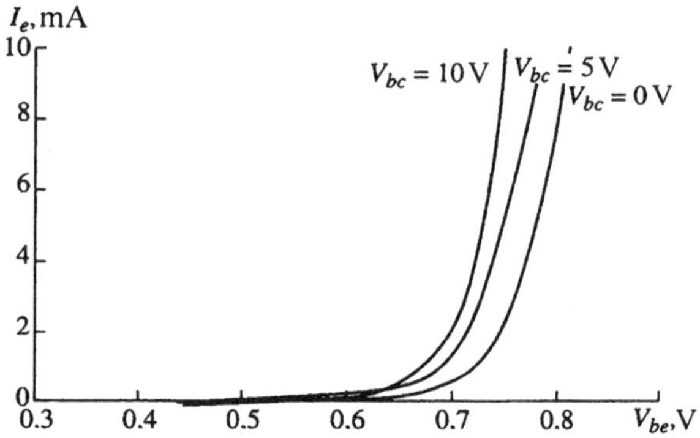

Fig. 5.6 The plots of the dependences obtained from (5.15) to (5.28)

$$K_{\mathrm{q1}} = \frac{1}{1 - \dfrac{V_{\mathrm{s12}}}{\mathrm{VA}} - \dfrac{V_{\mathrm{s1}}}{\mathrm{VB}}}, \tag{5.25}$$

$$K_{\mathrm{q2}} = \frac{I_{\mathrm{bf}}}{\mathrm{IKF}} + \frac{I_{\mathrm{br}}}{\mathrm{IKR}}, \tag{5.26}$$

$$R_{\mathrm{bb}} = \begin{cases} \mathrm{RBM} - \dfrac{\mathrm{RB} - \mathrm{RBM}}{K_{\mathrm{qb}}}, & \mathrm{IRB} = \infty, \\ \mathrm{RBM} - 3(\mathrm{RB} - \mathrm{RBM})\dfrac{\tan(x) - x}{x\tan^2(x)}, & \mathrm{IRB} > 0, \end{cases} \tag{5.27}$$

where

$$x = \frac{\sqrt{1 + \dfrac{144 \times I_{\mathrm{b}}}{\mathrm{IRB} \times \pi^2}}}{\dfrac{24}{\pi^2}\sqrt{\dfrac{I_{\mathrm{b}}}{\mathrm{IRB}}}};$$

V_{s1} is the internal base–emitter voltage; V_{s2} the internal base–collector voltage; $V_t = k\mathrm{TJ}/q$ (temperature voltage); k the Boltzmann's constant, $k = 1.380662 \times 10^{-23}\,\mathrm{J/K}$; q the electron charge; TJ is the absolute temperature of the transistor structure, $\mathrm{TJ} = 293.15\,\mathrm{K}$.

5.4 Calculation of Characteristics of Powerful Heteromagnetic Microwave Oscillators

The investigated generator will be described by the equivalent circuit shown in Fig. 5.7.

C_e, C_b, L_b are frequency driving elements. Resistance R_0 limits the displacement current of the base–emitter junction and, together with the power sources for the collector (E_c) and for base displacement, determine the mode of the transistor on direct current. Resistance R_0 is shunted on high frequencies by capacitor $C_0 \sim 1{,}000\,\text{pF}$, R_e, R_{ec} being the internal resistance of sources E_b and E_c Elements $R'_e, L'_e, R'_b, L'_b, R'_c, L'_c$ correspond to the conductors of terminals and assembly of the transistor, where C_{cb} is the parasitic capacity and Z is the load resistance. The element of coordination of the low target resistance of the transistor with the load is executed in the form of a P-shaped filter formed by elements C_f, L_f, C'_f, which are determined by the FMCR. Capacities C_e, C_b are formed by the contact platforms of the emitter and base terminals of the transistor. Inductance L_e is formed by a piece of the asymmetrical microstrip line.

The equivalent circuit describes the work of all the investigated HMT models. A number of models could be described by simpler equivalent circuits derived from that shown above.

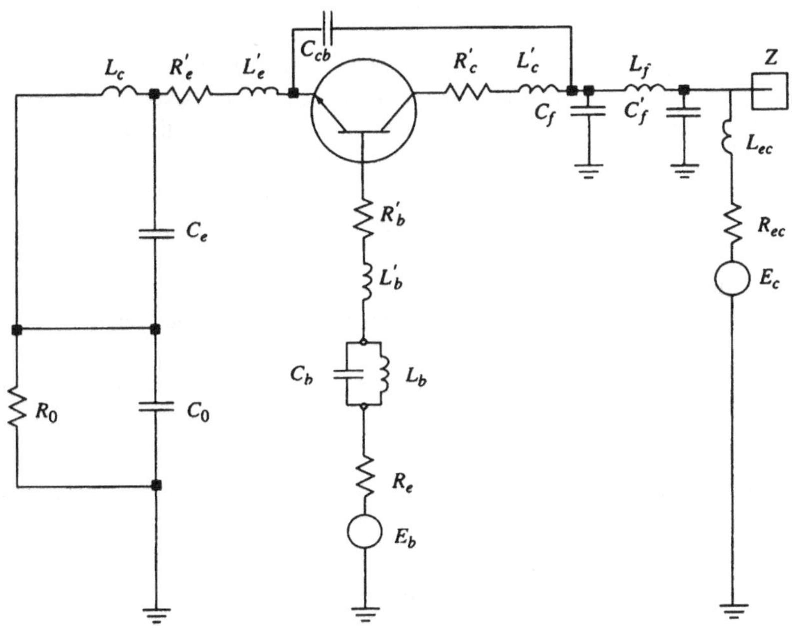

Fig. 5.7 The equivalent circuit of the investigated generator

Table 5.4 The parameters of the equivalent circuit of the investigated HMG

Element	L_b	C_b	L'_b	R'_b	C_0	L_{ec}	
Value	7.5 nH	16 pF	0.01 nH	0.01 Ω	10 nF	100 nH	
Element	E_b	Z	C_f	C'_f	L_e	C_e	
Value	6 V	50 Ω	0	0	7.5 nH	15.01 pF	
Element	L'_e	R'_e	R'_c	R_{ec}	E_c	L'_c	L_f
Value	0.03 nH	0.01 Ω	0.01 Ω	0.01 Ω	27 V	0.01 nH	0

The basic characteristics of the generator are: the frequency of generated oscillations – $\nu_{0,1,2}$, target integral power – $(P_{out})_{0,1,2}$, efficiency factor, spectral line width – $(\Delta\nu_{-3dB})_{0,1,2}$, and frequency change – $\Delta\nu_{0,1,2}/\Delta H$ and $\nu_{0,1,2}/\Delta V_{e,c}$. These characteristics depending on the feed voltages V_e and V_c and the parameters of elements of the equivalent HMT circuit in regular modes was studied by harmonic balance. The solution was sought as $\sum_{n=1}^{N} A_n e^{i\omega_n t}$, where A_n is the amplitude, $n = 0, 1, 2, \ldots$. By varying the number of harmonics N it is possible to achieve a preset accuracy of calculation. In practice, $N = 5$ was chosen, the accuracy of calculation lied within 10^{-3}–10^{-7}.

In a frequency band of 400–1,120 MHz, the output power of HMT oscillations was within 0.2–15.0 W, the efficiency factor is ≈20–50%. To change the frequency of HMG the parameters of the oscillatory systems in the emitter and base simultaneously changed.

In Table 5.4, the parameters of the equivalent circuit of the investigated HMG are shown.

The frequency of HMG generation was 450 MHz. The oscillatory system in the emitter circuit had a resonant frequency of 459 MHz, in the base circuit 474 MHz.

By fine tuning the elements of the oscillatory systems in the base and emitter circuits, the HMG frequency was 400–1,120 MHz. By selection of elements of the target matching filter and loading resistance, one could achieve a level of generated power within the limits of 7–15 W at an efficiency factor of 40–50%. Similar results have been obtained experimentally.

HMGs on powerful KT962 transistors (3–10 elementary transistors) were experimentally investigated. Each elementary transistor was described on the basis of the used model.

The equivalent circuit for the structure consisting of three elementary transistors is presented in Fig. 5.8.

Experimentally was investigated:

- Influence of the number of elementary transistors included in parallel in the structure on the generated frequency and power.
- Conditions under which it is possible to replace a complex (several transistors) structure by a model for one elementary transistor.

The terminals of the elementary transistors were experimentally disconnected from the structure. The equivalent circuit was numerically investigated. The resistance of the Z-loading was 75 and 50 Ω. The other parameters of the circuit are taken from Table 5.4.

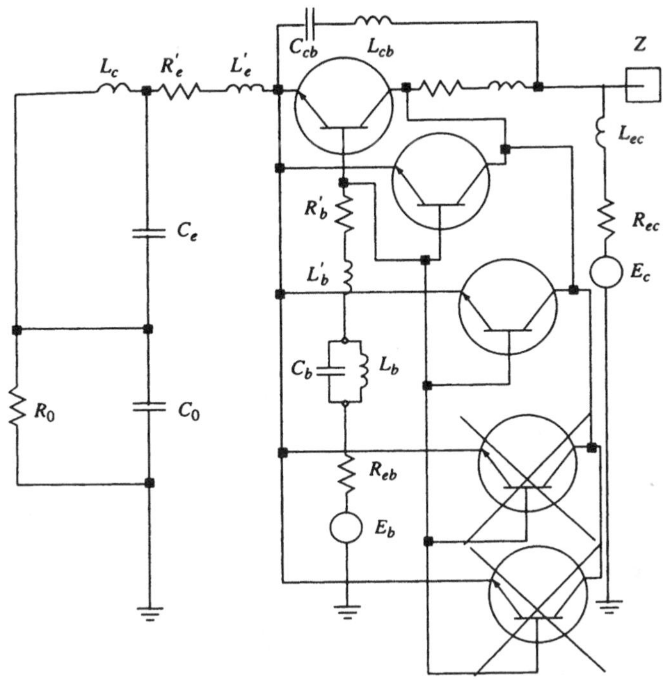

Fig. 5.8 The equivalent circuit for the structure consisting of three elementary transistors

At a load impedance of $Z = 75\,\Omega$, generation arose on structures made from 1, 2, and 3 elementary transistors (Fig. 5.8). At connection of a fourth transistor in the circuit, generation was killed. The frequency ν_0 of the generated oscillations changed slightly: for one transistor by 444,496 MHz; for two transistors by 443,702 MHz; for three transistors by 445,877 MHz. The failure of oscillations resulted from the change of the operating mode of the transistors by direct current and mismatches due to reduction of the target resistance of the structure of transistors.

At reduction of the load resistance down to $Z = 50\,\Omega$, generation renewed on a frequency of 435,618 MHz with an essential increase in the target power up to 7–10 W. For a load of $Z = 50\,\Omega$, a mode with two elementary transistors was explored. The target power has decreased slightly. The frequency of generation was 433,995 MHz. At working of one transistor with $Z = 50\,\Omega$ the frequency of generation was 435,735 MHz. The target power has decreased down to 5–7 W. Similar results have been obtained experimentally.

The results of our calculations of time realization of oscillations for the structure containing one and four transistors are presented in Figs. 5.9 and 5.10.

In Fig. 5.11 the time realization of oscillations of the generator is presented at working with a load $Z = 50\,\Omega$. The structure consisted of five identical elementary transistors. The target power $P_{out} \approx 7$ W at an efficiency factor of 40%.

At our experimental research of HMG, modes with slowly falling down amplitudes of harmonics have been obtained. In some cases, the reduction of the intensity

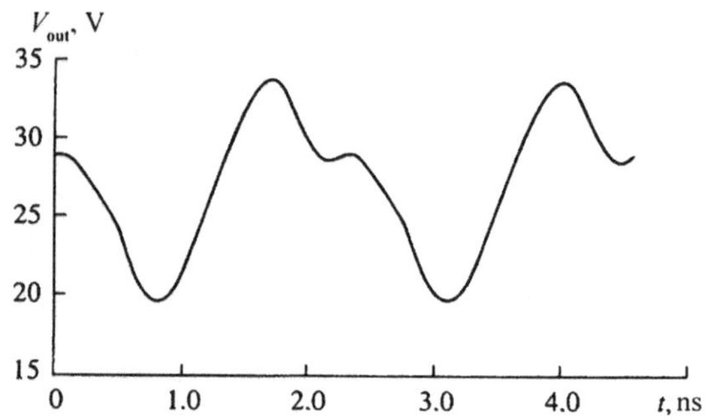

Fig. 5.9 The results of our calculations of time realization of oscillations for the structure containing one and four transistors

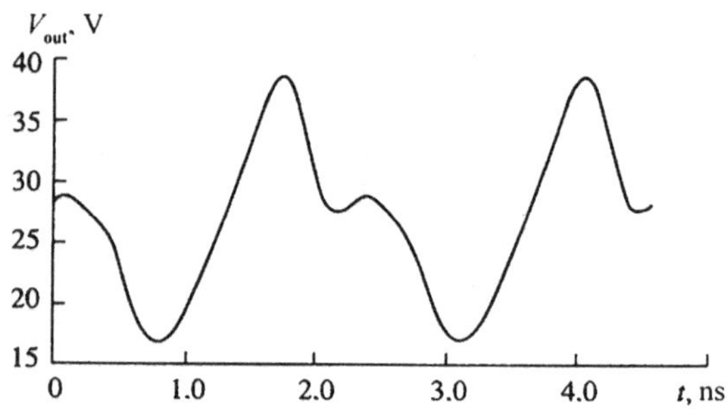

Fig. 5.10 The results of our calculations of time realization of oscillations for the structure containing one and four transistors

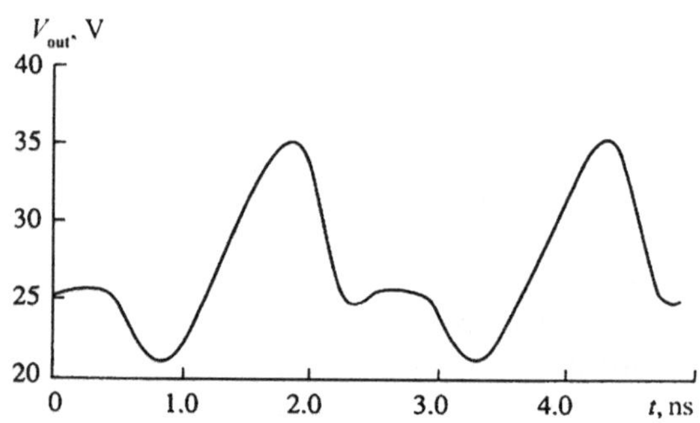

Fig. 5.11 The time realization of oscillations of the generator

of the higher harmonic amplitudes was not monotonous. Similar results have been obtained at numerical simulation. The results of calculation of the amplitudes of spectral components and time realization of oscillations are presented in Figs. 5.12 and 5.13, respectively.

The investigated HMG had a narrow width of the spectral line $\Delta\nu_{-3\text{dB}} \approx (15 \div 30)$ kHz, on frequencies $\nu_0 \approx 500$–750 MHz.

In the used model of HMG there is an opportunity to provide for the noisy properties of the transistor and determination of the generated spectral line width by the technique offered in [43].

The results of calculation of the phase noise $\mathbf{P}_{\varphi}$ for a generation frequency $\nu_0 = 450$ MHz are given in Fig. 5.14. The spectra of harmonic constituents are

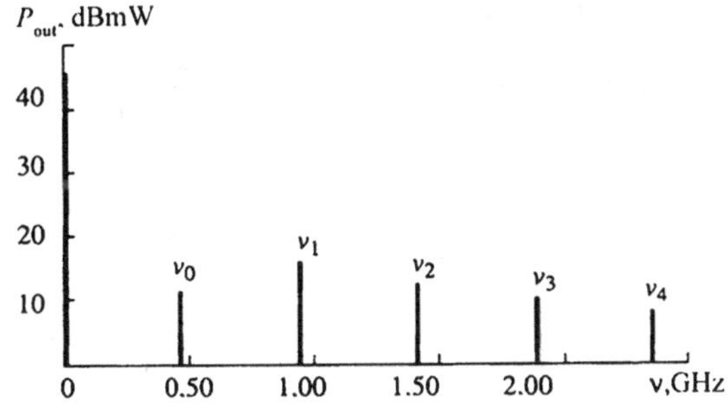

Fig. 5.12 The results of calculation of the amplitudes of spectral components

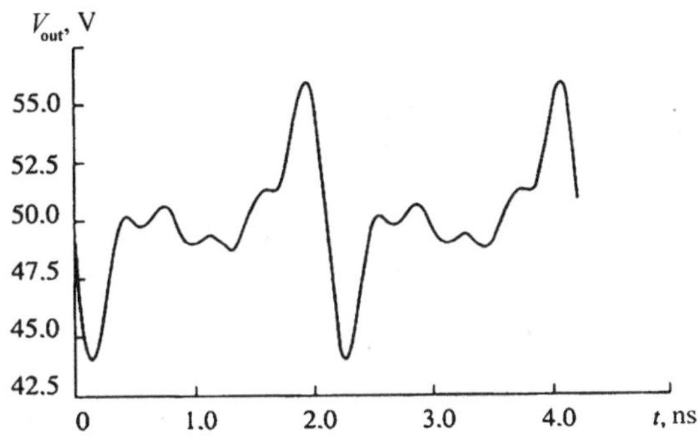

Fig. 5.13 Time realization of oscillations

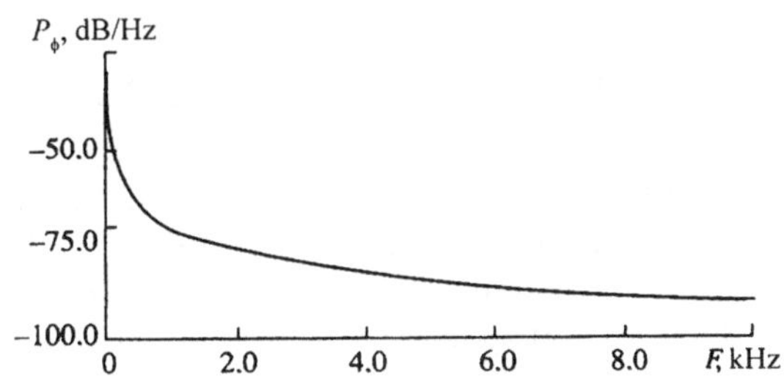

Fig. 5.14 The results of calculation of the phase noise P_φ for a generation frequency $\nu_0 = 450\,\text{MHz}$

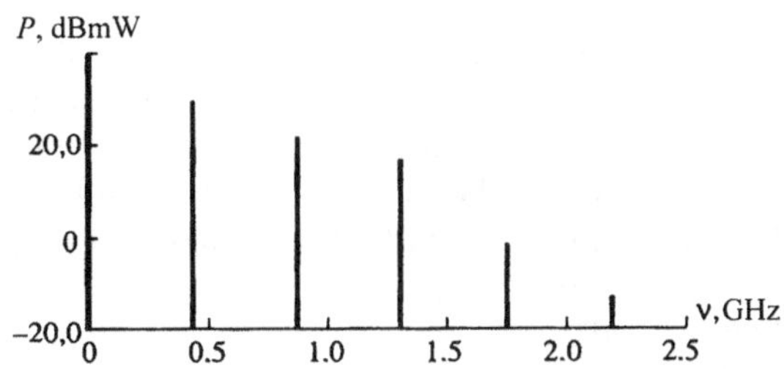

Fig. 5.15 The spectra of harmonic constituents

shown in Fig. 5.15. The data of calculations and experiment are in good conformity. The spectral line computed for the HMG model is narrower than that observed experimentally.

It is explained that only noise properties of the passive elements and transistor have been considered in the model. Technical noises caused by the influence of feed

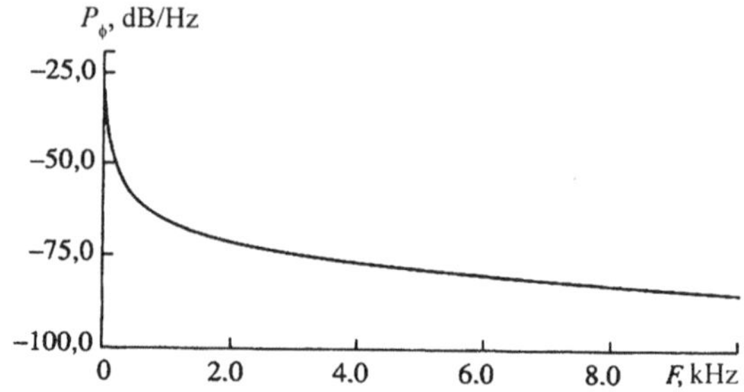

Fig. 5.16 The spectral line width in view of the instability of power sources

Table 5.5 The results of measurement and calculation of generation frequency and spectral line width depending on the number of the harmonics

Parameter	$n = 1$		$n = 2$		$n = 3$		$n = 4$	
	Exp	Calc	Exp	Calc	Exp	Calc	Exp	Calc
U_k (V)	3.0	3.0	3.0	3.0	4.0	4.0	9.0	9.0
U_e (V)	1.0	1.0	1.5	1.5	3.0	3.0	3.0	3.0
I_k (A)	0.2	0.22	0.34	0.37	0.6	0.61	0.66	0.65
P_{out} (mW)	200	250	100	135	3.0	15	4.0	7.0
ν_0 (MHz)	404	430	402	425	395	400	401	405
$(\Delta\nu_0)_{-3dB}$ (kHz)	30	15	100	30	30	30	30	20
$(\Delta\nu_0)_{-60dB}$ (kHz)	130	100	400	180	200	180	200	150
$(\Delta\nu_1)_{-3dB}$ (kHz)	200	62	200	70	100	78	50	44
$(\Delta\nu_1)_{-60dB}$ (kHz)	350	220	400	380	400	420	350	320
$(\Delta\nu_2)_{-3dB}$ (kHz)	300	95	400	120	200	134	120	74
$(\Delta\nu_2)_{-60dB}$ (kHz)	1,000	360	1,000	700	600	1,000	350	600

instability and breakthroughs in the model were not considered. For their account, pulsation generators in a loading mode of 2–10 mV were placed in the circuit of ideal power sources. In view of the instability of power sources the spectral line width increased up to the values observed experimentally (Fig. 5.16).

Our researches have shown that the structures consisting of several elementary transistors at modeling can be replaced by one transistor with the corresponding parameters that considerably simplifies calculations.

By experimental determination of the spectral lines width of signal harmonics of the generator in classical multiplication modes[1] their broadening was observed at transition to higher harmonics. The results of calculation of spectral line widths for the third harmonic of HMG signal on a frequency of 1,350 MHz are presented in Fig. 5.12. In comparison with the results in Fig. 5.15 broadening of the spectral line is noted.

The results of measurement and calculation of generation frequency and spectral line width depending on the number of the harmonics are shown in Table 5.5.

[1] Modes of parametrical multiplication and division of the fundamental frequency of a signal, for which the spectral line width $\nu_{n-1} = \nu_{-m+1} = \text{const}$, $n = 1, 2, 3, \ldots, m = 1, 2, 3 \ldots$.

The spectral line width was calculated at introduction of technical noise sources with a root-mean-square voltage of 10 mV into the collector and base circuits.

Analyzing the results of calculations and experimental data, it is possible to draw a conclusion on their good conformity that confirms the correctness of our choice of the equivalent HMG circuit and determination of its parameters.

5.5 Modeling of Complicated Regimes

The characteristics of a powerful HMG in the steady regular nonlinear mode are considered above. The analysis was made with the use of standard harmonic analysis. However, at such an approach it is impossible to solve the problem of stability of the obtained stationary solutions of the equations describing the generator and to pass to consideration of more complex operating modes, including irregular and complex multifrequency modes with modulation of the envelope of high-frequency oscillations. Such modes exist in HMG, as follows from experiments, both in the presence of ferrite microresonators built in the magneto transistor and in their absence.

A method of solution of ordinary differential equations describing its equivalent circuit was applied to description of such operating HMG modes. The mathematical model of the generator with heteromagnetic interactions was designed on the basis of the total equivalent circuit of Gummel–Poon's model for a powerful transistor.

For simplification of analysis the full equivalent circuit of the generator was separated into external and internal ones. The internal component represented the general model of Gummel–Poon for a powerful UHF transistor. The external one included a FMCR and elements of coordination and control over HMG parameters. Such an approach simplified deriving the equations describing the generator at change of the circuit's parameters of the external oscillatory system since some of the equations describing the behaviour of the transistor in this circuit remained unchanged.

For derivation of the equations describing the investigated generator, a method of central potentials with operational recording of capacitor and inductive conductances was used. The conductance of a nonlinear capacitor C can be presented as $-G_C = C(\mathrm{d}/\mathrm{d}t)$. For a nonlinear inductance L we have

$$G_L = \frac{1}{L}\int \mathrm{d}t.$$

The equivalent circuit of the transistor on the basis of Gummel–Poon's model (Fig. 5.1) has seven internal units $\varphi_0, \ldots, \varphi_6$ and three external ones φ_e, φ_b, φ_c (the emitter, base, and collector ones), the potentials of which will be considered are given. Three units are connected to each other by active resistances and their potentials are related algebraically. For assemblages amount we shall make the equations, including that for units the sum of all flowing currents is equal to zero, i.e., $\sum_i I_i = 0$. Equations (5.28–5.31) describe the HMG in the differential form for units $\varphi_0, \ldots, \varphi_6$:

$$C_{\text{bcl}}\frac{\mathrm{d}U_{S12}}{\mathrm{d}t} + C_{\text{bel}}\frac{\mathrm{d}U_{S1}}{\mathrm{d}t} + \frac{I_{\text{br}}}{\text{BR}} + I_{\text{lc}} + \frac{I_{\text{bf}}}{\text{BF}} + I_{\text{le}} - \frac{1}{R_{\text{bb}}}(\varphi_2 - \varphi_0) = 0, \tag{5.28}$$

$$\frac{1}{L_{\text{b}}}\int(\varphi_{\text{b}} - \varphi_1)\mathrm{d}t + \frac{1}{R_{\text{b2}}}(\varphi_2 - \varphi_1) + C_{\text{bc}}\frac{\mathrm{d}(\varphi_5 - \varphi_1)}{\mathrm{d}t} + C_{\text{be}}\frac{\mathrm{d}(\varphi_6 - \varphi_1)}{\mathrm{d}t} = 0, \tag{5.29}$$

$$\frac{1}{R_{\text{b2}}}(\varphi_1 - \varphi_2) - C_{\text{b}x}\frac{\mathrm{d}U_{23}}{\mathrm{d}t} + \frac{1}{R_{\text{bb}}}(\varphi_0 - \varphi_2) = 0, \tag{5.30}$$

$$C_{\text{b}x}\frac{\mathrm{d}U_{23}}{\mathrm{d}t} + C_{\text{bcl}}\frac{\mathrm{d}U_{S12}}{\mathrm{d}t} + \frac{I_{\text{br}}}{\text{BR}} + I_{\text{lc}} - I_{\text{cf}} + I_{\text{cr}} + \frac{1}{R_{\text{c2}}}(\varphi_5 - \varphi_3) = 0, \tag{5.31}$$

$$C_{\text{bel}}\frac{\mathrm{d}U_{S1}}{\mathrm{d}t} + \frac{I_{\text{bf}}}{\text{BF}} + I_{\text{le}} + I_{\text{cf}} - I_{\text{cr}} + \frac{1}{R_{\text{e1}}}(\varphi_6 - \varphi_4) = 0 \tag{5.32}$$

$$\frac{1}{L_{\text{c}}}\int(\varphi_{\text{c}} - \varphi_5)\mathrm{d}t + C_{\text{ce}}\frac{\mathrm{d}(\varphi_6 - \varphi_5)}{\mathrm{d}t} + C_{\text{bc}}\frac{\mathrm{d}(\varphi_1 - \varphi_5)}{\mathrm{d}t} + \frac{1}{R_{\text{c2}}}(\varphi_3 - \varphi_5) = 0, \tag{5.33}$$

$$\frac{1}{L_{\text{e}}}\int(\varphi_{\text{e}} - \varphi_6)\mathrm{d}t + C_{\text{be}}\frac{\mathrm{d}(\varphi_1 - \varphi_6)}{\mathrm{d}t} + C_{\text{ce}}\frac{\mathrm{d}(\varphi_5 - \varphi_6)}{\mathrm{d}t} + \frac{1}{R_{\text{e1}}}(\varphi_4 - \varphi_6) = 0, \tag{5.34}$$

where $U_{S1} = \varphi_0 - \varphi_4$, $U_{S12} = \varphi_0 - \varphi_3$, $U_{23} = \varphi_2 - \varphi_3$.

Let us make the following substitutions: $\varphi_4 = U_{S12} - U_{S1} + \varphi_3$ in (5.32) and (5.34); $\varphi_3 = \varphi_2 - U_{23}$ in (5.31) and (5.33); $\varphi_0 = U_{S12} - \varphi_2 + U_{23}$ in (5.28) and (5.30). Let us designate: $X_{\text{c}} = \int \varphi_{\text{c}}\mathrm{d}t$, $X_5 = \int \varphi_5\mathrm{d}t$, $X_6 = \int \varphi_6\mathrm{d}t$, $X_{\text{b}} = \int \varphi_{\text{b}}\mathrm{d}t$, $X_1 = \int \varphi_1\mathrm{d}t$, $X_{\text{e}} = \int \varphi_{\text{e}}\mathrm{d}t$. From (5.32) it follows that

$$C_{\text{bel}}\frac{\mathrm{d}U_{S1}}{\mathrm{d}t} = -\frac{I_{\text{bf}}}{\text{BF}} - I_{\text{le}} - I_{\text{cf}} + I_{\text{cr}} - \frac{1}{R_{\text{e1}}}(\varphi_6 - \varphi_4). \tag{5.35}$$

Let us substitute (5.35) into (5.28). From (5.30) it follows that

$$C_{\text{ce}}\frac{\mathrm{d}U_{23}}{\mathrm{d}t} = \frac{1}{R_{\text{b2}}}(\varphi_1 - \varphi_2) + \frac{1}{R_{\text{bb}}}(\varphi_0 - \varphi_2). \tag{5.36}$$

Then (5.28)–(5.34) can be rewritten as

$$C_{\text{bcl}}\frac{\mathrm{d}U_{S12}}{\mathrm{d}t} = -C_{\text{bel}}\frac{\mathrm{d}U_{S1}}{\mathrm{d}t} - \frac{I_{\text{br}}}{\text{BR}} - I_{\text{lc}} - \frac{I_{\text{bf}}}{\text{BF}} - I_{\text{le}} + \frac{1}{R_{\text{bb}}}(\varphi_2 - U_{S12} + \varphi_2 - U_{23}) = 0, \tag{5.37}$$

$$X_{\text{b}} - X_1 + \frac{L_{\text{b}}}{R_{\text{b2}}}(\varphi_2 - \varphi_1) + L_{\text{b}}C_{\text{bc}}\frac{\mathrm{d}\varphi_5}{\mathrm{d}t} - L_{\text{b}}C_{\text{bc}}Y_1 + L_{\text{b}}C_{\text{be}}\frac{\mathrm{d}\varphi_6}{\mathrm{d}t} - L_{\text{b}}C_{\text{be}}Y_1 = 0 \tag{5.38}$$

$$C_{\text{ce}}\frac{\mathrm{d}U_{23}}{\mathrm{d}t} = \frac{1}{R_{\text{b2}}}(\varphi_1 - \varphi_2) + \frac{1}{R_{\text{bb}}}(U_{S12} - \varphi_2 + U_{23} - \varphi_2), \tag{5.39}$$

$$C_{\text{b}x}\frac{\mathrm{d}U_{23}}{\mathrm{d}t} = -C_{\text{bcl}}\frac{\mathrm{d}U_{S12}}{\mathrm{d}t} - \frac{I_{\text{br}}}{\text{BR}} - I_{\text{lc}} + I_{\text{cf}} - I_{\text{cr}} - \frac{1}{R_{\text{c2}}}(\varphi_5 - \varphi_2 + U_{23})t, \tag{5.40}$$

$$C_{\mathrm{bel}}\frac{\mathrm{d}U_{\mathrm{S1}}}{\mathrm{d}t} = -\frac{I_{\mathrm{bf}}}{\mathrm{BF}} - I_{\mathrm{le}} - I_{\mathrm{cf}} + I_{\mathrm{cr}} - \frac{1}{R_{\mathrm{e1}}}(\varphi_6 - U_{\mathrm{S12}} + U_{\mathrm{S1}} - \varphi_3), \tag{5.41}$$

$$X_{\mathrm{c}} - X_5 + L_{\mathrm{c}}C_{\mathrm{ce}}\frac{\mathrm{d}(\varphi_6 - \varphi_5)}{\mathrm{d}t} - L_{\mathrm{c}}C_{\mathrm{bc}}\frac{\mathrm{d}\varphi_5}{\mathrm{d}t} + L_{\mathrm{c}}C_{\mathrm{bc}}Y_1 + \frac{L_{\mathrm{c}}}{R_{\mathrm{c2}}}(\varphi_3 - \varphi_5) = 0, \tag{5.42}$$

$$X_{\mathrm{e}} - X_6 - L_{\mathrm{e}}C_{\mathrm{be}}\frac{\mathrm{d}\varphi_6}{\mathrm{d}t} + L_{\mathrm{e}}C_{\mathrm{be}}Y_1 + L_{\mathrm{e}}C_{\mathrm{ce}}\frac{\mathrm{d}(\varphi_5 - \varphi_6)}{\mathrm{d}t} + \frac{L_{\mathrm{e}}}{R_{\mathrm{e1}}}(\varphi_4 - \varphi_6) = 0, \tag{5.43}$$

$$\frac{\mathrm{d}X_1}{\mathrm{d}t} = \varphi_1, \tag{5.44}$$

$$\frac{\mathrm{d}\varphi_1}{\mathrm{d}t} = Y_1, \tag{5.45}$$

$$\frac{\mathrm{d}X_5}{\mathrm{d}t} = \varphi_5, \tag{5.46}$$

$$\frac{\mathrm{d}X_6}{\mathrm{d}t} = \varphi_6, \tag{5.47}$$

$$\frac{\mathrm{d}X_{\mathrm{c}}}{\mathrm{d}t} = \varphi_{\mathrm{c}}, \tag{5.48}$$

$$\frac{\mathrm{d}X_{\mathrm{e}}}{\mathrm{d}t} = \varphi_{\mathrm{e}}, \tag{5.49}$$

$$\frac{\mathrm{d}X_{\mathrm{b}}}{\mathrm{d}t} = \varphi_{\mathrm{b}}, \tag{5.50}$$

Having substituted the expression for $\mathrm{d}\varphi_6/\mathrm{d}t$ from (5.38) into (5.42) and (5.43), we transform (5.37)–(5.50) to a form convenient for further analysis

$$\begin{cases} \dfrac{\mathrm{d}X_1}{\mathrm{d}t} = f_1(X_1, X_2, \ldots, X_n) \\ \vdots \\ \dfrac{\mathrm{d}X_n}{\mathrm{d}t} = f_n(X_1, X_2, \ldots, X_n) \end{cases}. \tag{5.51}$$

Let us supplement (5.37)–(5.50) with equations for the external units of the transistor $\varphi_{\mathrm{c}}, \varphi_{\mathrm{b}}, \varphi_{\mathrm{e}}$. The currents in these units are described by the equations

$$\frac{1}{L_{\mathrm{c}}}(\varphi_5 - \varphi_{\mathrm{c}}) = \frac{\mathrm{d}I_{\mathrm{c}}}{\mathrm{d}t}, \tag{5.52}$$

$$\frac{1}{L_{\mathrm{b}}}(\varphi_1 - \varphi_{\mathrm{b}}) = \frac{\mathrm{d}I_{\mathrm{b}}}{\mathrm{d}t}, \tag{5.53}$$

$$\frac{1}{L_{\mathrm{e}}}(\varphi_6 - \varphi_{\mathrm{e}}) = \frac{\mathrm{d}I_{\mathrm{e}}}{\mathrm{d}t}, \tag{5.54}$$

At solution of the equation set describing the HMG, the parameters for joining the equations describing the external oscillatory system and the transistor, are the potentials in the units $\varphi_c, \varphi_b, \varphi_e$ and input currents I_c, I_b, I_e.

Potentials $\varphi_c, \varphi_b, \varphi_e$ can be set as external conditions or be calculated as the solution of (5.37)–(5.54) coordinated with the external oscillatory system. The transistor is described by a set of 17 nonlinear ordinary differential equations.

The external equivalent circuit of the investigated generator is presented in Fig. 5.17.

In experiments for realization of effective interactions with the transistor structure, the FMCR was placed directly above the emitter area of the transistor. At a distance from the base, the multiple-parameter microfield interaction decreased. Therefore, at modeling of HMG a nonlinear magnetic-field-controlled contour L_f, C_f, R_f was placed in the equivalent circuit in the emitter area. The influence of the current emitter on the base current due to heteromagnetic interactions was described by elements L_f, L_m and connection M between them.

The equivalent circuit of HMG is presented in Fig. 5.18.

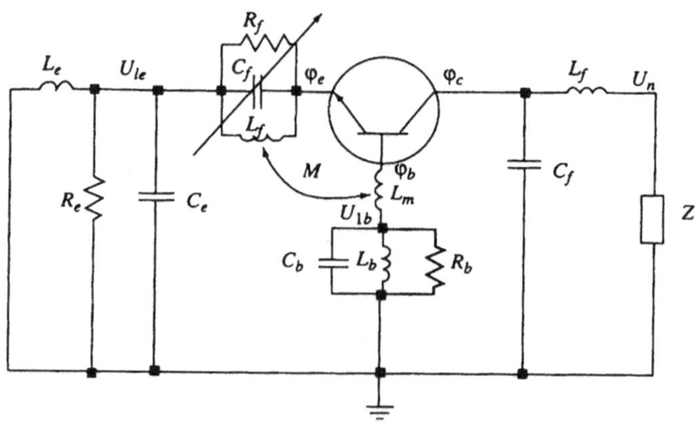

Fig. 5.17 The external equivalent circuit of the investigated generator

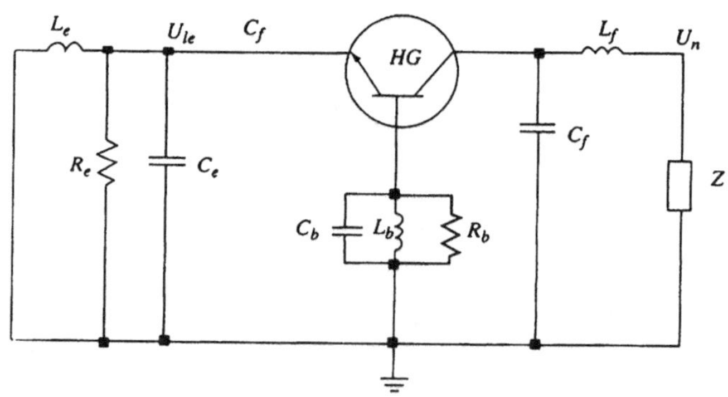

Fig. 5.18 The equivalent circuit of HMG

Equation set (5.55) describes the external HMG circuit:

$$\begin{cases} \int (U_{\mathrm{le}} - U_{\mathrm{e}})\mathrm{d}t + \dfrac{L_{\mathrm{f}}}{R_{\mathrm{f}}}(U_{\mathrm{le}} - U_{\mathrm{e}}) + L_{\mathrm{f}}C_{\mathrm{f}}\dfrac{\mathrm{d}(U_{\mathrm{le}} - U_{\mathrm{e}})}{\mathrm{d}t} + L_{\mathrm{f}}I_{\mathrm{e}} = 0, \\ \int (U_{\mathrm{le}} - MU_{\mathrm{lb}})\mathrm{d}t + \dfrac{L_{\mathrm{e}}}{R_{\mathrm{e}}}U_{\mathrm{le}} + L_{\mathrm{e}}C_{\mathrm{e}}\dfrac{\mathrm{d}U_{\mathrm{le}}}{\mathrm{d}t} = 0, \\ \int (U_{\mathrm{lb}} + MU_{\mathrm{le}})\mathrm{d}t + \dfrac{L_{\mathrm{b}}}{R_{\mathrm{b}}}U_{\mathrm{lb}} + L_{\mathrm{b}}C_{\mathrm{b}}\dfrac{\mathrm{d}U_{\mathrm{lb}}}{\mathrm{d}t} = 0, \\ \int (U_{\mathrm{lb}} - U_{\mathrm{b}})\mathrm{d}t + L_{\mathrm{m}}I_{\mathrm{b}} = 0, \\ \int (U_{\mathrm{n}} - U_{\mathrm{c}})\mathrm{d}t - L_{\mathrm{f}}C_{\mathrm{f}}\dfrac{\mathrm{d}U_{\mathrm{c}}}{\mathrm{d}t} + L_{\mathrm{f}}I_{\mathrm{c}} = 0, \\ \int (U_{\mathrm{c}} - U_{\mathrm{n}})\mathrm{d}t - \dfrac{L_{\mathrm{f}}}{R_{\mathrm{n}}}U_{\mathrm{n}} = 0. \end{cases} \tag{5.55}$$

The high frequency voltages on the transistor's terminals U_{e}, U_{b}, U_{c} are related to the full potentials $\varphi_{\mathrm{c}}, \varphi_{\mathrm{b}}, \varphi_{\mathrm{e}}$ of the transistor's model by $\varphi_{\mathrm{e}} = U_{\mathrm{e}} + V_{\mathrm{e}}$, $\varphi_{\mathrm{b}} = U_{\mathrm{b}} + V_{\mathrm{b}}$, $\varphi_{\mathrm{c}} = U_{\mathrm{c}} + V_{\mathrm{c}}$, where V_{e}, V_{b}, V_{c} are the voltage of displacement on the emitter, base, and collector of the transistor, respectively, which set a working point on a direct current.

Let us transform the equations to the form as in (5.51). For this purpose, we shall introduce new variables $X_{\mathrm{le}}, X_{\mathrm{lb}}, X_{\mathrm{e}}, X_{\mathrm{C}}, X_{\mathrm{B}}$.

$$\frac{\mathrm{d}X_{\mathrm{le}}}{\mathrm{d}t} = U_{\mathrm{le}}, \tag{5.56}$$

$$\frac{\mathrm{d}X_{\mathrm{lb}}}{\mathrm{d}t} = U_{\mathrm{lb}}, \tag{5.57}$$

$$\frac{\mathrm{d}X_{\mathrm{E}}}{\mathrm{d}t} = U_{\mathrm{E}}, \tag{5.58}$$

$$\frac{\mathrm{d}X_{\mathrm{C}}}{\mathrm{d}t} = U_{\mathrm{C}}, \tag{5.59}$$

$$\frac{\mathrm{d}X_{\mathrm{B}}}{\mathrm{d}t} = U_{\mathrm{B}}, \tag{5.60}$$

$$L_{\mathrm{b}}C_{\mathrm{b}}\frac{\mathrm{d}U_{\mathrm{lb}}}{\mathrm{d}t} = -X_{\mathrm{lb}} - MX_{\mathrm{le}} - \frac{L_{\mathrm{b}}}{R_{\mathrm{b}}}U_{\mathrm{lb}}, \tag{5.61}$$

$$L_{\mathrm{e}}C_{\mathrm{e}}\frac{\mathrm{d}U_{\mathrm{le}}}{\mathrm{d}t} = -X_{\mathrm{le}} - MX_{\mathrm{lb}} - \frac{L_{\mathrm{e}}}{R_{\mathrm{e}}}U_{\mathrm{le}}, \tag{5.62}$$

$$L_{\mathrm{f}}C_{\mathrm{f}}\frac{\mathrm{d}U_{\mathrm{e}}}{\mathrm{d}t} = X_{\mathrm{e}} - X_{\mathrm{le}} + \frac{L_{\mathrm{f}}}{R_{\mathrm{f}}}(U_{\mathrm{le}} - U_{\mathrm{e}}) + L_{\mathrm{f}}I_{\mathrm{e}} + L_{\mathrm{f}}C_{\mathrm{f}}\frac{\mathrm{d}U_{\mathrm{le}}}{\mathrm{d}t}, \tag{5.63}$$

$$L_{\mathrm{lm}}\frac{\mathrm{d}I_{\mathrm{B}}}{\mathrm{d}t} = U_{\mathrm{B}} - U_{\mathrm{lb}}, \tag{5.64}$$

$$\frac{L_{\mathrm{f}}}{R_{\mathrm{n}}} \times \frac{\mathrm{d}U_{\mathrm{n}}}{\mathrm{d}t} = U_{\mathrm{n}} - U_{\mathrm{C}}. \tag{5.65}$$

Expressions (5.56)–(5.65) describe the HMG parameters in the form of ordinary nonlinear differential equations. Considering (5.37)–(5.54) for the transistor, we derive a full set of 27 equations describing the HMG in general. For analysis, it is convenient to introduce dimensionless time $\tau = t/L_{\mathrm{b}}C_{\mathrm{b}} = \omega_{\mathrm{b}}^2 t$, where ω_{b} is the own frequency of the contour in the emitter circuit of the transistor.

The investigated equation set is complex enough, its solution demands significant computer time. No analysis of generation modes on the basis of the bifurcation theory has been made here. A research of the character of oscillations has been carried out at change of the contour mismatch $\Delta = \omega_{\mathrm{b}}^2/\omega_{\mathrm{f}}^2$, where ω_{f} is the own frequency of the contour $L_{\mathrm{f}}, C_{\mathrm{f}}, R_{\mathrm{f}}$ in the emitter circuit of the transistor of the generator. This parameter changes due to the magnetic field value in the HMG structure. A source modeling technical noise of the generator was placed into the collector circuit. The voltage of this source was set equal to 10 mV. At changes of parameter Δ, various modes of oscillations (from monochromatic unifrequent to multifrequency and noise-like ones) were observed. The spectral power density

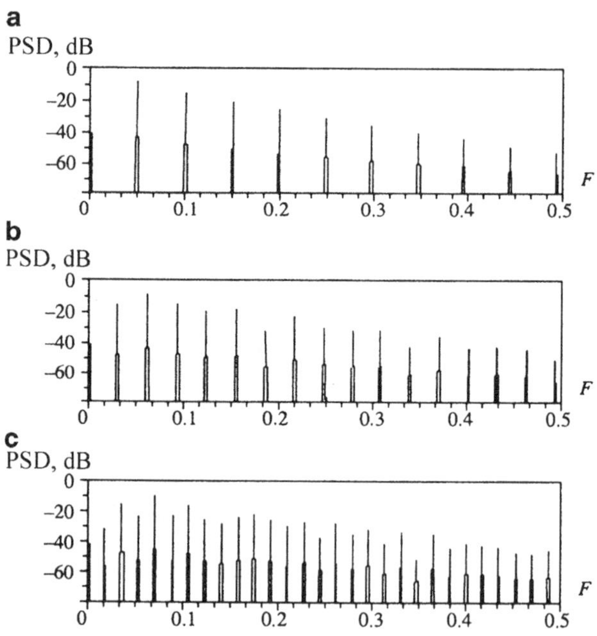

Fig. 5.19 The spectral power density distribution in the vicinity of the basic frequency of oscillations, typical for multifrequency and noise-like oscillations. (**a**) $\Delta = 1.20$, (**b**) $\Delta = 1.18$, (**c**) $\Delta = 1.15$

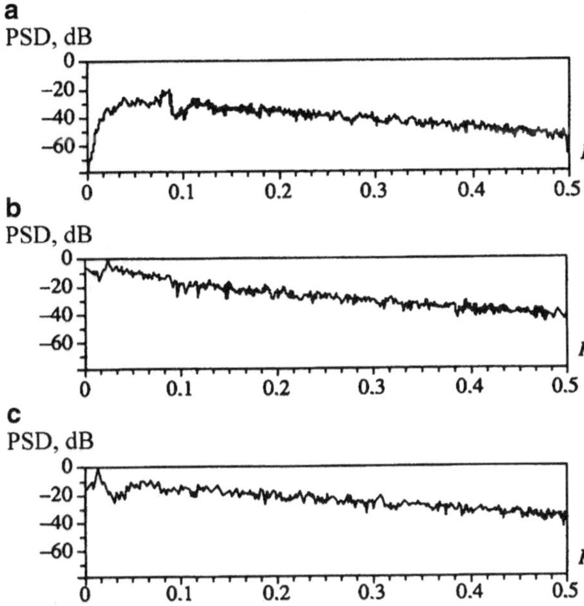

Fig. 5.20 The spectral power density distribution in the vicinity of the basic frequency of oscillations, typical for multifrequency and noise-like oscillations. (**a**) $\Delta = 1.23$, (**b**) $\Delta = 1.1$, (**c**) $\Delta = 1.09$

distribution in the vicinity of the basic frequency of oscillations, typical for multifrequency and noise-like oscillations, is presented in Fig. 5.19 ((a) $\Delta = 1.20$; (b) $\Delta = 1.18$; (c) $\Delta = 1.15$) and Fig. 5.20 ((a) $\Delta = 1.23$; (b) $\Delta = 1.1$; (c) $\Delta = 1.09$) as function of the dimensionless frequency $F = \nu_0/\nu_b$, where ν_b is the own frequency of the contour in the emitter circuit of the HMG transistor.

Chapter 6
Multicircuit Model of a Multifunctional Heteromagnetic Oscillator

6.1 Equivalent Circuit

Multipurpose oscillation modes of signals of various kinds and spectral compositions in a HMG arise at work of FMCR in an unsaturated nonlinear mode. In this case, the ferrite magnetization is nonuniform; the sample is separated into small ($\sim 10\,\mu m$) domains. The structure of these domains essentially depends on the material, shape, and sizes of the sample, an external magnetizing field [18]. Nonlinear effects in the ferrite are shown from power levels of the order of 0.1–1 mW. The resonant frequencies of FMCR depend on the saturation magnetization of the ferrite, the field of anisotropy, the orientation in an external constant magnetic field, the kind of polarization of the high-frequency magnetic field exciting oscillations of the magnetization vector in the ferrite. Modeling of interaction of FMCR with a high-frequency magnetic field of a semi-conductor structure in unsaturated nonlinear modes was conducted:

- On low (milliwatt) power levels.
- For various saturation magnetizations $4\pi M_s$ and orientations φ of FMCR in the magnetic field.
- At raised power levels (hundred mW–several W).
- The properties of FMCR in the modes of absorption and passage of signals were investigated.

In HMG, FMCR of small sizes in comparison with the lengths of waves arising in the ferrite at frequencies up to 1 GHz were used. Therefore, FMCR was considered as a concentrated oscillatory system without regard to wave effects. In the structures with monoaxial anisotropy or a cubic crystal the domain structure is [17, 18] presented in Fig. 6.1.

The ferrite sample is divided into sites with opposite magnetizations in the neighboring domains. For YIG, the domain width was $\sim 10^{-5}$ cm. In the model accepted by us, the domain structure is homogeneous and two domains with different magnetization orientations participate in the interaction with HF fields. At analysis of FMCR in unsaturated modes it is necessary to consider magnetization oscillations in domains and domain border oscillations simultaneously [43]. For a single domain

A.A. Ignatiev and A.V. Lyashenko, *Heteromagnetic Microelectronics: Microsystems of Active Type*, DOI 10.1007/978-1-4419-6002-3_6,

Fig. 6.1 The domain structure in the structures with monoaxial anisotropy or a cubic crystal

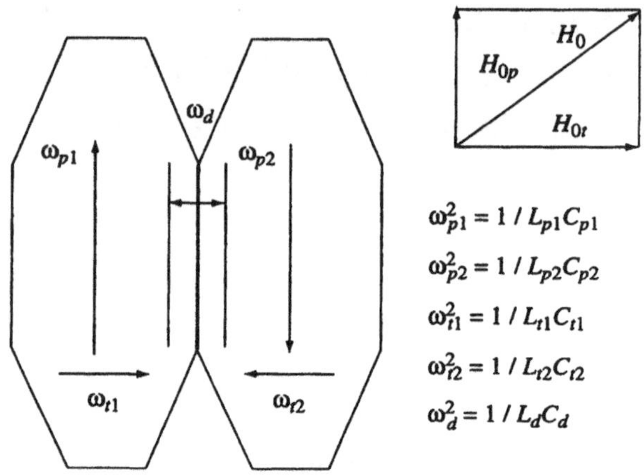

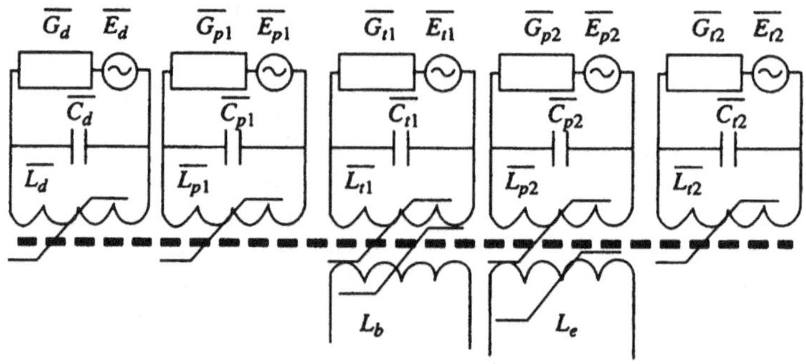

$$\overline{M} = \begin{vmatrix} 1 & M_{dp1} & M_{dt1} & M_{dp2} & M_{dt2} & M_{de} & M_{db} \\ M_{dp1} & 1 & M_{p1t1} & M_{p1p2} & M_{p1t2} & M_{p1e} & M_{p1b} \\ M_{dt1} & M_{p1t1} & 1 & M_{t1p2} & M_{t1t2} & M_{t1e} & M_{t1b} \\ M_{dp2} & M_{p1p2} & M_{t1p2} & 1 & M_{p2t2} & M_{p2e} & M_{p2b} \\ M_{dt2} & M_{p1t2} & M_{t1t2} & M_{p2t2} & 1 & M_{t2e} & M_{t2b} \\ M_{de} & M_{p1e} & M_{t1e} & M_{p2e} & M_{t2e} & 1 & M_{eb} \\ M_{db} & M_{p1b} & M_{t1b} & M_{p2b} & M_{t2b} & M_{eb} & 1 \end{vmatrix}$$

Fig. 6.2 The multivariable FMCR system

there are three normal oscillations, namely: the oscillation raised by the variable field perpendicular to the constant one (ω_t), an oscillation raised by the variable field parallel to the constant one (ω_p), and a domain wall oscillation (ω_d). In a more general case, the number of normal oscillations can exceed three.

In Fig. 6.2 an equivalent circuit of FMCR in an unsaturated mode with five normal frequencies is presented. The sources E_d, E_{p1}, E_{t1}, E_{p2}, and E_{t2} model the own temperature noise of FMCR. All the oscillations are coupled. Coupling between the oscillations in FMCR and its external circuits is set by a matrix $\overline{M}(7 \times 7)$. In the trivial case of three normal oscillations, the matrix degenerates into a 5×5 one. Generally, when the angle between constant field and the domain borders, $\varphi \neq 0, \pi/2$, all the types of oscillations existing in the sample are raised. Thus, in the heteromagnetic generator, FMCR is modeled by a multiconnected nonlinear oscillatory system. The frequencies of oscillations of the domain walls and the normal domain oscillation frequencies in the trivial case can be estimated theoretically [17]. This estimation was used as an initial approximation of the parameters of the equivalent circuit of the ferrite sample.

The equivalent parameters of the oscillatory contours were estimated on the basis of the results of our research of the absorption spectra of the used ferrite samples in the working ranges of a magnetic field $\overline{H}_0$ at various levels of high-frequency power. The equivalent parameters of HMG were optimized by the technique presented below.

The parameters of the ferrite sample and the equivalent circuit depend on the saturation magnetization $4\pi M_s$, the external magnetic field $\overline{H}_0$, and the orientation of the sample φ relative to the light magnetization axis. Therefore, the equivalent parameters of the ferrite with a certain value of $4\pi M_s$ were determined for each value of field $\overline{H}_0$, at a constant orientation φ. The halfwidth of the ferromagnetic resonance line (its nominal value) determined the parameter of equivalent GB product.[1]

By the amplitude maxima in the absorption spectra of FMCR the resonance frequencies in the domain modes $\omega_{p1}, \omega_{t1}, \omega_{p2}, \omega_{t2}$ were determined. At the use of low-Q measuring resonators, the spectral linewidth of signals $\Delta\nu_{-3dB}$ determines the GB products of the contours of the ferrite equivalent circuit.

At changes of the bias field H_0 for each oscillatory contour (Fig. 6.2) we have two equations of relation between the three (L, G, C) parameters:

$$\omega(H_0) = \frac{1}{\sqrt{L(H_0) \times C(H_0)}}, \tag{6.1}$$

$$\frac{\omega(H_0) \times C(H_0)}{G(H_0)} = \frac{\nu(H_0)}{\Delta\nu_{-3dB}}. \tag{6.2}$$

For determination of the coupling parameters between the contours, the dependence of the heteromagnetic generator frequency on the external field (which generally are sign-variable functions of $\overline{H}_0$), and the results of our research of generation modes of equidistant frequency spectra were used.

At designing of the equivalent circuit parameters of the ferrite in a nonlinear mode it is necessary to consider the experimental functional dependences of changes of the central frequency, the parameters of spectral lines $\Delta\nu_{-3dB}$ and $\Delta\nu_{-60dB}$, the amplitude levels of spectral components, i.e., the parameters L, G, C on the power level.

---

[1] Let us note that equivalent GB products of the spectral lines generated by heteromagnetic structures on frequencies ~1 GHz are ~10^5.

6.2 Model Equations

The used model of a multipurpose powerful HMG has undergone significant changes in comparison with the one-planimetric model.

In Fig. 6.3a structure constructed in a FMCR as a sphere on a powerful UHF transistor is schematically presented. FMCR is placed in the field of HF magnetic fields of the emitter and base junctions h_{be}. Due to the interactions of the HF magnetic fields of FMCR – h_{YIG} and the HF magnetic fields of the emitter and base currents h_{be}, additional inductances coupled with each other and with the multicoherent equivalent contours appear in the base and emitter circuits of the transistor. These multicoherent equivalent contours simulate nonlinear ferromagnetic oscillations and domain wall oscillations in the ferrite sample. Equivalent circuits of a multipurpose powerful HMG are presented in Fig. 6.4a,b. No direct current power elements of the transistor are shown in this figure. The elements C_{g} and Z are external in relation to the heteromagnetic transistor.

The equations of Gummel–Poon's model for the powerful transistor are presented below:

$$C_{\mathrm{bcl}}\frac{\mathrm{d}V_{\mathrm{S12}}}{\mathrm{d}t} = -C_{\mathrm{bel}}\frac{\mathrm{d}V_{\mathrm{S1}}}{\mathrm{d}t} - \frac{I_{\mathrm{br}}}{\mathrm{BR}} - I_{\mathrm{lc}} - \frac{I_{\mathrm{bf}}}{\mathrm{BF}} - I_{\mathrm{le}} + \frac{1}{R_{\mathrm{bb}}}(\phi_2 - V_{\mathrm{S12}} + \phi_2 - V_{23}), \quad (6.3)$$

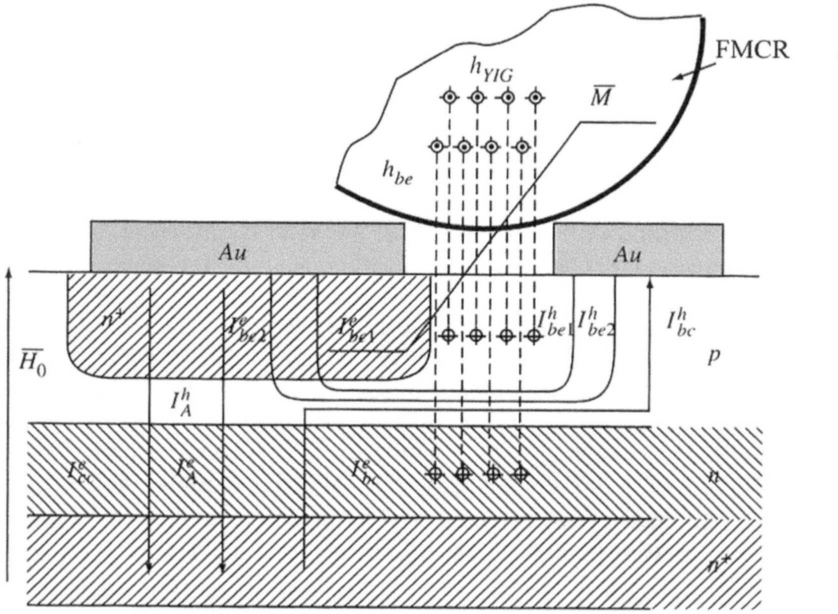

Fig. 6.3 A structure constructed in a FMCR as a sphere on a powerful UHF transistor

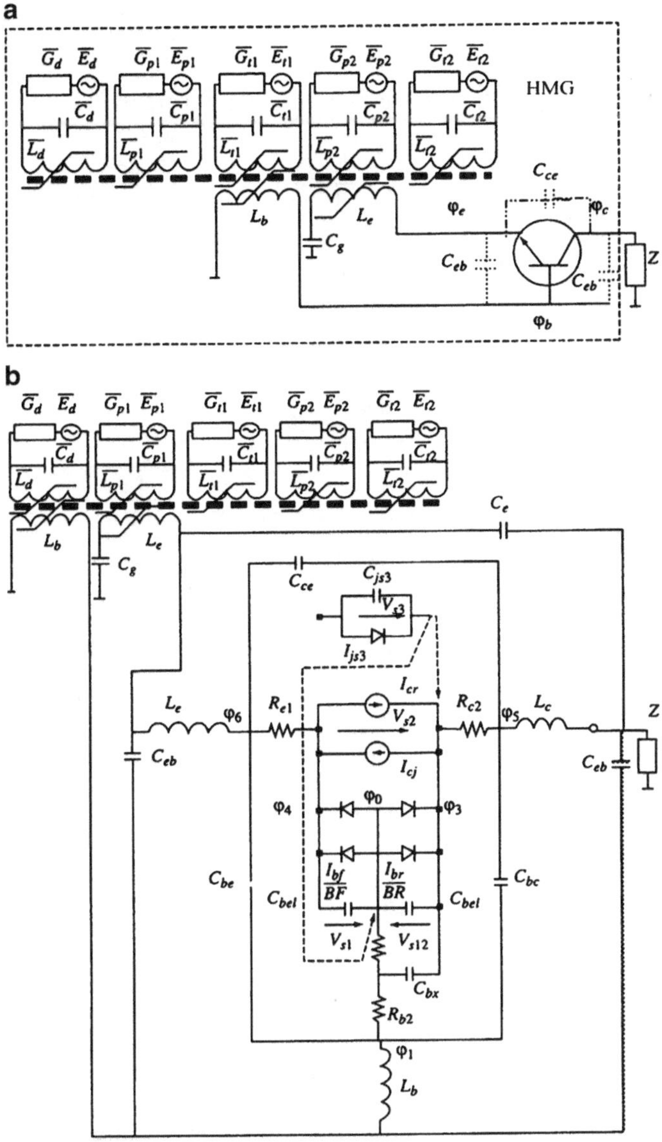

Fig. 6.4 Equivalent circuits of a multipurpose powerful HMG

$$X_{\mathrm{B}} - X_1 + \frac{L_{\mathrm{B}}}{R_{\mathrm{B2}}}(\varphi_2 - \varphi_1) + L_{\mathrm{B}}C_{\mathrm{bc}}\frac{\mathrm{d}\varphi_5}{\mathrm{d}t} - L_{\mathrm{B}}C_{\mathrm{bc}}Y_1 + L_{\mathrm{B}}C_{\mathrm{be}}\frac{\mathrm{d}\varphi_6}{\mathrm{d}t} - L_{\mathrm{B}}C_{\mathrm{be}}Y_1 = 0, \tag{6.4}$$

$$C_{\mathrm{CE}}\frac{\mathrm{d}V_{23}}{\mathrm{d}t} = \frac{1}{R_{\mathrm{b2}}}(\varphi_1 - \varphi_2) + \frac{1}{R_{\mathrm{bb}}}(V_{\mathrm{S12}} - \varphi_2 + V_{23} - \varphi_2), \tag{6.5}$$

$$C_{bx}\frac{dV_{23}}{dt} = -C_{bcl}\frac{dV_{S12}}{dt} - \frac{I_{br}}{BR} - I_{lc} + I_{cf} - I_{cr} - \frac{1}{R_{C2}}(\varphi_5 - \varphi_2 + V_{23}), \quad (6.6)$$

$$C_{bel}\frac{dV_{S1}}{dt} = -\frac{I_{bf}}{BF} - I_{le} - I_{cf} + I_{cr} - \frac{1}{R_{e1}}(\varphi_6 - V_{S12} + V_{S1} - \varphi_3), \quad (6.7)$$

$$X_c - X_5 + L_C C_{ce}\frac{d(\varphi_6 - \varphi_5)}{dt} - L_C C_{bc}\frac{d\varphi_5}{dt} + L_C C_{bc} Y_1 + \frac{L_C}{R_{c2}}(\varphi_3 - \varphi_5) = 0, \quad (6.8)$$

$$X_e - X_6 - L_E C_{be}\frac{d\varphi_6}{dt} + L_E C_{be} Y_1 + L_E C_{ce}\frac{d(\varphi_5 - \varphi_6)}{dt} + \frac{L_e}{R_{e1}}(\varphi_4 - \varphi_6) = 0, \quad (6.9)$$

$$\frac{dX_1}{dt} = \varphi_1, \quad (6.10)$$

$$\frac{d\varphi_1}{dt} = Y_1, \quad (6.11)$$

$$\frac{dX_5}{dt} = \varphi_5, \quad (6.12)$$

$$\frac{dX_6}{dt} = \varphi_6, \quad (6.13)$$

$$\frac{dX_c}{dt} = \varphi_c, \quad (6.14)$$

$$\frac{dX_e}{dt} = \varphi_e, \quad (6.15)$$

$$\frac{dX_b}{dt} = \varphi_b, \quad (6.16)$$

$$\frac{1}{L_C}(\varphi_5 - \varphi_C) = \frac{dI_C}{dt}, \quad (6.17)$$

$$\frac{1}{L_B}(\varphi_1 - \varphi_B) = \frac{dI_B}{dt}, \quad (6.18)$$

$$\frac{1}{L_E}(\varphi_6 - \varphi_E) = \frac{dI_E}{dt}. \quad (6.19)$$

The equations of the multipurpose powerful HMG supplementing the above ones of Gummel–Poon's model look as:

$$C_g\sum_{j\neq i}^{n} M_{1j}\frac{dU_j}{dt} - C_g\frac{dU_e}{dt} + C_{ce}\frac{d(\varphi_c - \varphi_e)}{dt} + C_{eb}\frac{d(\varphi_b - \varphi_e)}{dt} + I_e = 0, \quad (6.20)$$

$$C_{ce}\frac{d(\varphi_e - \varphi_c)}{dt} + C_{cb}\frac{d(\varphi_b - \varphi_c)}{dt} - \frac{1}{Z}\varphi_c + I_c = 0, \quad (6.21)$$

$$\frac{dI_b}{dt} - \frac{1}{L_b}\sum_{j\neq i} M_{2j}U_j + C_{eb}\frac{d^2\varphi_e}{dt^2} + C_{cb}\frac{d^2\varphi_c}{dt^2} = 0. \quad (6.22)$$

Equations (6.7)–(6.22) represent the generalized equations of the HMG.

6.3 Methods of Finalizing Equivalent Parameters of Transistor

Determination of the 26 equivalent parameters of the used model of the powerful transistor represents enough challenge, for which solution, specialized equipment and a series of measurements with their subsequent computer processing are required. The transistor is a component of the heteromagnetic structure that can be used in various generating, mixing, intensifying, and other modes.

All the unknown parameters of HMG were divided on static and dynamic ones. The static parameters were determined by means of measurements of the input and output characteristics families at direct current.

The dynamic parameters of HMG were determined from experiments with UHF signals in various generating modes from the dependences of the signal generation frequency on the collector voltage at various displacement voltages on the base of the transistor. The parameters of the equivalent circuit of HMG were optimized by the least squares method with the usage of groups of equivalent parameters and experimental characteristics. A group contained the parameters most essentially influencing a given family of characteristics.

Let a family of characteristics of the heteromagnetic structure be set by the functions $f_1(P_1, \ldots, P_n, V_1, U_1, \ldots, U_k), \ldots, f_m(P_1, \ldots, P_n, V_m, U_1, \ldots, U_k)$, where $P_1, \ldots, P_n$ are the optimized parameters of the model, $V_1, \ldots, V_m$ are consecutively set parameters which define the working point of the structure, $U_1, \ldots, U_k$ the parameters setting a certain curve in a given family of characteristics. For example, for static output characteristics these are the dependences of the collector current $I_{c1}, \ldots, I_{cm}$ on the base–collector voltage U_{bc} at fixed values of the emitter current $I_{e1}, \ldots, I_{em}$: $I_{c1}(P_1, \ldots, P_n, U_{be1}, I_{e1}), \ldots, I_{cm}(P_1, \ldots, P_n, U_{bem}, I_{em})$.

Optimization of the parameters was carried out by the minimum distinction of the areas between the experimental and calculated dependences summarized over all the curves of a family (Fig. 6.5).

Further, piecewise-linear approximation of the experimental and calculated curves was made and the total area of the trapezes was calculated (Fig. 6.6a,b).

$$U_0 = \frac{U_{e2}(f_{p1} - f_{p2}) + U_{e1}(f_{e2} - f_{e1})}{f_{e2} - f_{p1} - f_{e1} + f_{p1}},$$

$$f_{p1} = f_{p11} + (U_{e1} - U_{p11})\frac{f_{p12} - f_{p11}}{U_{p12} - U_{p11}},$$

$$f_{p2} = f_{p21} + (U_{e2} - U_{p21})\frac{f_{p22} - f_{p21}}{U_{p22} - U_{p21}}.$$

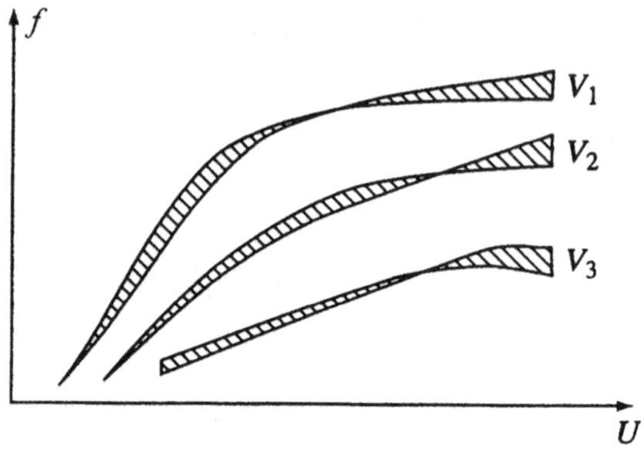

Fig. 6.5 Optimization of the parameters by the minimum distinction of the areas between the experimental and calculated dependences

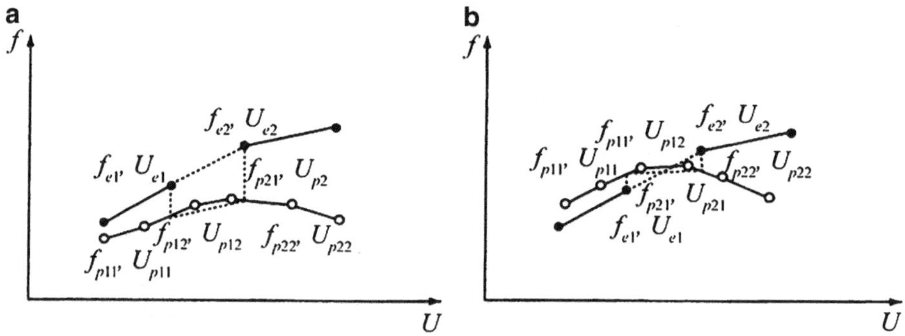

Fig. 6.6 Piecewise-linear approximation of the experimental and calculated curves

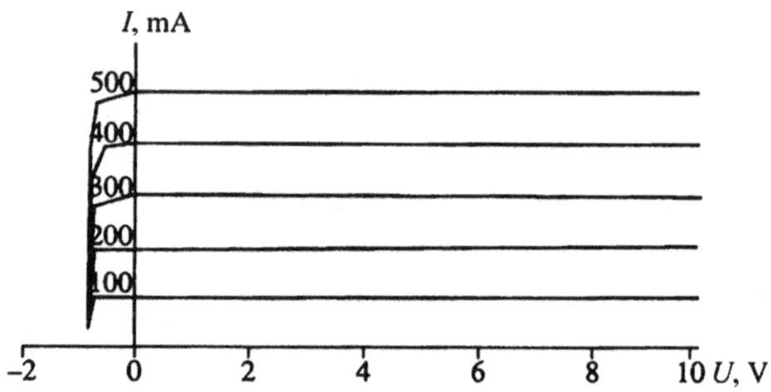

Fig. 6.7 The family of target static characteristics of the KT962B transistor

The areas calculated are summarized over all the experimental points of one curve and over all the curves of each family of characteristics.

At variations of certain parameters of the model, the area between the curves varied, which allows the error function ξ to be calculated.

From the family of target static characteristics of the KT962B transistor resulted in Fig. 6.7, dependences of the error functions ξ of the resistance in the base, emitter, and collector circuits R_{b}, R_{e1}, R_{b2} were calculated (Fig. 6.8).

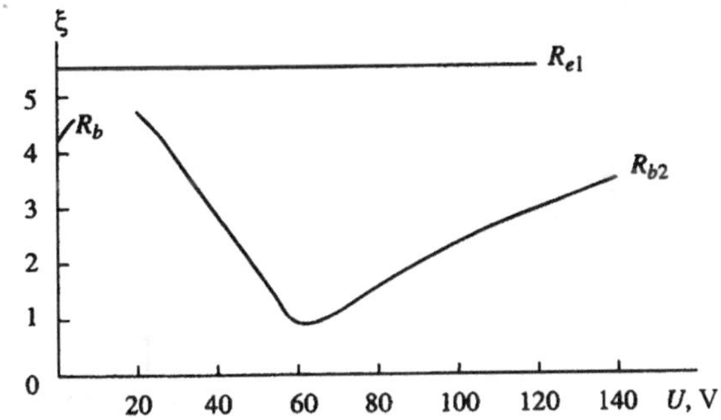

Fig. 6.8 Dependences of the error functions ξ of the resistance in the base, emitter, and collector circuits R_b, R_{e1}, R_{b2}

For the first case (Fig. 6.6a), the error function is calculated from the formula:

$$\xi = (U_{e2} - U_{e1})\frac{(f_{e2} - f_{p1}) + (f_{e2} - f_{p2})}{2},$$

where

$$f_{p1} = f_{p11} + (U_{e1} - U_{p11})\frac{f_{p12} - f_{p11}}{U_{p12} - U_{p11}},$$

$$f_{p2} = f_{p21} + (U_{e2} - U_{p21})\frac{f_{p22} - f_{p21}}{U_{p22} - U_{p21}}. \quad (6.23)$$

For the second case (Fig. 6.7b) we have:

$$\xi = (U_{e2} - U_0)\frac{f_{e2} - f_{p2}}{2} + (U_0 - U_{e1})\frac{f_{p2} - f_{e2}}{2},$$

where

$$U_0 = \frac{U_{e2}(f_p1 - f_{p2}) + U_{e1}(f_{e2} - f_{e1})}{f_{e2} - f_{p2} - f_{e1} + f_{p1}},$$

$$f_{p1} = f_{p11} + (U_{e1} - U_{p11})\frac{f_{p12} - f_{p11}}{U_{p12} - U_{p11}},$$

$$f_{p2} = f_{p21} + (U_{e2} - U_{p21})\frac{f_{p22} - f_{p21}}{U_{p22} - U_{p21}}. \quad (6.24)$$

Figure 6.8 shows that the parameter R_{e1}, R_b cannot be found from the family of output characteristics of the transistor. The parameter R_{b2} has the minimal error at $U = 65$ V. This result is caused by some specificity of measurements of target characteristics when the emitter current is fixed. The parameter R_{e1} can be found from the family of input characteristics of the transistor presented in Fig. 6.9 at various voltages.

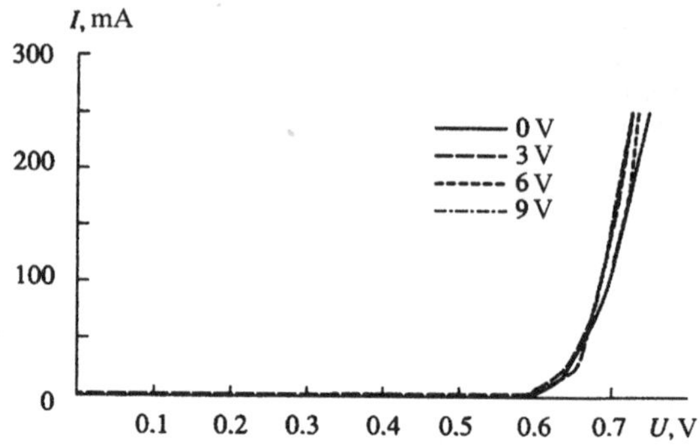

Fig. 6.9 The family of input characteristics of the transistor at various voltages

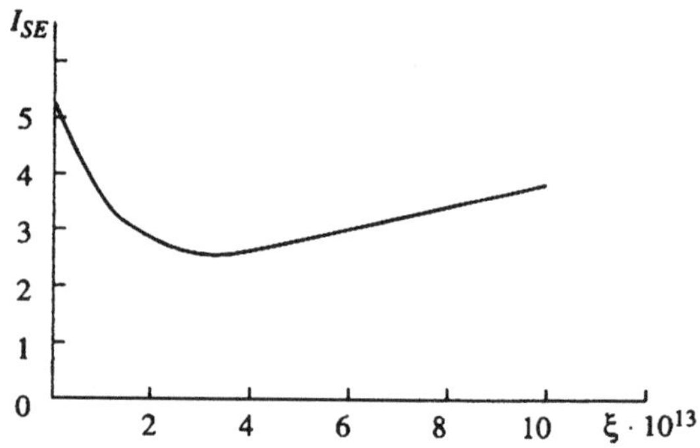

Fig. 6.10 Dependences of the error functions at changes of the parameters I_{SE}

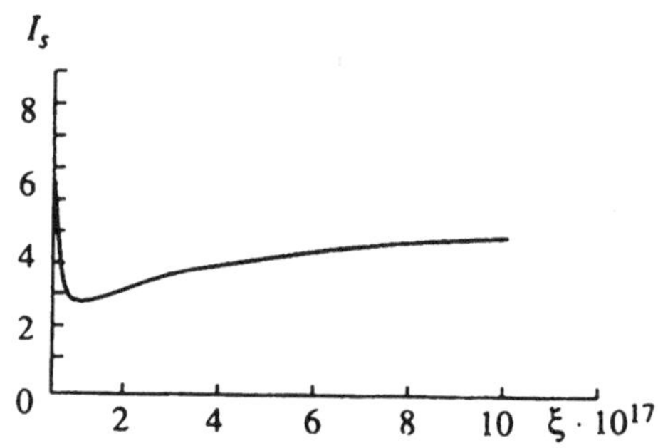

Fig. 6.11 Dependences of the error functions at changes of the parameters I_S

In Figs. 6.10–6.12 dependences of the error functions are presented at changes of the parameters I_{SE}, I_S, I_{SC}, calculated from the family of target static characteristics of the KT962B transistor, the parameters of cutoff currents of the diodes modeling the emitter and collector junctions of the transistor.

The dependences of the error functions of the parameters I_{SE}, I_S, I_{SC} have strongly pronounced minima. Therefore, at optimization, these parameters, together with the parameters N_f, N_R, N_E, N_C entering into the diode exponents, were united in a group to be processed first of all. Several dependences of the error functions at changes of the parameters N_f, N_R, N_E, N_C are presented in Figs. 6.13–6.16.

The dependences of the error functions on the parameters V_A and V_B, which enter in expression for source of current, are presented in Figs. 6.17 and 6.18. Here

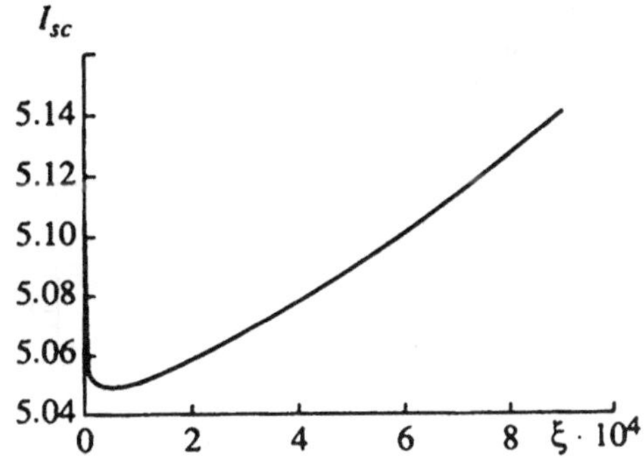

Fig. 6.12 Dependences of the error functions at changes of the parameters I_{SC}

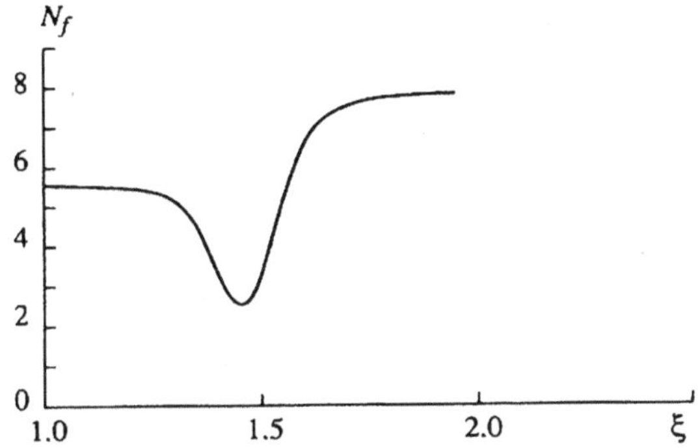

Fig. 6.13 Dependences of the error functions at changes of the parameters N_f

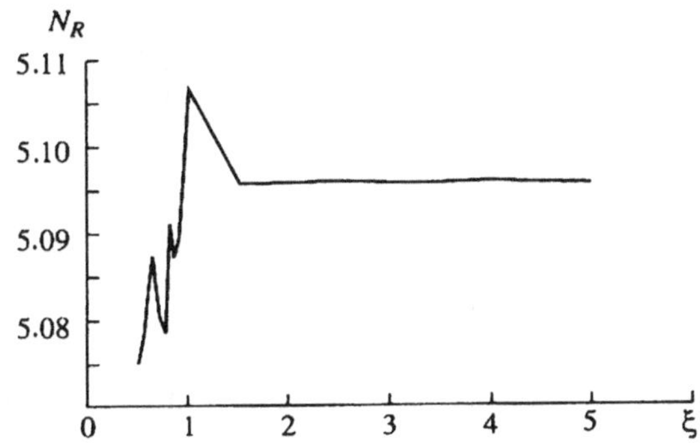

Fig. 6.14 Dependences of the error functions at changes of the parameters N_R

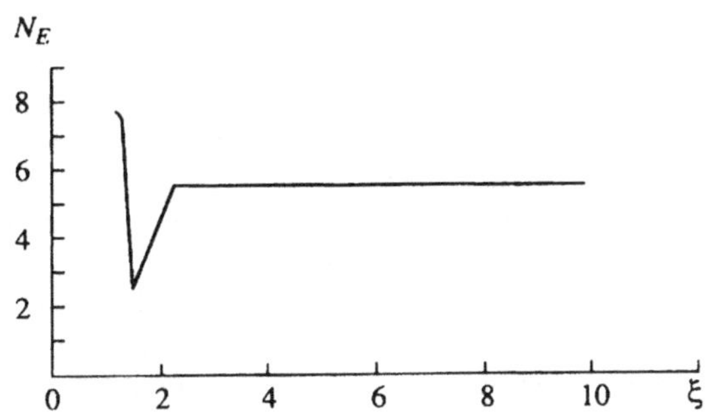

Fig. 6.15 Dependences of the error functions at changes of the parameters N_E

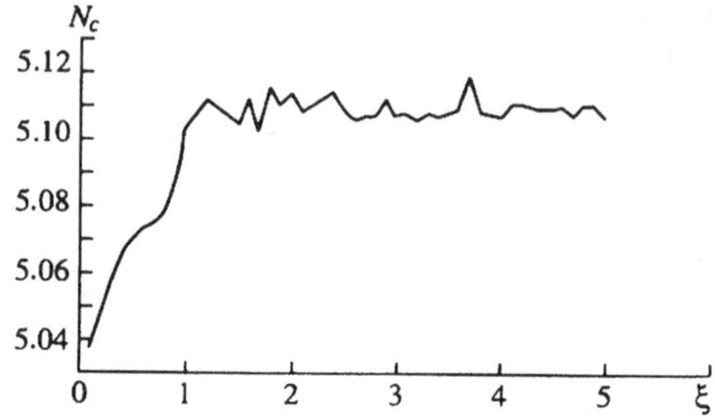

Fig. 6.16 Dependences of the error functions at changes of the parameters N_C

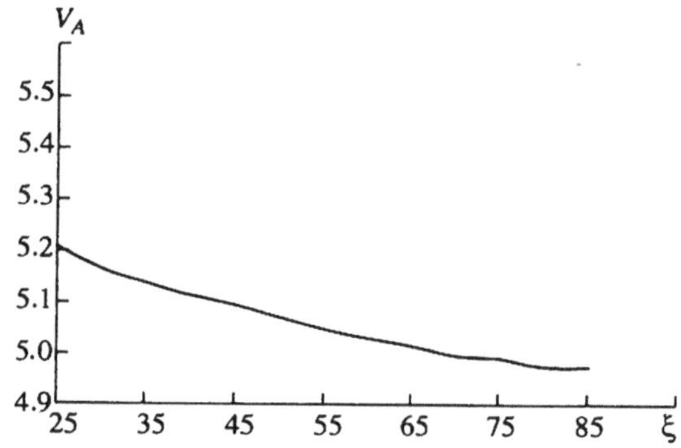

Fig. 6.17 The dependences of the error functions on the parameters V_A

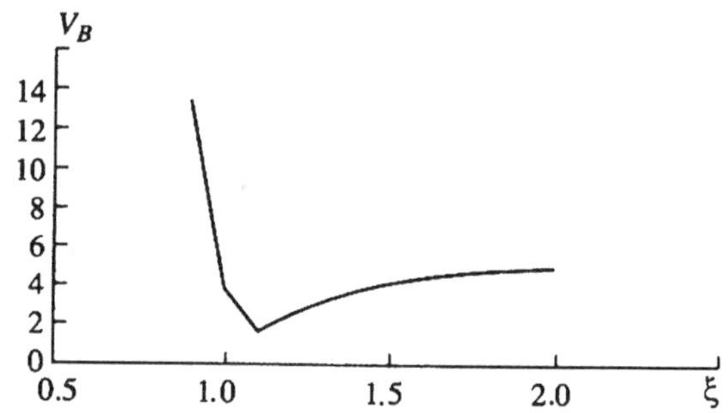

Fig. 6.18 The dependences of the error functions on the parameters V_B

by, target characteristics need to optimize only parameter V_B. The parameter V_A to act the immaterial part and appeared at inverse of transistor switching-on.

The dependences of the error functions on the parameters I_{KF}, I_{KR} and B_F, B_R are presented in Figs. 6.19 and 6.20. One can see that the parameters I_{KF} and B_R cannot be found from the family of target characteristics. The dependences I_{KR} and B_F have the corresponding minima.

The static equivalent parameters of the HMG model were sought by the gradient descent method. For calculation of the static characteristics, the set of nonlinear algebraic equations derived from (6.3) to (6.19) under the condition of zero derivatives by time was solved.

Fig. 6.19 The dependences of the error functions on the parameters I_{KF}, I_{KR}

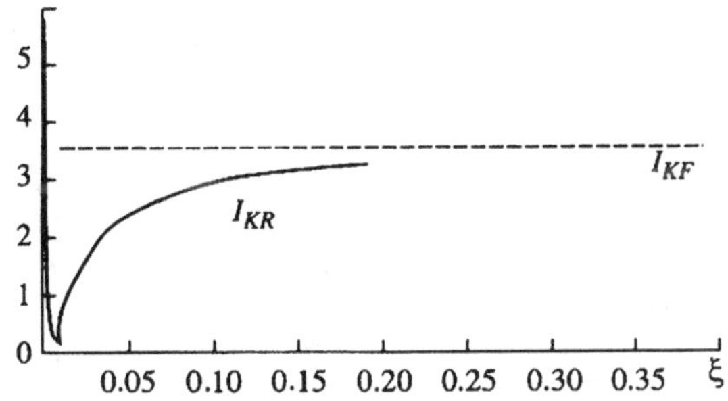

Fig. 6.20 The dependences of the error functions on the parameters B_F, B_R

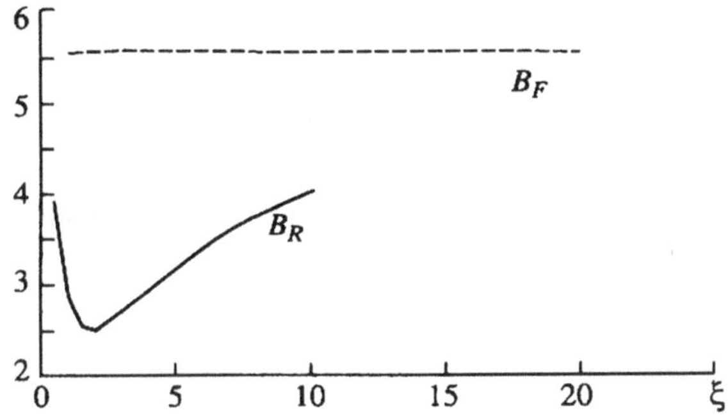

The results of calculation of the error function ξ at changes of the parameters of the HMG model confirm the expediency of preliminary splitting the parameters on groups. At a wrong choice of such groups, a loss of the physical sense of the optimized parameters (a negative resistance, etc.), weak monotonous change of the error function (Fig. 6.17), or failure of the gradient descent algorithm owing to overflow are possible.

The first group includes the parameters I_S, I_{SE}, I_{SC}, I_{SS}, N_S, N_F, N_E, N_R, N_C setting the exponential VACs of the diodes and most strongly influencing the shape of the input and output characteristics. At the first stage, the method of gradient descent was applied in the space of these parameters. The parameters of a foreign analog of the used transistor were taken as an initial approximation.

The second group included the parameters R_{b2}, R_{c2}, R_{e1}, and R_{bb} setting the base, collector, and emitter resistances.

The third group included the parameters B_F, B_R, V_A, I_{KR}, I_{KF}, R_B, and R_{BM} determining the properties of power sources.

After application of the method of gradient descent consistently on the three groups of parameters, the optimization was finished with minimization of the error function in the full space of the required parameters of the model.

Thus, the technique of determination of the 21 static parameters of Gummel–Poon's equivalent model has been developed.

6.4 Equivalent Circuit of a Multifunctional Heteromagnetic Oscillator

The considered HMG can be brought to the generating mode for recording the necessary families of characteristics. For determination of the dynamic parameters of the equivalent circuit, the family of experimental dependences of the generation frequency on the collector voltage was used at fixed displacement voltages on the base. The results of measurements of these dependences at various voltages on the base ($U_{BC} = 11, \ldots, 28$ V) are presented in Fig. 6.21. This family of characteristics was used for optimization of the dynamic parameters of HMG.

In Figs. 6.22–6.25 dependences of the error functions are presented at variation of the parameters CJC, CJE, MJC, and VJC determining the barrier capacities of the collector and emitter junctions.

From the figures, it follows that all these dynamic characteristics have expressed minima, which allows their optimization.

The parameter XCJC entering into the equations for the barrier capacities does not influence the error function (Fig. 6.26) and has not been used in our procedure of optimization.

The equivalent parameters of the HMG model were found by the method of gradient descent in the space of the parameters CJC, CJE, MJC, and VJC. The inductances of the terminals were taken as nominal data of the transistor.

Thus, the technique of finding the 26 parameters of the transistor model has been developed.

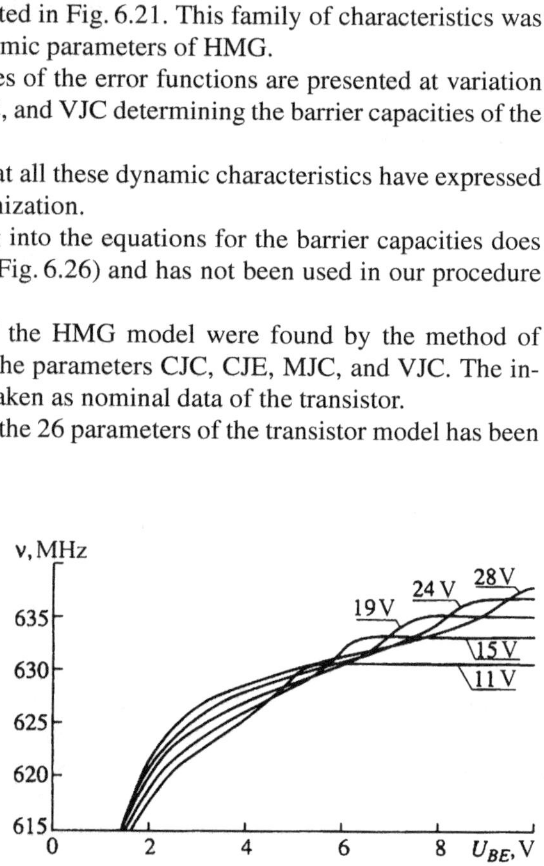

Fig. 6.21 The family of experimental dependences of the generation frequency on the collector voltage was used at fixed displacement voltages on the base

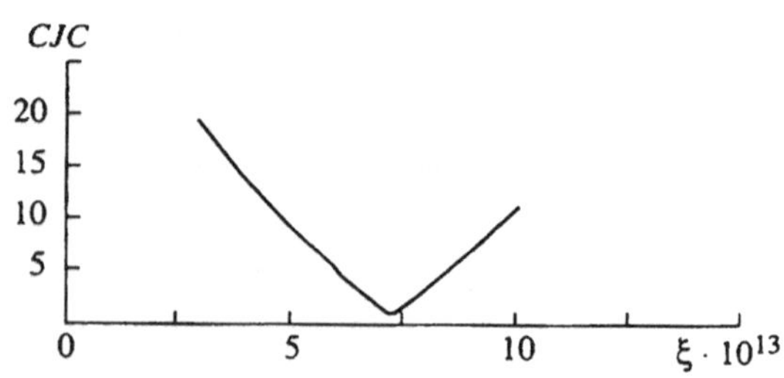

Fig. 6.22 Dependences of the error functions are resulted at variation of the parameters CJC

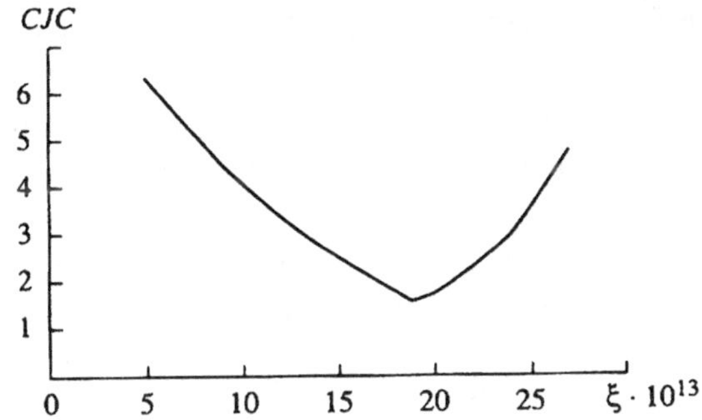

Fig. 6.23 Dependences of the error functions are resulted at variation of the parameters CJE

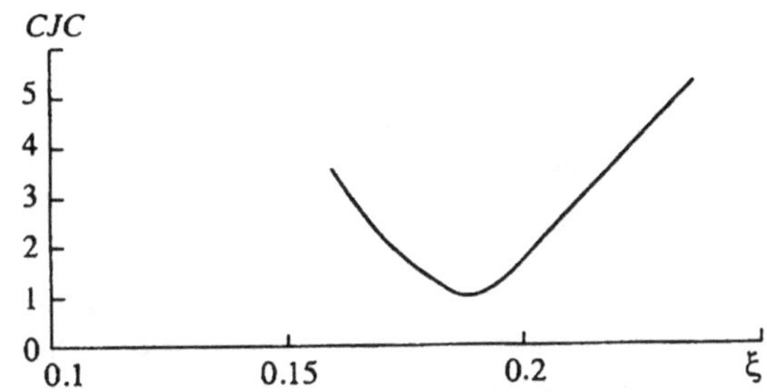

Fig. 6.24 Dependences of the error functions are resulted at variation of the parameters MJC

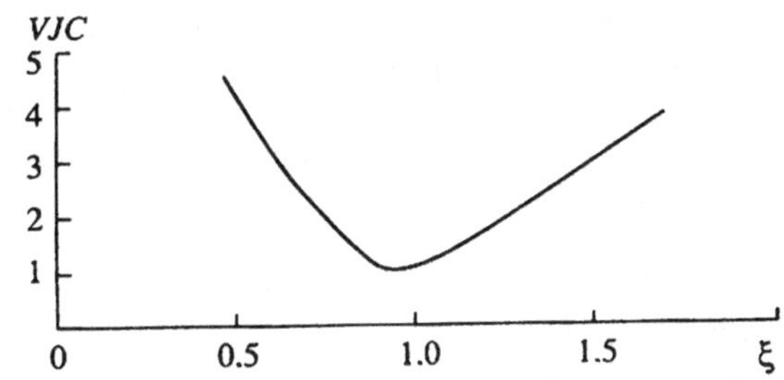

Fig. 6.25 Dependences of the error functions are resulted at variation of the parameters VJC

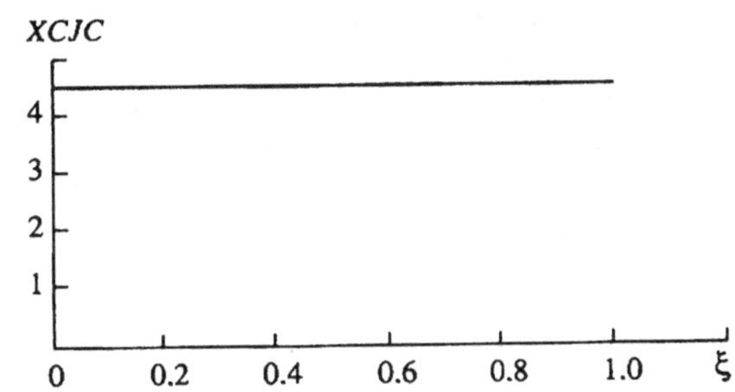

Fig. 6.26 The parameter XCJC entering into the equations for the barrier capacities does not influence the error function

6.5 Oscillating Modes of Subharmonic Constituents

Generation of spectral pure subharmonic components in HMG with a high number of subharmonics[2] at a constant width of the spectral line

$$(\Delta\nu_{-3\text{dB}})_0 = (\Delta\nu_{-3\text{dB}})_{-1} = (\Delta\nu_{-3\text{dB}})_{-2} = \cdots = (\Delta\nu_{-3\text{dB}})_{-m} = \text{const},$$

$$(\Delta\nu_{-60\text{dB}})_0 = (\Delta\nu_{-60\text{dB}})_{-1} = (\Delta\nu_{-60\text{dB}})_{-2} = \cdots = (\Delta\nu_{-60\text{dB}})_{-m} = \text{const}$$

and the constancy of the amplitudes $P_{\nu_0} = P_{\nu_{-1}} = P_{\nu_{-2}} = \cdots = P_{\nu_{-m}} = \text{const}$ determines the physics of processes as parametrical division of a signal of the basic frequency ν_0.

Another mode observable experimentally in HMG is associated with generation of equidistant array frequencies on the basic frequency ν_0 and in the field of each subharmonic component ν_{-m+1} $(m = 1, 2, 3, \ldots)$ with synchronous management of the frequency distances between the components, and the width of the spectral components of such signals also did not change. It confirms parametrical frequency modulation.

In both the modes on one crystal the following were observed: synchronous noisiness of the pedestal of bearing frequencies; broadening of the spectral lines; transition to broadband noisy signals by the envelope of the equidistant array frequency spectra; synchronous reorganization by a magnetic field; and electrical synchronous reorganization of all the central frequencies of the harmonic and subharmonic components and the frequency distances between them.

The indicated modes in heteromagnetic structures would be described by multicoherent parametrical interactions[3] in the ferrite and transistor subsystems possessing different nonlinear properties, including nonlinear resonance in the ferrite subsystem.

For description of the generation mode of subharmonic components (spectrally pure, noisy components, noisy signals), a model of nonlinear two-coherent transistor-magnetic structure with signal generation on the basic ν_0 and subharmonic $\nu_{-1} \approx \nu_0/2$ components has been developed. The method of slowly varying amplitudes used in analysis of this model does not allow research in the whole frequency range between ν_0 and ν_{-m} to be carried out. Therefore, for expansion of our analysis over the whole frequency range toward the area of the maximum subharmonic components with $m = 1, 2, \ldots$ the problem of generation of the lowest subharmonics in a two-coherent system must be solved with subsequent passage to the multiply connected generalized model of HMG.

---

[2] In the first experiments on the structures with the base KT9382A transistor at integrated power levels of the order of 40–60 mW, subharmonic components have been registered with $m = 71$ $\nu_{-70} = 17$ MHz at the basic frequency $\nu_0 = 1.2$ GHz.

[3] Multiplication (for harmonics), division (for subharmonics), simultaneous parametrical frequency modulation (frequencies spectra) of steady (for spectrally pure signals) and unstable (for noisy signals) oscillations.

The equivalent circuit of HMG represents an oscillatory system with many degrees of freedom. At a frequency rate of the resonant frequencies of domains ν_0 and ν_{-1} of the ferrite resonators, effective nonlinear interactions between the generated modes arise.

Let us present the transistor as a nonlinear element with active and reactive components. This simplifies the equation set, allows the usage of the method of slowly varying amplitudes and to obtain a number of important analytical and numerical results confirming the effects observable in our physical experiments.

Let us consider two oscillatory contours tuned to the basic ν_0 and subharmonic ν_{-1} frequencies. The equivalent scheme of HMG is presented in Fig. 6.27. The nonlinear elements R_t, C_t are determined by the transistor with a feedback in a dynamic mode. These elements have various WAC on the frequencies ν_0 and ν_{-1}.

The volt–ampere characteristics of the nonlinear element were calculated from the model of the KT962B transistor with feedbacks on the frequencies ν_0 and ν_{-1} (Figs. 6.28 and 6.29) ($\nu_0 = 635.0$ MHz, $\nu_{-1} = 317.5$ MHz).

On the nonlinear reactance L_0, the high-frequency current and voltage are shifted by phase, and this shift depends on their amplitude values.

At small amplitudes of oscillations, the relation between the basic and subharmonic types of fluctuations in FMCR is weak. Neglecting this relation we shall

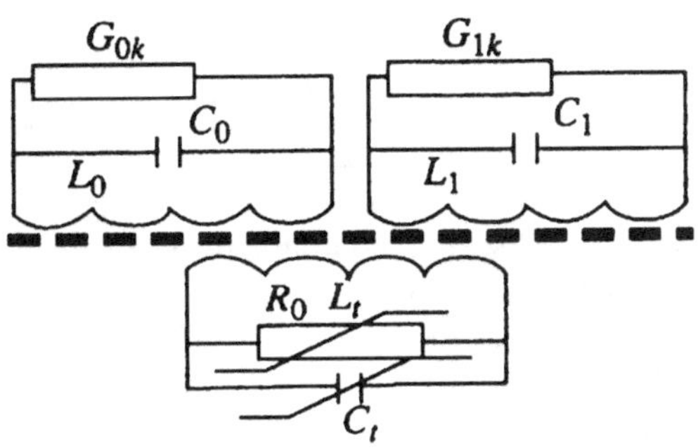

Fig. 6.27 The equivalent scheme of HMG

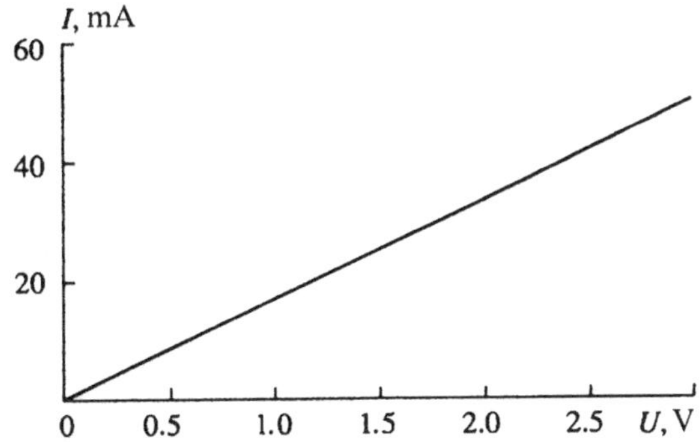

Fig. 6.28 The volt–ampere characteristics of the nonlinear element calculated from the model of the KT962B transistor with feedbacks on the frequency ν_0

Fig. 6.29 The volt–ampere characteristics of the nonlinear element calculated from the model of the KT962B transistor with feedbacks on the frequency ν_{-1}

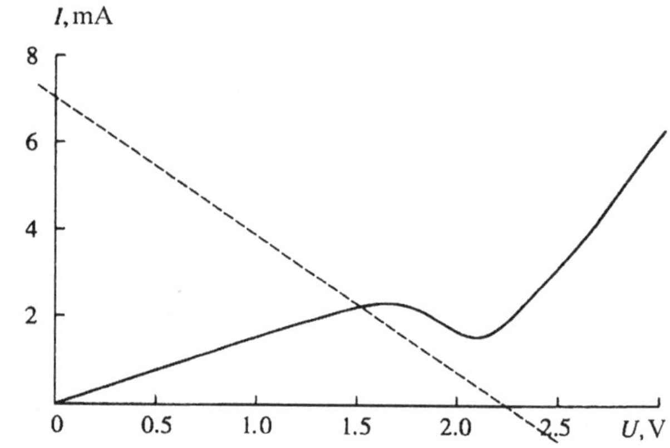

consider that a nonlinear relation between ν_0 and ν_{-1} arises in the transistor. In this case, the voltage on the nonlinear element will be equal to the sum of the voltages on the contours with the corresponding factors of transformation.

Having designated U_1 – the voltage on the first contour at ν_0 and U_2 – the voltage on the second contour at ν_{-1}, the current through the nonlinear element – I_t. Using the method of nodal potentials [45], we derive a set of two second-order differential equations:

$$\begin{cases} \dfrac{d^2U_1}{dt^2} + \omega_1^2 U_1 = \dfrac{1}{C_1 M_1} \dfrac{d(I_t + G_{1k} U_1)}{dt}, \\ \dfrac{d^2U_2}{dt^2} + \omega_2^2 U_2 = \dfrac{1}{C_2 M_2} \dfrac{d(I_t + G_{2k} U_2)}{dt}, \end{cases} \tag{6.25}$$

where $\omega_1 = 2\pi\nu_0$, $\omega_2 = 2\pi\nu_{-1}$ are the own cyclic frequencies of oscillations of the first and second contours, respectively, $M_{1,2}$ the coupling factors of the inductances $L_{1,2}$ and L_0.

Let us make the substitution of variables:

$$\begin{cases} U_1 = \dfrac{U_{10} \times e^{i\omega_1 t} + U_{10}^* \times e^{-i\omega_1 t}}{2}, \\ U_2 = \dfrac{U_{20} \times e^{i\omega_2 t} + U_{20}^* \times e^{-i\omega_2 t}}{2}. \end{cases} \tag{6.26}$$

Then (6.25) can be written in the form of:

$$\begin{cases} i\omega_1 \dfrac{dU_{10}}{dt} e^{i\omega_1 t} = -\dfrac{d}{dt}\left(\dfrac{I_t + G_{1k} U_{10}}{C_1 M_1}\right), \\ i\omega_2 \dfrac{dU_{20}}{dt} e^{i\omega_2 t} = -\dfrac{d}{dt}\left(\dfrac{I_t + G_{2k} U_{20}}{C_2 M_2}\right). \end{cases} \tag{6.27}$$

Let us truncate (6.27) by averaging [45,46]. Integrating the right-hand sides of (6.27) by parts, we obtain equations for the corresponding envelopes:

$$\begin{cases} \dfrac{\mathrm{d}U_{10}}{\mathrm{d}t} = \dfrac{1}{C_1 M_1 T_2} \displaystyle\int_0^{T} (I_t + G_{1k} U_1) \times \mathrm{e}^{-\mathrm{i}\omega_1 t} \mathrm{d}t, \\ \dfrac{\mathrm{d}U_{20}}{\mathrm{d}t} = \dfrac{1}{C_2 M_2 T_2} \displaystyle\int_0^{T_2} (I_t + G_{2k} U_2) \times \mathrm{e}^{-\mathrm{i}\omega_2 t} \mathrm{d}t. \end{cases} \tag{6.28}$$

The integrals in (6.28) are taken over the period of subharmonic oscillation $T_2 = 2\pi/\omega_2$. To get the equations for U_{10}, U_{20} in an explicit form, it is necessary to specify the VAC of the nonlinear element.

Let the sum of two voltages is applied to the transistor, namely: the basic one with a frequency ω_1 and a subharmonic one with a frequency $\omega_2 = \omega_1/2$. If the amplitudes of these voltages change slowly, the relation between the instant voltage and the total current will be functional and the VAC can be set as a polynomial

$$I_t = \sum_{j,k} G_{jk} U_1^j U_2^k + \sum_{p,q} C_{\mathrm{pq}} U_1^p U_2^q. \tag{6.29}$$

In our approximation $U_2 < U_1$, therefore let us limit to those members of the series (6.29) for which $j + 2k < 4, p + 2q < 5$. If we assume that generation on the frequency does not exist, the last relation means that we are limited to a cubic decomposition of the VAC of the nonlinear element.

Substituting the expression for the nonlinear element current (6.29) in (6.28) and taking integration in view of the slowness of change of U_{10}, U_{20} we derive a set of equations of the model in a complex form:

$$\begin{cases} C_1 M_1 \dfrac{\mathrm{d}U_{10}}{\mathrm{d}t} = \left(\alpha_1 + i\alpha_1^-\right) U_{10} + \left|G_1^-\right| \times \mathrm{e}^{\mathrm{i}\arg(G_1^-)} \\ \qquad \times U_{20}^2 + \gamma_1 U_{10}^2 U_{10}^* + i p_1^- \left|U_{20}\right|^2 U_{10}, \\ C_2 M_2 \dfrac{\mathrm{d}U_{20}}{\mathrm{d}t} = \left(\alpha_2 + i\alpha_2^-\right) U_{20} + \left|G_2^-\right| \times \mathrm{e}^{\mathrm{i}\arg(G_2^-)} \\ \qquad \times U_{10} U_{20}^* + p_2^- \left|U_{10}\right|^2 U_{20}, \end{cases} \tag{6.30}$$

where the factors are

$$\alpha_1 = -\frac{G_{10} + G_{1k}}{2}, \quad \alpha_1^- = -\frac{\omega_0 C_{00}}{4}, \quad G_1^- = -\frac{G_{02}}{4} - i\frac{\omega_2 C_{01}}{4},$$

$$p_1^- = -\frac{\omega_0 C_{02}}{4}, \quad \gamma_1 = -\frac{3}{8} G_{30},$$

$$\alpha_2 = \frac{G_{20} + G_{2k}}{2}, \quad \alpha_2^- = -\frac{\omega_2 C_{00}}{2}, \quad G_2^- = -\frac{G_{11}}{4} + i\frac{\omega_2 C_{10} - \omega_1 C_{01}}{4},$$

$$p_2^- = -\frac{\omega_2 C_{20}}{4}, \quad \gamma_1 = -\frac{3}{8} G_{30}.$$

In (6.30) at approximation of the reactive component of the current of the transistor, the components $p_2^- U_{10}^2$, $p_1^- U_{20}^2$ are considered as in the generation mode

$\arg\left(G_1^-\right) + \arg\left(G_2^-\right) = 0$. Despite the little nonlinear reactive conductance, their account leads to occurrence of qualitatively new modes in the system, including stochastic ones.

Let us lead the derived equations to a dimensionless form. Let us enter a generalized phase $\psi = \varphi_1 + \delta \times t - 2\varphi_2 + \arg\left(G_2^-\right)$, where φ_1 is the envelope phase on the frequency ν_0, φ_2 the envelope phase on the frequency ν_{-1}, $\delta = \nu_0 - 2\nu_{-1}$ the parameter of oscillation desynchronization. Let us pass from the equation set (6.30) to a set of equations for the envelope amplitudes and the generalized phase:

$$\begin{cases} C_1 M_1 \dfrac{\mathrm{d}A_1}{dt} = \alpha_1 A_1 + \left(G_1^{\mathrm{re}} \cos\psi + G_1^{\mathrm{im}} \sin\psi\right) \times A_2^2 + \gamma_1 A_1^3, \\ C_2 M_2 \dfrac{\mathrm{d}A_2}{dt} = \alpha_2 A_2 + \left(\left|G_2^-\right| A_1 A_2 \cos\psi\right), \\ \dfrac{d\psi}{dt} = \delta + 2\dfrac{\alpha_2^-}{C_2} + \left(\dfrac{G_1^{\mathrm{im}}}{C_1} \cos\psi + \dfrac{G_1^{\mathrm{re}}}{C_1} \sin\psi\right) \dfrac{A_2^2}{A_1} \\ \qquad -2A_1 \dfrac{\left|G_2^-\right|}{C_2} \sin\psi - \dfrac{2p_2^-}{C_2} + \dfrac{p_1^-}{C_1} A_2^2, \end{cases} \tag{6.31}$$

where $U_{10} = A_1 e^{i\varphi_1 + \delta \times t}$, $U_{20} = A_1 e^{i\varphi_2}$, $G_1^{\mathrm{re}} + iG_1^{\mathrm{im}} = |G_1^- \exp(\arg(G_1^-) + \arg(G_2^-))|$.

The physical sense of the factors in (6.30) is as following:

α_1 is the increment of the amplitude of oscillations of/at the basic frequency ($\alpha_1 > 0$); α_2 the decrement of the amplitude of oscillations of the subharmonic ($\alpha_2 > 0$); α_1^- and α_2^- the factors determining the constant shift of adjustment frequencies of the contours due to the capacity of the transistor; γ_1 the factor of nonlinear dissipation determining restriction of the amplitude of oscillations on the basic frequency due to the VAC nonlinearity ($\gamma_1 < 0$) ; G_1^- and G_2^- the factors of oscillation coupling in the first and second contours, respectively; p_1^- and p_2^- the factors reflecting the nonlinearity of the capacity at changes of the amplitudes of oscillations on the frequencies ν_0 and ν_{-1}.

Let us consider explicitly the assumptions concerning the signs of the factors considered above. For this purpose, change the variables:

$$A_1 = \frac{|\alpha_2|}{\left|G_2^{\mathrm{re}}\right|} a_1, A_2 = \sqrt{\frac{|\alpha_1 \alpha_2|}{\left|G_1^- G_2^-\right|}} \times a_2, t = \frac{C_2}{|\alpha_2|} \xi.$$

Then (6.31) can be written in the form of:

$$\begin{cases} \dfrac{d\alpha_1}{d\xi} = k \left(\alpha_1 + \left(\dfrac{G_1^{\mathrm{re}}}{\left|G_1^-\right|} \cos\psi + \dfrac{G_1^{\mathrm{im}}}{\left|G_1^-\right|} \sin\psi\right) \times \alpha_2^2 - \gamma_1 \alpha_1^3\right., \\ \dfrac{d\alpha_2}{d\xi} = -\alpha_2 + \alpha_1 \alpha_2 \cos\psi \Big), \\ \dfrac{d\psi}{d\xi} = \delta_s + p_2 \alpha_2^2 - p_1 \alpha_1^2 + \left(\dfrac{G_1^{\mathrm{im}}}{\left|G_1^-\right|} \cos\psi - \dfrac{G_1^{\mathrm{re}}}{\left|G_1^-\right|} \sin\psi\right) \times k \dfrac{\alpha_2^2}{\alpha_1} - 2\alpha_1 \sin\psi, \end{cases} \tag{6.32}$$

where the dimensionless factors are

$$k = \frac{\alpha_1 C_2 M_2}{\alpha_2 C_1 M_1},\ \gamma = \frac{\gamma_1 \alpha_2}{|G_2^-|^2}\frac{C_2 M_2}{C_1 M_1},\ p_1 = \frac{2p_2^- |\alpha_2|}{|G_2^-|^2},\ p_2 = \frac{p_1^- |\alpha_1|}{|G_1^- G_2^-|}\frac{C_2 M_2}{C_1 M_1}$$

$$\delta_s = \frac{C_2 \delta}{|\alpha_2|} - \frac{2\alpha_2^-}{|\alpha_2|} + \frac{\alpha_1 C_2}{|\alpha_2|\, C_1},\ \xi = \frac{|\alpha_2|}{C_2} t. \tag{6.32a}$$

For numerical analysis of (6.32) it is convenient to use the variables offered in Ref. [46]:

$$x = -\alpha_1 \cos\psi, \quad y = \alpha_1 \sin\psi, \quad z = \alpha_2^2.$$

This allows (6.32) to be driven to

$$\begin{aligned}
\frac{dx}{d\xi} &= kx + (\delta_s + p_2 z - p_1(x^2+y^2)) \cdot y - \cos\theta \cdot kz - 2y - \gamma \cdot x \cdot (x^2+y^2),\\
\frac{dy}{d\xi} &= ky - (\delta_s + p_2 z - p_1(x^2+y^2)) \cdot x + \sin\theta \cdot kz + 2xy - \gamma \cdot y \cdot (x^2+y^2),\\
\frac{dz}{d\xi} &= -2z(1+x),
\end{aligned} \tag{6.33}$$

where

$$\cos\theta = \frac{|G_1^{re}|}{|G_1^-|}, \quad \sin\theta = \frac{|G_1^{im}|}{|G_1^-|}.$$

Stationary points are the simplest solutions of the set of differential equations (6.33). They correspond to the modes of spectrally pure oscillations on the basic frequency ν_0 and the subharmonic frequency ν_{-1}. Studying of stationary points is important for understanding of the dynamics of the system investigated. Generally, determination of analytical stationers for (6.33) leads to bulky calculations, therefore, we shall consider special/particular cases:

1. Only oscillations on the basic frequency ν_0 exist (no oscillation on the half frequency ν_{-1} owing to the inefficiency of the interaction).
2. Simultaneous monochromatic generation on the two frequencies ν_0 and ν_{-1}.
3. Self-modulation of oscillations in connection with periodic energy swapping between the contours (equidistant frequency spectra in the vicinities of the generated modes ν_0 and ν_{-1}).
4. Stochastic generation of noisy oscillations in the surroundings of the frequencies ν_0 and ν_{-1}.

Let us consider the first case. Let the contour mismatch parameter be $\delta_s = 0$. Neglect also the displacement of the frequency of oscillations due to the influence of the nonlinear capacity $p_1^- = p_2^- = 0$. The stationary solution of (6.33) will be:

$$x = \pm\sqrt{\frac{k}{\gamma}},$$

$$y = 0,$$
$$z = 0. \tag{6.34}$$

The matrix of the linearized (6.33) generally looks like:

$$\begin{bmatrix} k - 2p_2xy - 3\gamma x^2 - \gamma y^2 & \delta_s + p_1z - p_2x^2 - 3p_2y^2 - 4y - 2\gamma xy & p_1y - \cos\theta \\ -\delta_s - p_1z + 3p_2x^2 + p_2y^2 & k + 2p_1xy + 2x - \gamma x^2 - 3\gamma y^2 & p_1x + \sin\theta \\ +2y - 2\gamma xy & & \\ -2z & 0 & -2 - 2x \end{bmatrix} \tag{6.35}$$

Solving the characteristic equation from (6.35), we find three values:

$$\lambda_1 = -2k,$$
$$\lambda_2 = \pm\sqrt{\frac{k}{y}},$$
$$\lambda_3 = -2 + 2\sqrt{\frac{k}{y}}. \tag{6.36}$$

For analysis of possible changes of the modes of spectrally pure oscillations[4] in (6.34), the first own value ($\lambda_1 = -2\,k$) represents no interest as it does not exceed zero. Analyzing the second own value $\lambda_2 = \pm\sqrt{k/y}$, it is possible to conclude that the point $x = \sqrt{k/y}$ is always unstable, and the point $x = -\sqrt{k/y}$ is always negative.

A zero third own value of (6.36) determines a bifurcation straight line $k = y$ in the space of parameters.

When $(k/y) > 1$, there occurs simultaneous generation on two frequencies. Note that the derived relationship is fair for the case of no mismatch between the adjustment frequencies of the contours. Relative mismatch of the contours appears because of the discrepancy of the doubled own frequency ν_{-1} of the second contour and the own frequency ν_0 of the first contour, due to the linear and nonlinear components of the nonlinear element capacity. The linear component is shown in the parameter δ_s, and the nonlinear one in the parameters p_1, p_2, θ. As when $\delta_s \neq 0$ the energy exchange between the contours is complicated, the two-frequency generation would occur at greater/higher values of the parameter $D = k/y > 1$. If we accept that the parameters δ_s, $p_1, p_2 \neq 0$, then the stationary solution of (6.33) will be:

$$x = \sqrt{\frac{k}{y} - y^2},$$

[4] Bifurcations of the stationary solution of (6.34).

$$y = \frac{\delta_s - p_1(k/y)}{2},$$
$$z = 0. \tag{6.37}$$

Thus, the phase shift of oscillations on the basic ν_0 and half ν_{-1} frequencies is $\psi \neq 0$. If the mismatch of the contours is not so great, it is possible to limit only to the third own value of the matrix (6.35) for bifurcational[5] analysis, believing the other two-ones to be negative. The bifurcational curve[6] in this case is set by the equation

$$\frac{k}{y} - \frac{(\delta_s - p_1(k/y))^2}{4} = 1 \tag{6.38}$$

Equation (6.37) shows that bifurcation always arises at values $D > 1$.

The energy exchange between the contours in (6.31) determines the depth of connection between the basic and subharmonic oscillations that depends on the factors G_1^{re}, G_1^{im}, G_2^-. Having multiplied the first equation in (6.31) by A_1 and the second one by A_2, we get equations for the oscillation energies in the contours.

There are two various energy exchanges between these oscillations on the frequencies ν_0 and ν_{-1}.

In the first case, occurrence of a subharmonic oscillation is associated with additional powering, and the energies of oscillations on the basic ν_0 and subharmonic ν_{-1} frequencies grow simultaneously. The oscillatory contours are not connected. No periodic energy swapping between the contours occurs.

In the second case, the oscillations on the frequencies ν_0 and ν_{-1} are strongly coupled, and the total power of oscillations in both contours is constant $P_{\nu 0} + P_{\nu 1} =$ const. The subharmonic oscillation exists at the expense of the basic one ν_0. The power lost by the basic frequency oscillation is equal to the power got by the subharmonic, and

$$G_1^{\mathrm{re}} \cos\psi + G_1^{\mathrm{im}} \sin\psi + |G_2^-| \cos\psi. \tag{6.39}$$

This equation holds for any value of ψ at

$$G_1^{\mathrm{re}} = -|G_2^-|, G_1^{\mathrm{im}} = 0.$$

In practice, both situations show to some extent, depending on the voltage of displacement of the transistor. In the second case, the energy exchange between oscillations may have an oscillatory character, therefore, it is of interest for further research. The modes of self-oscillations in (6.33) are important at violation of the condition (6.39).

---

[5] The analysis of qualitative change of a mode of fluctuations at change of operating parameters, for example, transition from one-private ν_{-1} generation to two-private ν_{-1}, ν_{-2} from spectral pure signals to equidistant to frequencies spectra.

[6] The curve in the parameter space whose crossing changes the mode of oscillations qualitatively. In this case, generation on the two frequencies ν_{-1} and ν_{-2} occur.

Let us put $\delta_s = 0$, $p_1 = 0$, $p_2 = 0$, $G_1^{re} = 0$, $z \neq 0$. In this case, (6.33) has two stationary solutions:

$$x = -1,$$
$$y^2 = \frac{k-2}{\gamma} - 1,$$
$$z = \frac{(4-2k)}{k\gamma}\cos\theta; \tag{6.40}$$

$$x = -1$$
$$y = \frac{\delta_s - p_1\frac{k}{\gamma}}{2},$$
$$z = \cos(\theta)\frac{k}{\gamma} - 1. \tag{6.41}$$

The first solution (6.40) exists when $k > \gamma + 2$ (since $y^2 > 0$) and is unstable. For existence of the second stationary solution (6.41), the condition $(\pi/2) < \theta < (3\pi/4)$ (since $z = \alpha^2 > 0$) is necessary. Exploring the stability of the stationary solution[7] (6.41), we derive three eigenvalues (6.35):

$$\lambda_1 = k - 2 - \gamma,$$
$$\lambda_{2,3} = \frac{k - 3\gamma \pm \sqrt{(k-3\gamma)^2 - 8k(1-(\gamma/k))}}{2}. \tag{6.42}$$

The first and the real parts of the second and third ones determine bifurcational straight lines in the space of parameters.

The bifurcational diagram of the stationary solutions of set (6.33) on the plane k, γ for the case $\delta_s = 0$, $p_1 = 0$, $p_2 = 0$, $\theta = \pi$ is presented on Fig. 6.30. The qualitative spectra of oscillations observed in HMG for each area/range are presented. The straight lines $\gamma = k$, $\gamma = k/3$ and $\gamma = k - 2$ divide the plane into four areas 1–4. Area 1 corresponds to the existence of oscillations on the basic frequency ν_0 only. The excitation parameter k equal to the ratio of the increment of oscillation on the basic frequency ν_0 and the decrement of oscillations on the subharmonic frequency ν_{-1} in this area is small. The parameter γ determining nonlinear energy dissipation on the basic frequency ν_0 is great enough. At a preset value of γ, the amplitude of oscillations on the basic frequency ν_0 is small and no oscillations on the subharmonic frequency ν_{-1} are supported.

With an increase of k, transition through the border $\gamma = k/3$ of areas 2 and 3 is accompanied by occurrence of simultaneous synchronous generation on two frequencies, namely, the basic one ν_0 and subharmonic one ν_{-1}. At transition through the border of areas 2 and 3, periodic energy swapping between the equivalent contours appears after Hopf's bifurcation, and the spectrum of the initial UHR

[7] The stationary solution corresponds to simultaneous generation of spectrally pure lines on the basic ν_0 and subharmonic ν_{-1} frequencies.

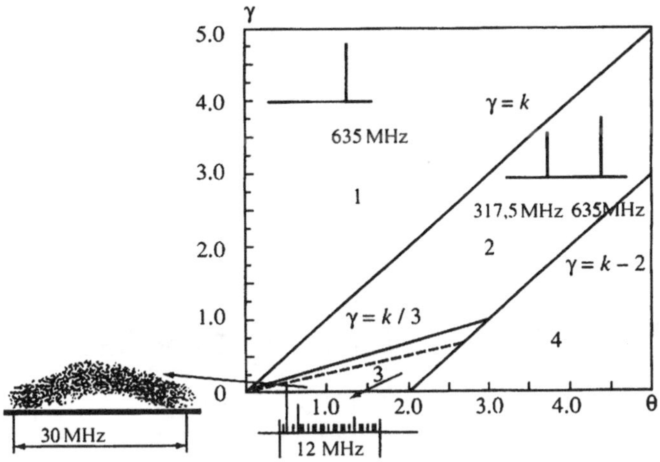

Fig. 6.30 The bifurcational diagram of the stationary solutions on the plane k, γ for the case $\delta_s = 0$, $p_1 = 0$, $p_2 = 0$, $\theta = \pi$

oscillations becomes a multifrequency one. The self-modulation frequency in the vicinity of the bifurcation straight line will be $\omega = \sqrt{4k/3}$.

At transition through the straight line $\gamma = k - 2$ dividing areas 2 and 4, 3 and 4, the stationary point (6.41) loses stability because of merging with the saddle value (6.40), and the representing point in the phase space extends to infinity. This situation has no physical sense and arises because of quadratic approximation of the transistor VAC on the subharmonic frequency ν_{-1}. In area 4, the values of parameter k are great and not realized in practice.

Thus, within the limits of our model and our analysis of the stationary solutions of (6.33), it is possible to explain the occurrence of various kinds of oscillations in HMG. These are the following modes:

- Spectrally pure oscillations on the basic frequency ν_0
- Two-frequency generation on the basic ν_0 and subharmonic ν_{-1} frequencies

To control the modes of oscillations in such generators is possible by changes of the adjustment parameters of the FMCR contours – i.e., by changes of the parameters of the external magnetic field, the parameters of excitation and nonlinear dissipation.

6.6 Oscillating Modes of Evenly Spaced Frequencies Spectra

Let us consider the range of existence of equidistant frequency spectra in the vicinity of the basic ν_0 and subharmonic ν_{-1} frequencies[8] of oscillations at values $\theta \neq \pi$.

In Fig. 6.31, bifurcation diagrams corresponding to the occurrence of multifrequency oscillations in HMG are presented. The curves have been obtained for

[8] The conditions of Hopf's bifurcation occurrence.

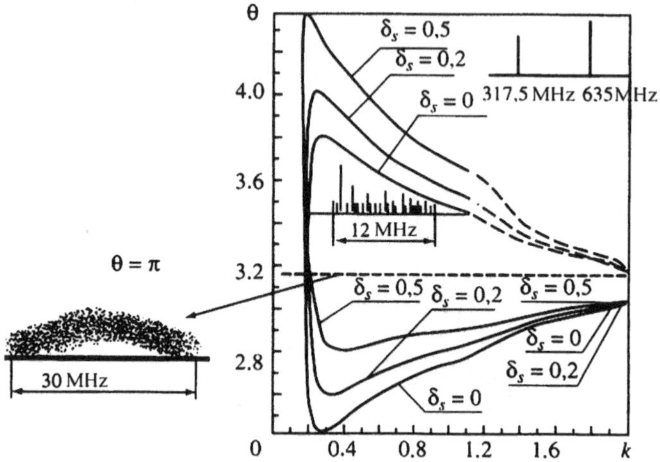

Fig. 6.31 Bifurcation diagrams corresponding to the occurrence of multifrequency oscillations in HMG

various values of the parameter δ_s. Inside of the areas limited by these curves, in the vicinity of the stationary point, a steady limiting cycle[9] "softly/gently appears." With increase in the parameter k, the cycle becomes unstable and the system passes in the modes of generation of noisy oscillations to be considered in paragraph/Sect. 6.7.

At $\delta_s = 0$, the bifurcation curve is symmetric relative to the axis $\theta = \pi$. At increasing of δ_s its symmetry disappears, and equidistant frequency spectra are observed at great values of θ. It occurs owing to compensation of mismatch of the adjustment frequencies of the contours because of the influence of the own nonlinear capacity of the transistor. When $\delta_s < 0$, similar results were observed, but the displacement of the bifurcation curve occurs toward lower θ. Thus, by changing the parameter δ_s it is possible to achieve the occurrence of periodic self-modulation of oscillations in HMG when $0 > \pi$ and $\theta < \pi$. The widest range of θ, in which Hopf's bifurcation is possible, lies in the field of small values of $k < 1$.

Earlier, it has been noted that at $\theta = \pi$, $\delta_s = p_1 = p_2 = 0$, Hopf's bifurcation arises at $k/\gamma = 3$. Hence, the range of γ can be bounded above by the value 1/3, that corresponds to Hopf's bifurcation at $k = 1$. At increasing k, the interval of θ values to satisfy the conditions of Hopf's bifurcation is narrowed. In Fig. 6.32, curves of Hopf's bifurcation, calculated for various values of δ_s, are presented. The parameter γ is chosen by 3.5 times greater/higher than for the bifurcation curves in Fig. 6.31, and Hopf's bifurcation occurs at great/high values of the parameter k. The interval of θ, in which the existence of limiting cycles is possible, is much less/narrower in Fig. 6.32 than in Fig. 6.31.

---

[9] Corresponds to periodic self-modulation of the basic and subharmonic oscillations and the occurrence of equidistant frequency spectra in the vicinity of the spectral lines of these oscillations.

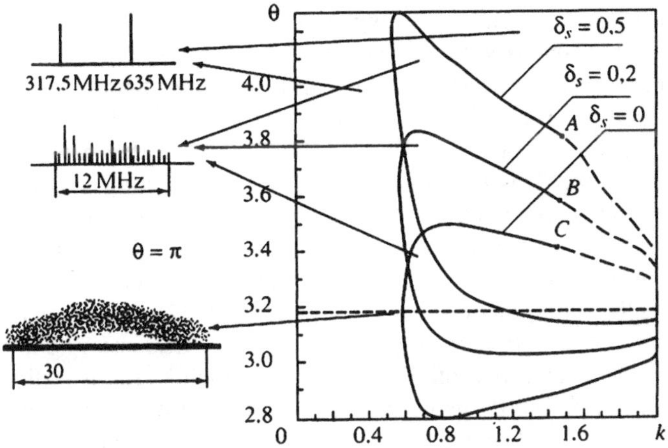

Fig. 6.32 Curves of Hopf's bifurcation, calculated for various values of δ_s

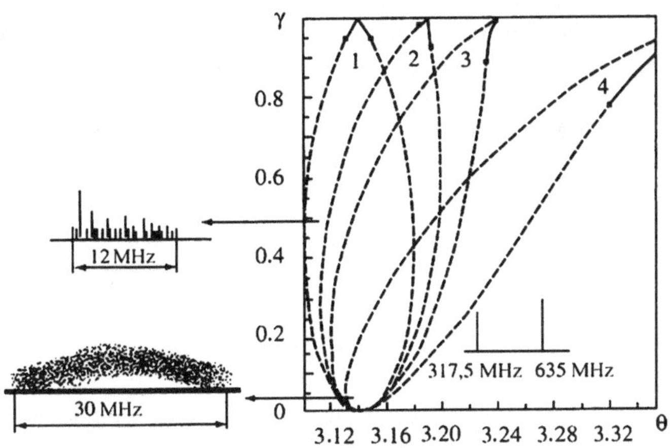

Fig. 6.33 Curves of Hopf's bifurcation are presented at various values of δ_s

In Fig. 6.33, curves of Hopf's bifurcation are presented at various values of δ_s. The parameter of excitation is fixed ($k = 3$).

Inside of areas 1, 2, 3, and 4 limited by the resulted curves, in the vicinity of the stationary point, a limiting cycle "softly/gently appears." At crossing the continuous curves, the cycle is steady, at crossing the dotted curves it is unstable (Fig. 6.33). At $k = 3$ the area/range of the appearance of such a steady limiting cycle is insignificant/small: $3.14 < \theta < 3.18$. With reduction of k this area/range increases.

To achieve Hopf's bifurcation at $\theta \neq \pi$ is also possible by means of frequency mismatch of the contours. However, such a way is inefficient. Really, at $\delta_s = 0.5$ the steady limiting cycle occurs within $3.33 < \theta < 3.39$. If $\delta_s < 0$, the bifurcation curves are displaced toward $\theta < \pi$, and the symmetry with similar curves obtained at $\delta_s > 0$ relative to the axis $\theta = \pi$ is kept.

From our analysis of the curves in Figs. 6.31–6.33 it follows that the greater value of the excitation parameter k is required for realization of Hopf's bifurcation, the narrower is the interval of the parameter θ within which the existence of limiting cycles in system (6.32) is probable. The value of θ significantly depends on the displacement voltage of the transistor, therefore, at its change the existence range of multifrequency and stochastic oscillations will be reduced with an increase of the GB product of the electrodynamic system, which was observed experimentally.

Investigate the evolution of limiting cycles in the (6.32). As earlier, for simplicity, shall put the parameters $\theta = \pi$, $\delta_s = p_1 = p_2 = 0$. After crossing the curves corresponding to Hopf's bifurcation (Fig. 6.31), in the vicinity of the stationary point, a steady limiting cycle "softly/gently appears." In the set of nontruncated equations, this situation corresponds to the occurrence of periodic self-modulation of oscillations on the basic ν_0 and subharmonic ν_{-1} frequencies. The spectrum of the UHF oscillations in HMG will be a multifrequency one in this case. The oscillations of the amplitude envelopes of the basic and subharmonic oscillations are in antiphase. At a distance from the bifurcation curve the amplitude of oscillations, which become relaxation ones, increases. Thus, the amplitude of oscillation on the subharmonic frequency ν_{-1} is practically equal to zero in the most part of the period [48]. This explains the smaller average intensity of oscillations on the subharmonic frequency.

Thus, within the framework of the model, the occurrence of equidistant frequency spectra in the vicinity of the basic ν_0 and subharmonic ν_{-1} fluctuations in HMG can be explained.

6.7 Regimes of Pseudonoise Signals

Explore the opportunity of the occurrence of noise-like oscillations in HMG in (6.37) at $\theta = \pi$, $\delta_s = p_1 = p_2 = 0$ on the basis of [47]. At increase of the k/γ ratio, the limiting cycle merges with the separatrices of the saddles $x = -\sqrt{k/\gamma}$ and $x = \sqrt{k/\gamma}$, then it disappears, the balance condition in the point $z = (k/\gamma) - 1$ remains an unstable focus, and the separatrix of the saddle $x = -\sqrt{k/\gamma}$ goes to the saddle $x = \sqrt{k/\gamma}$. Thus, at small values of k and $\gamma < k/3$, under any initial conditions, the movement shrinks to the vicinity of the three separatrices. One of them goes in the plane $y = 0$ from the saddle $x = -\sqrt{k/\gamma}$ to the saddle $x = \sqrt{k/\gamma}$, and the two others, symmetrized in the plane $z = 0$, go back from the point $x = \sqrt{k/\gamma}$ to the point $x = -\sqrt{k/\gamma}$. Figure 6.34 shows projections of the separatrices into the phase space of set (6.33) on the planes (a) xz, (b) yz, (c) zy. The letters mark the separatrices: AEB, BCA, BDA; A the saddle, $x = -\sqrt{k/\gamma}$; B the saddle, $x = \sqrt{k/\gamma}$; E the saddle, $x = 0$. The arrows mark the directions of movement of the representing point. In the xz plane any movement along the separatrices BCA and BDA is indiscernible and correspond s to movement along the straight line BA (Fig. 6.34b). Lying on the separatrixex AEB (Fig. 6.34b), the representing point gets in the saddle

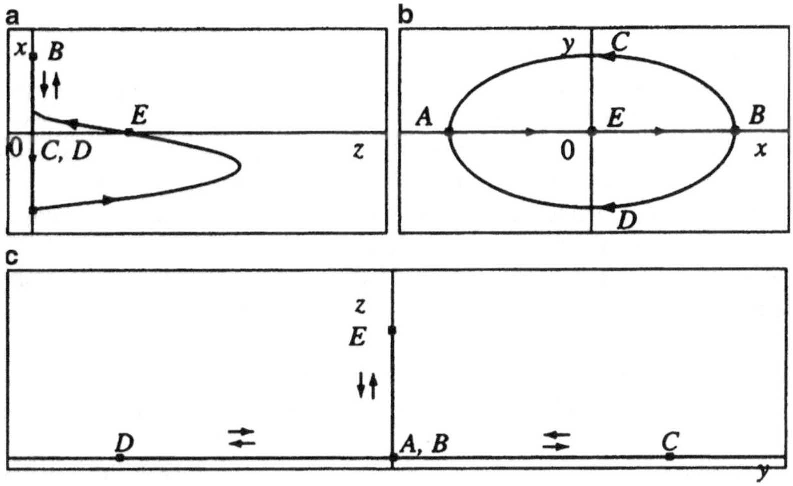

Fig. 6.34 Projections of the separatrices into the phase space of set (6.33) on the planes (**a**) xz, (**b**) yz, (**c**) zy

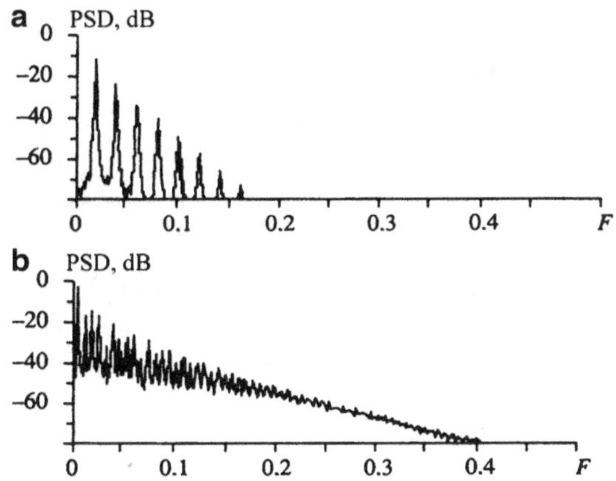

Fig. 6.35 (**a**) PSD as functions of the dimensionless frequency of tuning out F from the bearing frequency at reduction of the nonlinear dissipation parameter γ for peak and frequency modulations. (**b**) The chaotic oscillations at the movement in the vicinity of the separatrices

B and can go along either the separatrix BCA or the separatrix BDA. In Fig. 6.34c, this corresponds to the opportunity of movement from the points A, B to either the point D or the point G.

By analyzing Fig. 6.34c, it is possible to conclude that movement of the representing point occurs in the planes $y = 0$, $z = 0$. Hitting the point B (Fig. 6.34b), the representing point can go in the vicinities of either the top or bottom separatrix.

Thus, under the influence of external fluctuations, the movement of the representing point becomes chaotic.

In Fig. 6.35, PSD as functions of the dimensionless frequency of tuning out F from the bearing frequency at reduction of the nonlinear dissipation parameter γ are presented. Weak Gaussian noise with an intensity of 10^{-6} has been introduced into the equations of the model. If the movement occurs far from the separatrices, the external noise leads to insignificant peak and frequency modulations (Fig. 6.35a).

When the movement shrinks to the vicinity of the separatrices, the oscillations become chaotic (Fig. 6.35b). After Hopf's bifurcation, the amplitude of the arising limiting cycle quickly grows and after crossing the dashed line separating areas 2 and 3 on Fig. 6.31b the movement becomes random. Despite the global nonconservativeness, the system contains an area in the phase space where the behavior of the trajectories is not rough, which explains generation of noise-type signals.

With due account of the reactive square-law nonlinearity of the conductivity of the transistor in the field of small mismatches $\delta_s < 3$, both periodic and stochastic signals may arise in set (6.33). It is necessary to note that stochastic areas in the space of parameters of system exist at small/low values of the basic oscillation increment ($k < 1$).

In the field of small mismatches ($\delta_s < 0.5$) transition to chaotic fluctuations occurs through doubling of the period, and the universal relationships expressing scaling[10] on/by the parameters [50, 51] hold true. This corresponds to occurrence of equidistant frequency spectra of the higher orders in the vicinity of the basic ν_0 and subharmonic ν_{-1} oscillations with subsequent transition to noise-type oscillations.

For example, at $k = 0.2$; $\gamma = 0.02$; $p_1 = 0.02$; $\delta_s = 0$ and an increase in the parameter p_2 in set (6.33) a sequence of equidistant frequency spectra of the higher orders[11] arises. The values of $p_2^{n,2n}$ corresponding to bifurcations of the tactness cycle n[12] have been calculated: $p_2^{1,2} = 0.015461$, $p_2^{2,4} = 0.031089$, $p_2^{4,8} = 0.036968$, $p_2^{8,16} = 0.38251$. As early as for the second doubling bifurcation, the universal law expressing scaling by the parameter p_2 holds:

$$\delta_F = \frac{p_2^{4,8} - p_2^{2,4}}{p_2^{8,16} - p_2^{4,8}} = 4.5858.$$

The exact value of Feigenbaum's constant [49, 50] is $\delta_F = 4.6692$. By using the universality of δ_F, it is possible to calculate the values of the parameter for the subsequent doubling bifurcations.

The equality is satisfied:

$$p_2^{n,2n} \cong p_2^C - \bar{K}\delta_F^{-n}. \tag{6.43}$$

where p_2^C is the critical value of the parameter p_2, $\bar{K}$ a constant depending on a specific system and the parameters of the equations in which a doubling cascade arises. Using (6.43), it is possible to write:

$$\begin{aligned} p_2^{8,16} &\cong p_2^C - \bar{K}\delta_F^{-4}, \\ p_2^{4,8} &\cong p_2^C - \bar{K}\delta_F^{-3}. \end{aligned} \tag{6.44}$$

[10] Like modes of oscillations for various systems at changes of their operating parameters.

[11] The cascade of doubling period bifurcations.

[12] The number n can also be considered as the order of the frequency grid.

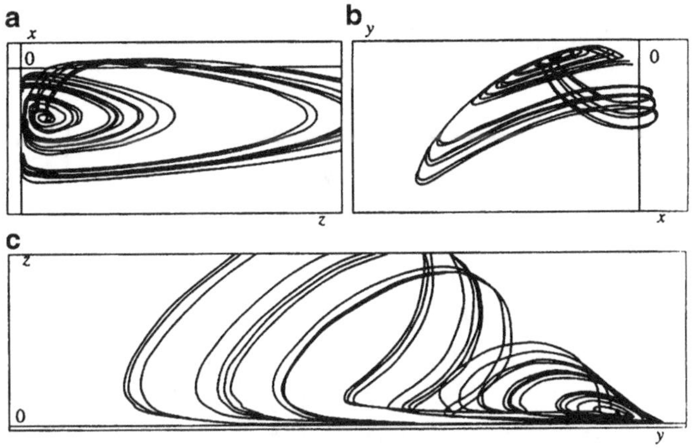

Fig. 6.36 Projections of the stochastic trajectory of the representing point in the phase space of (6.33) after transition to chaos through doubling period bifurcations in compliance with Fig. 6.34

Solving the set of linear equations (6.44), we find: $p_2^C = 0.0379, \overline{K} = 0.1660$. When $p_2 > p_2^C$ in set (6.33), stochastic fluctuations appear.

In Fig. 6.36, projections of the stochastic trajectory of the representing point in the phase space of (6.33) after transition to chaos through doubling period bifurcations are presented.[13] Stochasticity arises in the vicinity of the collapsed loop of the separatrix (the saddle-focus at point 0 (Fig. 6.34)).

In Fig. 6.37, distributions of PSD and HMG are presented at transition to chaos through oscillation period doubling for the parameters: $k = 0.2$; $\delta_S = 0$; $p_1 = 0.4$; $p_1 = 0$; $\theta = \pi$. The parameter γ was varied. In Fig. 6.37a, the spectra of relaxation oscillation power with the period ~5 is presented, which corresponds to an equidistant grid of frequencies with a step of 3.32 MHz in the basic frequency vicinity.

The spectrum is rich with harmonic components which slowly decrease with an increase in frequency. The self-modulation of microwave oscillations in the case of regular movement is close to rectangular. At reduction of γ the frequency of oscillations **a** little increases, their amplitude increases, and a period doubling bifurcation arises (Fig. 6.37b), i.e., the distances between the neighboring components of the grid decrease by 2 and are 1.66 MHz. Further, a next doubling bifurcation arises to determine the parameters of the three-order equidistant frequency spectra (Fig. 6.37c), and after their cascade, chaotic fluctuations appear (Fig. 6.38a). Such a sequence of transition to noise-type fluctuations was observed in our physical experiments with HMG.

[13] Sequence of occurrence equidistant frequency spectra of the maximum orders with transition to noise-type to fluctuations in vicinities of the basic ν_0 and subharmonic ν_{-1} frequencies.

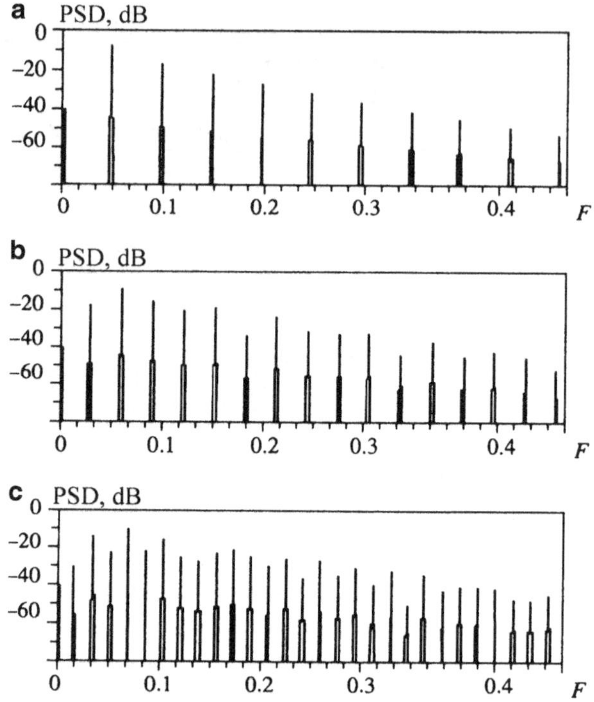

Fig. 6.37 Distributions of PSD and HMG at transition to chaos through oscillation period doubling for the parameters: $k = 0.2$; $\delta_S = 0$; $p_1 = 0.4$; $p_1 = 0$; $\theta = \pi$

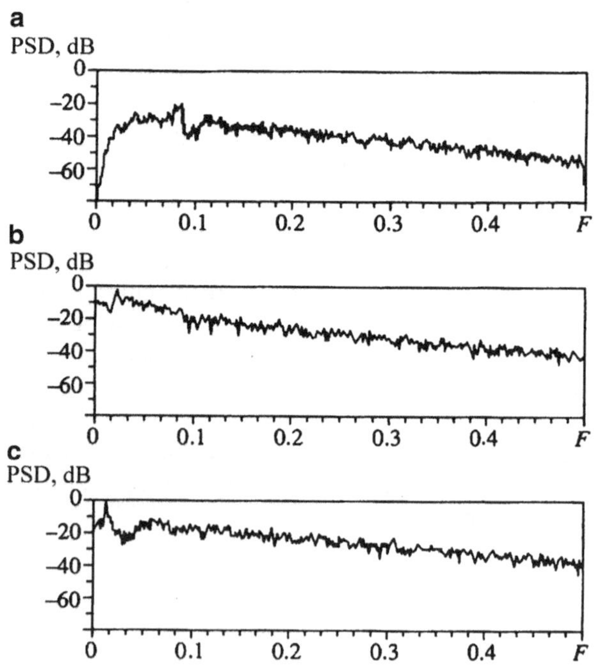

Fig. 6.38 PSD: (**a**) for oscillations with $x = -a_1 \cos \psi$; (**b**) $\nu_0 = a_1 \cos \varphi_1$; (**c**) $\nu_{-1} = a_2 \cos \varphi 2$

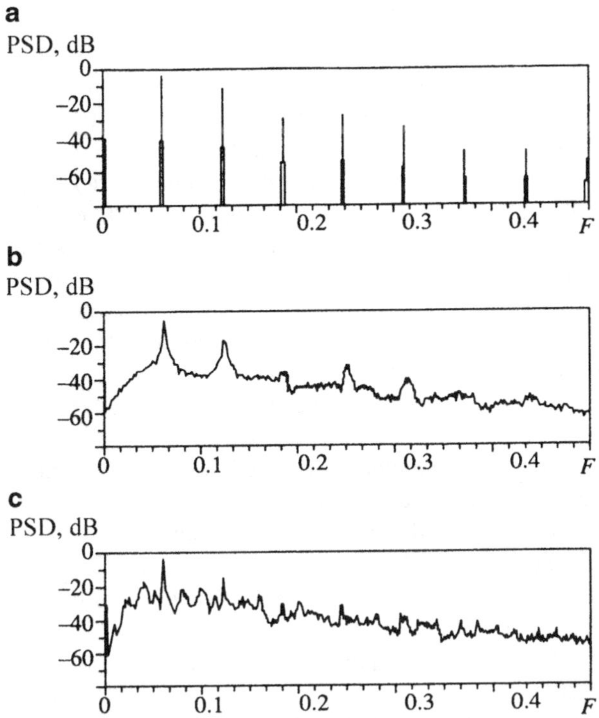

Fig. 6.39 The evolution of the power spectra is presented at transition to chaotic fluctuations through intermittence [53] due to changes of the parameter of mismatch of the contours

In Figs. 6.37–6.39 the results of our calculation of PSD in a dimensionless form are presented. With due account of the real/actual parameters of equivalent oscillatory contours and the coupling factors of the contours with the transistor structure – (6.31), (6.33), (6.35), a unit of dimensionless frequency corresponds to the frequency of 33.0 MHz and decreases at reduction of the coupling factors and at increase of the GB product of these contours.

Restoration of the real distribution of the spectral power density of oscillations on the basic ν_0 and subharmonic ν_{-1} frequencies in view of the generalized phase ψ is possible on the basis of solution of four equations. Add set (6.33) with the equations for the phase of the basic or subharmonic oscillation:

$$\begin{cases} \dfrac{\mathrm{d}\varphi_1}{\mathrm{d}\xi} = -kzy(x^2 + y^2) + p_2 z, \\ \dfrac{\mathrm{d}\varphi_2}{\mathrm{d}\xi} = y + p_1(x^2 + y^2)/2. \end{cases} \tag{6.45}$$

The right-hand sides of (6.45) are written up to a constant whose account would lead to a shift of the investigated spectra only.

In Fig. 6.38, PSD are presented: (a) for oscillations with $x = -a_1 \cos\psi$, (b) $-\nu_0 = a_1 \cos\varphi_1$; (c) $-\nu_{-1} = a_2 \cos\varphi_2$ (a_1 and a_2 are the normalized amplitudes of the oscillation envelopes on the basic and subharmonic frequencies).

The spectra in Fig. 6.38a–c have differences. The spectrum in Fig. 6.38a have a noise pedestal with a dip down to –70 dB near the carrier frequency. The spectrum in Fig. 6.38b, c smoothly deflates at tuning out from the carrier frequency.

Spectra similar to Fig. 6.38a–c were observed experimentally. Transition to chaotic fluctuations through doubling of the period in this system is not unique. Modes of transition to noise-type fluctuations at which no spectra of higher orders appeared were experimentally observed. At changes of the operating parameters (e.g., the external magnetic field or the working point of the transistor), noises in HMG arose at the pedestal spectral lines and increased up to the values equal to the intensity of the spectral line. Similar changes of the modes of oscillations were observed in the investigated model of HMG as well.

With an increase in the parameter of mismatch between the contours, the stochasticity in the system vanishes as a result of the return bifurcations of period doubling. In the field of high values of the parameter of mismatch $0.7 < \delta_s < 1.276$ there exists another stochastic area. In Fig. 6.39, the evolution of the power spectra is presented at transition to chaotic fluctuations through intermittence[14] [53] due to changes of the parameter of mismatch of the contours.

Stochastic fluctuations in the equations of the HMG model can arise at various values of the parameters of disalignment δ_s and the nonlinear shift of frequencies. At simultaneous reduction of the parameters k_s and γ_s determining the nonlinear energy dissipation in the system, the fluctuations remained stochastic. Only the width of the power spectra of oscillations decreased.

Thus, the effect of interaction with the subharmonic oscillation is one of the mechanisms of noise-type fluctuation occurrence in the investigated generator.

Estimate the band of generated frequencies in HMG. From the equivalent VAC of the transistor resulted in Figs. 6.28 and 6.29, we determine the linear parameter of excitation G_{10} on the basic frequency ν_0 and the dissipation parameter G_{01} on the subharmonic frequency ν_{-1}:

$$G_{10} = \frac{8 \times 10^{-3}}{2.5} = 3.2 \times 10^{-3}\ 1/\Omega,$$

$$G_{01} = \frac{50 \times 10^{-3}}{3} = 16.7 \times 10^{-3}\ 1/\Omega. \tag{6.46}$$

In (6.32a) for the dimensionless factor of excitation, put $G_2 = 2C_1$, $M_2 = M_1$, then

$$k = \frac{\alpha_1 C_2 M_2}{\alpha_2 C_1 M_1} = \frac{3.2 \times 2}{16.7} = 0.4 \tag{6.47}$$

[14] Casual transitions between two close attractors in the phase space of the investigated system.

In (6.43) no losses in the first contour are considered, therefore, assume $k < 0.4$ as an estimate. For determination of the real/actual band of generated frequencies, it is necessary to determine the factor of recalculation to dimensionless time $\xi = (|\alpha_2| / C_2)\, t$ in (6.35). Believing the wave resistance of the second contour to be $\sqrt{L_2/C_2} \cong 10$ and considering the subharmonic contour to be adjusted on a frequency of 317.5 MHz, we have:

$$t = \frac{C_2}{|\alpha_2|}\xi = \frac{\overline{2 \times 3.14 \times 317.5 \times 10^6}^{\,1}}{16.7 \times 10^{-3}}\xi = 3 \times 10^{-7}\xi, \tag{6.48}$$

i.e., a unit of dimensionless frequency F corresponds to 33.0 MHz.

In Fig. 6.37a, this corresponds to the distance between the neighboring components of the frequency grid –1.65 MHz, that agrees with the experimental data presented in Table 6.1 for fields $H_0 = 150$–185 Oe. The distance between the neighboring components of the frequency grid in Fig. 6.37c is 0.41 MHz, that corresponds to the modes of generation $H_0 = 190$–195 Oe in Table 6.1. The band of generated frequencies by a level $\Delta\nu_{-60\text{dB}} = 17$ MHz.

Thus, at changes of the operating parameters (the displacement voltage, the generator-loading coupling voltage, etc.) in the investigated generator (in HMG) the following modes are possible:

- Monochromatic generation on the basic frequency ν_0.
- Simultaneous monochromatic generation on the basic ν_0 and subharmonic ν_{-1} frequencies.
- Multifrequency generation on the basic ν_0 and subharmonic ν_{-1} frequencies.
- Chaotic generation.
- Hard changeover accompanied by discontinuous changes of the spectra of generated frequencies.
- A cascade of doubling of the envelope period.

Similar sequences of transition to noise-type fluctuations were observed experimentally $-\nu_0$.

Within the framework of the investigated HMG model it is possible to explain the occurrence of the following facts observed experimentally:

- Spectrally pure oscillations on the basic frequency ν_0.
- Spectrally pure oscillations on the basic ν_0 and subharmonic ν_{-1} frequencies.
- Equidistant frequency spectra with $\Delta\nu^{\text{fs}}$ in the vicinities of the basic ν_0 and subharmonic ν_{-1} frequencies.
- Equidistant frequency spectra of higher orders in the vicinities of the basic ν_0 and subharmonic ν_{-1} frequencies;
- Noise-type fluctuations in the vicinities of the basic ν_0 and subharmonic ν_{-1} frequencies.

The obtained results agree with our experimental data. The effect of interaction with the subharmonic oscillation is one of the mechanisms of occurrence of controlled equidistant frequency spectra and noise-type fluctuations in HMG.

Table 6.1 The experimental data for fields $H_0 = 150–185\,\text{Oe}$

H_0, Oe	Comment	Slope of reconstruction, kHz/Oe	Spectrum view
1	2	3	4
140	Noisy spectral line $(\Delta\nu)_{\text{-3dB}} = 20$ kHz	366.0	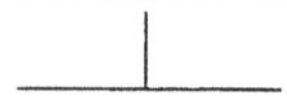
150	Central component with appearance of two lateral/side components distant by $\Delta\nu^{\text{ES FS}} = 1{,}75$ MHz from the central component	100.0	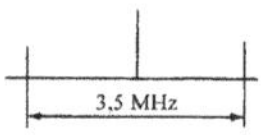
160	Equidistant frequency spectrum with $\Delta\nu^{\text{ES FS}} = 1{,}8$ MHz, occurrence of new components between the basic components $\Delta\nu^{\text{ES FS}} = 0.9$ kHz is observed	90.0	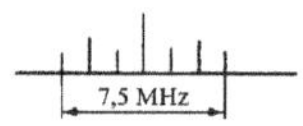
170	Equidistant grid with $\Delta\nu^{\text{ES FS}} = 2.25$ MHz		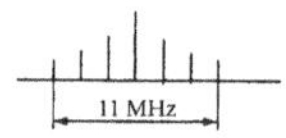
175	Equidistant grid with $\Delta\nu^{\text{ES FS}} = 2$ MHz		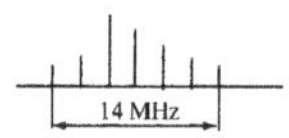
185	Occurrence of intermediate components in the equidistant grid to form an equidistant grid with $\Delta\nu^{\text{ES FS}} = 1$ MHz		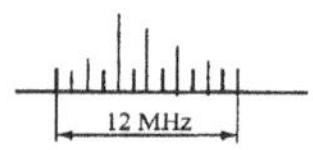
190	Equidistant grid with $\Delta\nu^{\text{ES FS}} = 1$ MHz		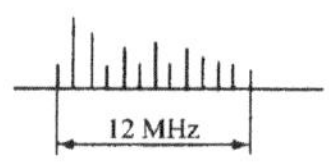
195	Occurrence of intermediate components in the equidistant grid to form an equidistant grid with $\Delta\nu^{\text{ES FS}} = 5$ MHz		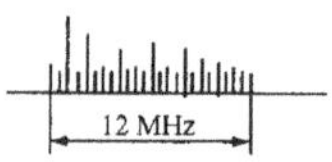
200	Transition to noisy mode		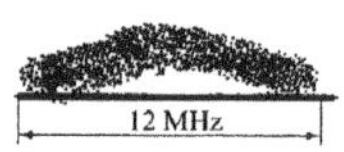
210	Noise signal		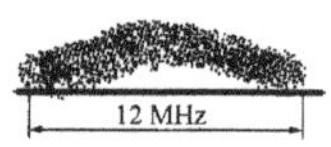
215	Noise signal		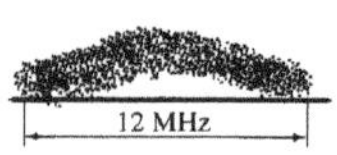
220	Transition from noise signal to an equidistant grid with $\Delta\nu^{\text{ES FS}} = 0.41$ MHz with non-uniformity of the components by amplitude		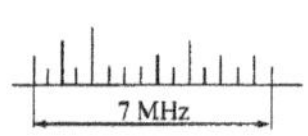
230	Transition to noise signal		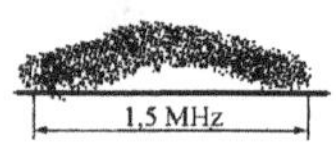
235	Transition to an equidistant frequencies spectrum with two side components $\Delta\nu^{\text{ES FS}} = 4.5$ MHz		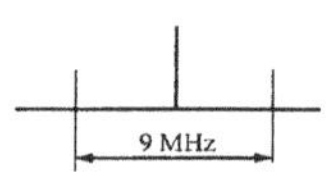
240	Equidistant frequency spectrum with $\Delta\nu^{\text{ES FS}} = 1.7$ MHz		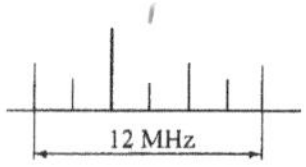

Part III
Calculation of Parameters of Heteromagnetic Structures

Programs for calculation of the parameters of heteromagnetic structures of various types of transistors (field and bipolar ones) and coupling elements, modes of strengthening/amplification, and generation of regular, semi-noisy, and noisy signals of low and high levels of continuous and pulse power in the VHF, UHF, MWF, and EHF ranges are discussed. For powerful heteromagnetic structures, thermophysical analysis of nonstationary and stationary thermal fields for magnetotransistors as a rectangular and as a multilayered cylinder was made.

Chapter 7
Calculation of Parameters of Transistors, Coupling Elements, Magnetotransistors in a Frequency Band Below 100 GHz

7.1 Bipolar Transistor in Omnirange, UHF Range

The bipolar transistor as a radio engineering element is a semiconductor crystal mounted in a case. The crystal is connected to the terminals of the transistor with boil soft wires; besides, a powerful transistor can include matching circuits. These additional elements (boil soft wires, the case, the transistor terminals, and matching circuits) bring essential distortions in the work of the semiconductor crystal and should be contained in the equivalent circuit of the transistor. At modeling of the bipolar transistor, Gummel–Poon's model was used.

The programs designed [53, 57] are intended for calculation of the static parameters of Gummel–Poon's model of the bipolar transistor and Materok's one of FET neglecting additional reactive elements.

Experimental families of the static characteristics of transistors serve as the initial data for calculation. The mathematical description of Gummel–Poon's static model of the bipolar transistor is presented in Chap. 5. The algorithm of the program is based on optimization of the static parameters of the models of transistors for the best calculation and experimental data fit.

The traditional technique of determination of the equivalent parameters of both linear and nonlinear models of transistors is based on carrying out numerous complex measurements with the usage of such vector circuit analyzers as, for example, N5250A.[1] In the developed programs, a simplified and effective technique allowing one to simulate the semiconductor structure of the crystal of the transistor from few simple measurements of static transistor characteristic families is realized.

In the programs [53], optimization algorithm of the additional reactive elements of the transistor (the inductances of boil soft wires and transistor terminals, matching element capacities) is realized. As input data for the program, reference data on the transistor (its boundary frequency, gain factor on the working frequency, matching capacities, inductance of the terminals, or experimentally measured S parameters in

[1] *Agilent Technologies*™ (USA) with a frequency band of 10 MHz–110 GHz, building-up of frequencies up to 325,500, and 1,000 GHz.

A.A. Ignatiev and A.V. Lyashenko, *Heteromagnetic Microelectronics: Microsystems of Active Type*, DOI 10.1007/978-1-4419-6002-3_7,

the working frequency range) are used. The program [53] allows simulation of the bipolar and FET transistors with a good accuracy.

The system requirements of the program are as follows:

- Operating system, Windows
- MFC Library
- CAD *Serenade* 8.0, MWO-2002 etc.

7.1.1 General Data on Programs

The programs [53] are written in the VC^{++} shell, use the library *MFC* 2, and contain a test amplifier for optimization of the frequency properties of the transistor.

The basic class of *CGummelPoon* contains the following functions, methods, and variables:

AreaTrapeze ($x1$, $x2$, $y1$, $y2$, $y3$, $y4$) – Calculation of the trapezoid area with the vertex coordinates (($x1$, $y1$), ($x2$, $y2$), ($x1$, $y3$), ($x2$, $y4$)). It is used for calculation of the error function.

Calculate() – Calculation of a curve family for the current values of the parameters *par* [].

DiffFuns (*x*[], *p*[], *type*) – Calculation of the coordinates of the residual vector in Newton's for independent variables *x*[] and the values of parameters *p*[]. The variable *type* defines the coordinate of the residual vector: a constant or an active measured quantity.

ERRFunction() – Calculation of the normed error function for the current values of parameters *par*[].

FindFirstPoint (*MinX*[], *MaxX*[], *BestRes*[]) – Calculation of an initial value within the interval (*MinX*[] – *MaxX*[]). In variable *BestRes*[], the found point is returned.

GradSpusk() – Carrying out gradient descent in the space of variables *par*[] for the error function *ERRFunction*().

InitPar(), *InitParValue*() – Initialization of *par*[] from the file *parameters.ini*.

Init() – Initialization of the global variables.

MethodNewton (*x*[]) – Newton's method for initial values of *x*[], the computed point is returned in the same variable.

Modfun (*x*[]) – Calculation of the module of the residual vector for Newton's method.

MotionToCurve (*x*0[], *i*, *type*) – Calculation of one (the *i*th) curve (movement along the curve with the initial values *x*0[]). The variable *type* defines the positions of the calculated points: through a set interval or strictly above the experimental points.

ReadMeassCurves() – The procedure of reading experimental curves (from the file *FileIn.ini*).

CalculatCurves[] – Calculated curves.

Eps – The accuracy of finding a point on the curve.
Funs – The reference to the calculated function (defines a set of equations for calculation of the transistor model).
MeassCurves[] – Experimental curves.
NActPar – A quantity set in measurements (varies in experiment).
NActParFun – A measured value (according to *NActPar*).
NConstPar – A quantity fixed in measurements. Variables *NActPar*, *NactParFun*, and *NConstPar* set families of the characteristics for numerical experiment.
Par[] – An array of measured values in the current point and the model's parameters (*par*[] – the measured values, *par*[] – the parameters).

The global functions are as follows:

CalculateInit(type) – Initialization of the variables for calculation of a certain family of curves.
GPFunction (*x*[], *p*[], *type*) – The function containing the equation set of Gummel–Poon's model of the bipolar transistor.
FETMaterkaFunction (*x*[], *p*[], *type*) – The function containing the equation set of Materok's model of the field transistor.

The information on the values of parameters of Gummel–Poon's model of the bipolar transistor is set in the *parameters.ini* file. Each parameter is described as:

> [ParName] [PN] [P0V] [PFV] [PLV] [dP] [Flag], where ParName is the displayed name of the parameter, *PN* the unique number of the parameter, P0V the initial value of the parameter to be set within the interval [PFV, PLV], PFV the first border of the interval of the parameter, PLV the second border of the interval of the parameter, dP the maximal increment of the given parameter for one step of optimization, and Flag is an activity flag of the parameter (its usage in the process of optimization).

The fields are separated with a symbol of tabulation. If a line begins with a sequence "//" it is ignored at reading the file.

Information on the experimentally measured points of the static characteristics is set in text files with arbitrary names. One file contains information on the experimental points of only one curve of the family of characteristics. The file format is:

- For the bipolar transistor:

$$[\mathrm{IE(mA)}]\ [\mathrm{UBE(V)}]\ [\mathrm{IC(mA)}]\ [\mathrm{UBC(V)}]. \tag{7.1}$$

- For the field transistor:

$$[\mathrm{Ig(mA)}]\ [\mathrm{Uds(V)}]\ [\mathrm{Id(mA)}]\ [\mathrm{Ugs(V)}]. \tag{7.2}$$

The fields are separated with a symbol of tabulation. If a line begins with a sequence "//" it is ignored at reading the file.

For inclusion of the files in the process of optimization they must be listed in the *FileIn.ini* file in the format:

[Family identifier 1]
[file 11]
[file 12]
...
[Family identifier 2]
[file 21]
[file 22]
...

Here [Family identifier] the bipolar transistor can accept the following values:

#BIPCBStatIn – The family of static input characteristics of the bipolar transistor in the circuit with a common base
#BIPCEStatIn – The family of static input characteristics of the bipolar transistor in the circuit with a common emitter
#BIPCBStatOut – The family of static output characteristics of the bipolar transistor in the circuit with a common base
#BIPCEStatOut – The family of static output characteristics of the bipolar transistor in the circuit with a common emitter
#MaterkaFETStatOut – The family of static output characteristics of the field transistor
#MaterkaFETStatIn – The family of static input characteristics of the field transistor

Preparation for calculation of the static parameters of transistor models contains the following steps:

1. Set all the measured curves in text files
2. List all these files in the file *FileIn.ini*
3. Set in the file *parameters.ini* the names, initial values, and intervals of each parameter

During optimization it is necessary:

1. To make sure that the experimental and calculated curves are displayed correctly.
2. To calculate the error functions for each parameter. For this purpose, it is necessary to specify the value "1" in the field *Flag* for each parameter in the file *parameters.ini* and to press the button with a curve family in the program shell.
3. To exclude the parameters not affecting the error function from optimization. For this purpose, it is necessary to specify *Flag* "0" for such a parameter in the file *parameters.ini*.
4. To select for the remaining parameters their initial and final range borders to have the minimum of the error function within.
5. To select for each parameter its maximal step to make the plot of the error function for the given parameter smooth enough.

6. To start searching the minimum of the error function. For this purpose, it is necessary to press the "*grad*" button in the program shell.
7. After finding the minimum it is necessary to correct the initial values of the parameters in the file parameters.ini in conformity with the found values, then go to step 2. If the error function has not reached a minimum by any parameters, restart the optimization (steps 2–7).

The found values of the parameters can be used for modeling bipolar and field transistors in CAD programs.

7.1.2 Test Task

To test the program [53], the static parameters of the bipolar MPSA92 transistor have been calculated. As an initial approximation of the model, the known parameters of the powerful BFR92 transistor were taken. The calculation was made on the basis of families of the static input and output characteristics.

As a second test task for the program [53] the reactive parameters of the equivalent circuit of the bipolar KT962B transistor have been calculated. The input data were: the boundary frequency $f_T = 750\,\text{MHz}$, the collector junction capacity $C_{col} = 35\,\text{pF}$, the power gain factor $K_{pg} = 4$.

The initial values of the parameters and optimization results for the MPSA92 transistor are presented in Table 7.1, for the KT962B transistor in Table 7.2.

The accuracy of modeling of static output characteristics for the MPSA92 transistor was 11 and 35% for output and input, respectively. The experimental and calculated families of output characteristics for the MPSA92 transistor are shown in Fig. 7.1.

The accuracy of modeling of static output characteristics for the KT962B transistor was 2 and 42% for output and input, respectively.

Table 7.1 The initial values of the parameters and optimization results for the MPSA92 transistor

Parameter	Initial value	Optimization result	Parameter change	Influencing family
IS	1.735×10^{-16}	6.349997×10^{-17}	−173%	Out., inp.
ISE	1.037×10^{-15}	3.142251×10^{-15}	67%	Out., inp.
NF	0.8858932	0.921912	4%	Out., inp.
NE	1.192	1.191512	−0.04%	Out., inp.
BF	141.4	142.974	1.1%	Out., inp.
IKF	1.023	0.2736885	−274%	Out.
VA	58.2	58.2	0	Out.
VB	50	50	0	Out.
RB	0.2	0.2	0	Inp.
RE1	0.2308085	0.01	100%	Inp.

Table 7.2 The initial values of the parameters and optimization results for the KT962B transistor

Parameter	Initial value	Optimization result	Parameter change	Influencing family
IS	6.349997×10^{-17}	5.5×10^{-16}	88.4%	Out., inp.
ISE	3.142251×10^{-15}	1.420222×10^{-14}	78%	Out., inp.
NF	0.921912	0.921912	0	Out., inp.
NE	1.191512	1.191512	0	Out., inp.
BF	142.974	500	72%	Out., inp.
IKF	0.2736885	0.2736885	0	Out., inp.
VA	58.2	2	−2810%	Out.
VB	50	40	−25%	Out.
RB	0.2	5.001548	96%	Inp.
RE1	0.01	0.185	95%	Inp.
CJC	1.0×10^{-11}	19×10^{-12}	90%	Power gain

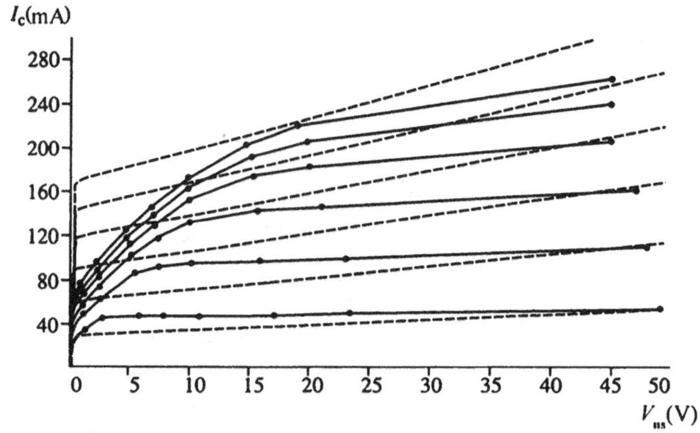

Fig. 7.1 The experimental and calculated families of output characteristics for the MPSA92 transistor

7.2 FET in Omnirange, UHF Range

7.2.1 Determination of Parameters of a FET Model with Schottky Gate

For modeling of powerful field-effect transistors with a Schottky gate, Materok's equivalent circuit (Fig. 7.2) is usually used.

In CADs, the nonlinear model of the transistor's active area in view of the active resistance of the electrodes is usually used. Unification of the model is necessary for simplification of further optimization by various classes of parameters and compatibility with various CADs; therefore, for further modeling, the equivalent circuit[2] shown in Fig. 7.3 will be used. For modeling of the transistor, modern CADs as Serenade, MatLab, Microwave Office, and others can be used.

[2] Designations of "MWO-2002" in used in scheme.

Fig. 7.2 Materok's equivalent circuit

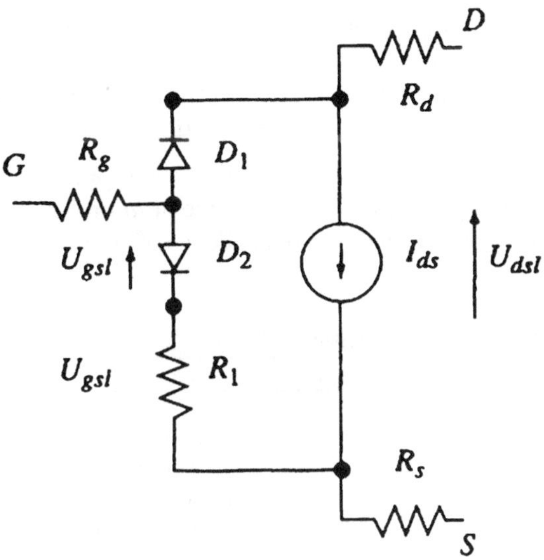

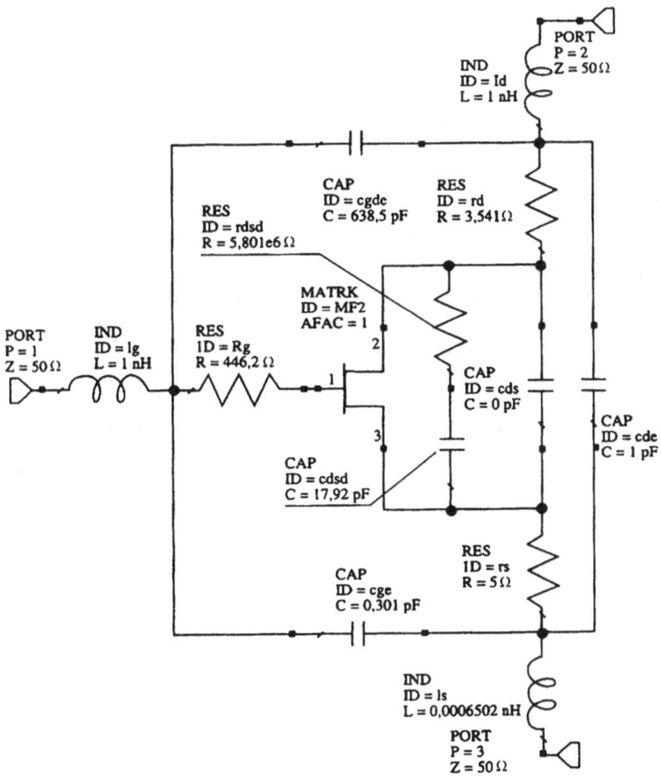

Fig. 7.3 The equivalent circuit

The transistor performance on direct current was defined by a set of 18 parameters (Table 7.3).

The field transistor connected in the circuit with a common source (Fig. 7.4) is characterized by three families of characteristics, namely:

- Input ones (the dependence of *Ig* on *Vgs* at *Vds* = const)
- Output ones (the dependence of *Id* on *Vds* at *Vgs* = const)
- Transfer ones (the dependence of *Id* on *Vgs* at *Vds* = const)

Table 7.3 The transistor performance on direct current

No.	Denomination	Description	Unit
1	IDSS	Saturation drain current at *Vgs* = 0	A
2	SS	Slope of the drain characteristic in the field of saturation	A/V
3	VP0	Cutoff voltage at *Vds* = 0	V
4	Gamma	Parameter of cutoff voltage slope	V
5	*E*	Constant component of the exponent for *Idsi*	
6	KE	Parameter of the dependence of exponent for *Idsi* on *Vgs*	1/V
7	SL	Parameter of the drain characteristics slope in the linear range	A/V
8	KG	Parameter of the dependence of the drain characteristics on *Vgs* in the linear range.	1/V
9	IG0	Saturation current of Schottky's diode	A
10	IB0	Return breakdown current of Schottky's diode	A
11	AFAG	Parameter of the slope of the forward current branch of the diode	1/V
12	AFAB	Parameter of the slope of the return current branch of the diode	1/V
13	VBC	Breakdown voltage of Schottky's diodes	V
14	*R*10	Internal resistance of the channel at *Vgs* = 0	Ω
15	KR	Parameter of the slope of the characteristics of the internal channel resistance	1/V
16	*R*s	Resistance of the source	Ω
17	*R*d	Resistance of the drain	Ω
18	*R*g	Resistance of the shutter	Ω

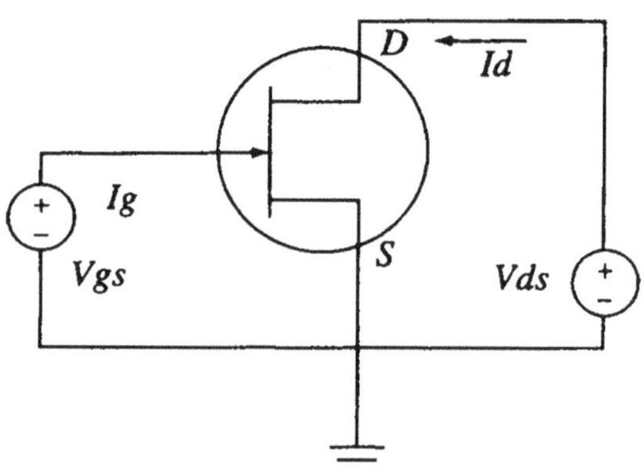

Fig. 7.4 The field transistor connected in the circuit with a commom source

Table 7.4 The parameters of the most significant reactive elements

No.	Denomination	Description	Unit
1	Lg	Shutter inductance	H
2	Ld	Drain inductance	H
3	Ls	Source inductance	H
4	$C10$	Shutter-source capacity at zero voltage	F
5	$K1$	Parameter of return internal shutter-source voltage	1
6	C1S	Constant component of the capacity Cgs	F
7	CF0	Shutter-drain capacity at zero voltage	F
8	KF	Parameter of return internal shutter-drain voltage	1

In the microwave-range, S parameters are usually used as characteristics of the transistor. For calculation of the S parameters of the transistor, the common-source circuit is used. The form of the frequency dependence of the S parameters is essentially influenced by the effects determined by reactive elements of the equivalent circuit. The parameters of the most significant reactive elements are shown in Table 7.4.

7.2.2 Method for Determination of Transistor Parameters

To determine the parameters of the transistor model as the error function for each ith, jth curve of each R-characteristics at a set point the following expression is used:

$$E = \sum_{i,j,p} \left(f^{\mathrm{m}}_{i,j,p} - f^{\mathrm{c}}_{i,j,p}\right)^2 , \tag{7.3}$$

where E is the error function value, $f^{\mathrm{m}}_{i,j,p}$ the measured value of the pth characteristic at the ith point of the jth curve, and $f^{\mathrm{c}}_{i,j,p}$ is the calculated value of the pth characteristic at the ith point of the jth curve.

Families of static characteristics and S parameters describe transistor processes weakly related to each other by some parameters and their various physical nature. For optimization we shall divide the error function into its components, each correspondent to some family.

$$E_p = \sum_{i,j} \left(f^{\mathrm{m}}_{i,j} - f^{\mathrm{c}}_{i,j}\right)^2_p , \tag{7.4}$$

where E_p is the error function for the pth characteristics.

Generally, the characteristics have the following functional dependence on its parameters:

$$f_{c,i,j,p} = f_{c,p}(z_1, \ldots, z_n), \tag{7.5}$$

For each function E_p, it is possible to resolve groups of strongly and weakly influencing parameters $\{z_j\} \in \Omega(g)$, optimization by which at initial iterations is run separately for each g. The sets of parameters $\Omega(g)$ can overlap for various g. Belonging of any parameter z_j to some group $\Omega(g)$ is defined from the physical sense of the parameter z_j. For reliable work and higher productivity, optimization is run consistently by the groups of parameters $\Omega(g)$. The iterative method of optimization essentially raises the productivity and stability of optimization because of decreasing the dimension of phase parameter space. The results of research of the influence of parameters of Materok's model on the output characteristics and possible distribution of the parameters by sets $\Omega(g)$ are shown in Table 7.5.

A program is developed for determination of the parameters of field transistors with the aid of optimization by static characteristics and S parameters. It contains schemes for calculation of characteristics families, an equivalent circuit of the field transistor, and a set of parameter optimizers by the error function (7.3) on the basis of experimental data.

The schemes are as follows:

- *Schematic* 1 – Calculation of the S parameters the model
- *Almaz Spar* – Input of experimentally measured characteristics
- *Static_in* – Calculation of the input static characteristics of the model
- *Static_Out*– Calculation of the output static characteristics of the model
- PTSH – The equivalent circuit of the transistor (see Fig. 7.3)

Files of experimental data:

- *Meas* – The measured static output characteristics of the transistor
- *Meas_In* – The measured static input characteristics of the transistor
- *3MS* – A matrix of the measured S parameters of the transistor

The error function optimizers:

- *Static_in* – Optimization of the input static characteristics of the model
- *Static_Out* – Optimization of the output static characteristics of the model
- *Err*11 – Optimization of the error function for S_{11}
- *Err*21 – Optimization of the error function for S_{21}
- *Err*22 – Optimization of the error function for S_{22}
- *Err*12 – Optimization of the error function for S_{12}

As initial values, the parameters of the transistor similar to that under study with a priori known integral characteristics, or the parameters of any reference transistor from CAD were taken. In the latter case, time expenses for calculation essentially increased owing to slow convergence of the algorithm outside of the range of physically adequate values of parameters.

The used technique of construction of a computer model of the transistor allows an effective machine-focused algorithm for development of a heteromagnetic field transistor to be designed.

7.2.3 Test Task

A test task is solved for the PTSh-600 transistor. The parameters of the model are shown in Table 7.6.

Table 7.5 The influence of parameters of Materok's model on the output characteristics

No.	Denomination	Description	Unit	Belonging to group g	Guaranteed range of parameter values
1	SS	Slope of the drain characteristic in the field of saturation	A/V	1,3,10	[−0.5; 0.5]
2	VP0	Cutoff voltage at $Vds = 0$	V	1,3,10	$[-Vg_{max}, 0]$
3	Gamma	Parameter of the cutoff voltage slope	V	2,3,10	[−5; +0.5]
4	E	Constant component of the exponent for $Idsi$	–	–	–
5	KE	Parameter of the dependence of the exponent for $Idsi$ onVgs	1/V	2,3,10	–
6	SL	Parameter of the slope of the drain characteristics in the linear range	A/V	2,3,10	[0, 10]
7	KG	Parameter of the dependence of the drain characteristics on Vgs in the linear range	1/V	2,3,10	[0, 10]
8	IG0	Saturation current of Schottky's diode	A	5,6,10	–
9	IB0	Return breakdown current of Schottky's diode	A	4,6,10	–
10	AFAG	Parameter of the slope of the direct current branch of the diode	1/V	5,6,10	–
11	AFAB	Parameter of the slope of the return current branch of the diode	1/V	4,6,10	–
12	VBC	Breakdown voltage of Schottky's diode	V	–	–
13	$R10$	Internal channel resistance at $Vgs = 0$	Ω	3,10	–
14	KR	Parameter of the slope of the internal channel resistance characteristics	1/V	3,10	–
15	Rs	Source resistance	Ω	3,10	[0, 100]
16	Rd	Drain resistance	Ω	3,10	[0, 100]
17	Rg	Shutter resistance	Ω	3,10	[0, 1000000]
18	Lg	Shutter inductance	H	9,10	–
19	Ld	Drain inductance	H	9,10	–
20	Ls	Source inductance	H	9,10	–
21	$C10$	Shutter-source capacity at zero voltage	F	7,8,9,10	–
22	$K1$	Parameter of the return internal shutter-source voltage	1	8,9,10	–
23	C1S	Constant component of the capacity Cgs	F	7,8,9,10	–
24	CF0	Shutter-drain capacity at zero voltage	F	7,8,9,10	–
25	KF	Parameter of the return internal shutter-drain voltage	1	8,9,10	–

Table 7.6 The calculation result of test task for the PTSh-600 transistor

No.	Denomination	Description	Unit	Initial value	Final value
1	2	3	4	5	6
1	IDSS	Drain saturation current at $Vgs = 0$	A	0	0.2287
2	SS	Slope of the drain characteristics in the field of saturation	A/V	0	0
3	VP0	Cutoff voltage at $Vds = 0$	V	−2	−5.26
4	Gamma	Parameter of the cutoff voltage slope	V	0	−0.085444
5	*E*	Constant component of the exponent for *Idsi*		2	2
6	KE	Parameter of the dependence of the exponent for *Idsi* on*Vgs*	1/V	0	0
7	SL	Parameter of the slope of the drain characteristics in the linear range	A/V	0.15	9.7734
8	KG	Parameter of the dependence of the drain characteristics on *Vgs* in the linear range	1/V	0	4.4958
9	IG0	Saturation current of Schottky's diode	A	0	0
10	IB0	Return breakdown current of Schottky's diode	A	0	0
11	AFAG	Parameter of the slope of the direct current branch of the diode	1/V	38.696	38.696
12	AFAB	Parameter of the slope of the return current branch of the diode	1/V	38.696	38.696
13	VBC	Breakdown voltage of Schottky's diode	V	1000000	1000000
14	*R*10	Internal channel resistance at $Vgs = 0$	Ω	0.001	0.09912
15	KR	Parameter of the slope of the internal resistance characteristics of the channel	1/V	0	0
16	*R*s	Source resistance	Ω	0	7.211
17	*Rd*	Drain resistance	Ω	0	1.211
18	*Rg*	Shutter resistance	Ω	0	48.069
19	*Lg*	Shutter inductance	H	0	2.3599×10^{-5}
20	*Ld*	Drain inductance	H	0	1.0427
21	*Ls*	Source inductance	H	0	0.5
22	*C*10	Shutter-source capacity at zero voltage	F	0	0.086272
23	*K*1	Parameter of the return internal shutter-source voltage	1	1.25	14
24	C1S	Constant component of the capacity *Cgs*	F	0	0.31055
25	CF0	Shutter-drain capacity at zero voltage	F	0	0.3667
26	KF	Parameter of the return internal shutter-drain voltage	1	1.25	7

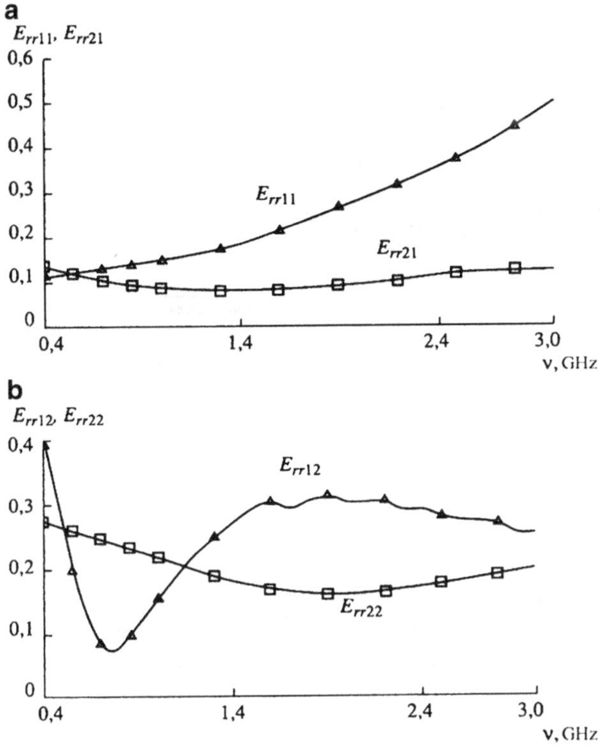

Fig. 7.5 The values of the error function E_{rr11}, E_{rr21} (**a**), E_{rr12}, E_{rr22} (**b**) at determination of the parameters S_{11}, S_{21}, S_{12}, S_{22} in a frequency band 0.4–3.0 GHz

The values of the error function E_{rr11}, E_{rr21}, E_{rr12}, E_{rr22} at determination of the parameters S_{11}, S_{21}, S_{12}, S_{22} in a frequency band 0.4–3.0 GHz are shown in Fig. 7.5. The calculated transfer factor is in agreement with its experimental value (the error does not exceed 15%).

7.3 Powerful FET in EHF Range

For modeling of field transistors in a frequency range up to 100 GHz, the base model of a transistor in the nonlinear operating mode designed (the HEMT-technology) is used. The flowchart of the computer program of calculation of the parameters of the model of EHF field transistors on the basis of its commercial prototype with locking to the working frequency range and the maximal output capacity is shown in Fig. 7.6.

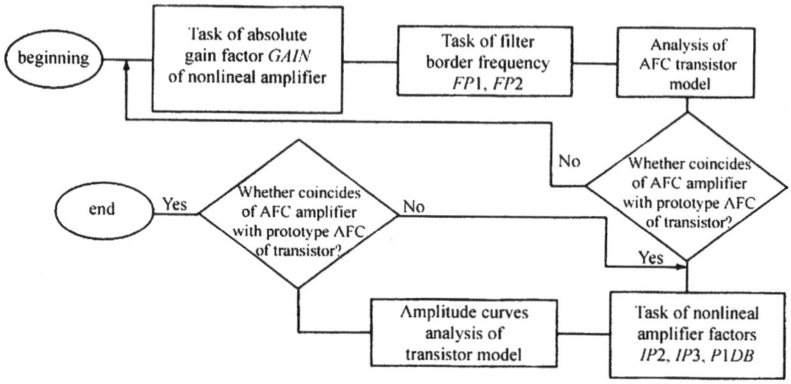

Fig. 7.6 The flowchart of the computer program [58] of calculation of the parameters of the model of EHF field transistors on the basis of its commercial prototype with locking to the working frequency range and the maximal output capacity

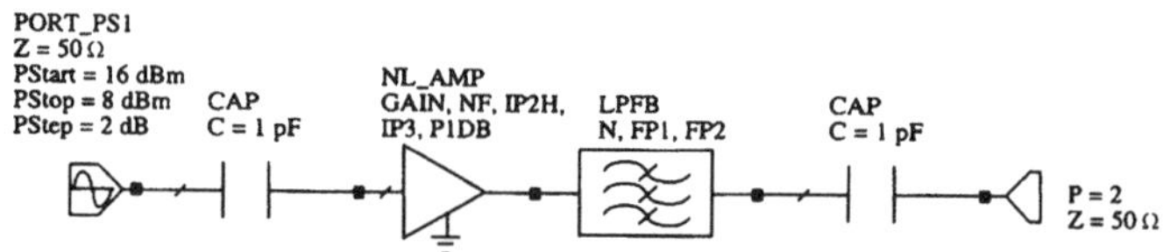

Fig. 7.7 The equivalent circuit of the base HEMT transistor

Modeling was run in the CAD MWO-2002 environment. A model of the nonlinear amplifier with a band filter of lower frequencies at output was used. In Fig. 7.7, the equivalent circuit of the base HEMT transistor[3] is shown.

The program uses the models of EHF field transistors of two types, namely, HEMT-1 and HEMT-2.

7.3.1 *Model of EHF Transistor of HEMT-1*

For modeling, the *S* parameters of dispersion and peak characteristics of the base HEMT transistor were used.

Test calculation for the HEMT-1 field transistor was run on a frequency of 65 GHz. The input parameters are as follows:

[3] MWO-2002 denomination was conserved.

```
!--------------------
!NET List file
!--------------------
DIM
     CAP  PF
     PWR  DBM
     ANG  DEG
    TIME  NS
    FREQ  GHZ
     RES  OH
CKT
CAP 1 2 ID = C2 C = 1
NL_AMP 2 3 ID = AM1 GAIN = 20 NF = 0 IP2H = 40 IP3 = 24 P1DB = 18 &
S11MAG = 0 S11ANG = 0 S22MAG = 0 S22ANG = 0 Z0 = 50 TDLY = 0
CAP 4 5 ID = C1 C = 1
LPFB 3 4 ID = LPFB1 N = 3 FP = 40 AP = 3.0103 RS = 50 RL = 50 QU = 1e + 012
PORT_PS1 1 P = 1 Z = 50 PStart = -16 PStop = 8 PStep  = 2 Ang = 0
PORT 5 P = 2 Z = 50
DEFOP circuit_1
```

The results are as follows: $f = 65$ GHz, $K_{gf} = 7.17$ dB, and $P_{max} = 3.6$ dBmW.

The calculation results of the gain factor K_{gf} of the HEMT-1 transistor in a frequency range of 1–85 GHz are presented in Table 7.7. Comparison of the gain factors, the maximal output capacity within various subranges of the frequency range for the HEMT-1 amplifier and its model are collected in Table 7.8. In Fig. 7.8, the dependencies of the output capacity on the input power on two frequencies are shown for the NEMT-1 amplifier: a – 4 GHz and b – 44 GHz.

Table 7.7 The calculation results of the gain factor K_{gf} of the HEMT-1 transistor in a frequency range of 1–85 GHz

f, GHz	1	3	5	7	9	11	13	15	17	19	21
K_{gf}, dB	9.01	17.77	19.07	19.46	19.62	19.70	19.74	19.76	19.76	19.75	19.72
f, GHz	23	25	27	29	31	33	35	37	39	41	43
K_{gf}, dB	19.67	19.58	19.46	19.29	19.05	18.75	18.36	17.90	17.35	16.73	16.04
f, GHz	45	49	51	53	55	57	59	61	63	65	67
K_{gf}, dB	15.30	13.72	12.90	12.07	11.24	10.41	9.60	8.79	8.00	7.23	6.47
f, GHz	69	71	73	75	77	79	81	83	85	–	–
K_{gf}, dB	5.72	5.00	4.29	3.59	2.92	2.25	1.61	0.97	0.35	–	–

Table 7.8 Comparison of the gain factors, the maximal output capacity within various subranges of the frequency range for the HEMT-1 amplifier and its model

Parameter, dB	Frequency band, GHz	Gain factor, dB HEMT-1	Gain factor, dB Model
$S21$	1–26	18–21	19.7
	26–45	15–17	15–19.6
	45–65	15–17	7.4–15

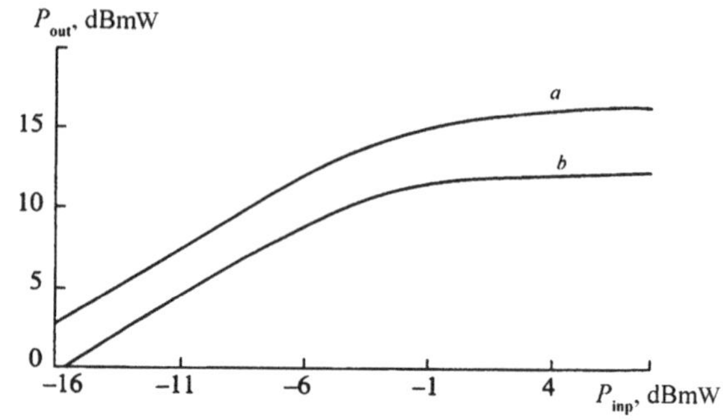

Fig. 7.8 The dependencies of the output capacity on the input power on two frequencies for the NEMT-1 amplifier: $a - 4$ GHz and $b - 44$ GHz

7.3.2 Model of EHF Transistor of HEMT-2

The model of the HEMT-2 transistor is designed similarly to HEMT-1. For modeling, the parameters of dispersion of the base HEMT transistor were used.

Test calculation for the HEMT-2 field transistor was run on a frequency of 65 GHz. The input parameters are as follows:

```
NET List file
DIM
     CAP  PF
    FREQ  GHZ
     RES  OH
     PWR  DBM
     ANG  DEG
CKT
CAP 1 2 ID = C2 C = 1
GAIN 2 3 ID = U2 A = 20 SL = 0 SH = 0 FL = 0 FH = 0 R = 50
BPFB 3 4 ID = BPFB1 N = 3 FP1 = 55 FP2 = 73 AP = 3.0103
  RS = 50 & RL = 50 QU = 1e+012
CAP 4 5 ID = C1 C = 1
PORT1 1 P = 1 Z = 50 Pwr = -30 Ang = 0
PORT 5 P = 2 Z = 50
DEFOP circuit_1
```

The results are as follows: $f = 65$ GHz, $K_{gf} = 20$ dB, and $SWR_e = 1.12$.

The carryover (gain) factor of the HEMT-2 transistor in a frequency range 45–85 GHz is presented in Fig. 7.9 and in Table 7.9.

The characteristics of HEMT-1 and HEMT-2 obtained in 7.3.1 and 7.3.2 were used for calculation of the parameters of powerful heteromagnetic field transistors in a frequency range up to 100 GHz.

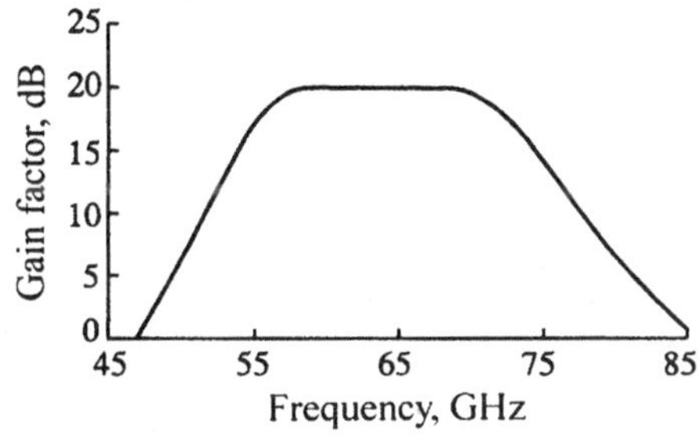

Fig. 7.9 The carryover (gain) factor of the HEMT-2 transistor in a frequency range 45–85 GHz

Table 7.9 The carryover (gain) factor of the HEMT-2 transistor in a frequency range 45–85 GHz

Parameter, dB	Frequency band, GHz	Gain factor, dB	
		HEMT-2	Model
*S*21	46–54	10–22	0–15
	54–72	22–23	15–20.6
	54–80	10–20	6.5–15

7.4 Magnetoelectronic Elements of LPL

A system of microstrip or coil electrodes of certain configurations creating HF magnetic fields in which FMCR is placed will be called as a magnetoelectronic element of communication (MECE).

The basic requirements to MECE are as follows:

- Small weights and dimensions
- Concentration of HF magnetic fields of a required polarization
- Maintenance of the maximum factor of filling
- Design simplicity
- Small transfer losses
- A high decoupling level
- A sufficient electric durability
- Control over the bias field, the resonant frequency, the total impedance, the phase
- Dynamic management of the microwave frequence power

Additional requirements to MECE are as follows:

- Planarity
- Work with no field bias (at autoresonance)
- A preset law of change of the magnetic parameters over thickness and the area of the multilayered ferrites
- Saturated (homogeneous), nonsaturated (multidomain), or transitive states
- Linear or nonlinear mode
- One- or the multiconnected systems of the passing or absorbing type

7.4.1 Coupling Element in Omnirange, UHF Range

Some types of planar coupling elements made on the microstrip technology on polybark and gallium arsenide (GaAs) substrates were simulated and experimentally investigated (Table 7.10).

The topology of coupling elements (see Table 7.10) is made on various substrates, namely:

- Gallium arsenide (No. 1–6 and No. 8–10) with a thickness of 0.1 mm, with a dielectric permeability 12.9 and the width of bringing strip conductors 0.1 mm

Table 7.10 The topology of coupling elements

No.	Coupling element topology	Description
1	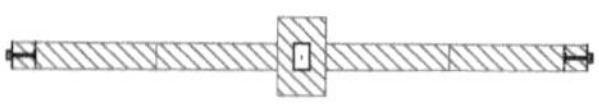	A strip short-circuited at the center, on a gallium arsenide substrate, the strip width is 100 μm
2	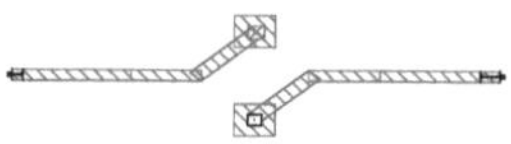	Strips short-circuited at their ends, on a gallium arsenide substrate, the strip width is 100 μm
3	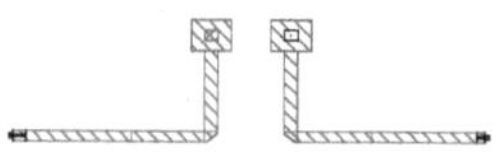	Strips short-circuited at their ends, on a gallium arsenide substrate, the strip width is 100 μm
4	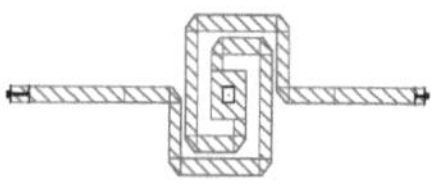	A rectangular spiral “360°” on a gallium arsenide substrate, the strip width is 100 μm
5	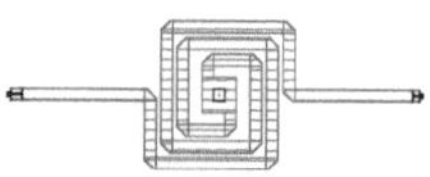	A rectangular spiral “540°” on a gallium arsenide substrate, the strip width is 100 μm.
6	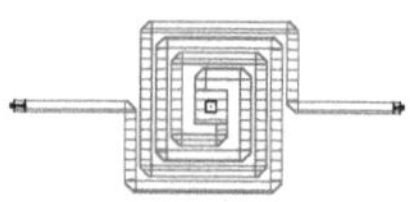	A rectangular spiral “720°” on a gallium arsenide substrate, the strip width is 100 μm
7	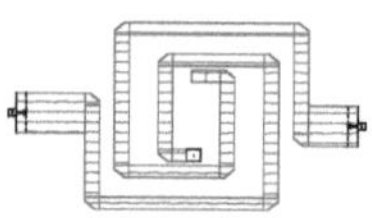	A rectangular spiral on a polybark substrate, the strip width is 0.5 mm
8	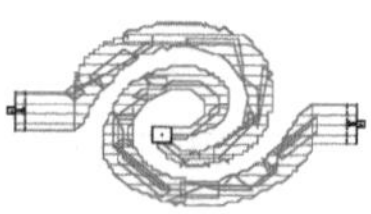	A circular spiral on a gallium arsenide substrate, the strip width is 0.5 mm.
9	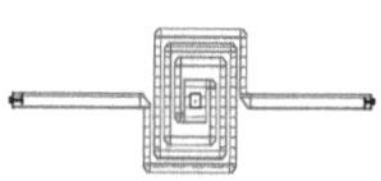	A rectangular spiral “720°” on a gallium arsenide substrate, the strip width is 60 μm
10	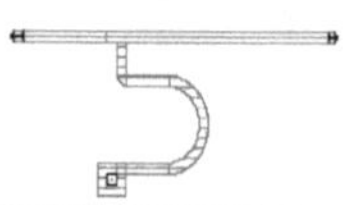	A power divider on a gallium arsenide substrate, the strip width is 100 μm

Table 7.11 Basic characteristics of various MECE (see Table 7.10) in a frequency range from 0.5 to 30 GHz with no external magnetic field

No.	MECE decoupling level, dB	Losses in MECE on FMR frequency, dB
1	27.5–11.0	4.0
2	26.8–10.7	5.5
3	24.6–8.6	3.2
4	30.0–18.5	0.7
5	22.7–9.9	0.9
6	17.5–8.9	1.2
7	19.5–16.5	11.5
8	21.6–9.7	0.6
9	20.0–8.76	0.85
10	17.0–3.3	0.96

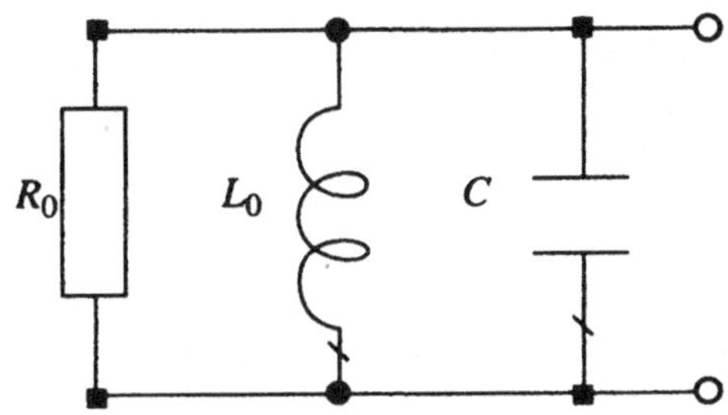

Fig. 7.10 The equivalent FMCR circuit as a single parallel oscillatory contour

- Gallium arsenide and polybark (No. 7–8) with a thickness of 0.5 mm, the width of bringing strips is 0.5 mm

In Table 7.11, basic characteristics of various MECE (see Table 7.10) in a frequency range from 0.5 to 30 GHz with no external magnetic field are presented.

The equivalent FMCR circuit is chosen as a single parallel oscillatory contour (Fig. 7.10).

The resonant frequency of FMCR is determined by

$$f_0 = \gamma H_{0i}, \tag{7.6}$$

where H_{0i} is the external magnetic field and $\gamma = 28\,\text{MHz/mT}$ is the gyromagnetic electron ratio.

The active resistance is determined by

$$R_0 = \mu_0 V K^2 \omega_{\text{m}} Q_{\text{nl}}, \tag{7.7}$$

where $\mu_0 = 4\pi \times 10^{-7}\text{H/m}$ is the magnetic constant, $V = \frac{1}{6}\pi d^3$ the ferrite sphere volume, and d its diameter, $\omega_{\text{m}} = 2\pi\gamma \times 4\pi M_{\text{s}}$ the FMR frequency, $4\pi M_{\text{s}}$ saturation magnetization, Q_{nl} – not loaded GB product of FMCR, ΔH is the resonant line width of FMR.

The value of the not loaded GB product is

$$Q_{\mathrm{nl}} = \frac{H_0 - \frac{1}{3}4\pi M_{\mathrm{s}}}{\Delta H}. \tag{7.8}$$

The contour inductance L_0 and capacity C_0 are determined by

$$L_0 = \frac{R_0}{\omega_0 Q_{\mathrm{nl}}}, \tag{7.9}$$

$$C_0 = \frac{1}{\omega_0^2 L_0}. \tag{7.10}$$

Modeling of the microstrip coupling elements with the ferrite spherical resonator in an external magnetic field can be implemented in a CAD loke MWO-2002, Serenade, etc.

The equivalent circuits of FMCR connection into coupling elements are presented in Fig. 7.11.

The circuit shown in Fig. 7.11 was used for MECE calculations from Tables 7.10 and 7.11: (a) coupling element No. 1 and (b) elements of communication No. 2–10. The three-pole elements in these figures represent investigated MECEs. The choice of the corresponding coupling element is made by changing the NET field value in [58]. The oscillatory contour simulates FMCR. The contour parameters are defined by (7.6)–(7.10) that allows changing the external magnetic field, resonant frequency, and not loaded GB product of the oscillatory contour. By the algorithm from [55], the frequency dependencies of the transfer factors for various magnetic fields H_0 (Fig. 7.12) and SWR (Fig. 7.13) were calculated for MECE. The resonant frequency of 1,400 MHz of the equivalent oscillatory contour in FMCR corresponds to the magnetic field of $H_{0i} = 50\,\mathrm{mT}$.

In Table 7.12, the calculation results in a frequency band 0.5–3 GHz for coupling element No. 6 from Table 7.10 are given.

The analysis of planar MECEs of various types in a real time mode is made strictly electrodynamically. The calculation program allows changing the topology

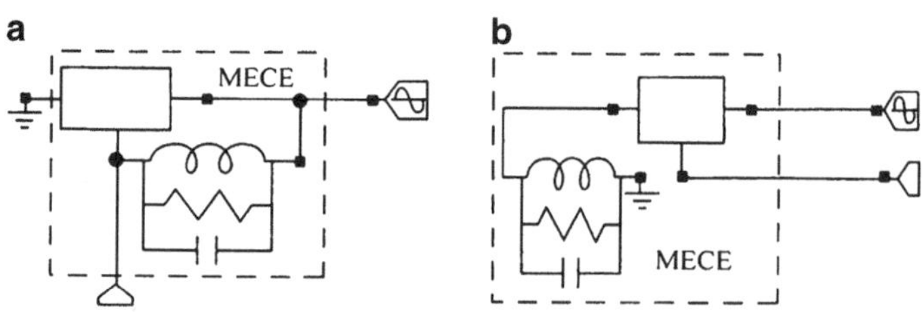

Fig. 7.11 The circuit for MECE calculations: (**a**) as flute, which is bridged in the middle (element 1 from Table 7.10); (**b**) in a various topologies of striplines (elements 2–10 from Table 7.10)

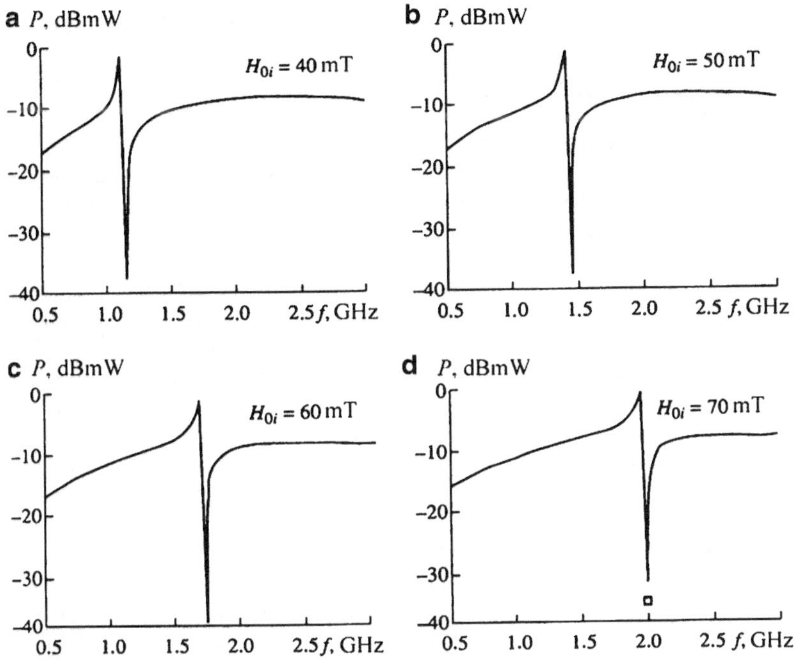

Fig. 7.12 Dependencies of the transfer factors for MECE for various magnetic fields H_0

Fig. 7.13 Dependencies of the transfer factors for MECE for various SWR

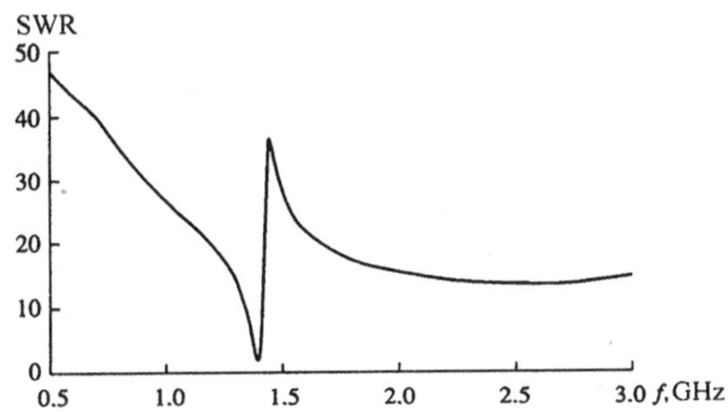

of the MECE conductors, the control modes over the magnetic field, including MECE the corresponding multiterminal network with a matrix of parameters into other projects and CADs.

The algorithm was used in programs of calculation of generation modes of regular and quasi-noise signals at high levels of continuous and pulse power in the VHF, UHF, microwave frequency, and EHF-ranges:

- Field magnetotransistors [54]
- Bipolar magnetotransistors [56]

Table 7.12 The calculation results in a frequency band 0.5–3 GHz for coupling element No. 6 from Table 7.10

f, GHz	P, dBmW	SWR	f, GHz	P, dBmW	SWR
0.50	−17.524	47.551	1.50	−13.604	28.104
0.55	−16.701	45.428	1.55	−11.829	24.018
0.60	−15.95	43.502	1.60	−10.958	21.750
0.65	−15.265	41.869	1.65	−10.402	20.235
0.70	−14.643	40.116	1.70	−9.9966	19.107
0.75	−14.071	37.738	1.75	−9.6788	18.212
0.80	−13.53	35.189	1.80	−9.4177	17.474
0.85	−13.015	32.804	1.85	−9.1973	16.849
0.90	−12.521	30.618	1.90	−9.008	16.312
0.95	−12.046	28.600	1.95	−8.844	15.846
1.00	−11.581	26.716	2.00	−8.7014	15.438
1.05	−11.119	24.928	2.05	−8.5774	15.083
1.10	−10.648	23.193	2.10	−8.4704	14.772
1.15	−10.148	21.451	2.70	−8.2717	13.678
1.20	−9.5858	19.604	2.75	−8.3439	13.782
1.25	−8.8877	17.464	2.80	−8.4312	13.918
1.30	−7.8641	14.583	2.85	−8.5341	14.086
1.35	−5.8123	9.6021	2.90	−8.6533	14.288
1.40	−1.1755	1.6138	2.95	−8.7896	14.525
1.45	−21.704	36.602	3.00	−8.9438	14.795

7.4.2 Coupling Element in Microwave Frequency, EHF Range

The developed algorithm of calculation is compatible with modern CADs (AWR Microwave Office, Ansoft HPFS, etc.). The various MECE types considered above had some restrictions at advancement into the frequency ranges from 10 up to 100 GHz, namely, increased introduced losses, a decreased decoupling level. In this connection, modified MECEs have been simulated, which allows obtaining effective interaction in the microwave and EHF frequency ranges.

The MECE presented in Fig. 7.14a has the best characteristics (the decoupling level and efficiency of interaction with FMCR in a range from 0.5 up to 10 GHz). However, on frequencies above 20 GHz the decoupling level in such a MECE decreases from 20 dB down to 2 dB, which considerably reduces its efficiency – see Fig. 7.14b in which the transfer factors for various magnetic fields are shown: $1 - H_{0i} = 5.4$ kOe; $2 - H_{0i} = 15.4$ kOe, $3 - H_{0i} = 25.4$ kOe, $4 - H_{0i} = 35.4$ kOe. For increasing the decoupling of this element at frequencies 20–100 GHz, the thickness of the dielectric layer in the field of coupling of the input and output line has been increased from 3 up to 15 μm.

The MECE topology presented in Fig. 7.15a is rather critical to the grounding hole resistance. On frequencies above 20 GHz the decoupling level considerably decreases (Fig. 7.15b). To increase the decoupling in a frequency up to 100 GHz it is necessary to minimize the grounding hole resistance, which may cause certain technological difficulties.

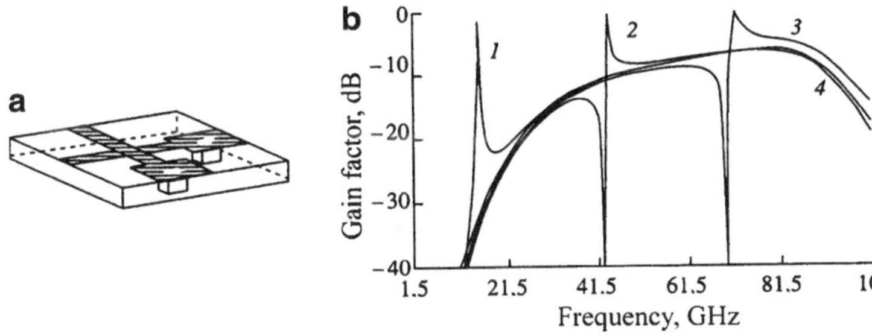

Fig. 7.14 (**a**) The MECE topology in the form of transposed bridged striplines. (**b**) The AFC of such MECE

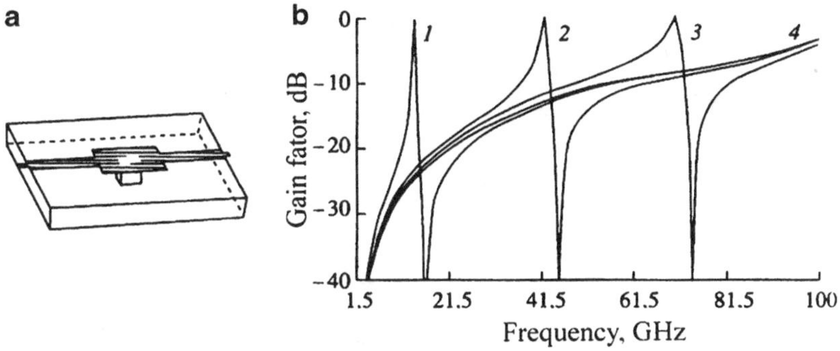

Fig. 7.15 (**a**) The MECE topology. (**b**) The MECE topology on frequencies above 20 GHz

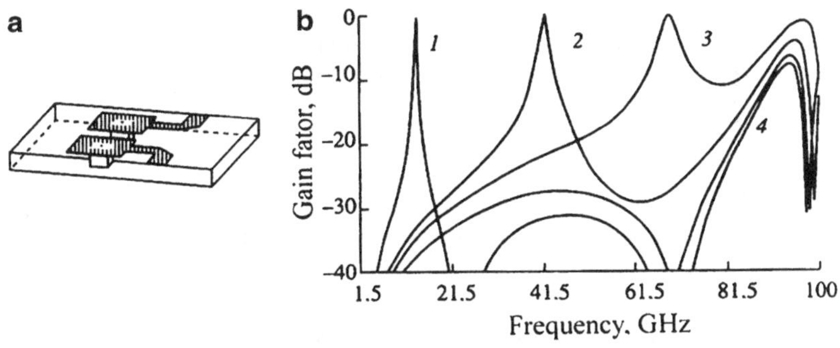

Fig. 7.16 (**a, b**) The coupling element and its characteristics in a frequency range up to 90 GHz

The coupling element presented in Fig. 7.16a possesses a high decoupling level in a frequency range up to 90 GHz (Fig. 7.16b).

The parameters of FMCR presented in Fig. 7.17 as an equivalent *RLC* contour were calculated with (7.6–7.10).

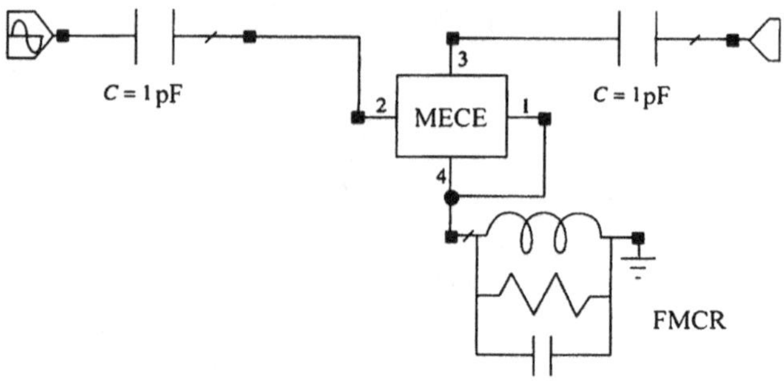

Fig. 7.17 The parameters of FMCR as an equivalent RLC contour

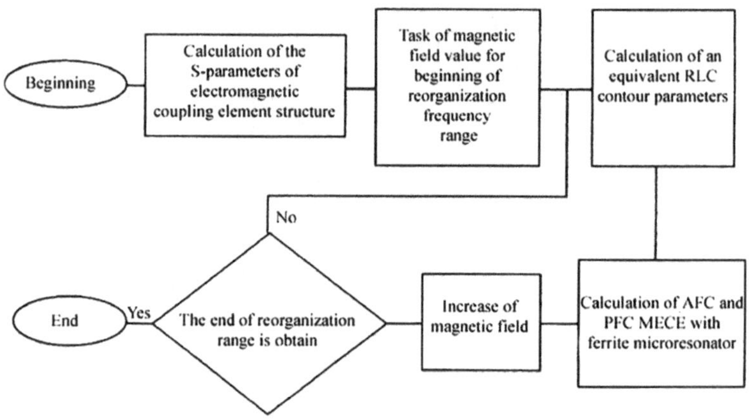

Fig. 7.18 The flowchart of the algorithm of calculation of the parameters of magnetoelectronic coupling elements with a ferrite microresonator with resonant frequency reorganization by a magnetic field in a frequency range up to 100 GHz

The flowchart of the algorithm of calculation of the parameters of magnetoelectronic coupling elements with a ferrite microresonator with resonant frequency reorganization by a magnetic field in a frequency range up to 100 GHz is shown in Fig. 7.18.

The algorithm of calculation of the parameters of magnetoelectronic coupling elements was used in development of a program of analysis of powerful bipolar magnetoelectronic transistors in the microwave frequency and EHF ranges.

7.5 Powerful Bipolar Transistor in Microwave Frequency Range

At designing of powerful bipolar transistors, cascading of transistor crystals in one case is applied. Separate cells of the compound transistor are incorporated in parallel that essentially reduces its input resistance. With the purpose to increase the resistance of the input circuit of the compound bipolar transistor, a matching MDP condenser was placed between the emitter and base near to the transistor crystals (Fig. 7.19). The task of modeling of the powerful transistor is divided into modeling of separate transistor crystals and of the whole transistor assembly in view of the inductances of the boiled conductors and the capacity of the matching condenser. As initial data, experimental families of the S parameters and of static characteristics of the bipolar transistor were used. Each cell of the 5-section bipolar transistor (Fig. 7.20) was modeled separately by iterative optimization of its parameters.

The calculated parameters of one cell of the compound transistor are as follows: $IS = 6.235 \times 10^{-13}$ mA, BF = 150, NF = 0.902033, VAF = 30, IKF = 1022.26 mA, NE = 1.192, BR = 8.01, NR = 1, VAR = 50 W, IKR = 3.514 mA, NC = 0.8, RB = 0.2 Ω, $IRB = 1.015 \times 10^{-10}$ mA, RBM = 0.503839 Ω, RE = 0.329702 Ω, CJC = 4 pF, and $LB = 1 \times 10^{-6}$ nH.

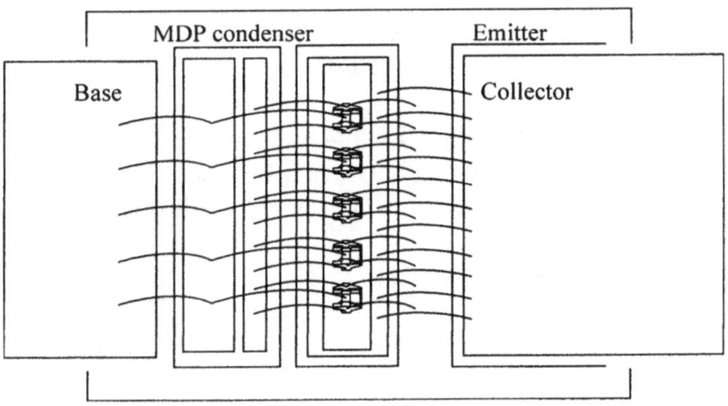

Fig. 7.19 Allocation of a matching MDP condenser between the emitter and base

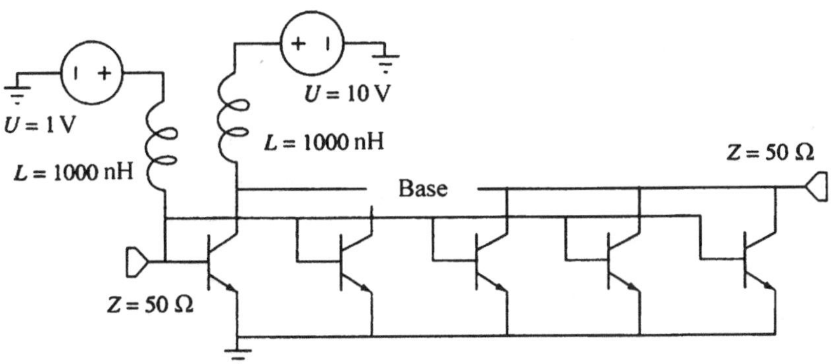

Fig. 7.20 The 5-section bipolar transistor

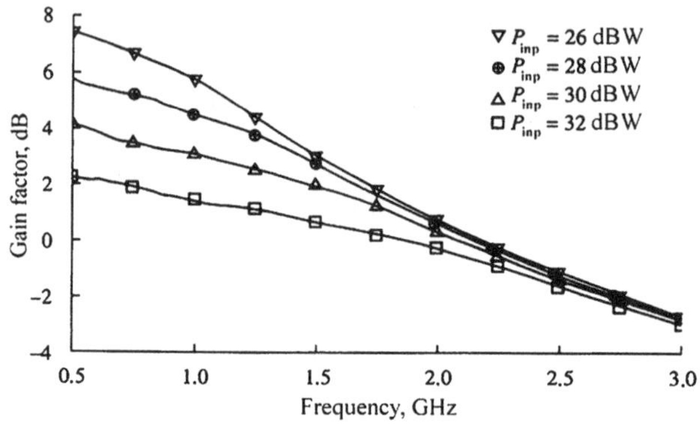

Fig. 7.21 The calculated AFC of the test powerful amplifier with no coupling element for various levels of the input power

Figure 7.21 shows the calculated AFC of the test powerful amplifier with no coupling element for various levels of the input power.

The initial static characteristics for one section of the compound transistor were calculated from the experimental static characteristics of the 5-section transistor, thus the terminal current was divided between all the sections equally, and the terminal voltage of each section was accepted equal to the terminal voltage of the whole transistor.

At parallel–parallel connection of four-pole elements the Y-matrix of the circuit can be obtained by addition of the Y-matrixes of all its components.

Let us pass from the Y-matrix to the S-matrix for the compound transistor by means of

$$Y = \sqrt{Y_0}(I - S)(I + S)^{-1}\sqrt{Y_0}, \tag{7.11}$$

$$Y_{11} = \frac{\frac{1}{Z_{01}}((1 - S_{11})(1 + S_{22}) - S_{12}S_{21})}{(1 + S_{11})(1 + S_{22}) - S_{12}S_{21}},$$

$$Y_{12} = \frac{\frac{1}{\sqrt{Z_{01}}}\frac{1}{\sqrt{Z_{02}}}(S_{21}S_{12} - (1 - S_{11})S_{12})}{(1 + S_{11})(1 + S_{22}) - S_{12}S_{21}},$$

$$Y_{22} = \frac{\frac{1}{Z_{02}}((1 - S_{22})(1 + S_{11}) - S_{21}(1 + S_{22}))}{(1 + S_{11})(1 + S_{22}) - S_{12}S_{21}},$$

$$Y_{21} = \frac{\frac{1}{\sqrt{Z_{01}}}\frac{1}{\sqrt{Z_{02}}}(S_{21}S_{12} - (1 - S_{22})S_{21})}{(1 + S_{11})(1 + S_{22}) - S_{12}S_{21}}, \tag{7.12}$$

where I is a unit matrix and Y_0 is the diagonal matrix of normalizing conductances of various inputs with diagonal elements equal to $1/Z_{01}$ and $1/Z_{02}$.

For one transistor structure of the compound transistor, the matrix is

$$Y' = \frac{Y}{N}, \tag{7.13}$$

where Y is the conductivity matrix of the compound transistor and N is the number of transistor structures in the compound transistor.

For calculation of the S-matrix of one transistor structure of the compound transistor we shall use the following expressions:

$$S = \sqrt{Z_0}\,(Y_0 - Y')\,(Y_0 + Y')^{-1}\,\sqrt{Y_0}, \tag{7.14}$$

$$S_{11} = \frac{\left(\frac{1}{Z_{01}} - \frac{Y_{11}}{N}\right)\left(\frac{1}{Z_{02}} + \frac{Y_{22}}{N}\right) + \frac{Y_{12}Y_{21}}{N^2}}{\left(\frac{1}{Z_{01}} + \frac{Y_{11}}{N}\right)\left(\frac{1}{Z_{02}} + \frac{Y_{22}}{N}\right) - \frac{Y_{12}Y_{21}}{N^2}},$$

$$S_{12} = \frac{\sqrt{\frac{Z_{01}}{Z_{02}}}\frac{Y_{12}}{N}\left(\frac{Y_{11}^2}{N} - \frac{1}{Z_{01}^2}\right)}{\left(\frac{1}{Z_{01}} + \frac{Y_{11}}{N}\right)\left(\frac{1}{Z_{02}} + \frac{Y_{22}}{N^2}\right) - \frac{Y_{12}Y_{21}}{N^2}},$$

$$S_{21} = \frac{\sqrt{\frac{Z_{01}}{Z_{02}}}\frac{Y_{21}}{N}\left(\frac{Y_{22}^2}{N^2} - \frac{1}{Z_{02}^2}\right)}{\left(\frac{1}{Z_{01}} + \frac{Y_{11}}{N}\right)\left(\frac{1}{Z_{02}} + \frac{Y_{22}}{N}\right) - \frac{Y_{12}Y_{21}}{N^2}},$$

$$S_{22} = \frac{\left(\frac{1}{Z_{01}} + \frac{Y_{11}}{N}\right)\left(\frac{1}{Z_{02}} - \frac{Y_{22}}{N}\right) + \frac{Y_{12}Y_{21}}{N^2}}{\left(\frac{1}{Z_{01}} + \frac{Y_{11}}{N}\right)\left(\frac{1}{Z_{02}} + \frac{Y_{22}}{N}\right) - \frac{Y_{12}Y_{21}}{N^2}}. \tag{7.15}$$

In Fig. 7.22, the flowchart of the algorithm of calculation for the powerful compound bipolar transistor, considering numerical experiment, coordination with experimental data, and optimization of the parameters of the external model is shown.

As a test task the AFC of the powerful bipolar 5-section KT962B transistor has been calculated, $P = 20\,\mathrm{W}$, $f_{\mathrm{T}} = 1{,}000\,\mathrm{MHz}$, and $K_{\mathrm{gf}} = 4$–5. The divergence with the experimental characteristics in a frequency range from 300 to 900 GHz does not exceed 15–20%.

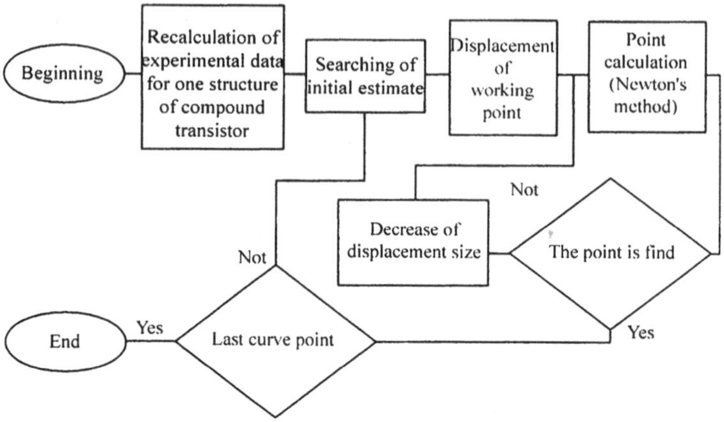

Fig. 7.22 The flowchart of the algorithm of calculation for the powerful compound bipolar transistor, considering numerical experiment, coordination with experimental data, and optimization of the parameters of the external model

7.6 Powerful Bipolar Heteromagnetic Transistor in Microwave Frequency Range

The equivalent circuit of the powerful bipolar HMT is a set of transistor structures connected by power summation with FMCR (Fig. 7.23).

The design of MECE with FMCR depends on the wavelength range and the used active element (transistor). In HMG samples up to 2 GHz, FMCR is placed directly into the current carrying conductor area of the powerful transistor. Therefore, for analysis the model in Fig. 7.24 was used. The current carrying system of the active element is presented by a piece of a two-wire line with a length $2R$ with a distance between conductors a and their diameter d. Near to the conductor located at $x = 0$, at a distance $h + R$ a ferrite sphere of a radius R was placed. The resonant characteristics of MECE and parameters of the equivalent oscillatory contour modeling FMCR were calculated.

For calculations the equations obtained in (5.4) and (5.5) were used in view of the temperature dependencies of the parameters of Gummel–Poon's model of the bipolar transistor.

Research of the characteristics of a powerful HMG in the mode of regular signal generation was made by the method of harmonic balance. The irregular modes of the generated signals were analyzed by Runge–Kutta's method.

The following were investigated: the frequency of generated oscillations ν_0, output integral power P_{out}, efficiency CE, the level of output power from frequency reorganization, thermal drift of generation frequencies, modes of superposition of complex multifrequency, and noise-like oscillations.

The equivalent circuit of the powerful HMT on the 2T9132Ac transistor includes five structures and allows getting pulse output power up to 400 W (see Fig. 7.23).

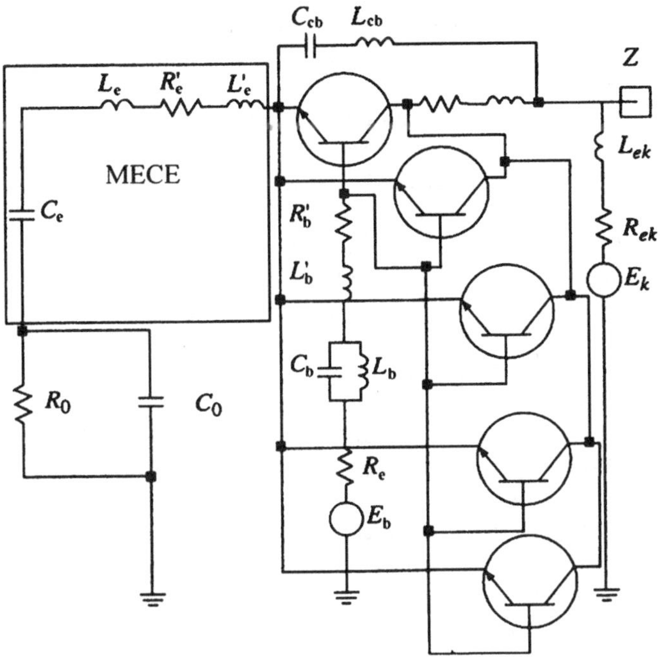

Fig. 7.23 The equivalent circuit of the powerful bipolar HMT

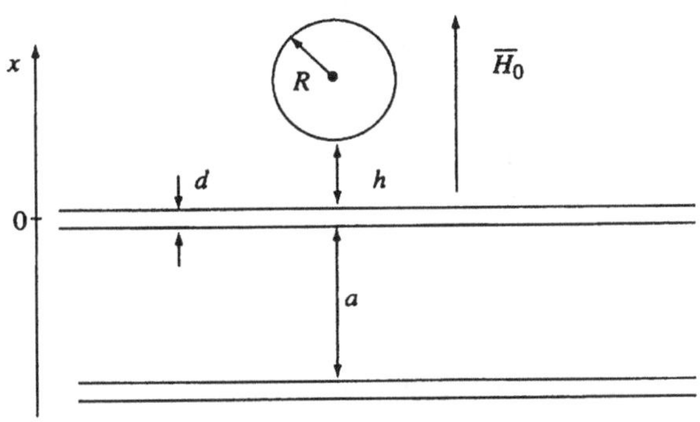

Fig. 7.24 FMCR in the current carrying conductor area of the powerful transistor

The Frequency-driving elements C_e, L_e, R'_e, L'_e determine the equivalent parameters of the FMCR located in the field of the emitter junction of the transistor. The resonant frequency is determined from $\omega_0 = \mu_0 \gamma \sqrt{H_0^2 - (\Delta H(\theta))^2}$, where $\Delta H(\theta)$ is the half-width of ferromagnetic resonance lines. At $\gamma = 1.76 \times 10^{11}$ C/kg and $\Delta H = 24$ A/m, the resonant frequency ω_0 is $f_0 = 471$ MHz; at $H_0 = 13{,}885$ A/m,[4] the linear resonant frequency is $f_0 = \omega_0/2\pi = 471$ MHz. The resistance R_0 limits

[4] 1 A/m = 79.6 Oe.

the displacement current of the base–emitter junction and together with the sources setting the power voltage (E_k) and the base displacement voltage (E_b), determines the mode of the transistor by direct current. The resistance R_0 on high frequency is shunted by the condenser C_0 with a high enough capacity (~1,000 pF); R_e, R_k being the internal resistance of the sources of base displacement and the power supply of the generator. The elements R'_e, L'_e, R'_b, L'_b, R'_k, L'_k are determined by the conductors of the terminals and the transistor assembly on the plate. C_{kb} is the parasitic capacity formed at installation of a semiconductor structure in the case of the transistor. Z is the load resistance. Coordination of the low output resistance of the transistor with the load is carried out with a P-shaped filter of their elements L_f, L'_f, C_f. The capacities C_e and C_b are formed by the contact platforms of the emitter and base terminals of the transistor. The inductance L_e is formed by a piece of an asymmetrical microstrip line.

For calculations, the equivalent parameters of Gummel–Poon's model of the following powerful bipolar transistors were used: 2T9132AC (a continuous mode up to 150 W), "Plotter B" (a pulse mode up to 500 W) [60]. The power voltage of the transistor "Plotter B" was 50 V, and that of the transistor 2T9132AC was 30 V.

In Table 7.13, the parameters of the equivalent circuit of a powerful HMT on the 2T9132AC transistor are shown.

Depending on the operating mode, the output capacity changed within the limits of 150–500 W, the basic frequency was within 400–500 MHz, DE was 20–40%.

For reorganization of the generator by frequency, the parameters of the oscillatory systems in the emitter and base areas simultaneously changed. Time realization of the oscillation amplitude of a powerful MES on the 2T9132AC transistor in the mode of continuous power generation is presented in Fig. 7.25.

For achievement of optimum values of the output power and DE of the generator at frequency changes, the elements of the equivalent circuit in the circuits of the emitter, base, and loading were tuned simultaneously. By means of selection of the elements of the output matching filter and the load resistance of HMG, a level of generated power of 150–400 W at DE 30–40% was reached.

Table 7.13 The parameters of the equivalent circuit of a powerful HMT on the 2T9132AC transistor

Element	Value	Element	Value
L_b	0.002 nH	L_e	7.3 nH
C_b	16 pF	C_e	15.01 pF
L'_b	0.01 nH	L'_e	0.03 nH
R'_b	0.01 Ω	R'_e	0.003 Ω
C_o	10 pF	R'_k	0.003 Ω
L_{ek}	100 nH	R_{ek}	0.01 Ω
E_b	6 V	E_k	45 V
Z	25 Ω	L'_k	0.01 nH
C_f	0	L_f	0
C'_f	0	–	–

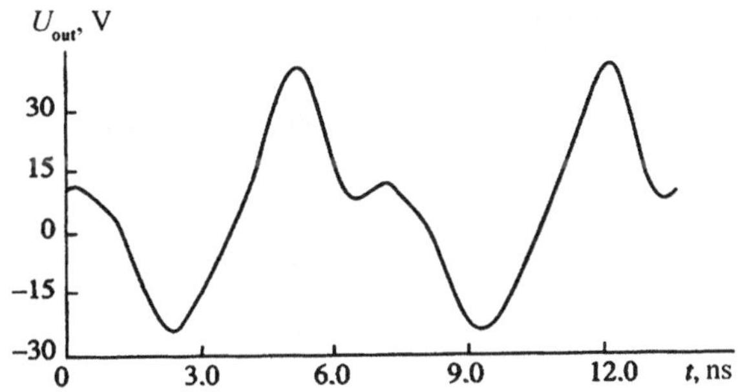

Fig. 7.25 Time realization of the oscillation amplitude of a powerful MES on the 2T9133AC transistor in the mode of continuous power generation

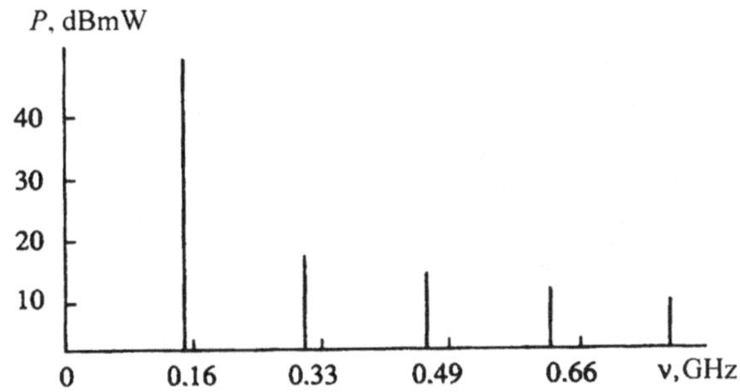

Fig. 7.26 The results of calculation of the power of spectral components of oscillations in HMG on the 2T9132AC transistor in the mode of continuous generation

The results of calculation of the power of spectral components of oscillations in HMG on the 2T9132AC transistor in the mode of continuous generation are presented in Fig. 7.26.

At those values of the equivalent parameters of FMCR as $C_e = 25$ pF, $L_e = 36$ nH, the magnetic bias $H_0 = 4{,}690$ A/m, the frequency of generation was 0.165 GHz at an output power 152 W. The other parameters of the circuit are taken from Table 7.13. At changes of the external bias field within the limits of $H_0 = 4{,}221$–$5{,}159$ A/m, the frequency of generation varied within the limits of 148–181 MHz.

The value of the external magnetic field $H_0 = 46{,}900$ A/m corresponds to the frequency of generation of 1.65 GHz. At adjustment of FMCR for this frequency, excitation of the second harmonic was realized in HMG. The spectrum of oscillations is presented in Fig. 7.27.

By the use of FMCR in the nonsaturated mode, multifrequency and noise-type oscillations arose in HMG. The mechanism of their occurrence is investigated in 6.7. The distribution of SPDN in the mode of noise-type oscillations with an integral output power 150 W is presented in Fig. 7.28 within a wave range 1.40–1.76 GHz and in Fig. 7.29 within a wave range 130–200 MHz.

The distribution of SPDN in HMG based on the "Plotter B" transistor in the mode of pulse noise-type oscillations with an integral output power 450 W is presented in Fig. 7.30.

From the figure, it follows that powerful HMTs can be used to develop noise generators in a range ±10% and a nonuniformity of SPDN not higher than 3 dB.

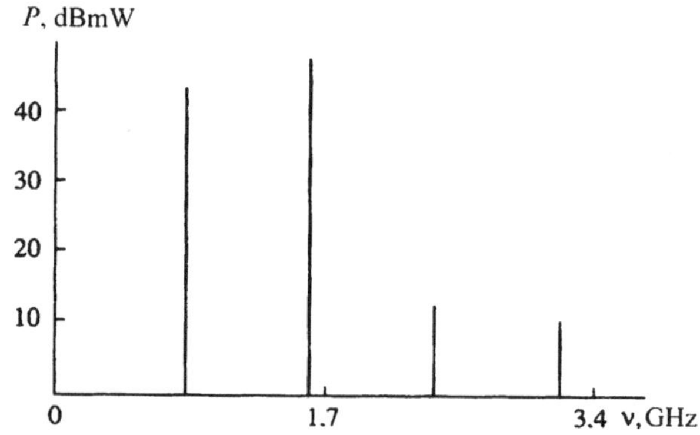

Fig. 7.27 The spectrum of oscillations on the second harmonic at adjustment of FMCR for frequency 1.65 GHz

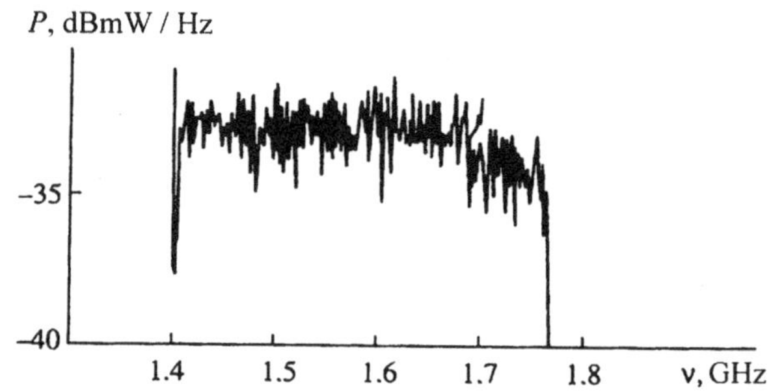

Fig. 7.28 SPDN in the mode of noise-type oscillations with an integral output power 150 W within a wave range 1.40–1.76 GHz

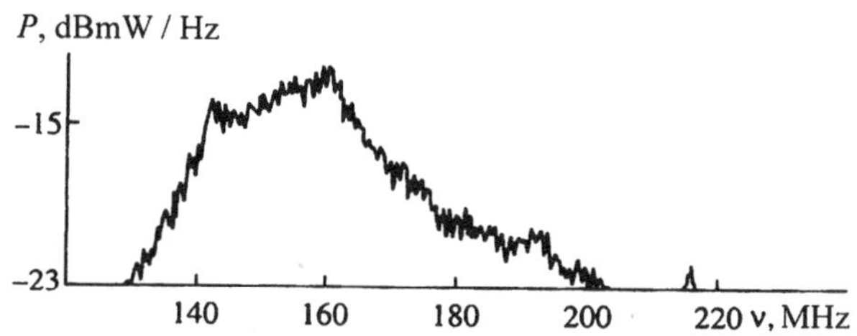

Fig. 7.29 SPDN in the mode of noise-type oscillations with an integral output power 150 W within a wave range 130–200 MHz

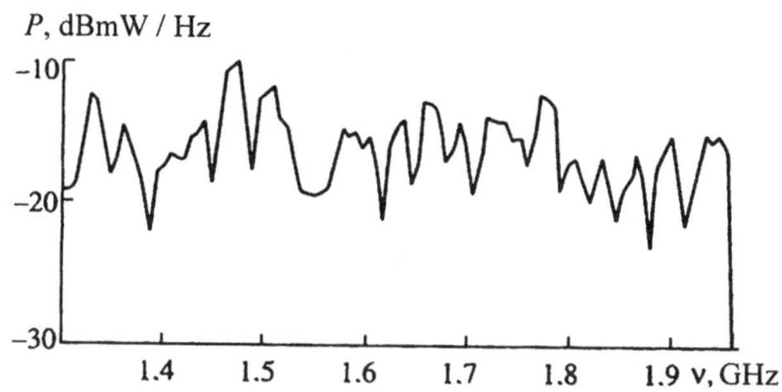

Fig. 7.30 The distribution of SPDN in HMG based on the "Plotter B" transistor in the mode of pulse noise-type oscillations with an integral output power 450 W

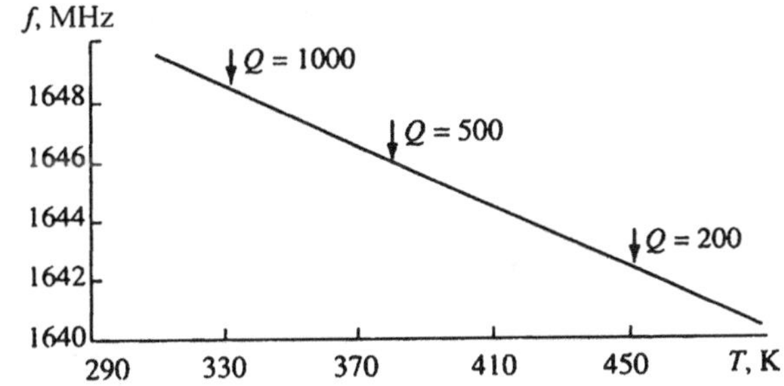

Fig. 7.31 The dependence of the frequency of a powerful HMG on the temperature of the "Plotter B" transistor in the mode of pulse noise-type oscillations with a pulse output power 450 W at pulse-to-pulse durations $Q = 1{,}000$, 500, 200 and a front duration of 8 ns

At output power levels above 150 W, HMT can work in a pulse mode only. Thus, the warming temperature of the crystal at the moment of pulse reaches 23°C. It is accompanied by a frequency change in the autogenerating mode. Calculations have shown that due to crystal warming in the transistor up to 23°C, the frequency of generation decreased by 1.12 MHz under the law close to a linear one. It corresponds to a relative reduction of frequency by 0.16%.

In Fig. 7.31, the dependence of the frequency of a powerful HMG on the temperature of the "Plotter B" transistor in the mode of pulse noise-type oscillations with a pulse output power 450 W at pulse-to-pulse durations $Q = 1{,}000$, 500, 200 and a front duration of 8 ns is presented. In HMG, introduction of compensation circuits of temperature drifts of frequencies with the use of a microprocessor control system is possible.

A powerful HMT consists of several coordinated transistors, connected in parallel. Effective frequency reorganization by magnetic field was reached at inclusion of an FMCR into each transistor. Owing to the small (in comparison with the powerful HMT) sizes of the spherical FMCR there was no possibility to provide effective communication between all the transistor structures and the microresonator. By the use of various microresonators in each transistor structure there is a mismatch of these structures because of a deviation from the face value of characteristics of the used microresonators. Calculations show a reduction of the generated power by 7–10 times at deviations of the parameters of FMCR used in various transistor structures, within the limits of 10% from the face value. For this reason, in powerful HMT it is necessary to use a ferrite monocrystal as a plate overlapping all the transistor structures. Technologically, it is possible by the use of a planar HMT design.

In Figs. 7.32–7.35, the curve of changes of the output power of a powerful HMT at reorganization by magnetic field in the modes of continuous and pulse generation of regular and noise-type signals is presented. The range of reorganization in the investigated powerful generators is much less than in low-power HMGs. It is explained by the necessity of simultaneous fine tuning of the matching circuits of the powerful transistor in HMG at changes of the frequency of generation by a magnetic field.

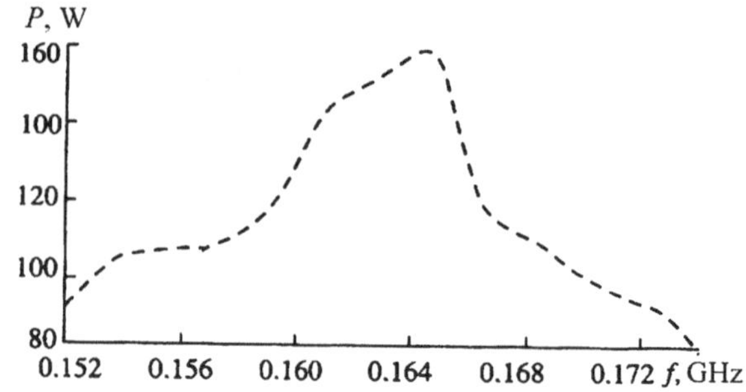

Fig. 7.32 The curve of changes of the output power of a powerful HMT at reorganization by magnetic field in the modes of continuous and pulse generation of regular and noise-type signals

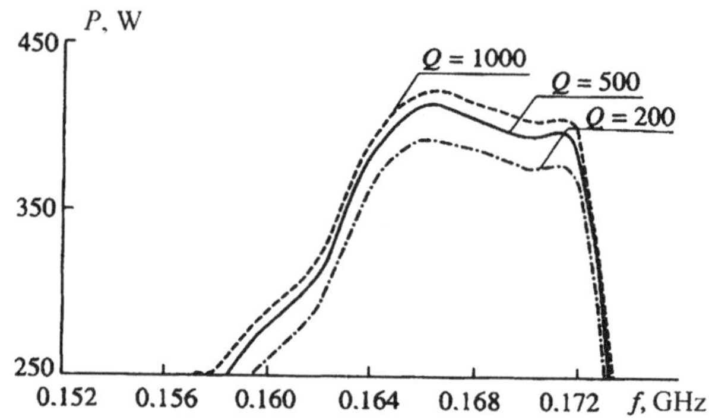

Fig. 7.33 The curve of changes of the output power of a powerful HMT at reorganization by magnetic field in the modes of continuous and pulse generation of regular and noise-type signals

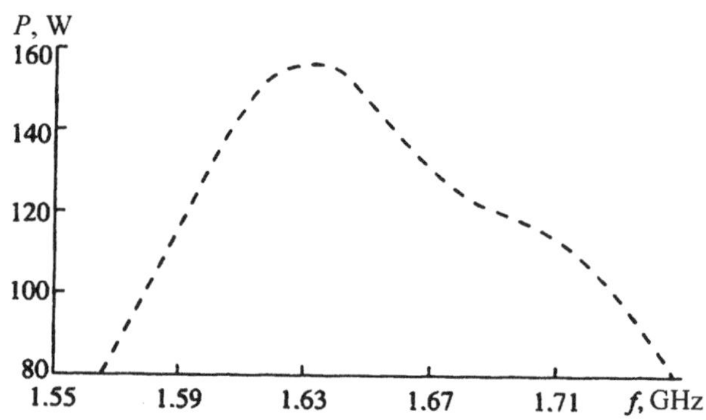

Fig. 7.34 The curve of changes of the output power of a powerful HMT at reorganization by magnetic field in the modes of continuous and pulse generation of regular and noise-type signals

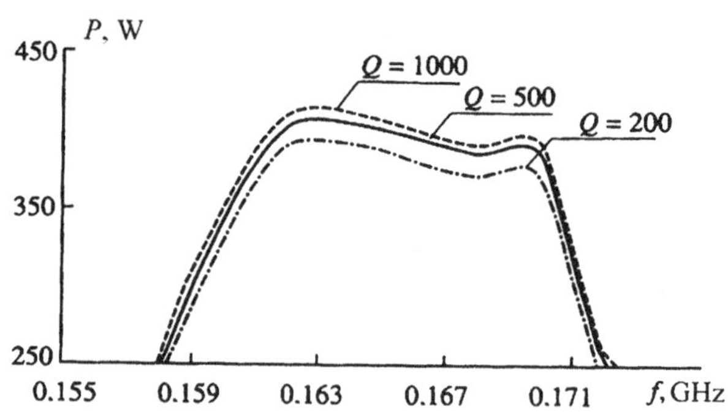

Fig. 7.35 The curve of changes of the output power of a powerful HMT at reorganization by magnetic field in the modes of continuous and pulse generation of regular and noise-type signals

From analysis of the curves, it follows that the relative range of reorganization of HMG by a magnetic field lays within the limits of 10%.

Thus, in a powerful HMG the following are possible:

- Reorganization of signal within a 10% frequency band
- Monochromatic generation of a signal
- Multifrequency generation of signals of equidistant frequency spectra
- Chaotic generation of signals at working of the ferrite microresonator in a non-saturated nonlinear mode

7.7 Powerful Magneto-FET in a Frequency Band Below 30 GHz

The elementary transistor cell with a Schottky shutter is presented (Fig. 7.36) in the form of Materok's equivalent circuit. The principle of cascade connection of elementary transistor cells is used for increase of the output power of a field HMT.

A nonlinear model of the active area of the transistor considering the total impedance of the electrodes was applied. For modeling of cascade connection, the elementary cells of the transistor were considered quasi-identical. This assumption essentially limited the number of the parameters necessary to solve the optimization problem. Updating of the transistor model is necessary for simplification of optimization by various classes of parameters and compatibility with various CADs. Therefore, further the equivalent circuit shown in Fig. 7.37, in which the elementary semiconductor structure was set by the circuit in Fig. 7.36, was used.

The full equivalent circuit of a cascading transistor can also be presented in the form of Materok's equivalent circuit (Fig. 7.38) for a single transistor.

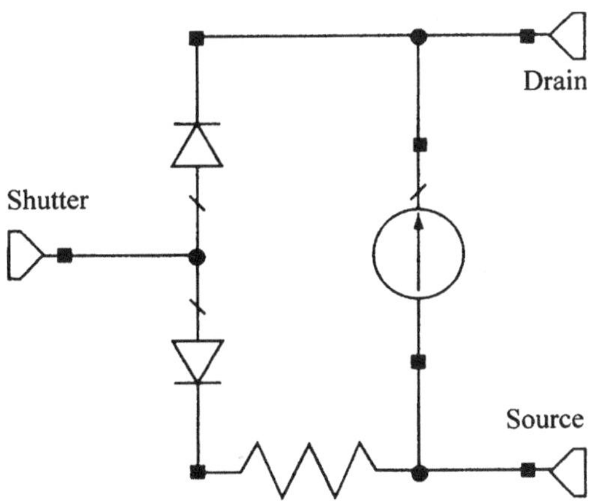

Fig. 7.36 The elementary transistor cell with a Schottky shutter in the form of Materok's equivalent circuit

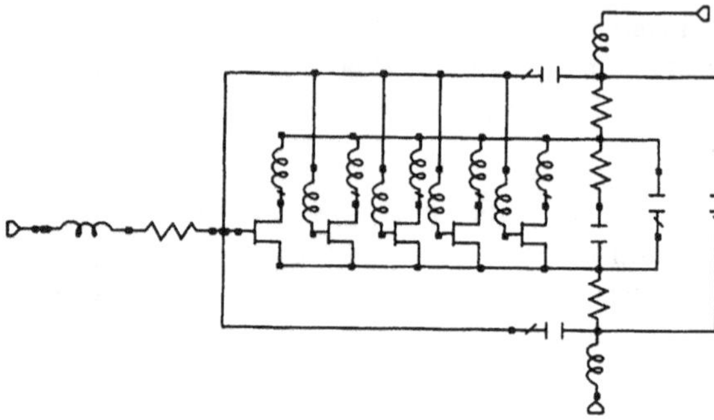

Fig. 7.37 The equivalent circuit of the transistor model

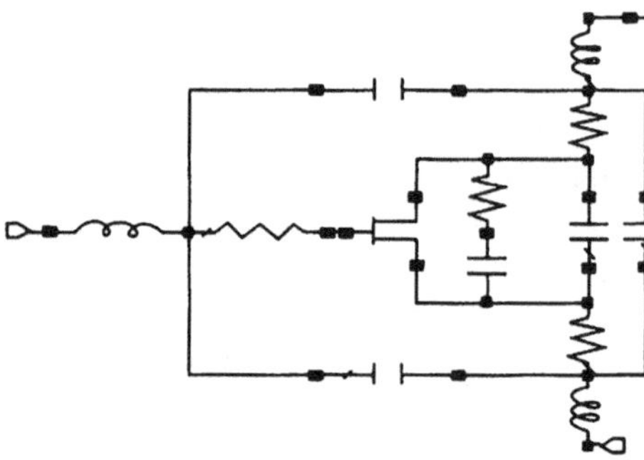

Fig. 7.38 The full equivalent circuit of a cascading transistor in the form of Materok's equivalent circuit for a single transistor

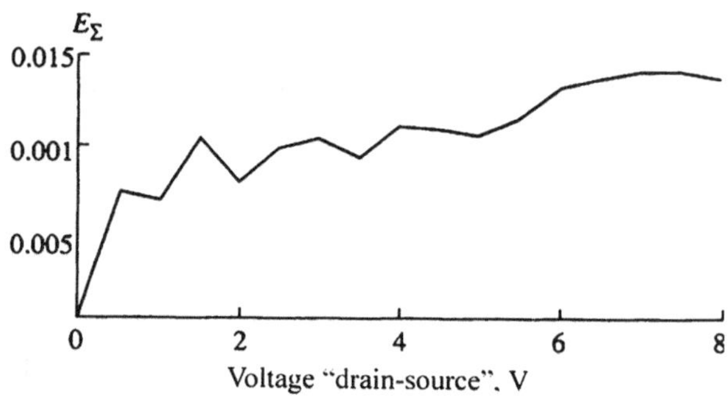

Fig. 7.39 The integral error by a family of static characteristics $E_\Sigma < 0.015$

At solution of a test task the following results have been obtained:

- The integral error by a family of static characteristics $E_\Sigma < 0.015$ (Fig. 7.39)
- The error function for the transfer factor in a range from 0.3 to 30 GHz: $E(S_{21}) < 0.5$, and in a range from 7 to 30 GHz $E(S_{21}) < 0.08$ (Fig. 7.40)

In the test task, the base cell on the basis of the NE27200 transistor was used. The choice of the transistor is caused by a high gain factor in a wide frequency band (up to 30 GHz).

In Fig. 7.41, the equivalent circuit of a powerful FMT is presented. It includes a coupling element, a ferrite structure, a powerful cascading field transistor in the

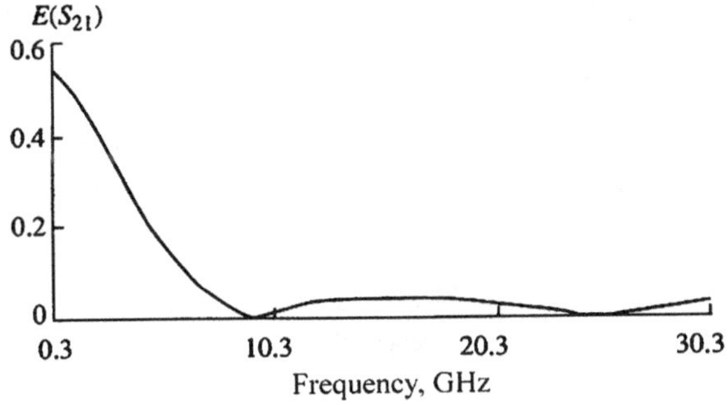

Fig. 7.40 The error function for the transfer factor in a range from 0.3 to 30 GHz: $E(S_{21}) < 0.5$, and in a range from 7 to 30 GHz $E(S_{21}) < 0.08$

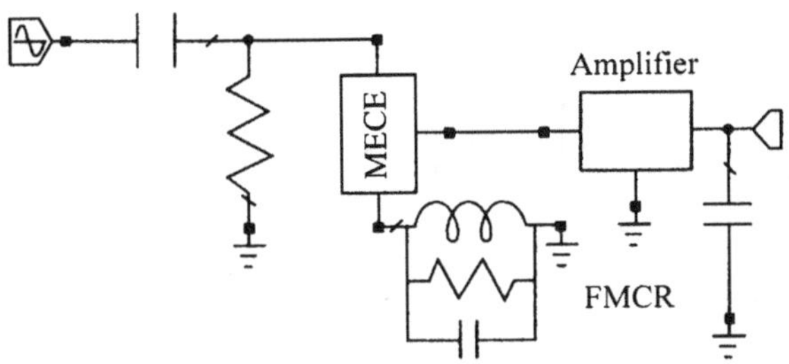

Fig. 7.41 The equivalent circuit of a powerful FMT is presented. It includes a coupling element, a ferrite structure, a powerful cascading field transistor in the form of a three-pole element (an amplifier, Fig. 7.38) with power supply elements (Fig. 7.42)

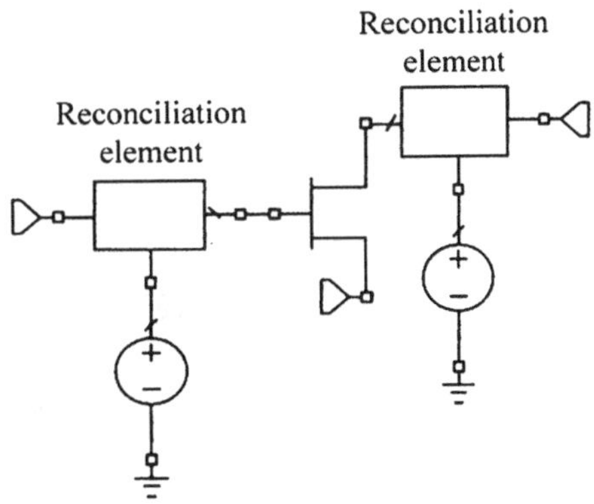

Fig. 7.42 The equivalent circuit of a powerful FMT with power supply elements

form of a three-pole element (an amplifier, Fig. 7.38) with power supply elements (Fig. 7.42).

The results of calculation of the frequency characteristics of the powerful FMT for several magnetic induction values are shown in Fig. 7.43: (a) 40 mT; (b) 61.5 mT; (c) 80 mT; and (d) 100 mT.

The dependencies of the gain factor of FMT on the magnetic induction are shown in Fig. 7.44: 1 – 0.04 T; 2 – 0.08 T; and 3 – 0.1 T.

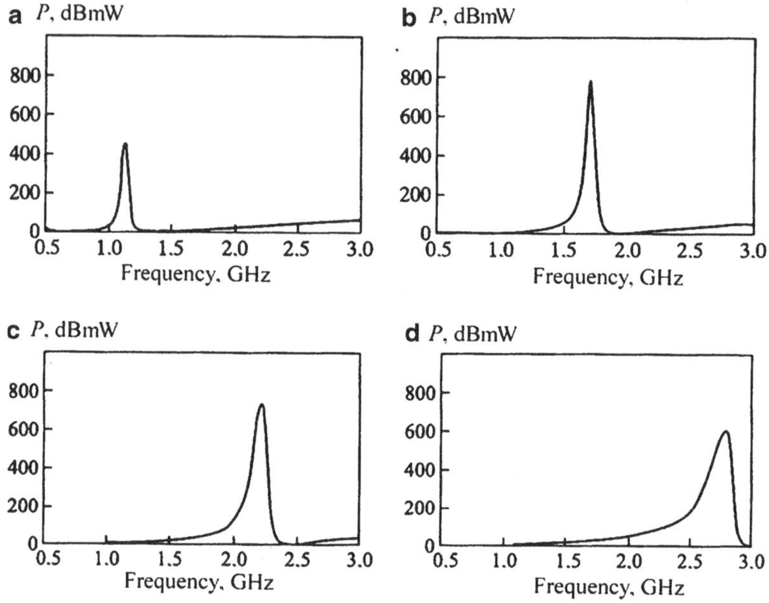

Fig. 7.43 The results of calculation of the frequency characteristics of the powerful FMT for several magnetic induction values: (**a**) 40 mT; (**b**) 61.5 mT; (**c**) 80 mT; and (**d**) 100 mT

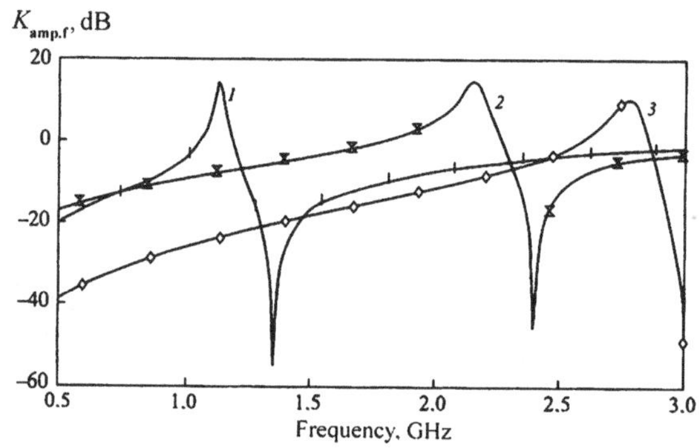

Fig. 7.44 The dependencies of the gain factor of FMT on the magnetic induction: 1 – 0.04 T; 2 – 0.08 T; 3 – 0.1 T

7.8 Powerful Magneto-FET in EHF Range

A mathematical model (HEMT-M) has been developed for calculation of a powerful FMT. The broadband UA1S65LM amplifier (USA, *Centellax Inc.*) has similar parameters. For modeling, the parameters of the dispersion matrix of the HEMT-M structure shown in Table 7.14 for a frequency of 65 GHz were used.

The program is written in the MWO-2002 CAD environment. A model of summation of the power of nonlinear amplifiers with a strip low-frequency filter at the output was used. The equivalent circuit of HEMT-M in the amplification mode is presented in Fig. 7.45.

Table 7.14 The parameters of the dispersion matrix of the HEMT-M structure

Parameter	Value
S_{11}	−10
S_{12}	−18
S_{21}	7.5
S_{22}	−12

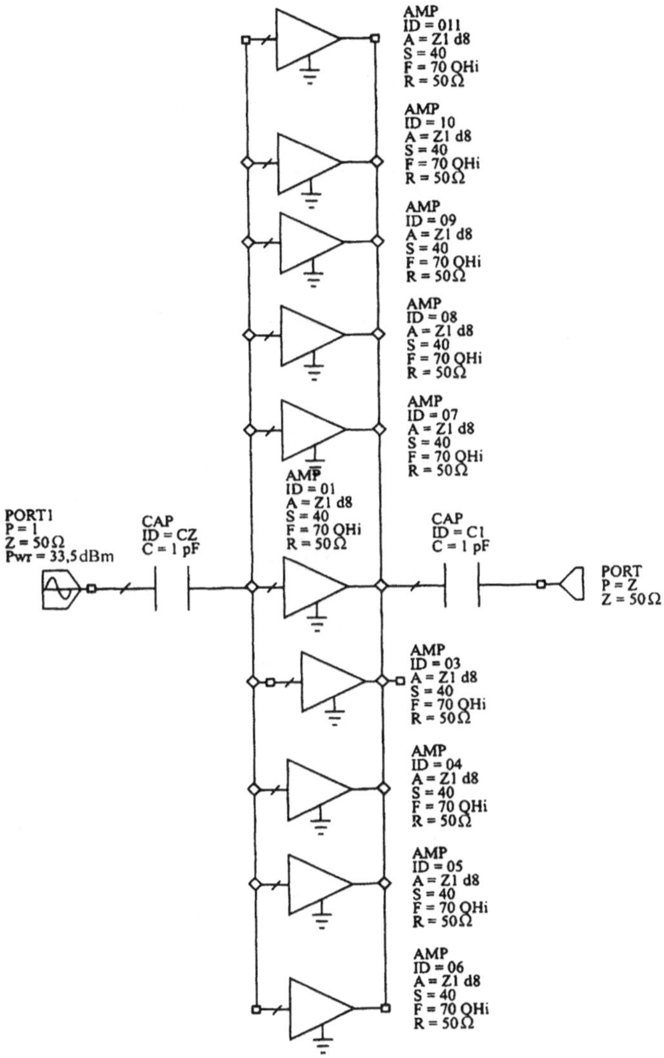

Fig. 7.45 The equivalent circuit of HEMT-M in the amplification mode

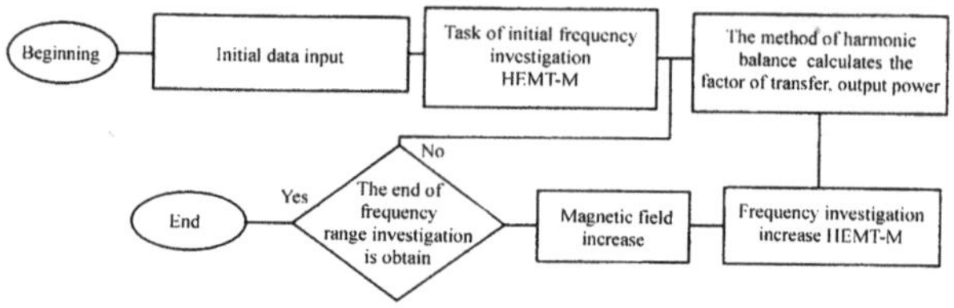

Fig. 7.46 The flowchart of the program

Table 7.15 The results of calculation of the frequency dependence of the module of the transfer factor and output power

Frequency, GHz	The module of the transfer factor, dB	Output power, W
50	3.1	4.7
55	2.8	4.4
60	2.6	4.2
65	2.4	4.0
70	0.9	2.8

The flowchart of the program is presented in Fig. 7.46. At the first step, the program inputs its initial data: frequency, power voltage, and the equivalent circuit parameters. At the second step, the method of harmonic balance calculates the factor of transfer, output power, and DE of the HEMT-M structure. At the third step, the frequency and constant magnetizing field increase, achievement of the higher border of the frequency range is checked, and the process of calculation either repeats with the new frequency or finishes.

The results of calculation of the frequency dependence of the module of the transfer factor and output power are presented in Table 7.15.

Chapter 8
Calculation of Thermal Conditions of Magnetotransistors in Continuous and Pulse Modes

8.1 General Remarks

The developed HMT can be used in modes of generation of both continuous and pulse power. The temperature field of the active semiconductor crystal in the HMT structure in these conditions reflects of a sequence of practically rectangular pulses of thermal power, whose peak value is determined by the transistor's CE. The quasi-stationary mode is most critical for HMT, being the reaction to a long sequence of identical power pulses or identical groups of pulses, so-called "trains" of pulses.

In this case, the temperature field of the semiconductor crystal can be presented as superposition of both the stationary and pulse components of temperature [59,61]. The stationary component of temperature is determined by the time-average thermal flux over all the elements of the HMT design (due to heat conductivity) to the environment and is calculated by means of the thermal scheme method [58].

The stationary temperature components (Fig. 8.1) were calculated on the basis of the following models with localized bulk thermal emission: (a) with rectangular multilayer elements and (b) with cylinder ones.

In a rectangular N-layer semiconductor structure (Fig. 8.1a), the contacts between different layers are considered ideal. In each layer, there can occur bulk thermal emission with a uniform distribution over thickness and any distribution in the plane of this layer. On the top and bottom surfaces, the conditions of convective heat exchange with the environment are set. Any thermal fluxes from the other surface of the structure were neglected.

The general problem equations for the devices shown in Fig. 8.1a look like:

$$\begin{aligned}
&\frac{\partial^2 T_i}{\partial x^2}+\frac{\partial^2 T_i}{\partial y^2}+\frac{\partial^2 T_i}{\partial z_i^2}+\frac{W_i(x,y)}{\lambda_i}=0, \quad i=1,2,\ldots,N,\\
&\left.\frac{\partial T_1}{\partial z_1}\right|_{z_1=0}=\frac{\alpha_1}{\lambda_1}\left[T_1(z_1=0)-T_{\mathrm{c}}\right],\\
&\left.T_i\right|_{z_i=h_i}=\left.T_{i+1}\right|_{z_{i+1}=0}, \quad \left.\frac{\partial T_i}{\partial z_i}\right|_{z_i=h_i}=\left.\frac{\lambda_{i+1}}{\lambda_i}\frac{\partial T_{i+1}}{\partial z_{i+1}}\right|_{z_{i+1}=0}, \qquad (8.1)\\
&i=1,2,\ldots,N-1,
\end{aligned}$$

A.A. Ignatiev and A.V. Lyashenko, *Heteromagnetic Microelectronics: Microsystems of Active Type*, DOI 10.1007/978-1-4419-6002-3_8,

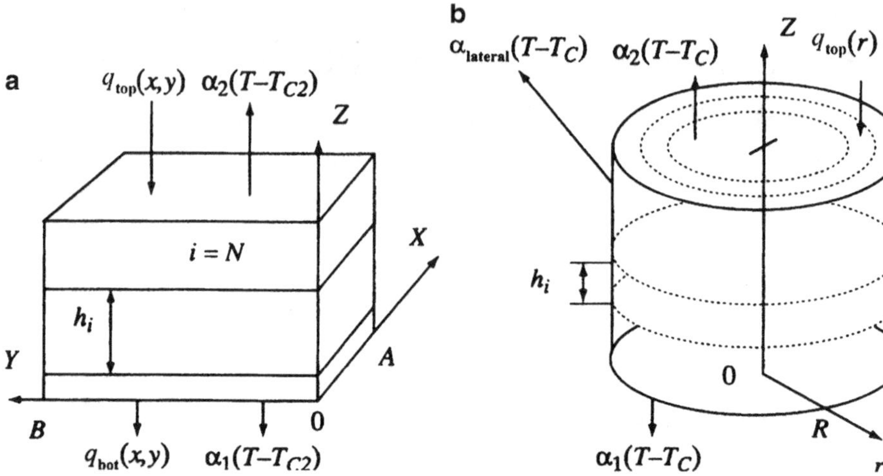

Fig. 8.1 Models of devises with localized bulk thermal emission: (**a**) with rectangular multilayer elements and (**b**) with cylinder ones

$$\left.\frac{\partial T_N}{\partial z_N}\right|_{z_N=h_N} = -\frac{\alpha_2}{\lambda_N}\left[T_n(z_N = h_N) - T_c\right] + \frac{q(x,y)}{\lambda_N},$$

$$\left.\frac{\partial T_i}{\partial x}\right|_{x=0,A} = \left.\frac{\partial T_i}{\partial y}\right|_{y=0,B} = 0 \quad \text{for all } i = 1, 2, \ldots, N,$$

where $T_i(x, y, z_i)$ is the temperature field in the ith layer; $0 \le x \le A, 0 \le y \le B$, and $0 \le z_i \le h_i$ are spatial variables; λ_i the heat conductivity of the ith layer; h_i the thickness of the ith layer; A and B the structure sizes in the plane of layers; α_1 and α_2 the convective heat transfer factors with the bottom ($z_1 = 0$) and top ($z_N = h_N$) surfaces of the structure, respectively; $W_i(x, y)$ the volume density of thermal emission in the ith layer; $q(x, y)$ the density of surface thermal emission on the top ($z_N = h_N$) side of the structure; and T_c is the ambient temperature.

The general problem has been solved analytically by finite integral transformations, which allows calculation of stationary temperature fields in planar and bulk microcircuits, field transistors, semiconductor lasers, etc.

A problem of calculation of stationary temperature differences in multilayered axisymmetric cylindrical objects with local bulk thermal emission (Fig. 8.1b) has been similarly formulated and solved. Its mathematical formulation looks like:

$$\frac{\partial^2 T_i}{\partial r^2} + \frac{1}{r}\frac{\partial T_i}{\partial r} + \frac{\partial^2 T_i}{\partial z_i^2} + \frac{W_i(r)}{\lambda_i} = 0, \quad i = 1, 2, \ldots, N,$$

$$\left.\frac{\partial T_1}{\partial z_1}\right|_{z_1=0} = \frac{\alpha_1}{\lambda_1}\left[T_1(z_1 = 0) - T_c\right],$$

$$\left.T_i\right|_{z_i=h_i} = \left.T_{i+1}\right|_{z_{i+1}=0}, \quad \left.\frac{\partial T_i}{\partial z_i}\right|_{z_i=h_i} = \left.\frac{\lambda_{i+1}}{\lambda_i}\frac{\partial T_{i+1}}{\partial z_{i+1}}\right|_{z_{i+1}=0},$$
$$i = 1, 2, \ldots, N-1, \tag{8.2}$$

$$\left.\frac{\partial T_N}{\partial z_N}\right|_{z_N=h_N} = -\frac{\alpha_2}{\lambda_N}\left[T_n(z_N = h_N) - T_c\right] + \frac{q(x,y)}{\lambda_N},$$

$$\left|T\right|_{r=0}, \quad \left.\frac{\partial T_i}{\partial r}\right|_{r=R} = 0 \quad \text{for all} \quad i = 1, 2, \ldots, N,$$

where $T_i(r, z_i)$ is the temperature field in the ith layer; $0 \le r \le R$ and $0 \le z_i \le h_i$ are spatial variables; R the radius of the cylindrical structure; $W_i(r)$ the volume density of thermal emission in the ith layer; $q(r)$ the density of surface thermal emission on the top ($z_N = h_N$) side of the structure, and the other designations are the same as in (8.1).

By means of the obtained general solutions of the problem, stationary thermal modes in the corresponding constructive elements of HMT were modeled at some preset forms of the functions of heat sources and heat exchange conditions.

The pulse modes of HMT were described by the heat conduction equation:

$$\frac{\partial T}{\partial \tau} = a\left(\frac{\partial^2 T}{\partial x^2} + \frac{\partial^2 T}{\partial y^2} + \frac{\partial^2 T}{\partial z^2}\right) + \frac{a}{\lambda} W(x, y, z, \tau),$$

$$\left.T\right|_{\tau=0} = T_0; \quad \left.T\right|_{z=0} = T_0; \quad \left.\frac{\partial T}{\partial z}\right|_{z=h} = 0; \quad \left.\frac{\partial T}{\partial x}\right|_{x=0,B} = 0, \qquad (8.3)$$

where $T(x, y, z, \tau)$ is the required temperature field; x, y, and z arespatial variables; τ current time; λ and α the factors of heat conductivity and temperature conductivity of the semiconductor, respectively; and $W(x, z, \tau)$ is thevolume density of thermal emission in the semiconductor structure depending on the coordinates and time.

Problem (8.3) has been solved in a general view by finite integral transformations. The spatial–temporal dependence of the volume density of thermal emission should be specified in each case.

8.2 Nonstationary and Temperature Field of Powerful Magneto-FET in Pulse Mode

Problem (8.3) for pulse HMTs (e.g., on the KT 9164 AC transistor) can be simplified essentially by ignoring one spatial coordinate, as the base areas in such a structure occupy practically all the width of the semiconductor crystal, and its temperature field can be considered as bidimensional.

The nonstationary temperature field of a powerful HMT in its pulse mode has been calculated at the following simplifications:

- The localized thermal sources of the rectangular shape generate thermal power in the form of a sequence of rectangular impulses, and the characteristics of this pulse sequence (the pulse duration, the period and amplitude of thermal power) can be individual for each thermal source.

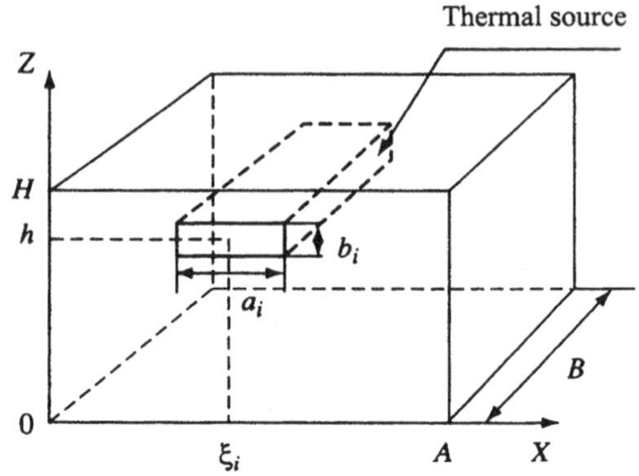

Fig. 8.2 The analyzable structure

- The whole thermal power released in the volume dissipates to a heat-conducting path through the bottom side of the structure, which is considered isothermal.
- The temperature field of the object is bidimensional.

The structure whose temperature field has been calculated is shown in Fig. 8.2.

The initial data for the program [62] are as follows:

- The sizes of the crystal $A \times B \times H$
- The heat conductivity and temperature conductivity of the material
- The number of thermal sources

Each thermal source is characterized by:

- The sizes of the source
- The coordinates of the centers of the source by axes X and Z
- The pulse duration
- The duration of the pulse sequence period
- The peak value of thermal power
- The borders of the spatial area of temperature calculation with the number of spatial points
- The borders of the time area of temperature calculation with the number of time points

As a result, the program outputs the spatial distributions of temperature during the instants of time specified by the user.

The program has been tested on several test models, including the case $\tau \to \infty$. The programming language is FORTRAN.

The flowchart of the program is shown in Fig. 8.3.

Test task 8.1. Monotonous warming up of the silicon semiconductor crystal with the following data:

- The sizes of the crystal are $A = 0.5$ mm, $B = 0.74$ mm, and $H = 0.12$ mm.
- The heat conduction of silicon is $\lambda = 120$ W/(m K).
- The temperature conductivity of silicon is $a = 0.5 \times 10^{-4}$ m^2/s.
- The number of thermal sources is 1.

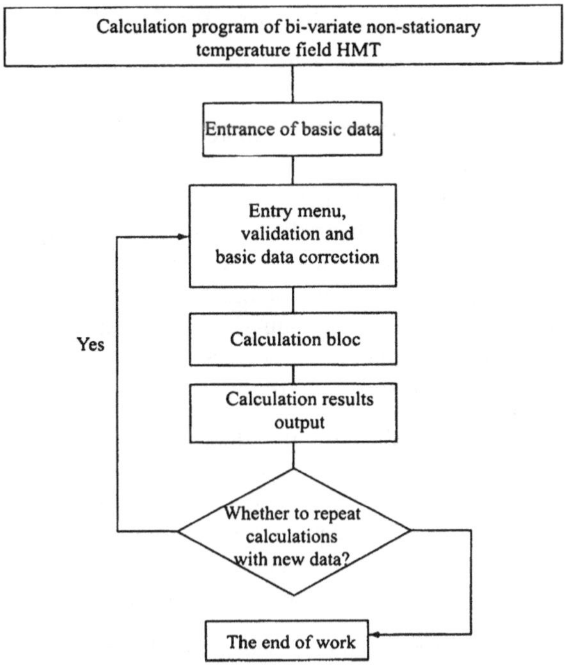

Fig. 8.3 The flowchart of the program

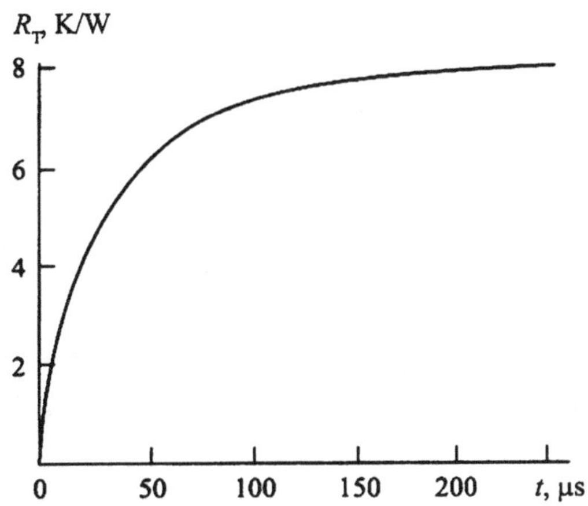

Fig. 8.4 The calculated transitive thermal resistance of the specified structure $R_T = T_S(t)/P_T$

- The sizes of a source are $a_1 = 0.080$ mm and $b_1 = 0.004$ mm.
- The coordinates of the center of a thermal source are $h = 0.118$ mm and $\xi = 0.37$ mm.
- The thermal power is $P_T = 1$ W in a continuous mode.

The calculated transitive thermal resistance of the specified structure $R_T = T_S(t)/P_T$ is shown in Fig. 8.4, where T_S is the nonstationary temperature change of the crystal in the field of a heat source.

The established value of the thermal resistance of the object is 7.98 K/W.

8.3 Stationary Thermal Resistance of Powerful Magneto-FET with Squared Shape

A local rectangular thermal source is located on the surface $z_N = h_N$, and there exists a heat-conducting isothermal side at $z_1 = 0$, while the other surface is isolated. All the parameters of the modeled structure are shown in Fig. 8.5.

A program has been developed [63] for analysis of the thermal modes of active semiconductor structures with a planar technology, and design elements of a similar geometry. The program contains a built-in subroutine for solution of sets of linear algebraic equations by Gauss' method with leading element choice. The limiting value of the number of layers is 5.

The initial data are as follows:

- The sizes of the structure along axes X and Y
- The number of heterogeneous layers
- The thickness (over axis Z) and heat conductivity of each layer
- The sizes of a superficial heat source along axes X and Y
- The coordinates of the center of a superficial heat source along axes X and Y
- The thermal emission power in a source

The values of the maximal and average temperature of each surface heat source were calculated. The analytical solution of test task 8.1 contains Fourier series by spatial variables. Their convergence was estimated by means of Leibniz's theorem of sign-variable series convergence. In the specified program, the error of Fourier series calculation did not exceed 1%. The programming language is FORTRAN.

Test task 8.2. The stationary thermal resistance of a semiconductor crystal with the parameters from test task 8.1 with the following initial data:

- The sizes of the crystal: $A = 0.5$ mm and $B = 0.74$ mm
- The number of layers is 1
- The thickness of a layer $h_1 = 0.12$ mm
- The heat conductivity of a layer $\lambda = 120\,\text{W}/(\text{m K})$

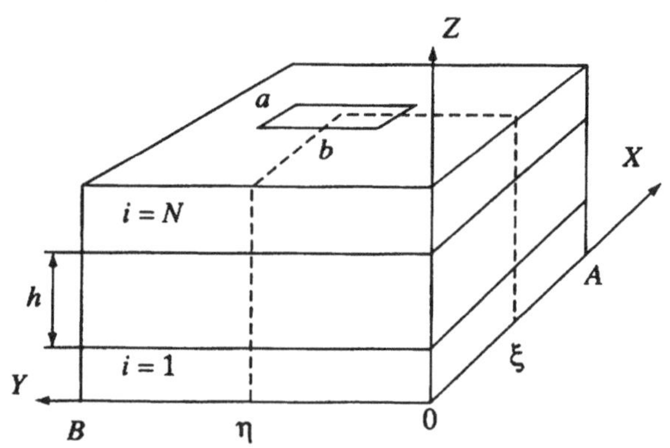

Fig. 8.5 The parameters of the modeled structure

Table 8.1 The results of calculation of power magnetotransistor

ξ, mm	1.0	2.0	3.0	4.0	5.0	6.0	7.0	8.0	9.0
max R_T, K/W	44.07	43.92	43.90	43.89	43.89	43.89	43.90	43.92	44.07
$(R_T)_s$, K/W	39.66	39.53	39.51	39.51	30.51	39.51	39.51	39.53	39.66

- The sizes of the source: $a_1 = 0.08$ mm and $b_1 = 0.74$ mm
- The coordinates of the center of the thermal source: $\xi = 0.25$ mm and $\eta = 0.37$ mm

The obtained value of the thermal resistance is 8.14 K/W. The insignificant difference of this result from that of the previous test task is explained by the heat source having no thickness (it is superficial) while in the previous example the source possessed a small thickness of 0.004 mm.

Test task 8.3. The dependence of thermal resistance R_T of a semiconductor structure on the location of a local thermal source with the following:

- The sizes of the crystal are $A = 10$ mm and $B = 10$ mm.
- The number of layers is 2.
- The thickness of the layers are $h_1 = 2.0$ mm and $h_2 = 0.12$ mm.
- The heat conductivities of the layers are $\lambda_1 = 200$ W/(m K) and $\lambda_2 = 120$ W/(m K).
- The sizes of a source are $a = 0.1$ mm and $b = 0.1$ mm.
- The coordinates of the center of the thermal source ξ varies from 1 to 9 mm, $\eta - 5.0$ mm.

The results of calculation are shown in Table 8.1.

8.4 Stationary Thermal Resistance of Powerful Magneto-FET in the Form of Multilayer Cylinder

In an HMT with an axial symmetry shown in Fig. 8.6, a local thermal source is located on the surface $z_N = h_N$, and there exists a heat-removing isothermal side $-z_1 = 0$, while the other surface is adiabatic (isolated).

The program in [64] contains the following blocks:

- Solution of a set of linear algebraic equations by Gauss's method with leading element choice
- Evaluation of Bessel's functions of the first sort of zero and first orders

The initial data are as follows:

- The radius of the structure R
- The radius of a heat source r_s
- The number of heterogeneous layers
- The thickness and heat conductivity of each layer
- The thermal emission power in a source

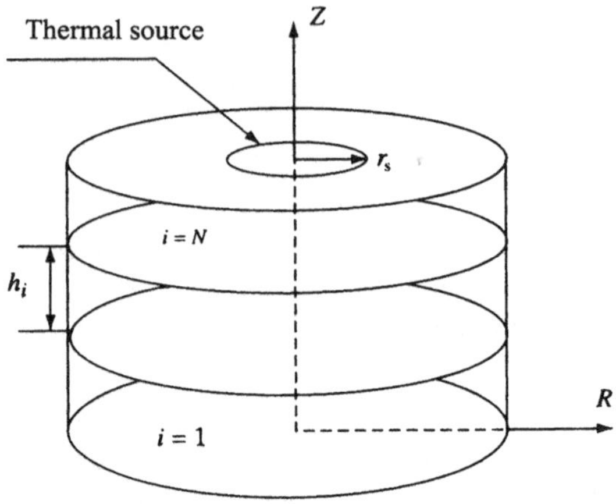

Fig. 8.6 HMT with an axial symmetry

In HMT, the values of the maximal and average temperatures of the surface heat source were calculated.

The solutions-series of Fourier and Bessel were calculated with an error not exceeding 1%. The programming language is FORTRAN.

The work of the program of calculation of the stationary thermal resistance of the constructive elements of a powerful HMT in the form of a multilayered cylinder is illustrated on an example of the thermal resistance of the own case of a KT 962 transistor.

Test task 8.4. The thermal resistance of a ceramics-copper case with the following initial data:

- The radius is 4.8 mm.
- The effective radius of a heat source is 0.66 mm.
- The thickness of the ceramic layer is 2.2 mm.
- The thickness of the copper layer is 1.5 mm.
- The heat conductivity of ceramics is 200 W/(m K).
- The heat conductivity of copper is 390 W/(m K).

The results of calculation are as follows:

- The maximal thermal resistance is 2.29 K/W.
- The thermal resistance averaged over the area of the source is 2.01 K/W.

Test task 8.5. The thermal resistance of a two-layered structure considered in test task 8.2 ($\xi = 5$ mm) with the following initial data:

- The sizes of the structure are $A = 10$ mm and $B = 10$ mm.
- The equivalent radius of the structure $R = 5.6419$ mm.
- The number of layers is 2.
- The thickness of the layers are $h_1 = 2.0$ mm and $h_2 = 0.12$ mm.
- The heat conductivities of the layers are $\lambda_1 = 200$ W/(m K) and $\lambda_2 = 120$ W/(m K).
- The sizes of a source are $a = 0.1$ mm and $b = 0.1$ mm.

- The equivalent radius of a heat source $r_s = 0.0564$ mm.

The results of calculation are as follows:

- The maximal thermal resistance is 44.75 K/W.
- The average (over the area of the source) thermal resistance is 39.84 K/W.

Part IV
Applied Aspects

Theoretical estimations of external influencing factors on heteromagnetic devices are made. Modes of multipurpose generation of regular semi-noise and noise signals of a raised level of continuous and pulse power in the VHF, UHF, microwave, and EHF ranges are analyzed. Modern principles of design of various types of frequency synthesizers are considered. Special attention is given to design of multipurpose operated frequency synthesizers on magnetotransistors in a frequency range up to 100 GHz, including the modes with a pseudorandom working frequency and phase manipulation of a noise-type signal on the basis of heteromagnetic structures with discrete phase shifters.

The results of physical researches of autogenerating magnetosensitive microcircuits and calculation methods of the basic characteristics for determination of small values of the magnetic induction vector with a raised accuracy and spatial resolution are given. Ways to decrease the noise factor in the amplifying cascades on magnetotransistors are investigated and ways of their designing for frequency ranges up to 40 GHz are suggested. The basic elements of magnetotransistors of various types at low and high power levels, including low-power amplification circuits up to 200 GHz and signal transformations up to 1,000 GHz are considered. The results of our development of the know-how of field magnetotransistors of low (mW) and high (W) power levels are presented.

Basic nonlinear effects in magnetotransistors and their role at power restrictions are discussed. The results of our theoretical research of the two-domain model of spherical microtransistors in unsaturated modes are presented.

Chapter 9
Influence of External Factors

9.1 General Remarks

All the factors affecting the radioelectronic equipment from the environment can be classified under the scheme in [65] (Fig. 9.1).

External influences can be subdivided into two groups, namely, those determined by natural – meteorological, climatic – factors and artificial factors. The conclusion about the stability of HMT to external influences is made on the basis of corresponding experimental tests. Preliminary theoretical analysis of the stability of HMT to external influences is important to estimate how optimal the design is.

Let us consider estimations of the influence of some EEFs on HMS.

The following types of mechanical EEFs can be discerned: sinusoidal vibration, casual vibration, mechanical impact of a single action, impacts of a repeated action, linear acceleration and acoustic noise, and raised or lowered atmospheric pressure.

At vibrations and impacts, distributed loadings are acted on the HMS design elements, and the peak values of the resultant forces are determined by the weight of the element and its peak acceleration. These forces, depending on their direction, aspire to shift or tear off the element from its place.

A preliminary theoretical estimation of the stability of HMS to mechanical influences was made on the basis of the quasistatic approach [66]. Deformations and pressures in the gauge elements were calculated under static loadings and were compared with the strength of the materials. For analysis of dynamic influences, a correction factor named as the dynamic factor [66] to reduce the effective strength has been introduced.

The investigated elements of the HMS design are schematically shown in Fig. 9.2: (a) the case and (b) the cover of the case.

In a brass HMS case, an assembly plate with two GaAs elements (a two-cascade amplifier on the basis of a field transistor with a Schottky barrier and a feedback element to FMCR as a standard ferrite sphere) was placed. The cover of the gauge was fixed with screws. The semiconductor crystals were soldered to the assembly plate. The plate was fixed with screws to the corresponding groove of the gauge case. The cover had an aperture for a brass screw, which has a constant magnet at its end in the form of a washer.

A.A. Ignatiev and A.V. Lyashenko, *Heteromagnetic Microelectronics: Microsystems of Active Type*, DOI 10.1007/978-1-4419-6002-3_9,

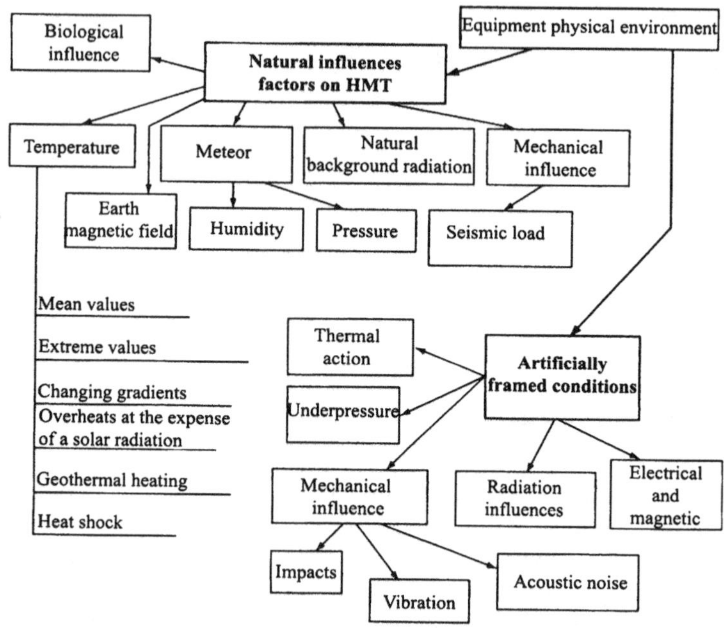

Fig. 9.1 All the factors affecting the radioelectronic equipment from the environment

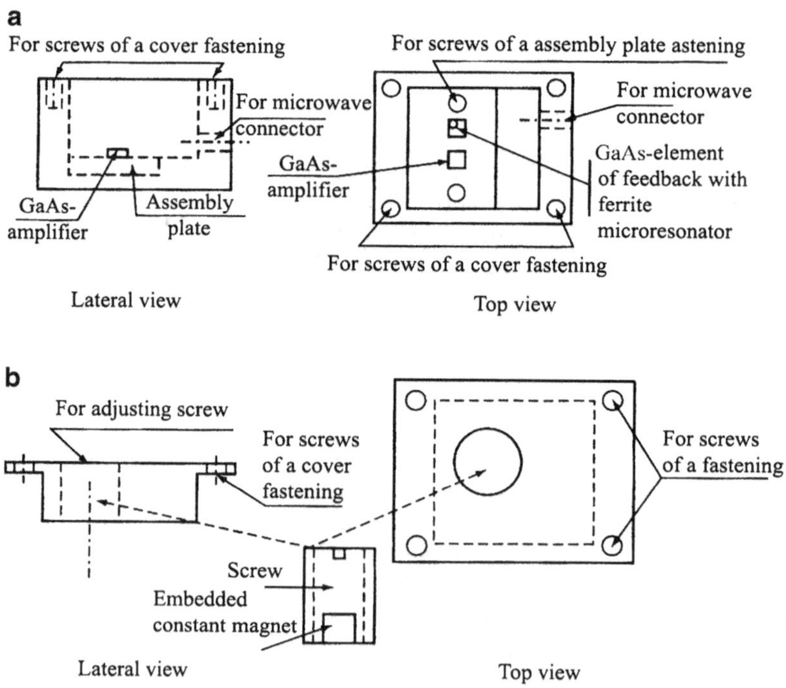

Fig. 9.2 The investigated elements of the HMS design: (**a**) the case, (**b**) the cover of the case

By mechanical influence on HMS, the following is possible: destruction of brazed connections, breakage of beam conductors, separation of the ferrite microresonator, and destruction of screw connections.

9.2 Estimation of Static Load

Depending on the direction of action of an external force on the crystal, two variants of solder layer destruction in HMS are possible, namely, at tangential shift and at normal loading. The external forces causing these loadings were expressed through the tangential a_τ and normal a_n accelerations of the crystal. Such a representation is shown in Fig. 9.3 (1 is a semiconductor crystal with the basis area S; 2 a solder layer with a thickness h; 3 a motionless base, and Δl is the shift deformation of the solder layer).

A typical value of the strength of soft solders (the Russian brands POS, POSV) is within $\sigma_B = 30$–40 MPa. The strength $\sigma_{B,\tau}$ in a tangential direction for plastic materials lies within $\sigma_{B,\tau} = 0.5$–$0.6\,\sigma_B$. More rigid solders have a bit higher σ_B values.

The tear distributed effort is estimated by a simple expression [66]

$$\sigma_n = \frac{m a_n}{S}, \tag{9.1}$$

where m is the weight of the semiconductor crystal, a_n the operating acceleration, and S is the area of the crystal base.

The value of the maximum normal acceleration destroying the solder was determined as

$$a_{n,\max} = \frac{\sigma_B \cdot S}{m}. \tag{9.2}$$

The weight of the $GaAs$ crystal used in HMS does not exceed 3.8×10^{-3} g with the base area of $5\,\text{mm}^2$. Hence $a_{n,\max} = 40 \times 10^6\,\text{m/s}^2 \cong 4 \times 10^6 \times g$, where $g = 9.81\,\text{m/s}^2$ is the acceleration of free fall on the sea level.

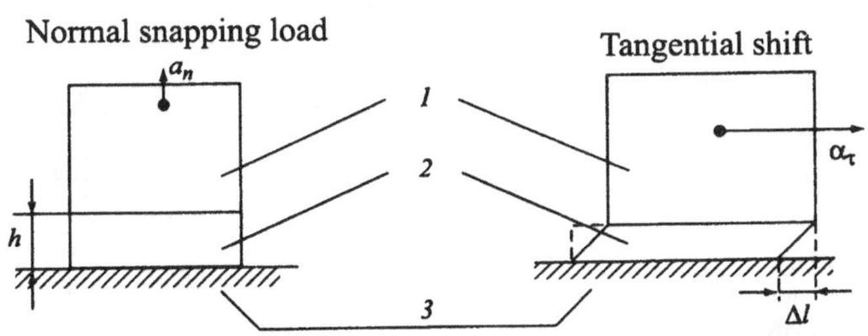

Fig. 9.3 Variants of solder layer destruction in HMS, at tangential shift and at normal loading

The tangential pressure at shift are defined by the expression

$$\sigma_\tau \cong 0.4E\frac{\Delta l}{h}, \tag{9.3}$$

where E is the tension modulus (Young modulus); Δl the absolute value of the shift of the crystal; and h is the thickness of the deformed layer.

At the thickness of the solder layer $h = 2\,\mu\text{m}$, the maximum destroying tangential acceleration is $a_{\text{n,max}} \approx 16 \times 10^3\,\text{m/s}^2 = 1{,}600 \times g$.

The resulted estimations are true provided that the durability of the interfaced HMS elements surpasses that of the connecting layer (solder).

9.3 Strength of Beam-Type Bonds

The conducting beam in the GaAs field transistor and in HMS was modeled by a half ring with a rectangular cross section (Fig. 9.4). Each beam is microwelded onto the corresponding contact platforms. Such welded possess a high durability and hold overloads of $16{,}000.0 - 40{,}000.0 \times g$. Therefore, consider the probability of a break of the block from its lateral bend.

The force σ arising in the beam near its welding places is estimated by the expression [66]

$$\sigma = \frac{ma_1 R_0}{8\left(R_1^2 - R_2^2\right)} + \frac{ma}{S_\text{n}}, \tag{9.4}$$

where $R_0 = h/\ln(R_1/R_2)$ is the radius of curvature of the beam across its central section, h the effective thickness of the beam, $S_\text{n} = h^2$ the area of the cross section of the beam, a the equivalent acceleration of the force acting, on the beam, and m is the mass of the beam.

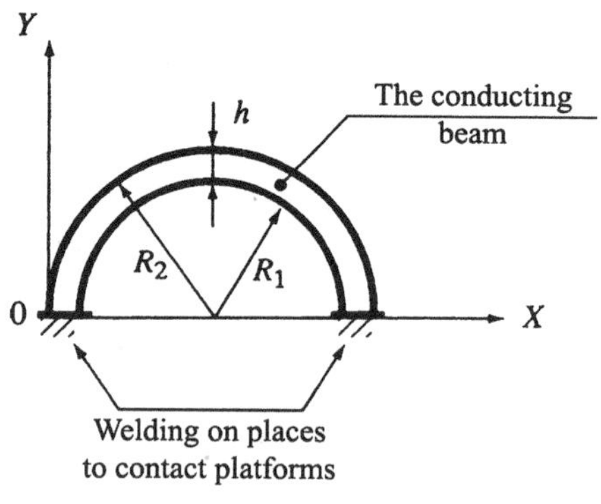

Fig. 9.4 The conducting beam in the GaAs field transistor and in HMS

Let the beam be made of gold wire with a diameter of 0.02 mm with a radius of $R_0 = 1$ mm. For gold, the tear strength is $\sigma_B = 150$ MPa, and the yield point is $\sigma_{02} = 40$ MPa. So the maximal acceleration acting on the beam and capable of bending it at the right angle near its welding places is about $\sim 1.1 \times 10^9\ \mathrm{m/s^2} \cong 1.1 \times 10^8 \times g$, and for break of the beam a linear acceleration of $0.8 \times 10^6\ \mathrm{m/s^2} \cong 0.8 \times 10^5 \times g$ is required.

9.4 Strength of Glue Fixation

A FMCR with a diameter of 0.4–0.5 mm was fixed on GaAs with a drop of epoxy resin glue (Fig. 9.5).

The mechanical properties of glues widely vary. To glue ferrite materials, the glues ETP-1, ETP-2, ETP-3 (The Russian brands developed by the Yekaterinburg Institute of organic synthesis together with a number of Moscow organizations such as Almaz Corp., etc.) can be used As an example, consider fastening of FMCR by ETP-2 glue.

The adhesive properties of ETP-2 glue are resulted in Table 9.1 (according to Institute of organic synthesis, Ekaterinburg, RF).

The density of epoxy resin glue noticeably varies depending on the used additive in a range of 1,300–2,000 kg/m^3. The glue contained no electrowire or heat conduction additives; therefore its density was $\rho = 1{,}500\ \mathrm{kg/m^3}$. Take the radius of a glue drop to be $R = 0.5$ mm. Then the volume of glue (without a ferrite sphere)

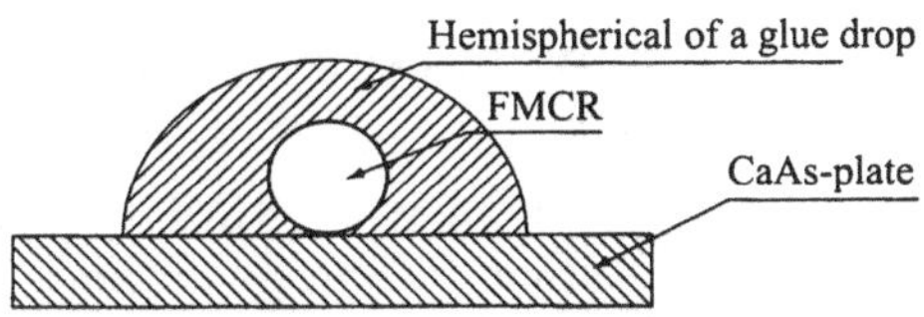

Fig. 9.5 A FMCR with a diameter of 0.4–0.5 mm was fixed on GaAs with a drop of epoxy resin glue

Table 9.1 The parameters of used glues

Characteristics of glue	Numerical values
Strength limit σ_B of ferrite rod joints at static console bend, MPa	
In initial state, MPa	82–110
After metallization, MPa	116–128
After climatic tests, MPa	80
Strength of glue joint of 30HGSA steel at shift, MPa	
In initial condition, MPa	20–30
After long (12 years) storage test, MPa	20–25
Temperature linear expansion (TEC) factor $\times 10^4 (\mathrm{K}^{-1})$, at	
$25 \pm 5°\mathrm{C}$	0.4
$50 \pm 5°\mathrm{C}$	0.8
$85 \pm 5°\mathrm{C}$	3.5

is $0.2265 \times 10^{-9}\,\mathrm{m}^3$. The total weight of the glue drop-spherical microresonator system is 738.7×10^{-9} kg. The brake-off force created by the normal (to the surface of the GaAs plate) acceleration a_n is related to the strength of adhesion σ_B by

$$\frac{ma_\mathrm{n}}{S} \leq \sigma_\mathrm{B}, \tag{9.5}$$

where m is the weight of glue with the ferrite sphere, S the glue–base contact area, and $\sigma_\mathrm{B} = 82\,\mathrm{MPa}$. Hence, for destruction of ETP-2 glue the maximal acceleration $a_{\mathrm{n,max}} = 87 \times 10^6\,\mathrm{m/s}^2 \cong 87 \times 10^5 \times g$ is needed.

In the case of shift, the strength of ETP-2 glue is one-fifth larger; therefore, the glue-destroying tangential acceleration is $a_{\tau,\mathrm{max}} \approx 17 \times 10^6\,\mathrm{m/s}^2 \cong 17 \times 10^5 \times g$.

The basic demerit of epoxy resins is their high TEC (see Table 9.1). Therefore, temperature changes in HMS essentially intensify development of defects in glue layers. Such defects originally arise owing to glue shrinkage at polymerization.

The durability of FMCR connections by means of epoxy resin glues is essentially influenced by long vibrations with a frequency from 5 up to 5,000 Hz at accelerations up to $40 \times g$, which leads to fatigue failures in the glue layer.

In glue hardening, the residual solvent creates porosity in the bulk of glue, which is another cause of the occurrence of internal tensions and a decrease in the durability of glue joints.

With growth of the glue layer thickness, the number of defects increases, and the durability of joints falls. Usually, it is recommended to limit the thickness of the glue layer to values from 0.05 to 0.1 mm.

9.5 Strength of Screw Connection

All screw connections in the HMS case at any direction of external acceleration counteract to tension fractures only, since the fixed elements are made in the form of fixed inserts. All the elements of the case, including the screws, are made of brass.

The force to break the screw of fastening is determined by the external acceleration:

$$a_{\max} = \sigma_\mathrm{B} \times \frac{\pi \times d^2}{2 \times m}, \tag{9.6}$$

where σ_B is the break strength of the screw material and d is the diameter of the continuous part of the screw.

The break force of the basic material is determined by

$$\sigma_\mathrm{B} = \frac{ma}{4lh}, \tag{9.7}$$

where h is the thickness of the basic material.

Let us consider a screw cover–case connection of the sensor. The weight of the case (the brass density is $8.8\,\mathrm{g/sm^3}$) is 28.2 g, and the weight of the cover is 11.4 g. At general dispersed influence on the case the inertial forces are due to the superfluous weight $\Delta m = 16.8\,\mathrm{g}$. Therefore, the estimated value of acceleration to break off the screws is $a_{\max} = 212 \times 10^3\,\mathrm{m/s^2} \cong 21{,}200 \times g$. The thickness of the case cover in the places of its fastening to the case is 0.5 mm only. For this, the acceleration to break the basic material is considerably smaller as follows:

$$\alpha_{\max} = \frac{4lh\sigma_{\mathrm{B}}}{m} = 33{,}333\,\mathrm{m/s^2} \cong 3{,}330 \times g.$$

9.6 Resistivity to Dynamic Forces

In mechanics, within the limits of validity of Hooke's law, it is accepted that the maximal dynamic stress will exceed the static one so much as how much the dynamic deformation exceeds the static one [66]:

$$\sigma_{\mathrm{d}} = k_{\mathrm{d}} \times \sigma_{\mathrm{st}}, \tag{9.8}$$

where σ_{d} is the dynamic stress, σ_{st} the static stress, and k_{d} is the dynamic factor at impact.

When a body undergoes a dispersed external force, a value of $k_{\mathrm{d}} = 2$ is used for estimation. Therefore, for estimation of the stability of the HMS elements to a short-term (below 2 ms) single impact, it is necessary to halve the value of critical accelerations obtained earlier.

At repeated mechanical impacts, a residual deformation is accumulated in HMS. Therefore, the stability was estimated not by the tensile strength but by the yield stress.

The reaction of HMS to an external sine oscillation with a circular frequency was determined by a dynamic factor

$$k_{\mathrm{d}} = 1 + \frac{P_{\mathrm{f}}}{P_{\mathrm{st}}} \cdot \frac{1}{1 - \left(\frac{\omega}{\omega_0}\right)^2}, \tag{9.9}$$

where P_{f} is the maximal value of the external periodically varying force, P_{st} the value of the static force acting on the system, and ω_0 is the own frequency of the mechanical oscillations of the system.

Usually, it is supposed that a "deviation" from resonance will be provided, if the relative mismatch is:

$$\left|\frac{\omega - \omega_0}{\omega_0}\right| \approx 30\%.$$

In this case, for the system not loaded statically $k_d \leq 2$. Besides, it is necessary to consider that under a sine wave loading the sign of stress constantly varies. Therefore, for theoretical estimation of the HMS stability to sine wave vibrating influence, it is possible to take advantage of repeated impact data, having toughened them twice. It is fair provided that the own frequency of the system noticeably differs from the frequency of the external sine wave influence.

Let us estimate the own frequencies of the polycoric assembly plate of HMS of the sensor. The static deflection δ_{st} of the plate under the action of its own weight is determined by

$$\delta_{st} = 0.1336 \frac{mg}{lb} \frac{l^4}{Eh^2}, \tag{9.10}$$

where l, b are the length and width of the plate ($l > b$), h the thickness of the plate, and E is the modulus of elasticity of the plate's material.

The own frequency of mechanical oscillations of the plate is determined by

$$f_{own} = \frac{1}{2\pi}\sqrt{\frac{g}{\delta_{st}}}. \tag{9.11}$$

Hence, for the considered plate $\delta_{st} = 0.17 \times 10^{-12}$ m and $f_{own} = 1.21 \times 10^{-3}$ kHz.

Under the action of inertial forces determined by acceleration

$$a = n \times g,$$

where $n = 1, 2, 3, \ldots$, the own frequency of the plate will change under the law

$$f_{own}(n) = \frac{1.21 \times 10^3}{\sqrt{n}} \text{ (kHz)}. \tag{9.12}$$

For $n = 600$, the frequency $f_{own}(n) = 49.39$ kHz, which is far from the typical frequency values of mechanical oscillations.

Doing similar calculations for the thin walls ($h = 2$ mm) of the cases of HMS, we will get

$$f_{own}(n) = \frac{1.05 \times 10^3}{\sqrt{n}} \text{ (kHz)}.$$

9.7 Resistivity to Pressure Changes

Let us write b for the height, l for the width, and h for the thickness of the thinnest wall of the gauge case ($l > b$). At a sudden decrease in the external atmospheric pressure, the deflection δ_{st} of the middle part of the wall, caused by gas pressure inside the case (in a static approach) is determined by [69]

$$\delta_{st} = 0.1106 \frac{Pl^4}{Eh^2}, \tag{9.13}$$

where P is the homogeneous loading (in Pa) and E is the modulus of elasticity of the material of the wall.

In our example, the deflection of the wall of the case at a sharp drop of the external pressure will be 21×10^{-10} m. Estimate the force bending the wall as

$$\sigma = 0.61 \times \frac{Pk_{\mathrm{d}}l^2}{h^2}, \tag{9.14}$$

where $k_{\mathrm{d}} = 2$. The upper estimate of such a force is 2.2 MPa that is two orders of magnitude lower than the yield stress of brass. Hence, it is possible to approve that the case of the HMS sensor is not sensitive to external pressure differences. In the case of a sharp increase in the pressure up to three atmospheres (2,207 mm Hg), this statement remains true.

9.8 Resistivity to Temperature Excitations

The value of temperature, the rate of its change in time, and temperature drops between separate elements of the construction are the major factors influencing the reliability of a product.

A change of the ambient temperature will result not only in a change of the thermal mode of the HMS sensor (in the conditions of its functioning) and a respective alteration of its working parameters, but also in possible occurrence of significant thermotensions in the joint places of heterogeneous elements. Hence, the interfaced elements of the design should be coordinated by TEC.

In the mode of HMS functioning the calculated value of the thermal resistance of the magnetometric gauge without the use of any heat-conducting paths is equal to 74 K/W (for conditions natural convective heat exchange between the gauge and its environment at a room temperature). At a thermal emission power of 30 mW, the own overheat of the amplifier of the gauge above the ambient temperature will be 2.3 K. So, in view of the value of limiting working temperature for GaAs devices being 140°C, it is possible to note that a rise in the temperature of the environment up to 85°C will not lead to a failure of the sensor.

At the use of HMS in the conditions of a strongly lowered temperature (down to −70°C), its workability will be determined by the reliability of the semiconductor components and by the temperature change of the saturation magnetization of FMCR. In exclusive situations, it is necessary to provide a system cure, an active semiconductor structure of the sensor maintaining the working temperature at the required level.

Let us consider a situation with a sharp change of the ambient temperature in the conditions of storage (transportation) of the sensor and estimate the time of its cooling (or heating). Within the limits of our estimation, we shall consider that the conditions of heat exchange with the environment on the surface of the gauge will not vary eventually and that the thermophysical characteristics of the material of the case of the gauge also do not vary. Let T_0 be the reference temperature of the sensor, T_1 the new ambient temperature, and α be the factor of convection heat exchange between the surface of the case of the sensor and the environment. Then the volume-average temperature will vary in time under the law [67]

$$T(\tau) = T_0 + (T_1 - T_0)\left\{1 - \exp\left[-4a\tau\left(\frac{\mu_x^2}{h_x^2} + \frac{\mu_y^2}{h_y^2} + \frac{\mu_z^2}{h_z^2}\right)\right]\right\}, \quad (9.15)$$

where τ is current time, $a = 0.3049 \times 10^{-4}\,\mathrm{m^2/s}$ the factor of temperature conductivity of brass, h_x, h_y, h_z the sizes of the sensor, and μ_x, μ_y, μ_z are the first roots of the characteristic equation. The roots μ_x, μ_y, μ_z are from the solution of the equation

$$\mathrm{ctg}(\mu_{x,y,z}) = \frac{\mu_{x,y,z} \cdot \lambda}{\alpha \cdot h_{x,y,z}}, \quad (9.16)$$

where $\lambda = 110\,\mathrm{W/(m\,K)}$ is the factor of heat conductivity of brass.

The values of α were calculated by the semiempirical techniques stated in [74]. Generally, the factor of convective heat exchange depends on both the surface and the temperature difference between the surface and the environment. For our example, the calculated values of α in the conditions of natural convection for various temperature differences with a reference point of 293 K are presented in Fig. 9.6.

In connection with that the temperature of the sensor changes in time exponentially (asymptotically coming nearer to a finite value), we accept for the moment of time of acceptance by a temperature body T_1 with such value at which the temperature of the sensor is equal to $0.95T_1$. Then the time τ_h of heating or cooling of the gauge from the temperature T_0 up/down to temperature T_1 is determined by

$$\tau_\mathrm{h} = \frac{1}{4a_\mathrm{T}\left(\frac{\mu_x^2}{h_x^2} + \frac{\mu_y^2}{h_y^2} + \frac{\mu_z^2}{h_z^2}\right)} \cdot \left|\ln\left(\frac{0.05T_1}{T_1 - T_0}\right)\right|. \quad (9.17)$$

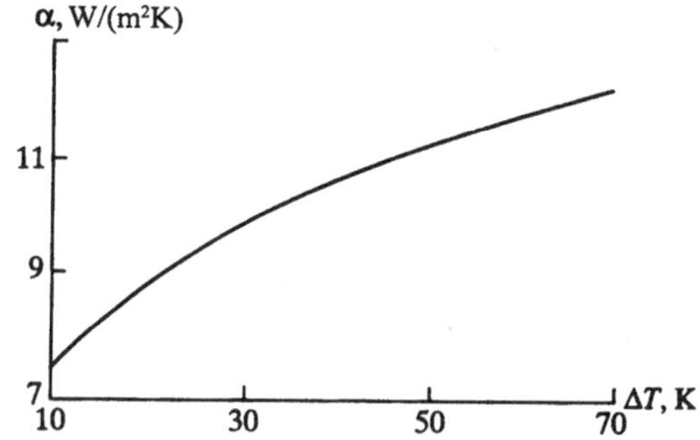

Fig. 9.6 The calculated values of α in the conditions of natural convection for various temperature differences with a reference point of 293 K

Our calculations show that the estimated time of heating of the gauge from the room temperature up to +60°C is not less than 14.3 min, and the time of cooling of the gauge from the room temperature down to −70°C is not less than 23.8 min; thus, it is necessary to note that in the latter case the value of heat emission factor α was taken equal to 16.6 W/(m^2 K) by extrapolation of the data presented in Fig. 9.6, to a temperature range $\Delta T = 90$ K toward lower temperatures.

The time-average rate τ_h of temperature change is given by

$$\left\langle \frac{\partial T}{\partial \tau} \right\rangle = \frac{T_1 - T_o}{\tau_h} \left\{ 1 - \exp\left[-4a\tau_h \left(\frac{\mu_x^2}{h_x^2} + \frac{\mu_y^2}{h_y^2} + \frac{\mu_z^2}{h_z^2} \right) \right] \right\}. \tag{9.18}$$

The rate of temperature change of the gauge at its heating from room temperature up to +60°C is ∼3 K/min and at cooling from room temperature down to −70°C is ∼4 K/min. Hence, no critical "temperature impacts" in the sensor will be observed.

In our example, only the passive elements of the sensor were considered. It is connected with that the stability of the active semiconductor module of the sensor to mechanicoclimatic and temperature influences was specified by its manufacturer and, consequently, was considered as known.

Protection against other kinds of influences (dust, aggressive environments, mould fungi, etc.) is provided with the presence of special casings, screens, hermetic sealing of modules, and the use of proof coverings of the case.

The problem of a high stability of the design of the gauge to EEF should be solved by an integrated approach in view of the conditions of its prospective operation, electromagnetic compatibility, and the stability to other influences, for example, to ionizing radiation.

9.9 Resistivity of HMS to External Factors

The basic requirements of the stability of HMS to mechanicoclimatic and temperature influences are resulted in Table 9.2.

9.10 Estimation of Jam Protection

The noise protection of a radioelectronic device is included into a wide class of the problems of EC of radioelectronic equipment. The EC of radioelectronic equipment is its ability to function jointly and simultaneously with other devices having their electromagnetic properties, under the possible action of electromagnetic interferences and not creating inadmissible interferences to other radioelectronic equipment [68, 69].

The solution of the EC problem of radioelectronic equipment has a complex character and directly depends on the purposes of the use of a specific device

Table 9.2 The requirements of the stability of HMS to mechanicoclimatic and temperature influences

External factor	Requirements to parameters	Results of theoretical estimations	Note
Sine wave vibration	The range of frequencies from 1 up to 5,000 Hz, the amplitude of acceleration 40 × *g* or 100 × *g* at short-term influence	Complies The amplitude of acceleration 10 × *g*	Restriction of stability of the semiconductor subsystem is maintained
Single mechanical impact	Peak shock acceleration >3,000 × *g* at a duration shorter than 2 ms	The design withstands a peak acceleration 1,600 × *g*	Restriction of stability of the semiconductor subsystem Restriction on the durability of screw connection of the cover of the case to the case
Repeated mechanical impacts	Peak shock acceleration 150 × *g* at a duration below 5 ms	Complies	
Acoustic noise	The range of frequencies from 20 to 10,000 Hz, a level of sound pressure 175 dB (rather 0.2 Pa)	Complies Level of sound pressure 150 dB	Restriction of stability of the semiconductor subsystem
Linear acceleration	Value of acceleration >500 × *g*	Complies Acceleration of 100 × *g* is withstood	Restriction of stability of the semiconductor subsystem
Raised environment temperatures	The maximal value of operation >125°C	Complies The maximal value at operation +85°C	Restrictions on the limiting working temperature of the semiconductor subsystem
Lowered environment temperatures	The minimal value at transportation and storage −60°C The minimal value at operation −60°C	Complies	
Temperature changes	From the maximal value at operation to the minimal value at transportation and storage	Complies	Restrictions on the semiconductor subsystem A mismatch on the factor of thermal expansion of the design elements

(continued)

Table 9.2 (continued)

External factor	Requirements to parameters	Results of theoretical estimations	Note
Raised air humidity	Relative humidity of 100% at temperature +35°C	Complies at a hermetic sealed case	
Lowered air humidity	Dew-point at temperature −40°C	Complies at a hermetic sealed case	
Lowered atmospheric pressure	Value at operation less than 5 mm Hg	Complies	
Raised atmospheric pressure	Value at operation of 2,207 mm Hg	Complies	

and the local electromagnetic properties of the place of its operation. The EC of radioelectronic equipment at the level provides:

- Separate elements of the device
- Devices and blocks of devices
- Groups of means and systems
- Spatial, time, and frequency factors

The provision of EC at the level of separate elements of the device is connected, first of all, with loss of the interferences caused by the elements of the equipment and reduction of the level of external fields, directed on these elements.

By noise immunity, we shall understand the ability of the device to resist to external and internal interferences, and this ability depends on specially applied additional circuit constitutive receptions and ways, which do not break the main principles of the device's design. In this sense, the term "noise immunity" is not equivalent to the concept of noise stability, which is understood as the ability of the device to resist to internal and external interferences only on the basis of the principles of its design, that is, without the use of additional ways and means.

Natural electromagnetic interferences are generated by the following causes [69]:

- Atmospheric electric processes
- Thermal radiation of the terrestrial surface, troposphere, and ionosphere
- Noise radio emissions of space (cosmic) sources

Interferences of artificial origin are subdivided on deliberate and inadvertent ones. Inadvertent interferences, in turn, are divided into those caused by the radiation of radiodevices and industrial ones.

The internal noise of the device is also classified as inadvertent interferences. They are caused by various fluctuating processes and always exist in real circuits along with the useful signal.

The degree of influence of inadvertent interferences on the quality indicators of radioelectronic devices is determined by the level of interferences, their spectral structure, statistical characteristics, and way of information processing.

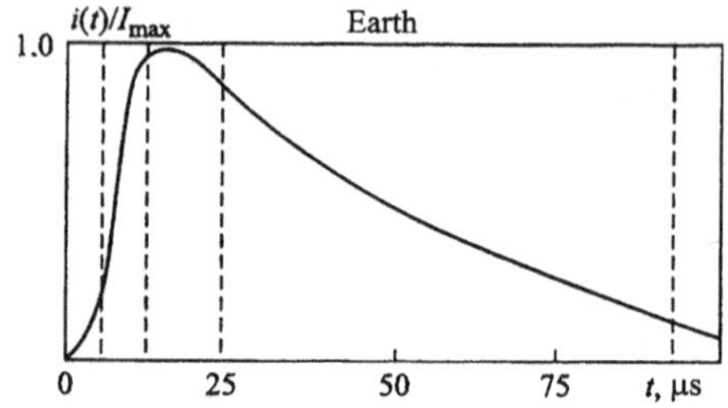

Fig. 9.7 The time distribution of the current of a lightning at a typical lightning discharge

Atmospheric electric processes create powerful electromagnetic interferences. In Fig. 9.7, the time distribution of the current of a lightning is shown at a typical lightning discharge. The discharge duration is ~100 μs, and the peak value of current is $I_{max} \approx (1 \div 20) \times 10^4$ A.

In [70], a detailed technique of theoretical analysis of the values of the electromagnetic fields created by discharge current in the surrounding space is presented, which is important for preliminary theoretical estimation of the electromagnetic conditions for solving the question about the gauge's EC.

Other sources of powerful electromagnetic interferences are high-voltage installations, pulse sources of high currents, and contact networks. An essential feature of such interferences is high values of the intensity of the magnetic field, which will render the corresponding influence on the parameters of HMS.

In detection and detailed characterization of interferences, laboratory experiments with model and pilot HMS sample should take an important place.

The basic ways of protection of radioelectronic means from electromagnetic interferences are well developed. These ways include various screening, optimized grounding, active and passive filtration, and computer (mathematical) processing of the useful signal by means of algorithms cutting its fluctuating distortion.

The most complex part of the problem of noise immunity of any radioelectronic means, including the magnetometric sensor, is protection against powerful pulse interferences – pulses of great energy with a duration from 50 to 100 μs. Such interferences affect both the electric and magnetic component of the interference bearing fields. The last circumstance should compel special attention by virtue of the functional applicability of the sensor.

The most effective means of protection against powerful interferences are continuous screens. So, a nonferromagnetic case-screen effectively weakens the pulse and high-frequency components of the magnetic field of interferences (Fig. 9.8).

In Fig. 9.8a, a model of the closed continuous screen placed in an external interference-bearing magnetic field is shown.

The degree of shielding is determined by the conductance of the material of such a screen, its walls' thickness, and the ratio of its external geometrical sizes. The last parameter is referred to as the shape factor. For a screen as a rectangular parallelepiped the value of the shape factor n is resulted in Fig. 9.8b, c as well.

Let us consider a specific example of the influence of a high-energy pulse interference on the gauge. As it was already noted, the case of HMS represents a

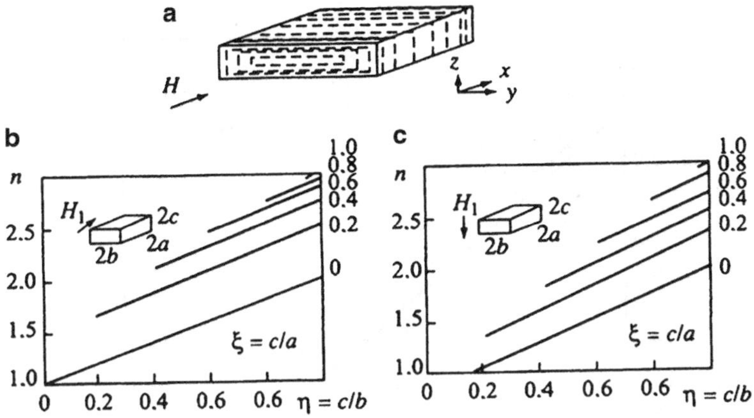

Fig. 9.8 A model of the closed continuous screen placed in an external interference-bearing magnetic field (**a**) and the value of the shape factor n (**b**, **c**)

continuous brass screen with the thickness of its wall of 4 mm. Assume that the interference pulse envelope corresponds to a lightning discharge current (see Fig. 9.7).

In this case, the intensity of the magnetic component H_i of the interference is approximated by a set of two exponential time dependences [70]

$$H_{\mathrm{i}}(t) = H_{\max}\left[\exp(-a_1 t) - \exp(-a_2 t)\right], \tag{9.19}$$

where $H_{\max}$ is the peak value of the magnetic field intensity and a_1 and a_2 are the constants describing the steepness of the front and cut of a interference pulse.

For the continuous rectangular screen, the intensity of the magnetic field inside the screen cavity is determined by

$$\begin{aligned} H_{\mathrm{scr}}(t) &= H_{\max}\sum_{i=1}^{2}(-1)^{i+1} \\ &\times\left\{\frac{\exp(-a_i t)}{\cos\left(k\sqrt{a_i}d\right) - \left(k\sqrt{a_i}R/n\right)\sin\left(k\sqrt{a_i}d\right)}\right. \\ &\left.-\sum_{m=0}^{\infty}\frac{2\beta_m\exp\left(-\beta_m^2 t/k^2 d^2\right)}{\left(\beta_m^2 - k^2 a_i d^2\right)\left[(1+R/nd)\sin\beta_m + (R\beta_m/nd)\cos\beta_m\right]}\right\}, \end{aligned} \tag{9.20}$$

where R is the least linear size of the screen, d the wall thickness of the screen, n the factor of shape, $k = \mu_0\sigma, \sigma$ the specific conductance of the material of the screen, μ_0 the magnetic constant, and β_m are the roots of the characteristic equation $\cos\beta_m - (\beta_m R/nd)\sin\beta_m = 0$.

The resulted expression describes the change in time of the pulse magnetic field inside the screened area depending on the peak-time parameters of the interference pulse and the constructive and electrophysical parameters of the screen.

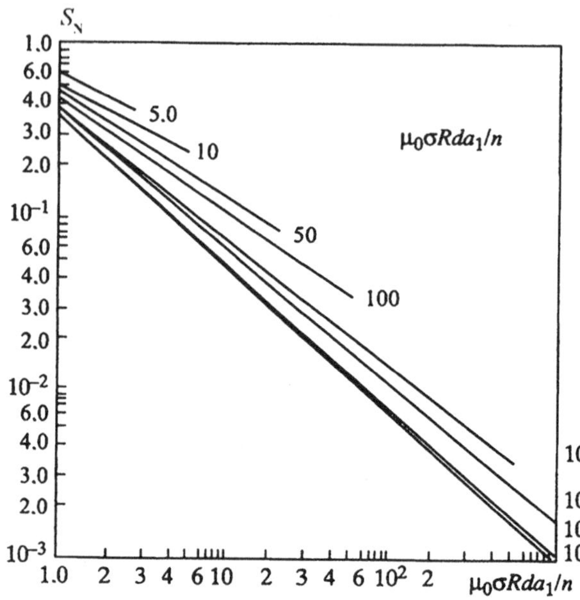

Fig. 9.9 Nomograms for determination of the factor of magnetic shielding

Table 9.3 The expressions by means can be estimated the values of the factor of magnetic shielding, the pulse front duration in the shielded cavity, and the pulse duration

Parameter	Analytical expression	Note
Factor of magnetic shielding S_N	$\frac{(a_2 - a_1)n}{\mu_0 \sigma R d a_1 a_2}$	The analytical expressions hold when $\mu_0\sigma Rd/n > \tau_i$
Pulse front duration τ_f(s)	$0.33\ \mu_0\sigma d^2$	
Pulse duration τ_i(s)	$0.69\ \mu_0\sigma Rd/n$	

In Fig. 9.9, nomograms for determination of the factor of magnetic shielding $S_N = (H_{scr})_{max}/H_{max}$ as a function of the dimensionless parameter $\mu_0\sigma Ra_2\, d/n$ are presented.

According to the techniques from [78], the values of the factor of magnetic shielding, the pulse front duration in the shielded cavity, and the pulse duration can be estimated by means of the expressions presented in Table 9.3.

Analysis shows [69] that for the frequencies at which the sizes of the continuous screen are essentially less than the wavelength, an appreciable distinction in the screening of the electric and magnetic fields is characteristic.

On low frequencies, with increasing frequency the efficiency of electric shielding first decreases and then starts to increase. The efficiency of magnetic shielding always grows with frequency, and the thicker the walls and the higher the magnetic permeability of the material of the screen, the higher the efficiency is.

On high frequencies, when the sizes of the screen become comparable with the wavelength, the distinction in shielding the electric and magnetic components of the field disappears. Owing to the small penetration depth of an electromagnetic field, the efficiency of shielding by continuous screens is high and improves with frequency growth.

From our calculations, it follows that the used brass case of the gauge can weaken the magnetic field of powerful pulse interference by approx. 1,800 times if this interference has a duration 100 μs, and the duration of its front is ~10 μs. Therefore, for improvement of screening from a pulse high-energy interference, additional shielding is required.

A powerful electromagnetic interference can have a regular character. In this case, it is modeled by a sine wave signal. If the field of a interference is low frequency, the factor of magnetic shielding depending on frequency ω is approximated by the following expression

$$S_{\mathrm{N}}(\omega) = \frac{1}{\sqrt{1+\tau_{\mathrm{scr}}^2\omega^2}}, \tag{9.21}$$

where $\tau_{\mathrm{scr}} = \mu_0\sigma Rd/n$.

For a high-frequency interference

$$S_{\mathrm{HF}}(\omega) = \frac{1}{R}n\delta\sqrt{2}\exp(-d/\delta), \tag{9.22}$$

where $\delta = \sqrt{2/(\mu_0\sigma\omega)}$ is the thickness of the skin layer.

In Fig. 9.10, the dependences of the factor of magnetic shielding of continuous nonferromagnetic screens from external factors for the sinewave field of interference [70] are shown.

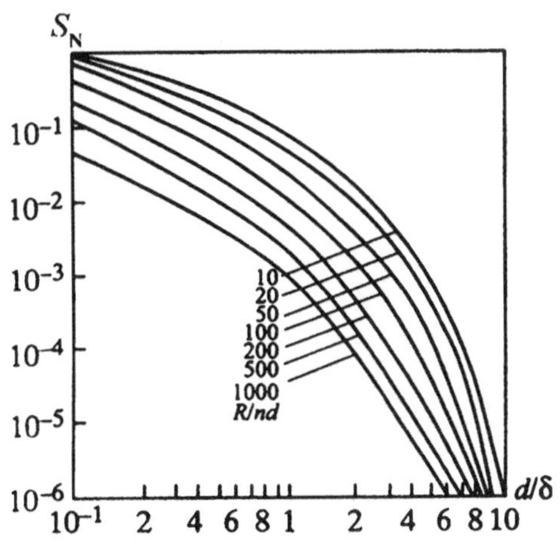

Fig. 9.10 The dependences of the factor of magnetic shielding of continuous nonferromagnetic screens from external factors for the sine wave field of interference [70]

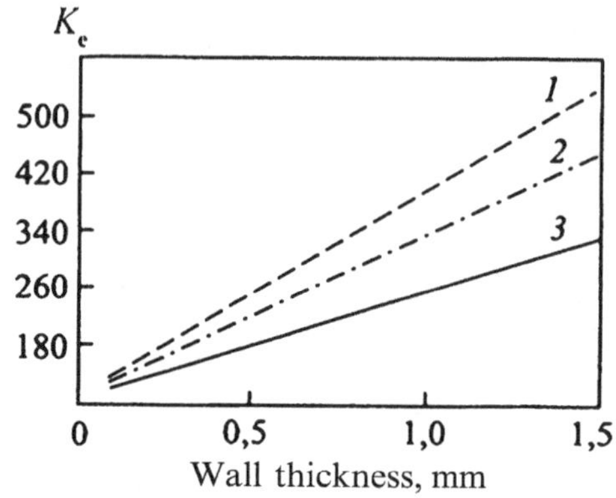

Fig. 9.11 The dependences of the calculated values K_{scr} for an ideal continuous screen of the cubic shape

In the case of higher frequencies, weeding such plots becomes inconvenient because of the exponential coefficient tending to zero. A concept of the efficiency of shielding is introduced

$$K_{scr} = 20 \times \lg \frac{H_{max}}{H_{scr\ max}} = 20 \times \lg \left(\frac{E_{max}}{E_{scr\ max}} \right). \quad (9.23)$$

In Fig. 9.11, the dependences of the calculated values K_{scr} for an ideal continuous screen of the cubic shape with the edge size of 10 cm on the thickness of the wall are presented at the frequency of a regular sine wave interference of 800 MHz for screens made of: copper – 1, aluminum – 2, and brass – 3. This screen is intended for protection of the HMS and the block of data processing.

Thus, high-frequency and pulse interferences in HMS can be effectively reduced due to the use of shielding, filtration in the circuits of power, and transfer of the useful signal. A low-frequency magnetic interference is shielded poorly, and its influence can be considered only by means of mathematical processing of the useful signal.

Chapter 10
Multifunctional Generation and Boosting

10.1 Generation of Increased Continued and Pulse Power Levels in Omnirange, UHF Ranges

The results of calculation of the HMT parameters in the modes of generation of regular and semi-noise signals with a band of 1–5% are presented in this chapter at a nonuniformity of 3–5 dB on the central frequencies of the VHF (0.165 GHz) and UHF (1.65 GHz) ranges.

The equivalent circuit of a powerful HMT contained a set of transistor structures placed for power summation. According to the aforementioned statement, such a structure is replaced by one transistor (Fig. 10.1). Power voltage of the transistor is 50 V.

A numerical research of the parameters of the powerful HMT in the mode of regular signals was made by harmonic balance. For the research of irregular signals by using Runge–Kutta's method, the set of nonlinear ordinary differential equations was solved. FMCR is modeled by a contour with the equivalent parameters C_{ap} and L_{nd}.

The key parameters were investigated: namely, the frequency of generated oscillations ν_0, the output integral power P_{out}, and CE.

The results of our calculations of the key parameters for HMG on the KT9164AS transistor with a technical efficiency of $\cong$ 45% on continuous power are shown in Table 10.1.

The equivalent circuit of the powerful generator on a bipolar HMT is shown in Fig. 10.1.

The continuous thermal power is $P_{\text{T}} = 185.8$ W. The maximal overheating of the semi-conductor crystal is $\Delta T = 78.97$ K. The limiting working temperature of the collector joint to standard is 200°C. Therefore, the limiting temperature of the base of the semiconductor crystal is 121°C. Value of thermal resistance for the case of the device in view of thermal energy dispersion in the environment is $R_{\text{TR}} = \frac{121°\text{C} - t_{\text{env}}}{P_{\text{T}}}$, where t_{env} is the temperature of the environment (centigrade). Even for the normal conditions ($t_{\text{env}} = 0°\text{C}$) $R_{\text{TR}} = 0.65\,\text{K/W}$, that is, it is very difficult to provide.

A.A. Ignatiev and A.V. Lyashenko, *Heteromagnetic Microelectronics: Microsystems of Active Type*, DOI 10.1007/978-1-4419-6002-3_10,

Fig. 10.1 The equivalent circuit of a powerful HMT

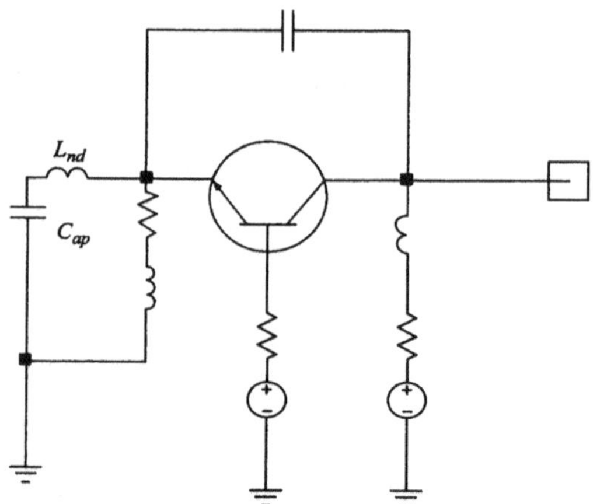

Table 10.1 Results of our calculations of the key parameters for HMG on the KT9164AS transistor

No.	C_{ap}, pF	L_{nd}, nH	ν_0, MHz	P_{out}, W	CE, %
1	7.0	4.0	165.0	152.0	52
2	0.7	0.4	1,650.0	147.0	47

Fig. 10.2 The equivalent circuit of a powerful HMG in the pulse mode

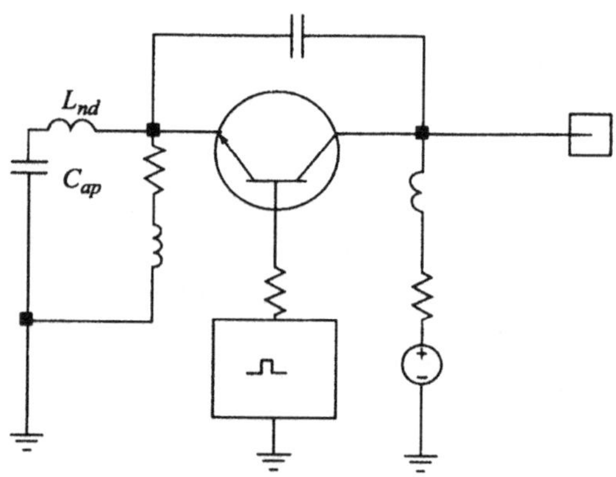

The equivalent circuit of a powerful HMG in the pulse mode is shown in Fig. 10.2.

Results of our calculations of the pulse modes of HMG at various relative pulse durations (Q) and the duration of forward pulse front $\tau_p = 8$ ns are given in Table 10.2.

The thermal mode of HMG will be estimated for a pulse power $P_p = 472$ W at a relative pulse duration of $Q = 1{,}000$. At a pulse thermal power of $P_T = 577$ W and a pulse duration of $\tau_p = 50\ \mu$s, the maximal pulse overheating of a single transistor structure at the end of the impulse action of current reached a value of 197.54 K, which is inadmissible. When dual or ternary semiconductor structures are used, the density of thermal emission in HMG decreased. At reduction of the density

Table 10.2 Results of our calculations of the pulse modes of HMG at various relative pulse durations Q, the duration of forward pulse front $\tau_p = 8$ ns

No.	C_{ap}, pF	L_{nd}, nH	ν_0, MHz	P_{out}, W	Q	CE, %	τ_p, ns
1	7.0	4.0	165.0	472.0	1,000	45	8
	7.0	4.0	163.0	432.0	500	42	8
	7.0	4.0	160.0	410.0	200	40	8
2	0.7	0.4	1,650.0	457.0	1,000	38	8
	0.7	0.4	1,620.0	432.0	500	35	8
	0.7	0.4	1,607.0	410.0	200	32	8

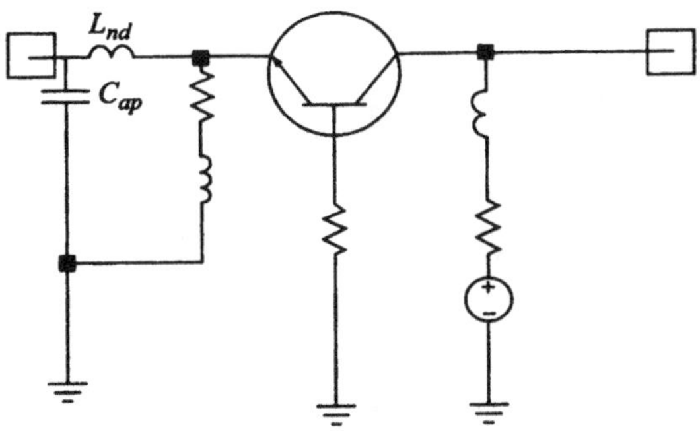

Fig. 10.3 The equivalent circuit of the powerful amplifier on BMT

of thermal emission twice, the maximal pulse overheating of the semi-conductor crystal will be 98.77 K, in which the average thermal power dissipated by the generator is from $P_{aver} = 0.6$ W for $Q = 1{,}000$ up to 2.5 W for $Q = 200$. The admissible value of the whole thermal resistance of the case can change over a wide range. The thermal resistance of the case with a standard convection cooler in the conditions of natural convection is not more than 5 K/W. This will ensure the functioning of the device at ambient temperature.

10.2 Signal Multiplication in Omnirange, UHF Range

The equivalent circuit of the powerful amplifier on BMT is shown in Fig. 10.3.

The results of calculations of the parameters of the amplifier on BMT are presented in Table 10.3.

The equivalent circuit of a powerful BMT in a mode of pulse signal amplification is given in Fig. 10.4.

The results of calculations of the parameters of a powerful BMT in the mode of pulse signal amplification of various relative pulse durations are given in Table 10.4.

Table 10.3 Results of calculations of the parameters of the power amplifier on BMT

No.	C_{ap}, pF	L_{nd}, nH	ν_0, MHz	P_{inp}, W	P_{out}, W	K_{gf}, dB	CE, %
1	7.0	4.0	165.0	50.0	157.0	5.0	37
2	0.7	0.4	1,650.0	50.0	142.0	4.5	34

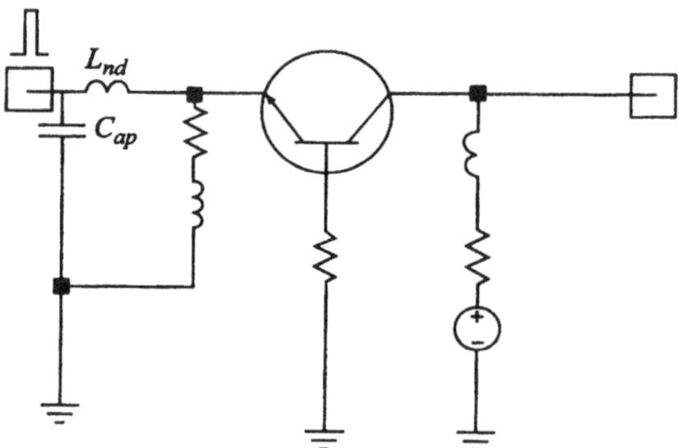

Fig. 10.4 The equivalent circuit of a powerful BMT in a mode of pulse signal amplification

Table 10.4 Results of calculation of the parameters of a powerful BMT in the mode of pulse signal amplification of various relative pulse durations

No.	C_{ap}, pF	L_{nd}, nH	ν_0, MHz	Q	P_{inp}, W	P_{out}, W	K_{gf}, dB	CE, %
1	7.0	4.0	165.0	1,000	170.0	469.0	4.4	40
	7.0	4.0	165.0	500	170.0	460.0	4.3	38
	7.0	4.0	165.0	200	170.0	442.0	4.1	35
2	0.7	0.4	1,650.0	1,000	200.0	472.0	3.7	34
	0.7	0.4	1,650.0	500	200.0	456.0	3.5	33
	0.7	0.4	1,650.0	200	200.0	441.0	3.4	32

10.3 Generation and Multiplication of Signals of Low and High Power Levels in UHF and Microwave Frequency Ranges

Calculated values of the key parameters of MECE on a low level are presented in Table 10.5 for the power from 20 to 100 mW in the range of frequencies from 0.3 to 18.0 GHz.

At high power levels in the UHF and at lower part of MWF ranges, FMCR saturation occurs, which limits the level of passing power.

In Table 10.6, calculated values of the basic MECE characteristics at power levels from 0.1 up to 2.0 W are presented.

A pulse HMG has been developed on the basis of the two-cascade amplifier, whose scheme is given in Fig. 10.5. In the feedback circuit of the amplifier, MECE 10.1 from Table 10.5 with a FMCR on KG-30 ferrite is included.

Table 10.5 Calculated values of the key parameters of MECE on a low level for the power from 20 to 100 mW in a range of frequencies from 0.3 to 18.0 GHz

No. of MECE	Working range of frequencies, GHz	Level of decoupling without external magnetic field, dB	Transfer losses in the passband with an external magnetic field, dB
4.1	12.0–18.0	−(12.0−15.0)	−(2.0–4.0)
8	1.0–18.0	−(23.0−30.0)	−(3.0–5.0)
9.1	1.0–2.5	−(8.0−2.0)	−(3.0–5.0)
10.1	0.3–3.0	−(25.0−30.0)	−(3.5–4.5)
10.2	2.0–18.0	−(25.0−30.0)	−(3.0–5.0)

Table 10.6 Calculated values of the basic MECE characteristics at power levels from 0.1 up to 2.0 W

No. of MECE	Working range of frequencies, GHz	Level of decoupling without external magnetic field, dB	Transfer losses in the passband with an external magnetic field, dB
8/M	from 1.0 up to 8.0	−(15.0–20.0)	−(5.0–7.0)
9.1/M	from 1.0 up to 2.0	−(8.0–10.0)	−(4.0–5.0)
10.1/M	from 0.5 up to 3.0	−(20.0–25.0)	−(5.0–6.5)

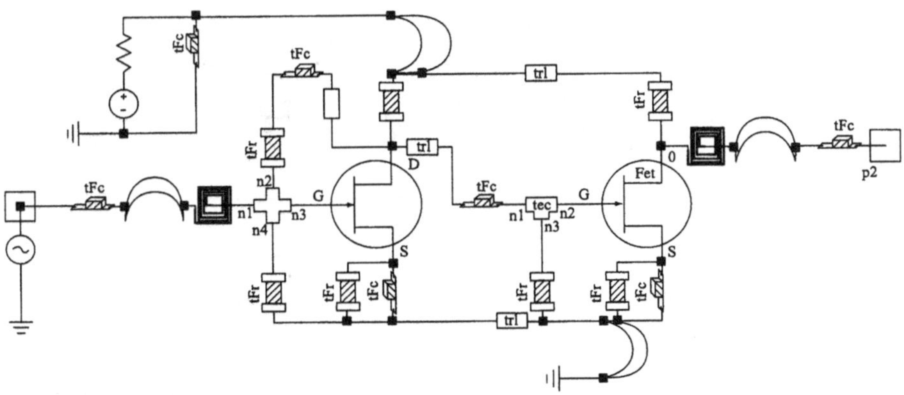

Fig. 10.5 A pulse HMG on the basis of the 2-cascade amplifier

The results of our modeling of the two-cascade amplifier on the basis of the field PTSh-800 and PTSh-1000[1] transistors are presented further. The model can be used in modern CADa such as *Serenade, Microwave Office*, etc. The peak-frequency

[1] Russian brands.

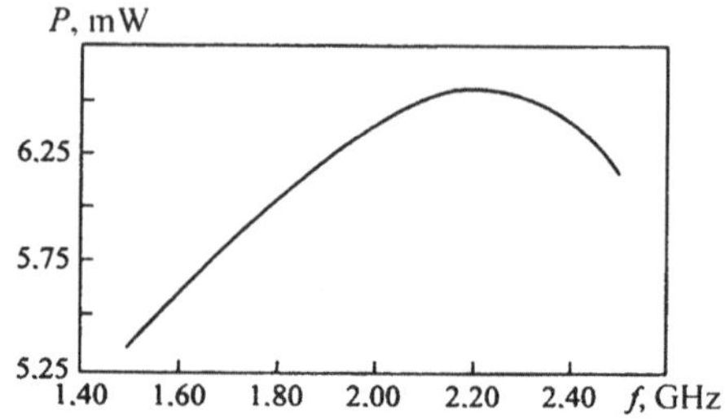

Fig. 10.6 The peak-frequency characteristics of the amplifier without magnetic field

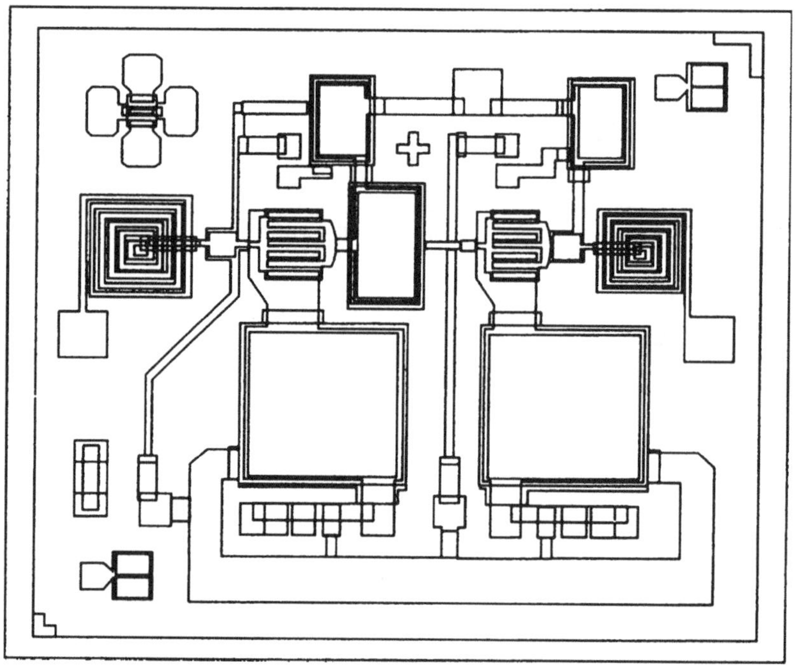

Fig. 10.7 The amplifier topology on a GaAs substrate of size 2.5 × 2.5 mm^2

characteristics of the amplifier without magnetic field management and its topology on a GaAs substrate of size 2.5 × 2.5 mm^2 are shown in Figs. 10.6 and 10.7.

In Fig. 10.8, the AFC of the coupling element (curve 1 – without magnetic field, curve 2 – with a magnetic field) is shown.

For getting the generating mode the output terminal of the amplifier was connected to the input through MECE (Fig. 10.9).

For supplying rectangular pulses on the HMT shutters, regular and noise signals were generated.

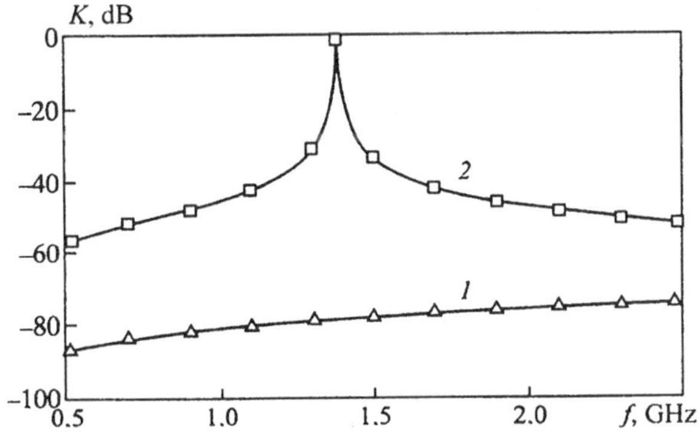

Fig. 10.8 The AFC of the coupling element (curve 1 – without magnetic field, curve 2 – with a magnetic field)

Fig. 10.9 For getting the generating mode the output terminal of the amplifier was connected to the input through MECE

10.4 Generation of Powerful Signals in the EHF Range

Theoretical research of the BMT parameters on the principles of power summation in the mode of generation of monochromatic oscillation signals on the central frequency of the EHF in the range of 65 GHz at an output power of below 4 W was conducted on the basis of the program from [53].

A mathematical model of the powerful generator on the bipolar HEMBIP-C magnetotransistor has been developed. The equivalent circuit of HEMBIP-M represents a set of active BS connected for power summation (Fig. 10.10). For every BS, Gummel–Poon's model was used.

In a range of frequencies above 40 GHz, FMCR t was brought directly into the area of the microstrip coupling element of the powerful transistor.

For calculations the equations describing the magnetoelectronic generator were used. A numerical research of the characteristics of a powerful BMT in the amplification mode was done by harmonic balance.

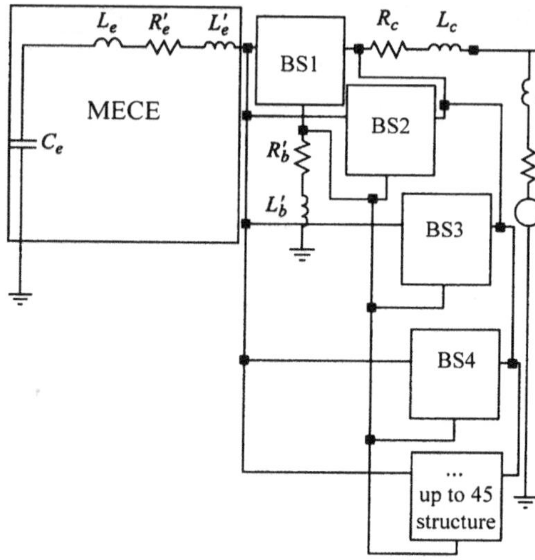

Fig. 10.10 The equivalent circuit of HEMBIP-M

As equivalent parameters, those of the bipolar transistor with the upper boundary frequency of 80 GHz were taken.

The output power P_{out}, its dependence on the signal frequency of the generator, and efficiency were investigated.

The equivalent circuit of the powerful generator on BMT included up to 45 BS and allowed to generate an output power up to 4 W on a frequency of 65 GHz. The frequency driving elements – C_e and L_e – represented the FMCR located in the field of the emitter junction of the powerful BMT. The resonant frequency was determined by

$$\omega_0 = \mu_0 \gamma \sqrt{H^2{}_0 - (\Delta H(\theta))^2}. \qquad (10.1)$$

At $\gamma = 1.76 \times 10^{11}$ C/kg, $H_0 = 2.19$ A/m, and $\Delta H = 38 \times 10^{-4}$ A/m, the resonant frequency is $f_0 = 47{,}100$ GHz. The resistance R_0 limits the current of displacement of the base–emitter junction and, together with the sources setting the voltage of feed E_c and the voltage of displacement of the base, determines the mode of the transistor on a direct current. The resistance R_0 on a high frequency is shunted by the condenser C_0 having a high enough capacity of $\sim$10 pF; R_e, R_c being the internal resistance of the displacement sources of the base and that of the power of the generator. The elements R_e', L_e', R_b', L_b', R_c', L_c' are the parameters of the conductors of conclusions and installation of the transistor on the plate; C_{cb} is the parasitic capacity formed at installation of a semiconductor structure inside the case of the transistor; and Z is the resistance of loading.

The dependence of the output power of BMT on the principles of power summation in the mode of monochromatic oscillation generation on the number of BS was investigated. The results are presented in Table 10.7, from which it follows that the output power of BMT grows not proportionally to the number of BS.

Table 10.7 The dependence of the output power of BMT on the principles of power summation in the mode of monochromatic oscillation generation on the number of BS

P_{out}, W	No. of BS
2.5	10
4.0	23
4.3	30
4.0	40
1.1	45

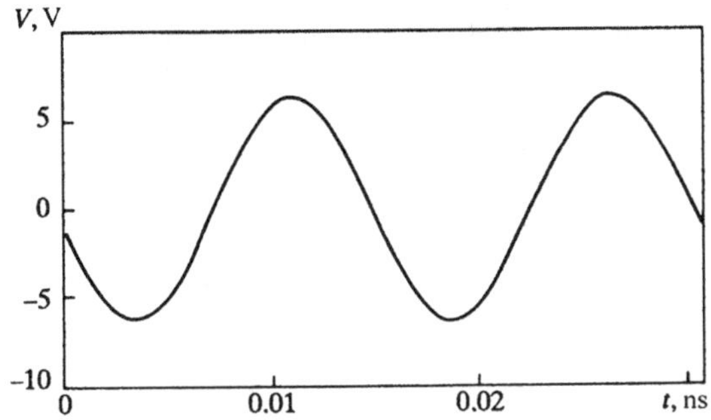

Fig. 10.11 Time realization of oscillations on the output of the powerful generator

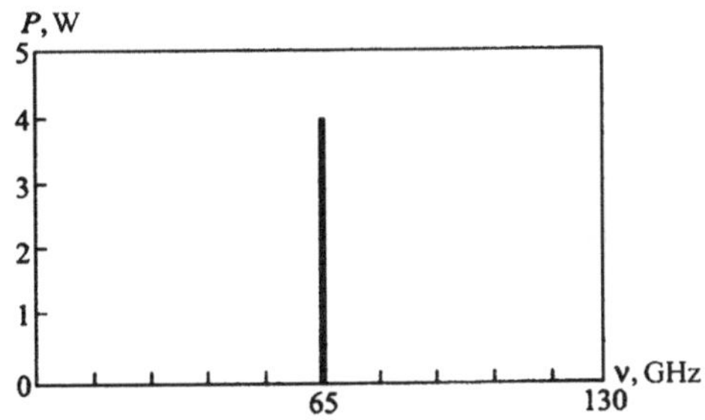

Fig. 10.12 The result of calculation of the power of the spectral component of oscillations on a frequency of 65 GHz

With an increase in the number of BS above 23–40, the growth of the output power of the generator was retarded down to its decrement, that is, due to the prevalence of BS mismatches and their work. At q 23 BS, the output power of the generating BMT is $P_{out} = 1$–4 W in a range of frequencies 40–70 GHz at CE = (5–7)%.

Time realization of oscillations on the output of the powerful generator is presented in Fig. 10.11.

For achievement of optimum values of the output power and CE at change of frequency in HMG, it is necessary to arrange the parameters of the equivalent circuit in the circuits of the emitter, base, and loading simultaneously.

The result of calculation of the power of the spectral component of oscillations on a frequency of 65 GHz is presented in Fig. 10.12. The equivalent parameters of FMCR are $C_e = 0.025$ pF and $L_e = 0.036$ nH.

Theoretical research of the parameters of a powerful generating FMT on the principles of power summation on the central frequency of the EHF in the range of 65 GHz was made on the basis of the program from [55]. In the mode of power summation with 10 BS included under positive feedback with MECE, an output power of $P_{out} = 4$ W on a frequency of $\nu_0 = 65$ GHz was obtained. The equivalent circuit of the generator on ten active structures is presented in Fig. 10.13.

The power distribution on the frequencies $\nu_0 = 65$ GHz and $\nu_1 = 130$ GHz in the generator on FMT is presented in Fig. 10.14.

The power of a signal on the basic frequency is $P_{\nu 0} = 4.1$ W, on the first harmonic $P_{\nu 1} = 0.1$ W. The integral output power of the generating FMT in a range of frequencies of 50–75 GHz is shown in Fig. 10.15.

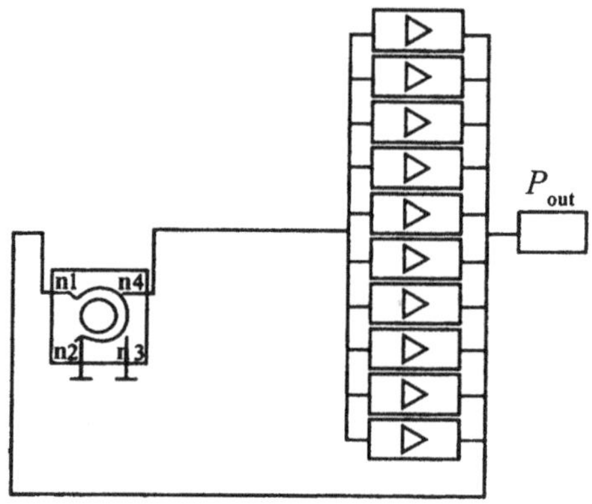

Fig. 10.13 The equivalent circuit of the generator with 10 BS included under positive feedback with MECE

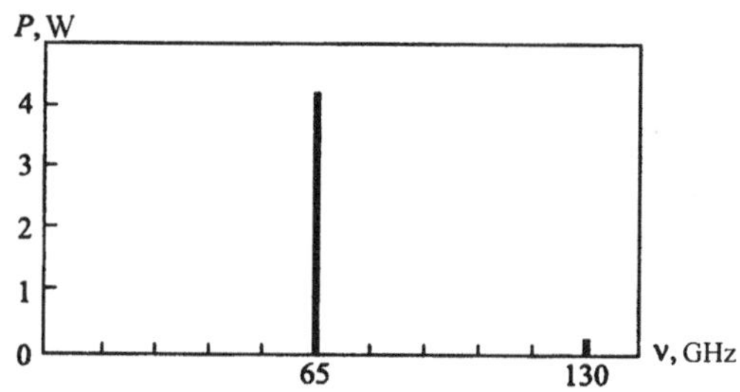

Fig. 10.14 The power distribution on the frequencies $\nu_0 = 65$ GHz and $\nu_1 = 130$ GHz in the generator on FMT

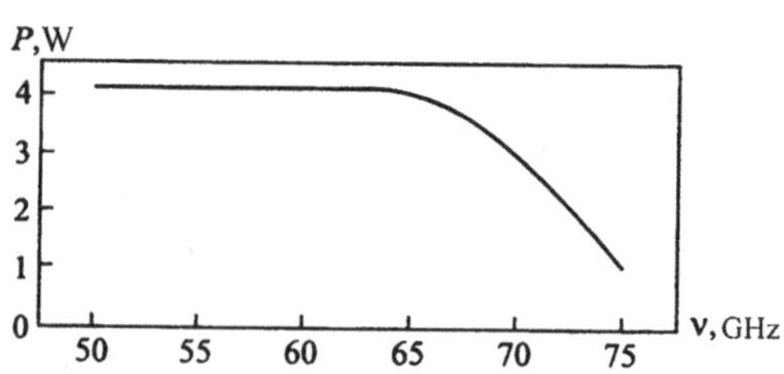

Fig. 10.15 The integral output power of the generating FMT in a range of frequencies of 50–75 GHz

Chapter 11
Multifunctional Frequency Synthesizers

11.1 General Data

The frequency synthesizer is understood as an electronic device forming a signal of a demanded frequency or a set of frequencies, changeable with a finite step by operating signal-commands [72]. The frequency synthesizer of the MWF range is a multimodular device uniting the following basic elements and units: a highly stable quartz generator, a voltage-controlled generator, a VCG, stable MWF generators with a frequency trim, frequency dividers and multipliers, amalgamators, amplifiers, MWF filters, and a control device for all the elements of the synthesizer.

The basic electric parameters of the frequency synthesizer, determining its quality and possible scopes, are as follows:

- Range of reorganization
- Rate of reorganization
- Step of reorganization (the frequency resolution) and accuracy of frequency setting
- Number of generated frequencies
- Clearance of the output signal spectrum (the level of side components and the level of phase noise)
- Long-term frequency instability
- Possible realization of various kinds of modulation, calibration of output power, etc.
- Continuous phase of output signal at reorganization (at return to the former frequency)

The operational parameters are as follows:

- Weight and dimensions
- Power consumption and power supply system
- Range of external working temperatures
- Stability to vibrations, etc.

A.A. Ignatiev and A.V. Lyashenko, *Heteromagnetic Microelectronics: Microsystems of Active Type*, DOI 10.1007/978-1-4419-6002-3_11,

All the known frequency synthesizers can be conditionally divided into two groups by their set of electric parameters, namely:

- Universal synthesizers having a wide frequency band, a small step of reorganization, a low level of side components and phase noise, possible realization of various kinds of modulation
- Specialized synthesizers having a narrower frequency band, a greater step of reorganization, a higher level of side components and phase noise

Universal frequency synthesizers have greater weights and dimensions and are intended for work in composite automated measuring complexes as stationary sources of signals at the development, manufacturing, control, checking of communication facilities, and radio-electronic equipment.

Specialized frequency synthesizers have small enough weights (few kg) and dimensions and, consequently, can be used in mobile communication systems, monitoring systems for equipment, etc.

The basic electric and operational parameters of some domestic and foreign frequency synthesizers are presented in Tables 11.1 and 11.2. The majority of companies produce frequency synthesizers as a series overlapping a wide range of frequencies. In the tables, data for the most low-frequency and high-frequency devices are cited.

For designing of frequency synthesizers, three basic methods are now used, namely:

- Direct analog synthesis
- Indirect synthesis on the basis of an APLC ring
- Direct digital and hybrid synthesis (a combinations of several methods)

In the MWF range, the method of indirect synthesis is most effective and widespread, when a demanded output frequency is formed by means of a reconstructed generator stabilized with a loop APLC. In such frequency synthesizers, the working range of frequencies and the reorganization band are determined by their reconstructed MWF generator. In domestic frequency synthesizers, VCG on transistors and Hannah diodes with varactor frequency reorganization are applied. In foreign frequency synthesizers, VCG with YIG filters or YIG resonators are applied.

The experience accumulated in SSU[1] and the results obtained in other organizations show that the MCG and HMG essentially surpass VCG with varactor reorganization by a number of parameters (the bandwidth of reorganization, linearity of the frequency characteristics). Besides, an HMG allows to adjust the parameters of the output signal spectrum.

---

[1] In the 1980–1990s in Saratov state university (SSU), linear and nonlinear processes in various electrodynamic structures containing monocrystal ferrite films, in the ranges of frequencies up to 60 GHz, were intensively investigated. There have been developed: methods of diagnostics of film ferrites, methods of designing of filters, delay lines, receivers, including devices of a high power level up to 5 kW in the 8 mm a range [1], and MCG.

Table 11.1 The basic electric parameters of frequency synthesizers

Parameter	Trademark, manufacturer								
	P46–05 "Meridian"	P46–03 "Meridian"	VMK2401 "Kvartz"	VMK2401 "Kvartz"	"Kvartz"	MSPLER 4506–05 "Micron"	MSPLER 1726–01 "Micron"	Frequency synthesizer 7883 "Izhevsk radio factory"	"Istok"
Range of output frequencies	0.3–1,200 MHZ	8.15–17.85 GHz	0.1–8.15 GHz	8.15–17.85 GHz	37.5–178.4 GHz (4 subranges)	450–600 MHz	1.7–2.6 GHz	7.8–8.3 GHz	Series of 0.4–8 GHz with band reconstruction up to 20%
Step of reorganization	1–2 Hz	20 kHz	1 kHz	2 kHz	100 kHz	5 kHz	1 kHz	28 kHz	1–5 MHz
Rate of reorganization	–	–	–	–	40 ms	0.5 ms	1 ms	–	–
Time of transients at frequency set-up	–	<40 ms	–	–	–	–	–	–	–
Level of phase noise at tuning out	–(96–90) dB/Hz on 20 kHz	–	−90 dB/Hz on 100 kHz	−70 dB/Hz on 100 kHz	–	−115 dB/Hz on 100 kHz	−115 dB/Hz on 100 kHz	−90 dB/Hz on 100 kHz	−90 dB/Hz on 100 kHz

(continued)

Table 11.1 (continued)

Parameter	Trademark, manufacturer								
	P46–05 "Meridian"	P46–03 "Meridian"	VMK2401 "Kvartz"	VMK2401 "Kvartz"	"Kvartz"	MSPLER 4506–05 "Micron"	MSPLER 1726–01 "Micron"	Frequency synthesizer 7883 "Izhevsk radio factory"	"Istok"
Attenuation of side components	70–60 dB	Harmonics 25 dB –(40–60 dB)	50 dB on >100 kHz	−40 dB on >100 kHz	–	Harmonics <−40 dB, collateral <−70 dB	Harmonics <−40 dB, collateral <−70 dB	−50 dB	<−60 dB
Output power	–	3–5 mW	10 mW	5 mW	2–20 mW	10 mW	10 mW	2 mW	To 10 mW
Kinds of modulation	–	–	AM, FM, PM	AM, FM, PM	AM, FM, PM	–	–	–	–
Instability of frequency	$\pm 5 \times 10^{-10}$ for 24 h	–	–	–	$\pm 5 \times 10^{-10}$ for 24 h	–	–	2×10^{-6}	–
Power supply (voltage, frequency, power)	220 V; 50 Hz; 350 VA	220 V; 50 Hz	–	–	220 V; 50 Hz; 500 VA	–	–	40 V; 1.2 A	–
Weight	45 kg	40 kg	4 kg	4.2 kg	47 kg			2 kg	0.006 kg
Range of working temperatures	–	–	−10...+50°C	−10...+50°C	5 ÷ 40°C	–	–	0 ÷ 40°C	–

Table 11.2 The basic operational parameters of frequency synthesizers

Parameter	Trademark, manufacturer "Saturn" Ukraine	MTS1500 Verti Com.	SNP-0710 "Microsource Inc"	EDRS "CT1"	MLSN-2040 "Micro Lambda Wireless, Inc"	MLSN-8010 "Micro Lambda Wireless, Inc"
Range of output frequencies, GHz	4.3–5.1	11.8–20.0	6.5–20.0	13.8–15.0	2.0–4.0	8.0–10.0
Step of reorganization	1 MHZ	0.8 MHz		From 1 Hz to 10 MHz	1 Hz	1 Hz
Rate of reorganization	–	–	25 ms	–	12 ms	12 ms
Level of phase noise at tuning out, dB/Hz, on 100 kHz	−100	−114	−115	−115	−120	−115
Attenuation of side components	−70 dB Harmonics <−30 dB	<−64 dB Harmonics <−15 dB	<−−50 dB Harmonics <−25 dB	<−70 dB Harmonics <−20 dB	<−60 dB Harmonics <−12 dB	<−60 dB Harmonics <−20 dB
Output power	10 mW	5 mW	15 mW	20 mW	20 mW	5 mW
Instability of frequency	–	–	–	$\pm 1 \times 10^{-11}$	–	–
Power supply	24 V	12 V, 0.75 A	15V, 1 A	5.25 V, 0.9 A	15 V, 0.9 A	15 V, 0.9 A
Size, mm^3	420 × 199 × 99	146 × 70 × 48	172 × 76 × 36	147 × 102 × 30	–	–
Range of working temperatures, °C	–	−35… + 70	−40… + 70	−10… + 80	−40… + 70	−40… + 70
Special conditions	–	YIG oscillator design	YIG oscillator design	–	YIG oscillator design	YIG oscillator design

VCG users usually refer to a lower rate of frequency reorganization in generators with FMCR, hysteresis phenomena occurrence, and temperature dependencies. However, the rate of frequency reorganization in the synthesizers of indirect synthesis is determined not only by the rate of reorganization of the generator itself, but also by the time of transients at frequency capture in the APLC system. The hysteresis phenomena in MCG can be almost eliminated by using combined adjustment: the magnetic one for greater changes of frequency, and the varactor one for small frequency changes in the APLC system.

The main advantages of HMG in comparison with the known VCG and MCG are as follows:

- The possibility of creation of a supertiny operated multipurpose synthesizer on a single crystal
- A high technical CE for frequency synthesizers of a high power level.

11.2 Oscillators Operated by Magnetic Field in Frequency Synthesizers

The simplest frequency synthesizer is based on direct analog synthesis (*DAS*). In such a synthesizer the input frequency is produced from the basic frequency by means of several operations: mixing, filtration, multiplication, and division. The structural block diagram of *DAS* is shown in Fig. 11.1, where MCG with frequency switches of f_1 and f_2 on *pin*-diodes (keys) can be used as the basic ones.

For frequency reorganization in such a synthesizer, a set of continuously working basic generators connected to the amalgamator by means of high-speed *pin*-keys is used. It provides an important advantage of the synthesizers of direct synthesis, namely, a short time of working frequency switching (less than 1 μs). The second important advantage of such synthesizers is the possibility of return on the former frequency and continuation of their work in the same phase as though no frequency switching occurred (this effect is referred to as "phase memory").

Direct synthesis is most optimally realized in the decimeter range. For design of synthesizers in the MWF range by the principle of direct synthesis, now two

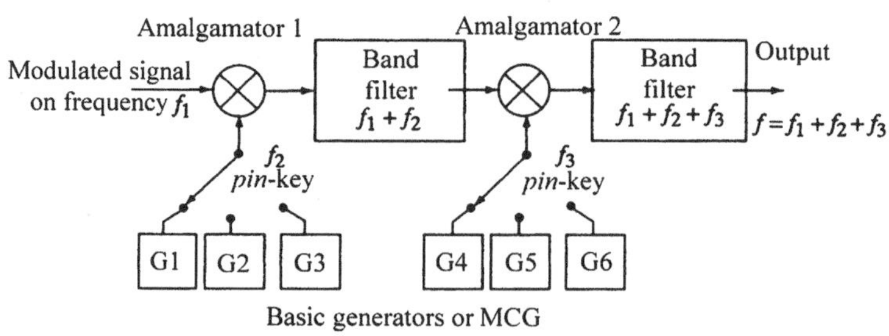

Fig. 11.1 The structural block diagram of DAS

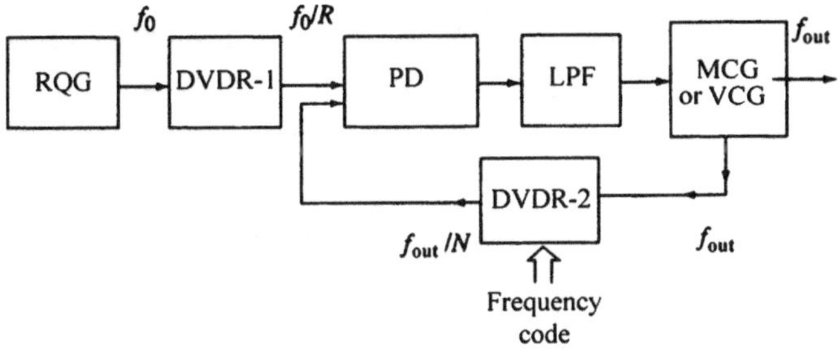

Fig. 11.2 A simplified structural flowchart of the frequency PLL

basic ways are used, namely: direct signal synthesis of the decimeter range with subsequent frequency multiplication, and shifting of the frequency of a decimeter wave synthesizer to the MWF range by its summation with the signal of a highly stable low-noise MWF generator. However, in any of these cases the level of phase noise increases in comparison with the noise of the basic signal.

The basic demerit of direct synthesis synthesizers is few output frequencies limited by the number of basic generators.

The number of synthesized frequencies can be increased by using frequency dividers; however, filtration of the output signal becomes complicated.

The method of indirect synthesis on the basis of phase-locked loop (PLL) is most widely spread now, at which the demanded output frequency is formed by means of an additional frequency-changed generator captured by the APLC loop.

A simplified structural flowchart of the frequency PLL is presented in Fig. 11.2.

A highly stable signal of a frequency f_0 from the reference quartz generator (RQG) is applied on the divider with a variable division ratio (DVDR-1) on R. From the output of DVDR-1 the signal of the frequency f_0/R gets on one of the inputs of the phase detector (PD). On the second input of PD the signal from the DVDR-2 divider with a variable factor of division by N from the frequency-changed VCG or MCG is applied. The target voltage of the error signal from PD, proportional to the frequency difference of f_0/R and f_{out}/N, is filtered by LPF and is applied on the element controlling the frequency of the basic generator (VCG or MCG). The frequency of the basic generator will be driven toward its nominal value until the frequencies f_0/R and f_{out}/N are equal. The step of frequency change of the synthesizer will be determined by the frequency of comparison f_0/R, and the working frequency range by the frequency-change band of the basic generator.

The basic demerit of the synthesizer with indirect synthesis is a low rate of frequency change (more than 1 ms).

In recent years, the method of direct digital synthesis of frequencies (*DDS*) has gained spread. In such a synthesizer, the output signal is created by digital methods with a high accuracy peculiar to digital systems and unambiguity. The structural flowchart of the elementary digital frequency synthesizer is shown in Fig. 11.3.

A signal from the reference clock-frequency generator (GCR) goes to an input of the divider with a variable factor of division (DVDR) on N. The obtained sequence

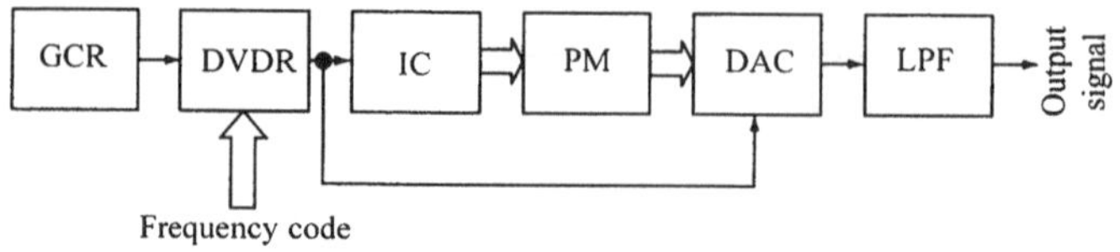

Fig. 11.3 The structural flowchart of the elementary digital frequency synthesizer

of pulses goes to an input of the impulse counter (IC) to form the phase of a harmonious signal as a linear function of time. Readouts of the sine function of the output signal with a frequency of digitization are formed in the PM with the help of a conversion table. The output signal in the digital form in the digital-to-analog converter (DAC) is translated into the analog form and filtered with the help of a filter of bottom frequencies (LPF). Change of the output frequency in *DDS* occurs due to change of the factor of division *N*.

DDS have a high-frequency resolution, reaching hundredths and even thousandths of Hz. Frequency and phase control occurs digitally that allows any signal modulation. Transition from one frequency to another one is carried out with a high rate. Frequency change occurs without breakage of phase and other anomalies associated with transients in analog systems. The parameters of digital synthesizers are almost not subject to temperature drift and ageing.

At the same time, some application restrictions of the systems of direct digital synthesis are due to some features of digitization and DAC processes. So, the maximum frequency of output signal should be essentially below the clock frequency of the processor. This limits the area of application of *DDS* to frequencies of few hundreds MHz. Isolated side components of the output signal in *DDS* can be considerably higher than in an analog one. The power consumed from the power supply in *DDS* synthesizers is higher than in *DAS* and *PLL*.

11.3 Frequency Synthesizers of Indirect Synthesis Based on APLC

Let us consider in more detail the design principles of the most widely spread synthesizers of indirect synthesis on the basis of APLC. In a similar synthesizer, two conflicting functions are incorporated, namely: generation of an MWF, signal with a wide band of frequency change; and generation of an MWF signal, stable by frequency and having a low level of phase noise.

The ways of separate realization of each of these functions are well known. Continuous frequency change in generators on transistors, Hannah diodes, or IMPATTD is usually carried out by means of varactors in VCG or MCG (Fig. 11.4a). Expansion of the generated frequency band toward lower frequencies is reached by mixing of the signals of the frequency-changing generator and oscillator with resolution of a

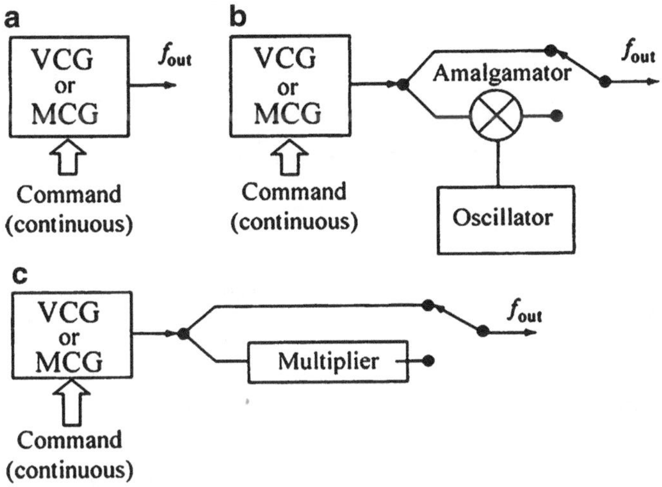

Fig. 11.4 (**a**) The control format by the signal frequency. (**b**) The expansion circuit of the generated frequency band toward lower frequencies. (**c**) The expansion circuit of the generated frequency band toward higher frequencies

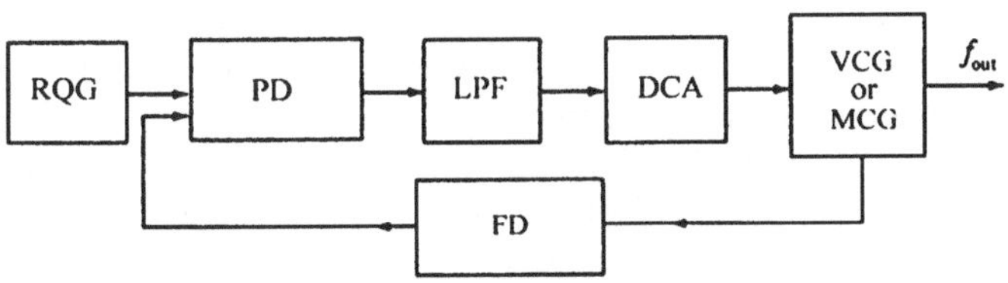

Fig. 11.5 A possible flowchart of highly stable MWF generators with an APLC with direct frequency division in the feedback channel

difference frequency (Fig. 11.4b). Expansion of the frequency change range toward higher frequencies is carried out by frequency multiplication (Fig. 11.4c).

Frequency stabilization in MWF generators is usually reached by application of various types of high-Q or superhigh-Q resonators (parametrical stabilization), or basic highly stable quartz generators with various circuits of the APLC system. A possible flowchart of highly stable MWF generators with an APLC with direct frequency division in the feedback channel is shown in Fig. 11.5. To avoid higher factors of frequency division in the MWF range, a preliminary frequency divider with a constant factor of division, prescaler (PR), is placed before the basic divider.

To convert the last circuit into a frequency synthesizer, a broadband MWF generator (VCG or MCG) and a frequency divider with a controllable variable factor of division are necessary (Fig. 11.6).

In such a synthesizer, the output frequency is determined by $f_{out} = f_0 N$ and the frequency grid step, $\Delta f_{out} = f_0$, where f_0 is the frequency of RQG and N is the factor of division in DVDR.

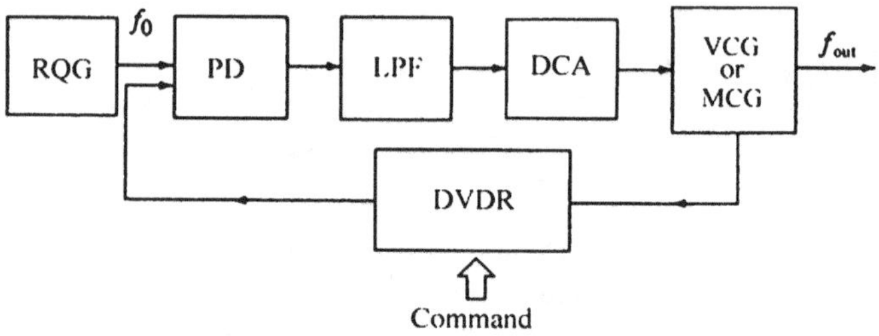

Fig. 11.6 A frequency synthesizer with a broadband VCG or MCG and a frequency divider with a controllable variable factor of division

11.4 Oscillator Operated by Magnetic Field

Any frequency synthesizer contains a VCG or MCG frequency-changing in an electrical way [71, 73].

The basic advantage VCG is the high rate of frequency change. The demerits are as follows: the narrow wide of electric frequency change (not higher than 20%), the nonlinear voltage dependence of the generated frequency, high enough Ohmic losses that leads to expansion of the resonant curve of the oscillatory contour and increase in the level of phase noise, and the GB product of the contour with a varactor decreases at increase in the working frequency [79].

The basic advantages of MCG are the wide band of electric frequency change (an octave or more), the high degree of linearity of the dependence of the generated frequency on the external magnetic field induction (up to 0.05%), and the high GB product of FMCR provide a good spectral frequency of regular signals and a low level of phase noise. Figure 11.7 shows the dependencies of the equivalent GB product of FMCR Q_0 on frequency f for various saturation magnetizations. The FMCR GB product Q_0 increases with frequency, which makes MCG attractive for use in the MWF. These advantages of MCG are used in MWF panoramic measuring instruments.

The demerit of MCG is the low rate of frequency change (tens ms from the beginning of the band up to its end) due to the high inductance of the bias coil. Besides, the MCG is characterized by a raised sensitivity to vibrations and external electromagnetic interferences.

MCG is usually based on a transistor generator with a YIG sphere as the frequency-changing resonator, providing (at the correct design) an octava band of frequency change at a high linearity, and a small magnetic backlash and a small volume of the area of the working magnetic field. For connection with the external electric circuits, the YIG sphere (FMCR) is located at the center of the coupling coil (or the half-coil). The electric circuit of a generator with the coupling half-coil with FMCR is shown in Fig. 11.8.

The inductance L_b in the circuit of the transistor base provides a positive feedback. For improvement of coordination of the target resistance of the transistor with the resistance of loading, the generator's output is connected to the loading through

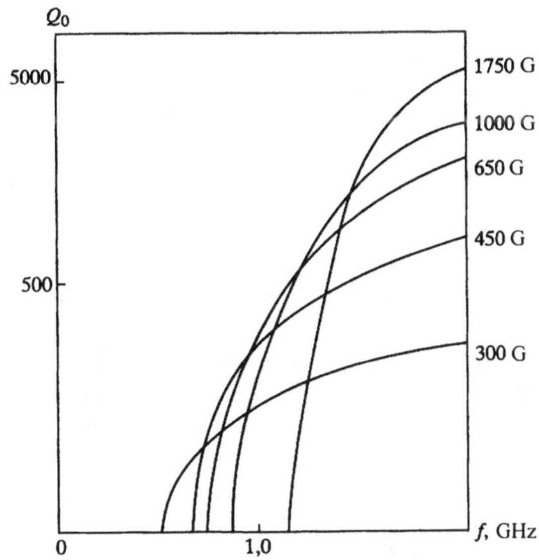

Fig. 11.7 The dependencies of the equivalent GB product of FMCR Q_0 on frequency f for various saturation magnetizations

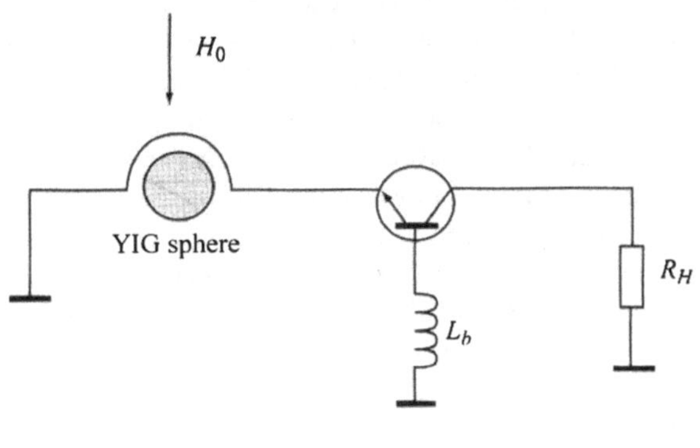

Fig. 11.8 The electric circuit of a generator with the coupling half-coil with FMCR

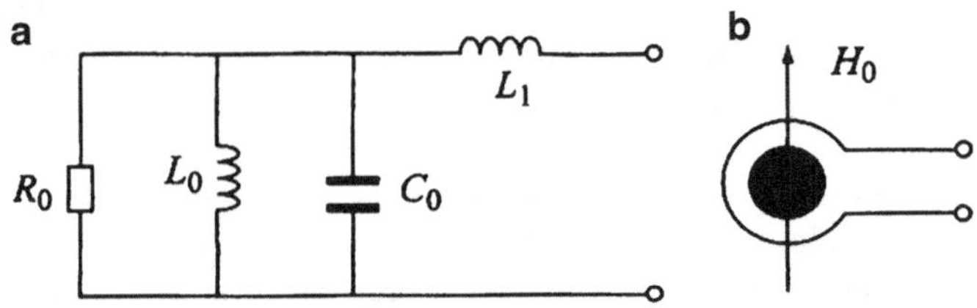

Fig. 11.9 The equivalent circuit (**a**) of FMCR with a coupling coil (**b**)

a buffer amplifier. Generation of a wide (octava) band of frequency change requires careful orientation of FMCR within the coupling coil, selection of matched elements and the value of inductance L_b. In Fig. 11.9, the equivalent circuit (a) of FMCR with a coupling coil (b) is shown.

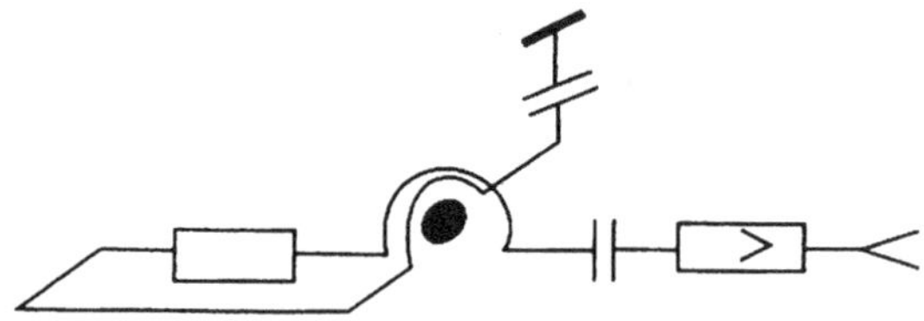

Fig. 11.10 A circuit of inclusion of FMCR in crossed coupling coils into the circuit of external feedback of the amplifier

A circuit of inclusion of FMCR in crossed coupling coils into the circuit of external feedback of the amplifier (Fig. 11.10) that complicates the design of MCG [74] is also known.

The basic design demerit of the considered circuits of MCG is due to spatial splitting of the active element (transistor) and FMCR with the coupling coils. This causes the necessity of use of additional connecting elements, their careful coordination with the basic circuits, and increases the sizes of the whole device. For overcoming this demerit, it is offered to place the YIG resonator together with the coupling coils directly on the transistor crystal, i.e., to create a magnetotransistor.

Ferrite YIG-based materials are manufactured in the form of FMCR of various diameters and various saturation magnetizations, and thin epitaxial films on a gallium-gadolinium substrate. The overwhelming majority of MWR devices with thin YIF films are based on the effects of excitation and distribution of MSW and hybrid electromagnetic waves [1]. The chair of general physics (SSU) has contributed much to the development of the magnetoelectronic theory of the MWF and EHF-ranges and design of devices on their basis [1, 75, 76]. They have designed a lot of devices, including those for the millimeter range, with no analogs; they are patented.

Narrow-band MSW filters frequency-changing by magnetic field, included in the amplifier's feedback circuit, have been used for design of pilot models of MCG [77]. However, similar MCG have not gained wide spread because of their "inconvenient" tangent of the external magnetic field orientation relative to the YIG film surface and its weight-dimensions. Now, OmniYIG ® has anew released MCG on thin YIG films, working on frequencies from 0.5 up to 18 GHz.

One of the basic elements of MCG is the magnetic system creating an electrically changed magnetic field. The generated frequency of MCG is determined by the induction of the magnetic field B in which FMCR is placed, by the known equation f (MHz) $\cong 2.8$ (MHz/G) $\times$ B (G).

The magnetic system of MCG is a small-sized electromagnet, in which the electric winding, as a rule, is located on the iron magnetic circuit only on one side of the magnetic gap. The closing magnetic circuit usually has a cylindrical form, simultaneously acting as a tight case. In typical MCG, the sensitivity of frequency adjustment is from 10 up to 20 MHz/mA. The magnetic system of MCG, as a rule, has an additional winding near the magnetic gap for thin frequency adjustment, for example, at phase synchronization. The sensitivity of frequency tuning is usually $300 \div 400$ kHz/mA. The rather long time of frequency change in a wide band is due, first of all, to the high inductance of the electromagnet winding. The typical values of the parameters of the electromagnet winding are $L = 0.1$ H and $R = 10\,\Omega$.

In this case, the constant of the circuit is $\tau_0 = L/R \approx 0.01\,\mathrm{s} = 10\,\mathrm{ms}$. The current relaxation period in the inductance is usually taken equal to $3\tau_0$. However, this period decreases down to τ by application of pulse voltage exceeding the required value to the electromagnet winding at frequency change. Thus, MCG enable getting the rate of frequency change below 10 ms.

For high-frequency MCG, the magnetic system is made as a constant magnet or an electromagnet. For reduction of the influence of external electromagnetic interferences on the generated frequency the magnetic screen is separated from the core magnetic circuit.

11.4.1 Experimental MCG Research

An MCG with its electric circuit presented in Fig. 11.11 has been made and tested. Figure 11.12 shows a view of the breadboard model of MCG.

The generator is assembled on a bipolar KT371A transistor under the common–base circuit. FMCR locates at the center of the coupling coil with a 12 mm diameter. The FMCR diameter is 0.8 mm and its saturation magnetization is $4\pi M_0 = 100\,\mathrm{G}$.

The generator was tested on the installation whose flowchart is shown in Fig. 11.13: 1 – a MCG; 2, 3 – power supplies; 4 – electromagnet windings; 5 – power supplies of the electromagnet; 6 – a directed coupler; 7 – an attenuator; 8 – a measuring instrument of power; 9 – an attenuator; 10 – a spectrum analyzer; 11 – the probe of the measuring instrument of magnetic induction; 12 – a measuring instrument of magnetic induction.

The external adjustable magnetic field was created by a laboratory electromagnet and measured near FMCR by a teslameter with an accuracy ±0.5 G. The device

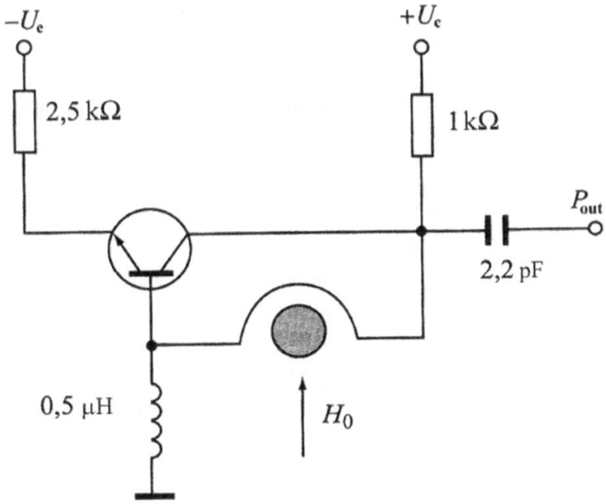

Fig. 11.11 Electric circuit of MCG

Fig. 11.12 A view of the breadboard model of MCG

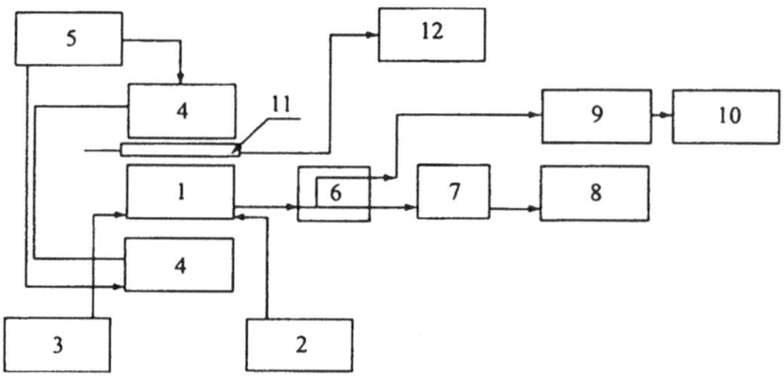

Fig. 11.13 Flowchart of measuring equipment

allowed observation of the spectrum of the generated signal to measure its frequency with an accuracy ±0.5 MHz, and to determine the output power.

The results of measurements in Fig. 11.14 show that in a range of 750–1000 MHz the dependence of the frequency in MCG on the induction of the magnetic field is practically linear. The output power decreases with frequency growth, which is due to mismatching of the output of the transistor with the loading. For elimination of this defect, a buffer amplifier must be placed between the generator and the loading.

The MCGs have an asymmetrical loop of hysteresis, i.e., the frequency distinction of generation at setting of the same value of the external magnetic field when approaching it from lower or higher fields. At the rating value of magnetic field of 404 G, the maximal value of hysteresis is 4 MHz in good agreement with the known data.

Our research of the spectrum of MCG signals has shown that no parasitic signals are found within the limits of the dynamic range of the C4–60 spectrum analyzer (60 dB).

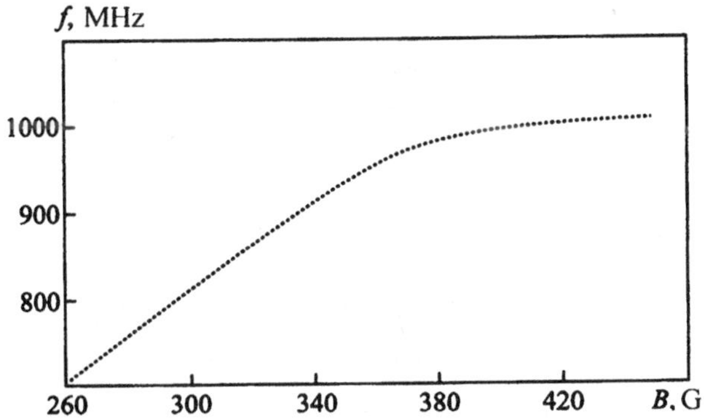

Fig. 11.14 The dependence of the frequency in MCG on the induction of the magnetic field

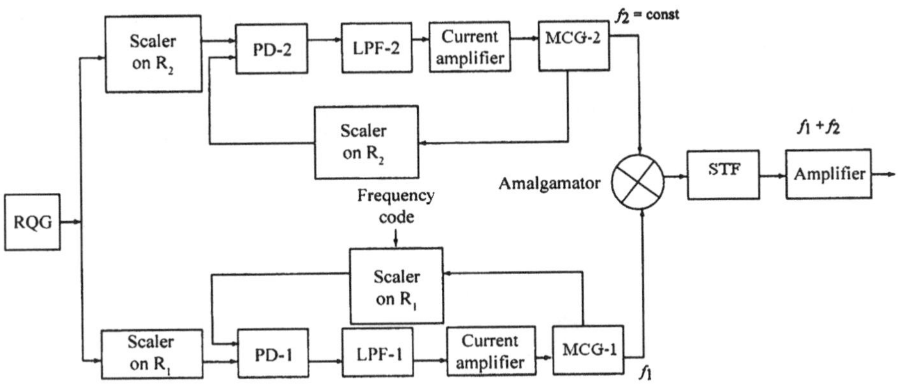

Fig. 11.15 The flowchart of a broadband multipurpose magnetoelectronic frequency synthesizer

11.5 Multifunctional Frequency Synthesizers Based on APLC Using GSM

For combination of a wide frequency-change band and management of the target signal spectrum in a frequency synthesizer, two types of MCG are offered to be used:

- MCG-1, with a wide frequency-change band, generating a monochromatic frequency-changed signal of the frequency f_1
- MCG-2 generating a signal of the frequency $f_2 = \text{const}$ with a controlled spectrum

The signals from MCG-1 and MCG-2 go to an amalgamator, on which output the total signal controlled by frequency and spectrum is isolated.

The basic flowchart of a broadband multipurpose magnetoelectronic frequency synthesizer is shown in Fig. 11.15. In this scheme, the use of traditional ways of target signal frequency modulation and working in pulse modes, doubling of the target frequency, the use of MCG for management of the target signal spectrum, and VCG for getting a frequency-changed signal are possible. The rate of frequency change will depend on the time of transients in the APLC system (1–2 μs).

In APLC synthesizers, the stability of the target signal frequency is determined by RQG and amounts to 10^{-7}–10^{-9}.

11.6 Multifunctional Operated Frequency Synthesizer Based on Transistor *BFR* 90

For design of a frequency synthesizer in a range of 0.5–3.0 GHz, a serial bipolar *BFR* 90 transistor (the working frequency range up to 5 GHz, the amplification factor not less than 14 dB on a frequency of 800 MHz) is chosen. The manufacturers of this transistor give detailed tables of its parameters at various power levels and in various parts of the frequency ranges, which facilitates the task of its computer modeling. The parameters of the transistor by Gummel–Poon's model are as follows:

$IS = 4.11877 \times 10^{-16}$ A	$RB = 10\ \Omega$	$ITF = 0.213776$ A
$BF = 102.639$	$IRB = 1 \times 10^{-6}$ A	$CJC = 5.46563 \times 10^{-13}$ F
$NF = 0.997275$	$RBM = 10\ \Omega$	$VJC = 0.380824$ V
$VAF = 62.6719$ V	$RE = 1.1645\ \Omega$	$MJC = 0.202935$
$IKF = 3.20054$ A	$RC = 2.32\ \Omega$	$MJS = 0.33$
$ISE = 4.01062 \times 10^{-15}$ A	$EG = 1.11$	$VJS = 0.75$ V
$NE = 1.57708$	$XTI = 3$	$Lb = 6 \times 10^{-10}$ H
$BR = 18.1086$	$CJE = 8.90512 \times 10^{-13}$ F	$Le = 6 \times 10^{-10}$ H
$NR = 0.996202$	$VJE = 0.6$ W	$L1 = 3.4 \times 10^{-10}$ H
$VAR = 3.36915$ V	$MJE = 0.25857$	$L2 = 1 \times 10^{-10}$ H
$IKR = 1.28155$ A	$TF = 1.54973 \times 10^{-11}$ s	$L3 = 3.4 \times 10^{-10}$ H
$ISC = 2.79905 \times 10^{-16}$ A	$XTF = 39.1402$	$Ccb = 1 \times 10^{-13}$ F
$NC = 1.07543$	$VTF = 2.15279$ V	$Cbe = 2 \times 10^{-15}$ F

The topology of the transistor's conductors measured with an ocular micrometer with a least division of 0.05 mm is presented in Fig. 11.16. The dotted line specifies the location of the semiconductor crystal.

With the purpose of creation of a magnetocontrolled element suitable for inclusion in the transistor, a design of the coupling element on crossed lines has been developed. The isometric projection of the coupling element's conductors is presented in Fig. 11.17. The coupling element is made in the form of conductors with a 1 mm width, located perpendicularly to each other in parallel planes.

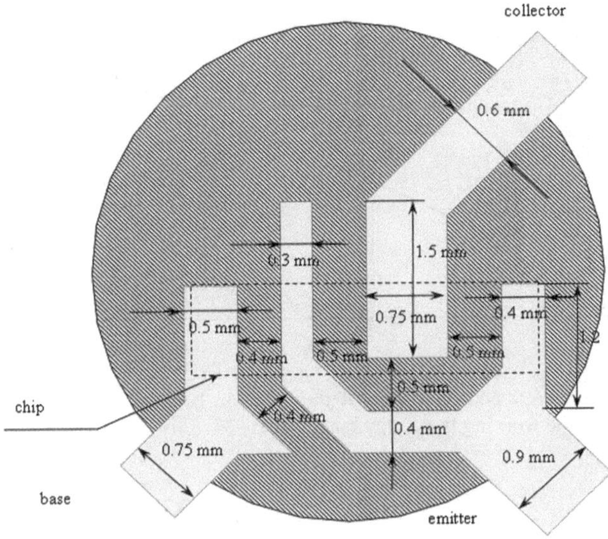

Fig. 11.16

Fig. 11.16 The topology of the transistor's conductors

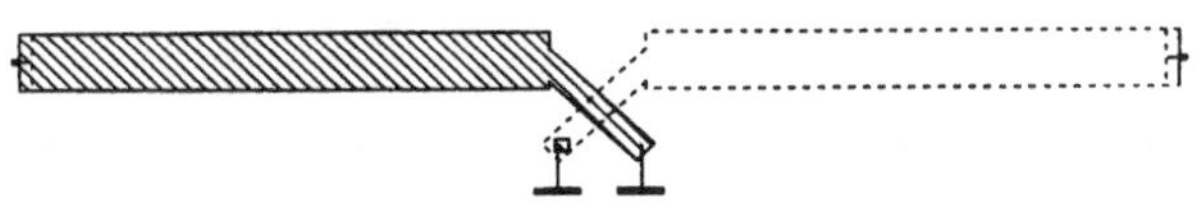

Fig. 11.17 The coupling element's conductors

The distance between the planes of the conductors is 0.1 mm, the structure is dipped into a dielectric material with a permeability 4.0. The outputs of the primary and secondary transfer lines "are short-circuited on the ground" for maintenance of effective FMCR interaction with the conductors of the coupling element. The ferrite resonator can be placed near one of the corners formed by the crossed conductors.

In Fig. 11.18, the frequency characteristics of the isolation between the input and output of the coupling element in the working frequency range are presented. The nominal level of isolation is reached by selection of the corresponding geometrical sizes of the conductors and the gap between the conductors.

In Figs. 11.19 and 11.20, various variants of the changed topology of the *BFR* 90 transistor are presented, allowing its usage in the mode of multipurpose generation with frequency management depending on the value of the applied external constant magnetic field.

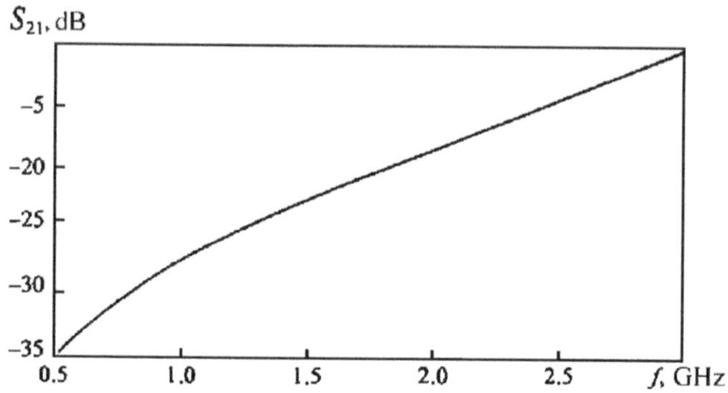

Fig. 11.18 The frequency characteristics of the isolation between the input and output of the coupling element in the working frequency range

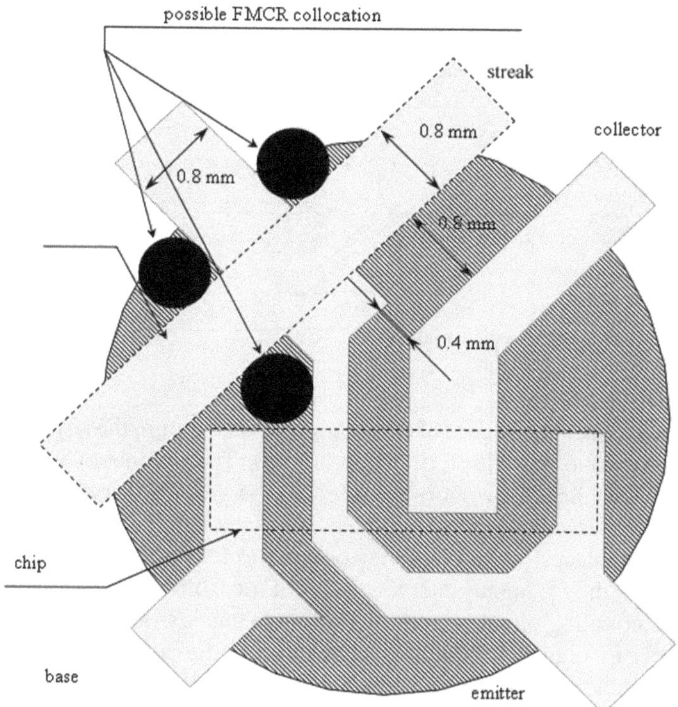

Fig. 11.19 Variants of the changed topology of the BFR 90 transistor, allowing its usage in the mode of multipurpose generation with frequency management depending on the value of the applied external constant magnetic field

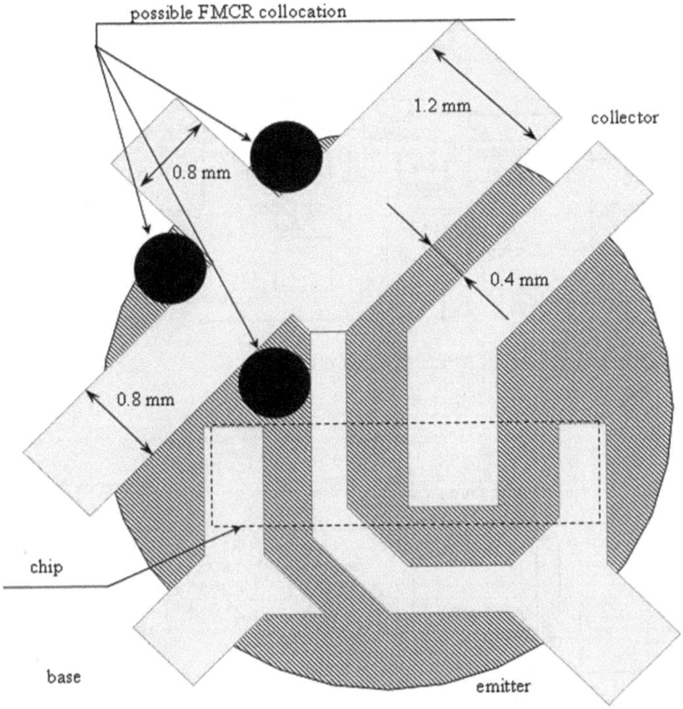

Fig. 11.20 Variants of the changed topology of the BFR 90 transistor, allowing its usage in the mode of multipurpose generation with frequency management depending on the value of the applied external constant magnetic field

11.7 Transient Processes Inside Synthesizers with APLC

The minimal time of frequency change in synthesizers with APLC is determined not only by the rate of frequency change in the MWF-generator, but also by the time of transients. A feature of these processes is that the change of frequency and target signal phase is characterized by fading oscillations of a finite duration even when the MWF generator frequency exactly follows the change of the operating voltage.

The duration of these processes can be evaluated by numerical methods. Calculation was made by ADI SimPLL ver. 2.0 (Applied Radio Labs) for a typical model of an APLC synthesizer, presented in Fig. 11.21. In Fig. 11.22, a specific circuit of a synthesizer with an APLC system based on an *ADF* 4007 microcircuit with a low-frequency *RC* filter is shown. As the MWF generator, VCG (*UMX*-612-*B*14) was used, the time of frequency setup at change of the operating voltage in the used program was considered negligibly small.

The results of calculations of the dependencies of frequency, the errors of the frequency and phase of the output of the generator as functions of time (in μs) at frequency changes within the intervals: 30–230 MHz, 300–800 MHz, 3.0–3.3 GHz, and 3.0–7.5 GHz are presented in Figs. 11.23–11.26 and generalized in Table 11.3.

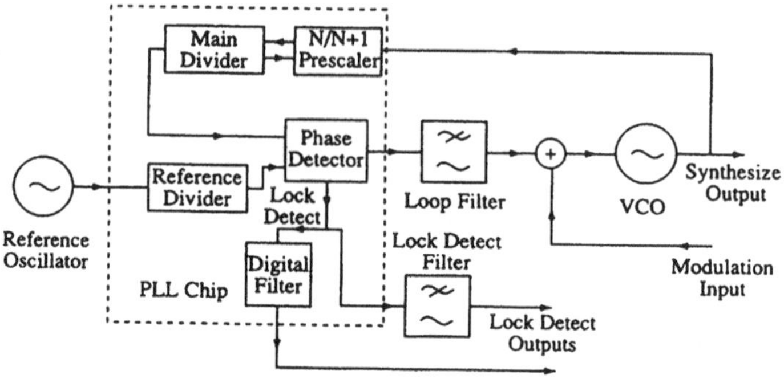

Fig. 11.21 A typical model of an APLC synthesizer

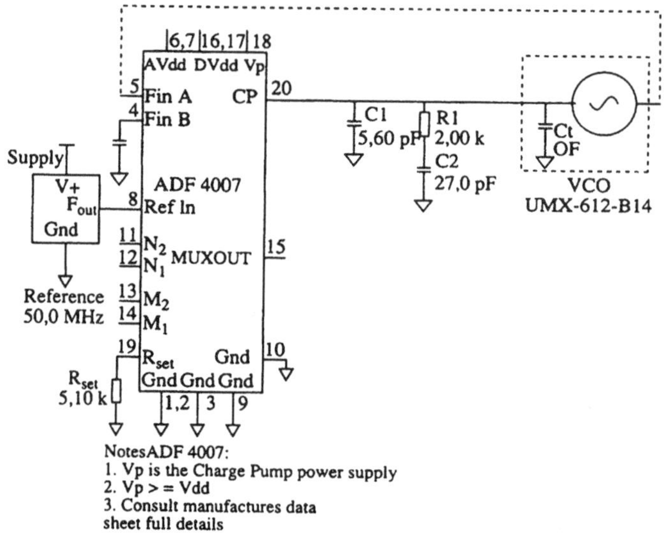

Fig. 11.22 A specific circuit of a synthesizer with an APLC system based on an ADF4007 microcircuit with a low-frequency RC filter

11.8 Output Characteristics of GSM

Consider a transistor generator with MECE on the basis of two crossed coupling coils in the feedback circuit between the collector and emitter (Fig. 11.27). The resistances, capacities, and inductances are adjusted to value so as to provide the maximal value of output power in the working band of frequency change.

In Fig. 11.28, the spectrum of harmonic components f_0, f_1, f_2, f_3, f_4 of MCG in a frequency range below 50 GHz is shown. Figure 11.29 shows the amplitude V_0 of output power as a function of time for target frequency change in MCG within an octave.

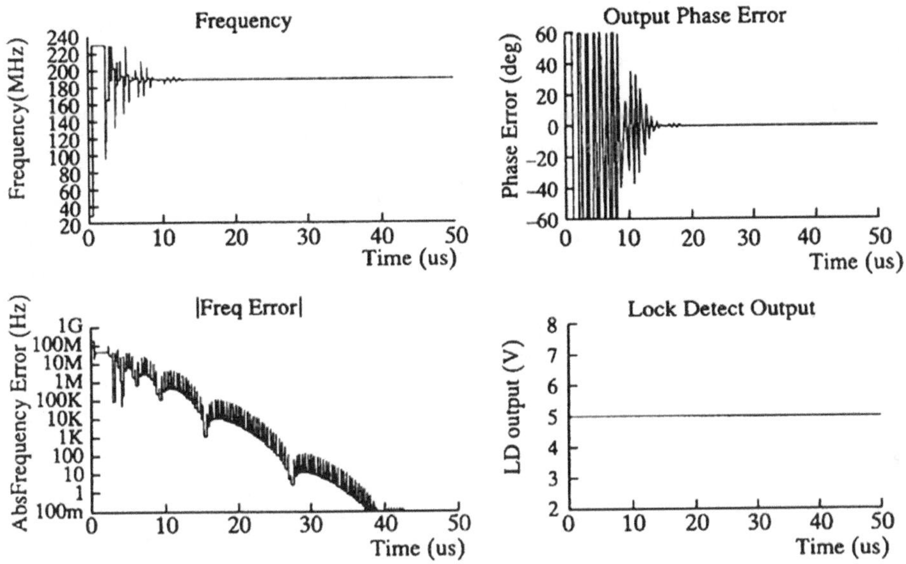

Fig. 11.23 The results of calculations of the dependencies of frequency, the errors of the frequency and phase of the output of the generator as functions of time (in μs) at frequency changes within the intervals: 30–230 MHz, 300–800 MHz, 3.0–3.3 GHz, and 3.0–7.5 GHz

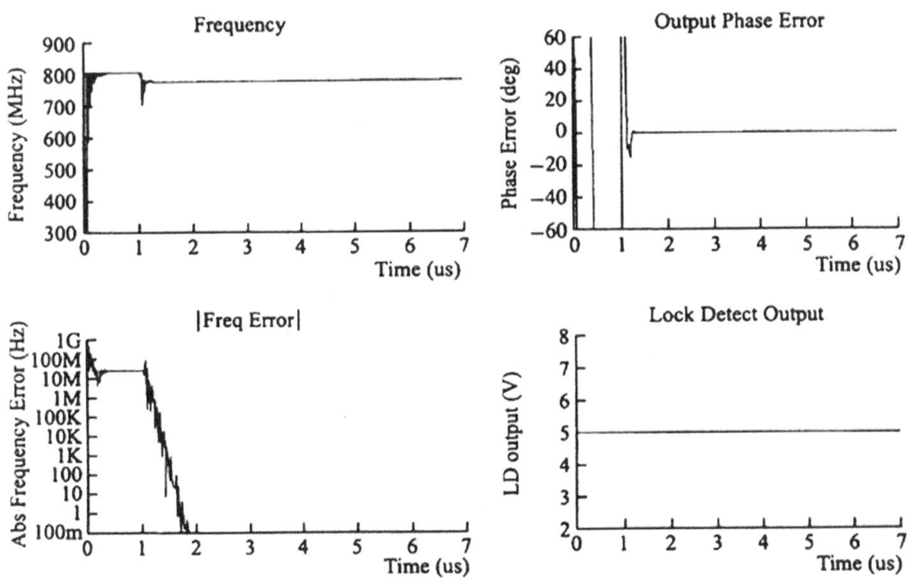

Fig. 11.24 The results of calculations of the dependencies of frequency, the errors of the frequency and phase of the output of the generator as functions of time (in μs) at frequency changes within the intervals: 30–230 MHz, 300–800 MHz, 3.0–3.3 GHz, and 3.0–7.5 GHz

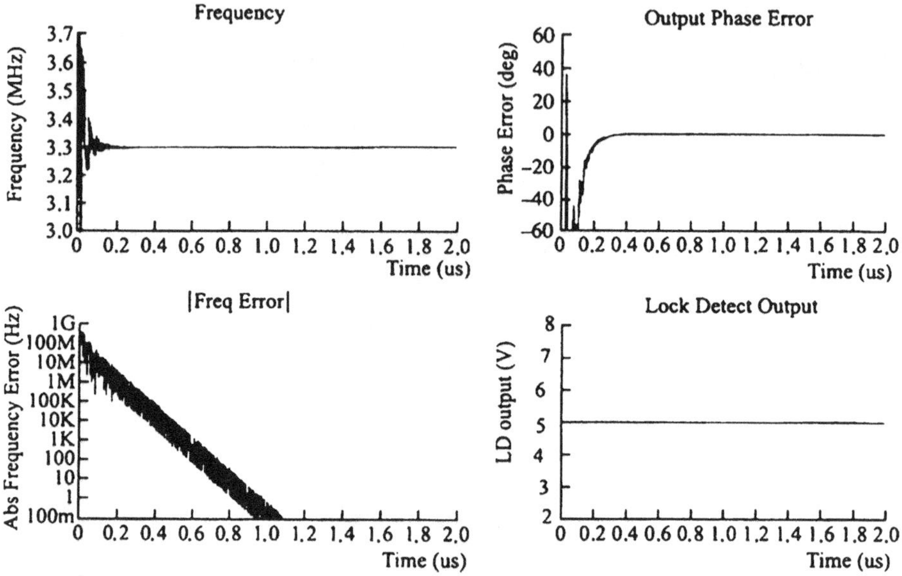

Fig. 11.25 The results of calculations of the dependencies of frequency, the errors of the frequency and phase of the output of the generator as functions of time (in μs) at frequency changes within the intervals: 30–230 MHz, 300–800 MHz, 3.0–3.3 GHz, and 3.0–7.5 GHz

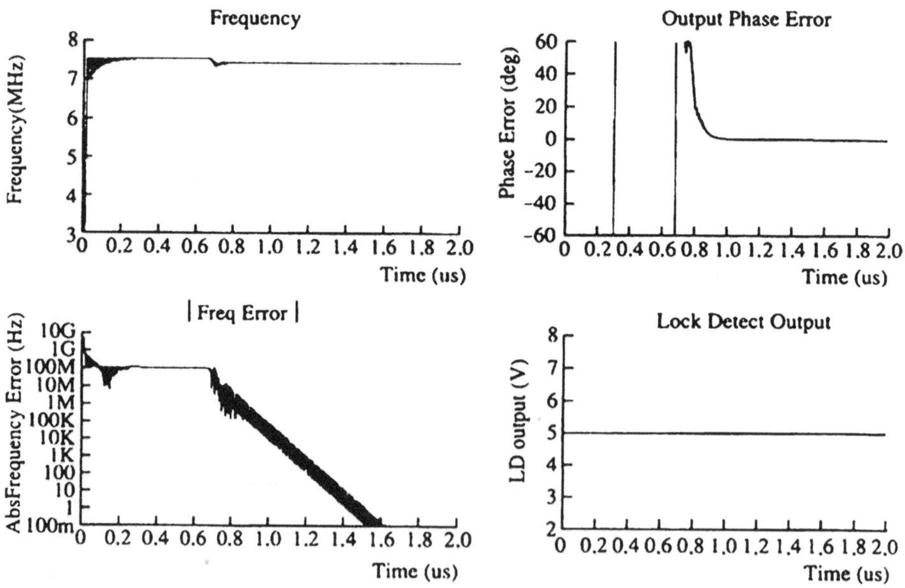

Fig. 11.26 The results of calculations of the dependencies of frequency, the errors of the frequency and phase of the output of the generator as functions of time (in μs) at frequency changes within the intervals: 30–230 MHz, 300–800 MHz, 3.0–3.3 GHz, and 3.0–7.5 GHz

Table 11.3 The results of calculations of the dependencies of frequency, the errors of the frequency and phase of the output of the generator as functions of time (in μs) at frequency changes within the intervals: 30–230 MHz, 300–800 MHz, 3.0–3.3 GHz, and 3.0–7.5 GHz

Microcircuit and its parameters	Lower frequency, MHz	The step on a frequency, MHz	Steepness (max), MHz/V	Temporary frequency, MHz	Frequency stabilization time, μs	Phase stabilization time, μs	Frequency stabilization time to 1 Hz, μs
ADF4007				3,050	0.40	0.50	1.20
$F_{min} = 0$ MHz			200	3,250	0.45	0.80	1.55
$F_{max} = 7{,}500$ MHz				3,950	1.30	1.70	2.60
Noise level =				3,050	0.25	0.30	0.75
−216 dB/Hz			250	3,600	0.40	0.55	1.20
$V_{min} = 0$ V		50		4,200	1.35	1.55	2.30
$V_{max} = 5$ V	3,000		300	4,450	1.45	1.55	2.25
			400	4,950	1.60	1.75	2.20
			500	5,450	1.75	1.85	2.25
			600	5,950	1.90	2.05	2.35
		60	800	6,960	2.45	2.60	3.00
		75	900	7,430	1.20	1.50	1.90
			10	325	2.40	2.50	5.50
			25	400	1.05	1.50	2.80
			50	525	1.00	1.10	1.90
	300	25	100	775	1.03	1.30	1.75
			200	1,275	1.95	2.10	3.00
			400	2,250	1.95	2.30	3.00

(continued)

Table 11.3 (continued)

Microcircuit and its parameters	Lower frequency, MHz	The step on a frequency, MHz	Steepness (max), MHz/V	Temporary frequency, MHz	Frequency stabilization time, μs	Phase stabilization time, μs	Frequency stabilization time to 1 Hz, μs
			550	3,000	2.65	3.00	4.00
			1	32.5	80	95	225
			5	52.5	27	30	70
			10	77.5	22	30	60
	30	2.5	20	127.5	30	35	60
			30	157.5	20	25	45
			40	227.5	60	67.5	87.5
			54	297.5	65	70	87.5

The control voltage changes frequency under the law: $f = f_0 + K_V \times V$, where $V = \{0; 5\}$

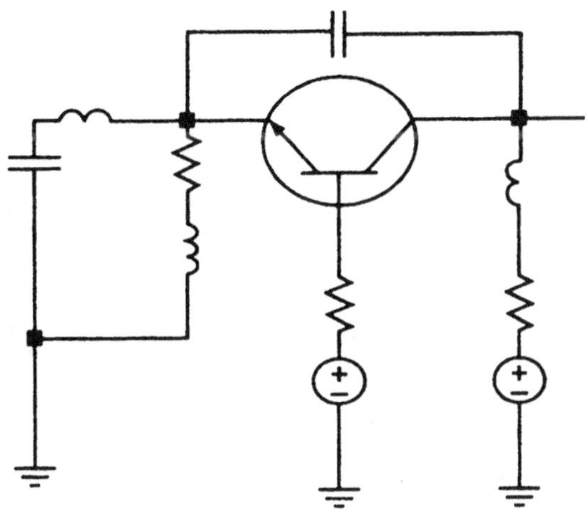

Fig. 11.27 A transistor generator with MECE on the basis of two crossed coupling coils in the feedback circuit between the collector and emitter

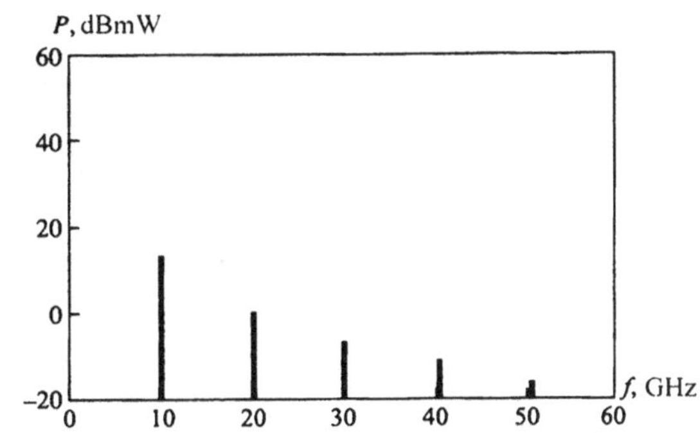

Fig. 11.28 The spectrum of harmonic components f_0, f_1, f_2, f_3, f_4 of MCG in a frequency range below 50 GHz

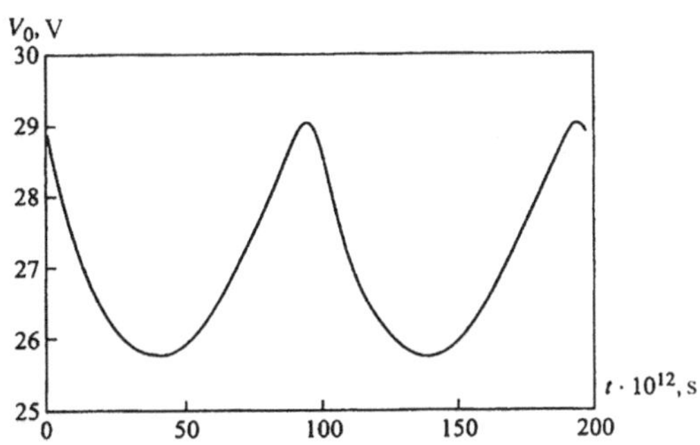

Fig. 11.29 The amplitude dependence V_0 of output power as a function of time for target frequency change in MCG within an octave

In Fig. 11.30, the results of calculation of the spectral characteristics on frequencies f_0, f_1, f_2, f_3, f_4 MCG in a frequency range from 8 up to 95 GHz for parameters $L_{nd} = 0.03$ nH and $C_{ap} = 0.025$ pF are presented. Figure 11.31 shows the time characteristics of the amplitude $V_0(t)$ for a frequency of 18 GHz. Figure 11.32 shows the spectral components f_0, f_1, f_2, f_3, f_4 in a frequency range below 204 GHz for $L_{nd} = 0.015$ nH and $C_{ap} = 0.01$ pF. Figure 11.33 shows the time realization of signal $V_0(t)$ on a frequency of 51 GHz. FMCR was modeled as an

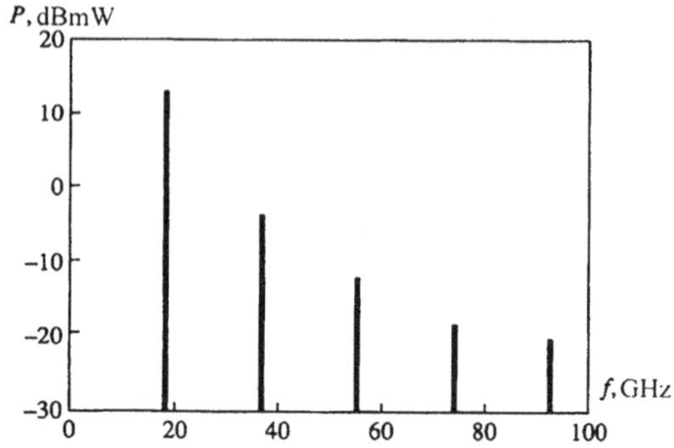

Fig. 11.30 The results of calculation of the spectral characteristics on frequencies f_0, f_1, f_2, f_3, f_4 MCG in a frequency range from 8 up to 95 GHz for parameters $L_{nd} = 0.03$ nH and $C_{ap} = 0.025$ pF

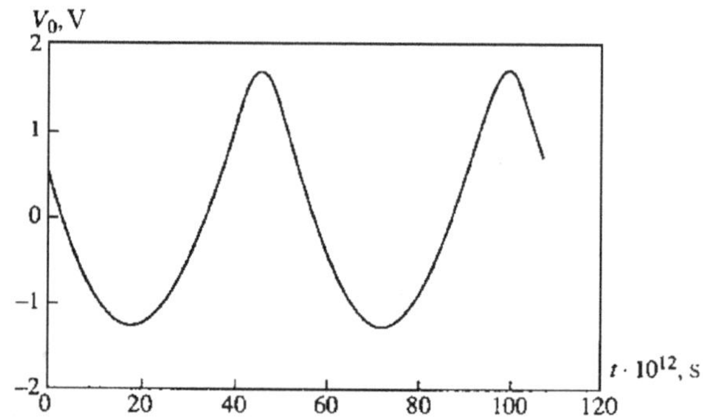

Fig. 11.31 The time characteristics of the amplitude $V_0(t)$ for a frequency of 18 GHz

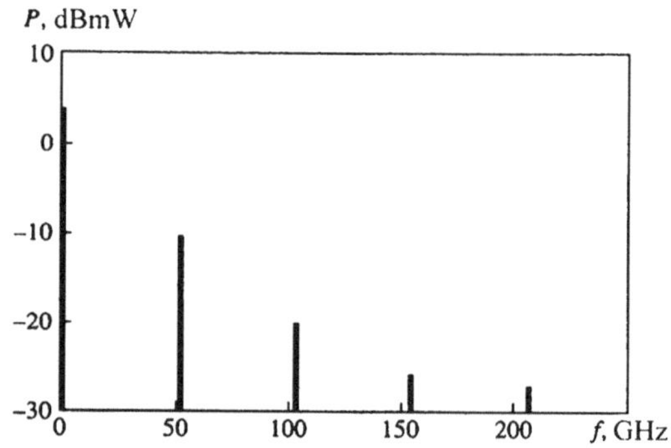

Fig. 11.32 The spectral components f_0, f_1, f_2, f_3, f_4 in a frequency range below 204 GHz for $L_{nd} = 0.015$ nH and $C_{ap} = 0.01$pF

equivalent oscillatory contour with its parameters depending on the external field bias. Calculation was done by the method of harmonic balance. The GB product of the equivalent oscillatory contour was determined by the width of ferromagnetic resonance of the used sphere. The wave resistance depends on the design of MECE.

Figure 11.34 shows the characteristics of phase noise S_φ in the Doppler frequency range F of tuning out from the carrier of MCG signal on a frequency of

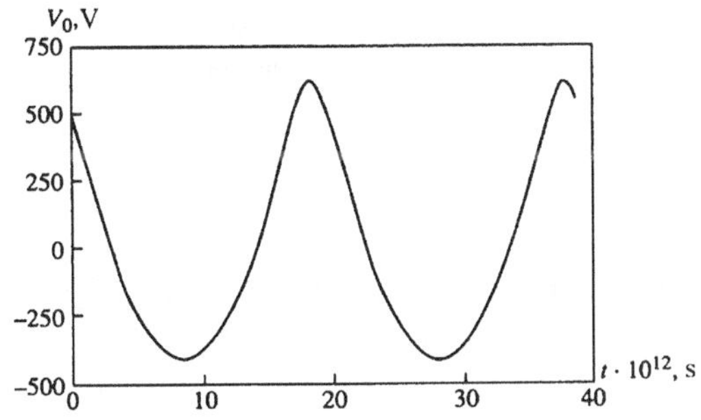

Fig. 11.33 The time realization of signal $V_0(t)$ on a frequency of 51 GHz

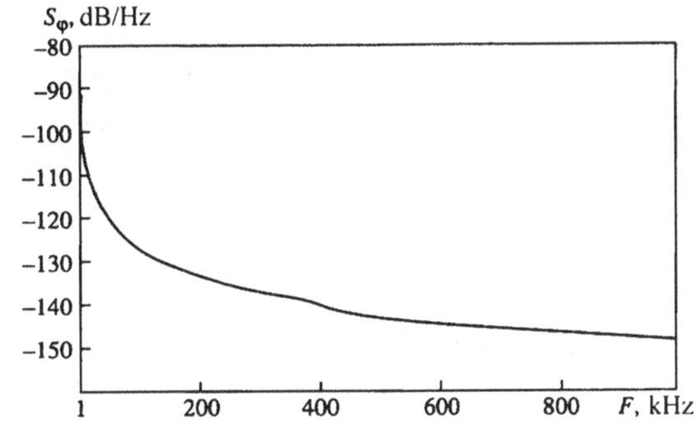

Fig. 11.34 The phase noise S_φ in the Doppler frequency range F of tuning out from the carrier of MCG signal on a frequency of 800 MHz

Table 11.4 The numerical values of S_φ are resulted at tuning out F from the carrier within the limits of 1 kHz to 1 MHz

F, kHz	1.0	10.0	1000.0
$S_\varphi = \text{dB/Hz}$	−89	−107	−148

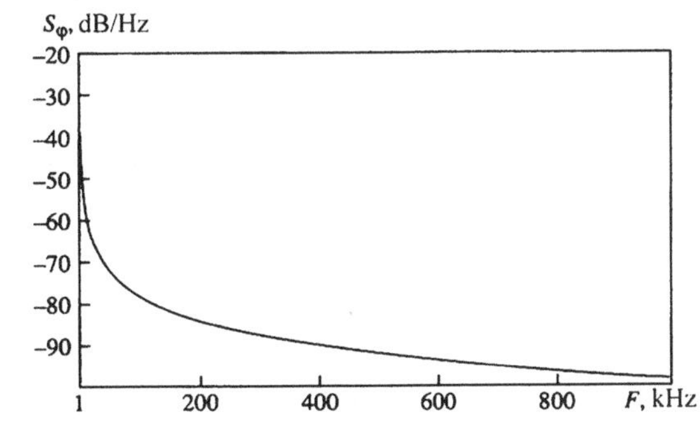

Fig. 11.35 Phase noise $S_\varphi(F)$ of MCG signal on a frequency of 51 GHz

800 MHz, and in Table 11.4 the numerical values of S_φ are presented at tuning out F from the carrier within the limits of 1 kHz to 1 MHz.

Figure 11.35 shows the characteristics of phase noise $S_\varphi(F)$ of MCG signal on a frequency of 51 GHz. Table 11.5 shows the values of phase noise S_φ at tuning out F from the carrier within the limits of 1 kHz to 1 MHz.

Table 11.5 The values of phase noise S_φ at tuning out F from the carrier within the limits of 1 kHz to 1 MHz

F, kHz	1.0	10.0	1000.0
S_φ = dB/Hz	−39	−59	−99

11.9 Pseudorandom Working Frequency Tuning and Phase-Shift Keying of Pseudonoise Signal Using GSM

A new promising lead in telecommunication is application of PM PS and PWFT, which, in comparison with the usual narrow-band systems, possess a number of advantages. Consider the possibility of MCG application in such systems.

The basic characteristics of a signal with PWFT are the range of frequency change Δf and the duration of a session of communication on a programmable frequency.

11.9.1 GSM with PWFT Function

The operating element of MCG is the inductance L creating a magnetic field in MECE, into which FMCR is placed. In an ideal MCG for maintenance of the preset frequency of generation, a field with $B = f/\gamma\mu\mu_0$, where γ is the gyromagnetic ratio, is necessary. If the field in the plane of the core S is considered homogeneous and the induction relates to the inductance L as $B = \mathrm{LI}/S$, then the current in the MCG bias coil is

$$I = \frac{fS}{\gamma L\mu\mu_0} \tag{11.1}$$

Let us estimate the transit time of the frequency-change circuit of MCG. In Fig. 11.36, a simplified scheme of frequency change in MCG is presented, where CVS is a source of a steady voltage U, and CE* is an operating element in the form of a changeable resistance R.

The steady current is

$$I = \frac{U}{R + R_\mathrm{L}}, \tag{11.2}$$

where R_L is the resistance of the bias coil. The transition of the necessary current $\tilde{I}$ to a steady state in such a system is described by the expression:

$$\tilde{I} = I \cdot \left(1 - \mathrm{e}^{-\frac{(R+R_\mathrm{L})\cdot t}{L}}\right) + I_0, \tag{11.3}$$

where I_0 is the initial current in the bias coil.

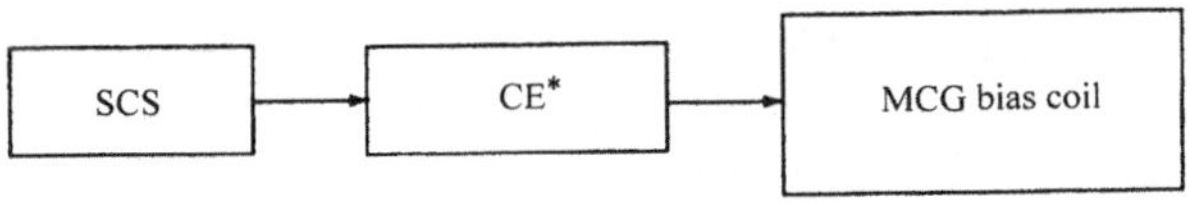

Fig. 11.36 A simplified scheme of frequency change in MCG

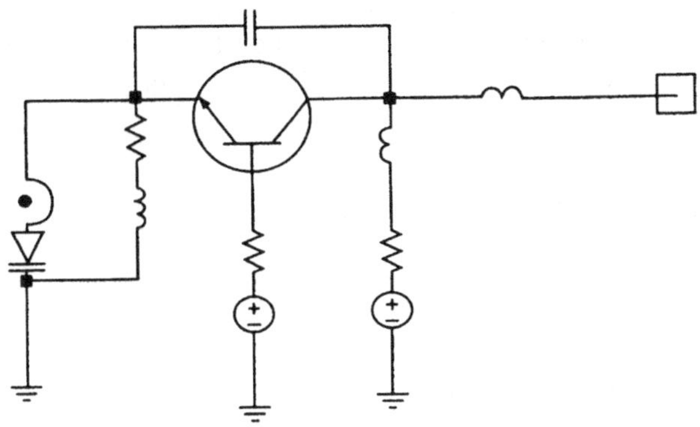

Fig. 11.37 The equivalent circuit of the combined MCG with a varactor

In view of (11.1) and (11.2), (11.3) can be rewritten as

$$f_{\text{tun}} = f\left(1 - \mathrm{e}^{-\frac{U\gamma\mu\mu_0 t}{fS}}\right) + f_0, \tag{11.4}$$

where f_0 is the initial frequency of generation of MCG.

Then for an estimation of the transit time of frequency change of MCG we have

$$\tau = \frac{fS}{U\gamma\mu\mu_0}. \tag{11.5}$$

With typical values: $U = 12\,\text{V}$, $\gamma = 28\,\text{Hz/nT}$, $f = 1\,\text{MHz}$, $\mu = 1000$, $\mu_0 = 1{,}257 \times 10^{-6}\,\text{H/m}$, and $S = 0.0001\,\text{m}^2$, we get a constant $\tau = 0.237\,\mu\text{s}$.

To reduce the time of frequency change is possible by an increase in the voltage of power supply of the coil; however, this would worsen the energetical characteristics of the MCG block. The energy heating the bias coil, in view of (11.1), is

$$W = UI = U\frac{fS}{\gamma L\mu\mu_0}. \tag{11.6}$$

The time of change of the MCG parameters proportionally decreases at reduction of the bias core area. Therefore, the bias coil should be made tiny.

The number of coils of the bias coil is great, and the intercoil capacity cannot be neglected in the equivalent circuit of the combined MCG with a varactor (Fig. 11.37), which additionally increases the transit time of frequency change. To reduce this effect, two bias coils are applied, and one of them has a considerably smaller number of coils and is intended for realization of the PWFT function.

Thus, the transit time of MCG is great enough, and to essentially improve this characteristics for due maintenance of the PWFT function, a combined scheme including a varactor can be used. A possible variant of the MCG circuit with double management of frequency by means of FMCR and a varactor is presented in Fig. 11.37. The consecutive contour *LC* models an FMCR with frequency change in a wide range. Fast frequency change of the generator is carried out by the varactor *D*. Such a combined circuit of the generator has its transit characteristics of frequency change similar to the traditional VCG on varactors [79].

11.9.2 GSM with PSK PS Function

For realization of the PM PS function, a phase shifter on *pin*-diodes is used on the output of MCG. A possible circuit of such a phase shifter is shown in Fig. 11.38.

At change of the polarity of applied voltage the phase of oscillations turns by 180°. The rate of manipulation is determined by the type of the applied *pin*-diodes, being few nanoseconds.

The characteristics of the switching *pin*-diodes made by Salute Corp. [82] are presented in Table 11.6.

The results of calculations of the target signals of the phase shifter at phase shift by 180° are presented in Fig. 11.39.

The phase shifter works in a frequency range of 5%. If necessary, it is possible to use broadband phase shifters [82] whose characteristics are presented in Table 11.7.

Thus, at use of MCG in a frequency synthesizer, the functions PM PS and PWFT can be implemented. The specific design of a synthesizer depends on the technical requirements on realization of these functions. Combined circuits of MCG possess the rate of frequency change similar to varactor VCG at preservation of their advantages on a low level of phase noise and a wide range of frequency change.

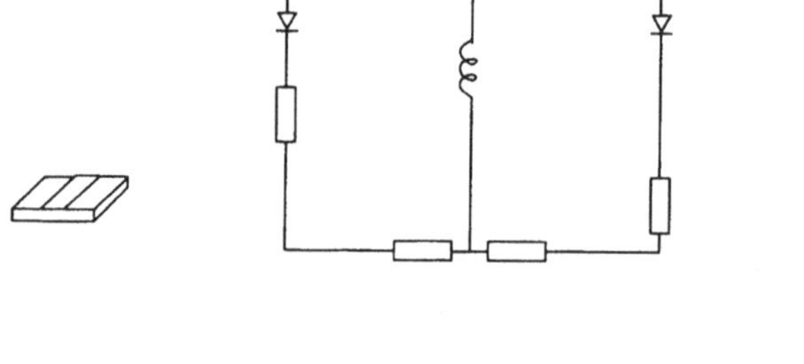

Fig. 11.38 A possible circuit of a phase shifter

Table 11.6 The characteristics of the switching *pin*-diodes made by Salute Corp. [93]

Parameter	Model SKP211404	SKP211501
Band of working frequencies, GHz	1–12	1 ÷ 18
Introduced losses, max., dB	−2	3
Decoupling, min, dB	−45	50
Time of switching, max., ns	50	100
$SWR_{e, inp./out.}$, max.	2	2
Input power, max., W	2.0	1.5
Power voltage, V	–	±5
Current of management, mA		
Open channel	−10	+15
Closed channel	+30	+30
Range of working temperatures, °C	From −60 to +70	From −60 to +70
Overall dimensions, mm	36 × 40 × 11	52 × 46 × 37
Weight, g	50	100

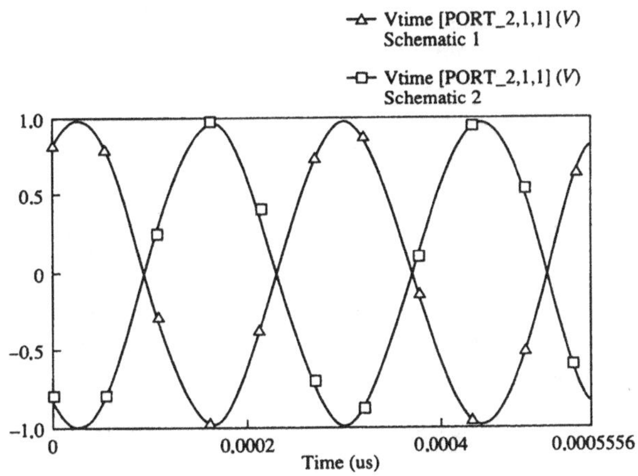

Fig. 11.39 The results of calculations of the target signals of the phase shifter at phase shift by 180°

11.10 Discrete Phaser for PSK PS

For realization of the PM PS function, a phase shifter "0–π" on *pin*-diodes can be used on the output of a standard MCG.

There are various kinds of such phase shifters.

The reflective phase shifter (RPS) is made in the circuit shown in Fig. 11.40. The electric length of the transfer line piece, characterized by the parameter $\beta l (\beta = 2\pi / l$, l being the piece length, λ the wavelength in the transfer line), is selected so that at switching of the *pin*-diode D the phase jump $\Delta\varphi$ on the output of the circuit has the preset value. Thus, the equivalent parameters of the *pin*-diode in these two states are considered.

Table 11.7 The characteristics of phase shifters

Parameter	Model SKFV1201	SKFV2101	S1KFV2102
Range of frequencies, GHz	0.5 − 4.0	4 − 8	8 − 12
Band of working frequencies, GHz	One octave	One octave	40
Categories, deg	11; 22.5; 45; 90; 180		
Accuracy of phase setting, deg	±4 (±6 for 180°)		
Losses on the discharge, max., dB	−1.5		
Time of phase switching, max., n	5	10	10
Input power, max., W	0.1	0,3	0,3
Management voltage, V	−7 ± 1	±3	±3
Current of management on the discharge, max., mA	0.01	15	15
Weight, g	20	20	20
Working temperature range, °C	From −60 to +85		
Input and output of MWF signals	Coaxial (rosette), 3.5/1.52 mm		

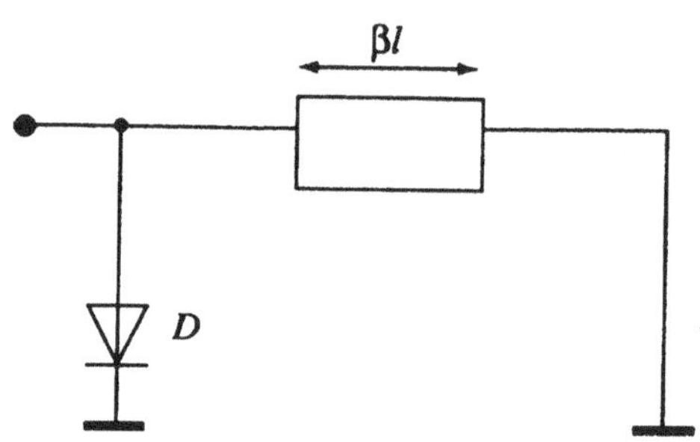

Fig. 11.40 The circuit of the reflective phase shifter

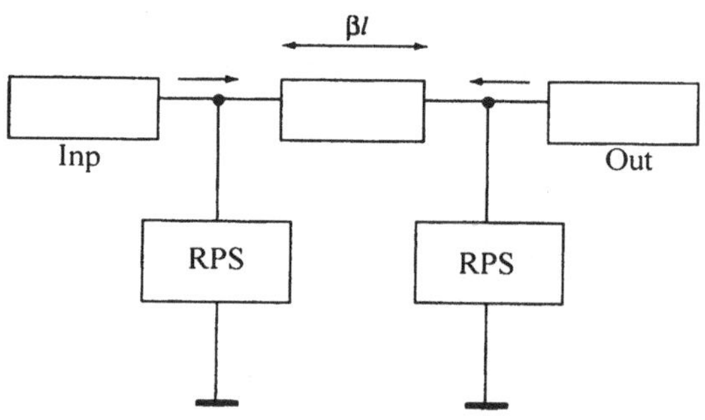

Fig. 11.41 The through-passage phase shifter in the form of a loaded line

The through-passage phase shifter in the form of a loaded line is shown in Fig. 11.41.

In this circuit, two identical RPS connected in the circuit in Fig. 11.40 are placed in the transfer line at a distance βl from each other. Phase shifters of this type cannot be used advantageously at great discrete phase values ($\Delta\varphi > \pi/2$), since the introduced losses will be higher than in other kinds of phase shifters.

A through-passage diode phase shifter on switched pieces of the transfer line (Fig. 11.42) is often used.

The diodes switch the signal passage way, and $\beta l_2 - \beta l_1 = \Delta\varphi$. Such a phase shifter, on the contrary, cannot be advantageously used at small discrete phase values ($\Delta\varphi$).

Through-passage phase shifters of a bridge type are most widespread in MWF paths. They are made by inclusion of two identical reflective phase shifters on switching *pin*-diodes in the mutually decoupled outputs of an MSF bridge (Fig. 11.43).

The basic element determining the parameters of the phase shifter is a *pin*-diode which changes its state at change of the input voltage polarity. Simplified equivalent circuits of the *pin*-diode ignoring the parameters of the case at direct r_+ and return r_- connections are shown in Fig. 11.44.

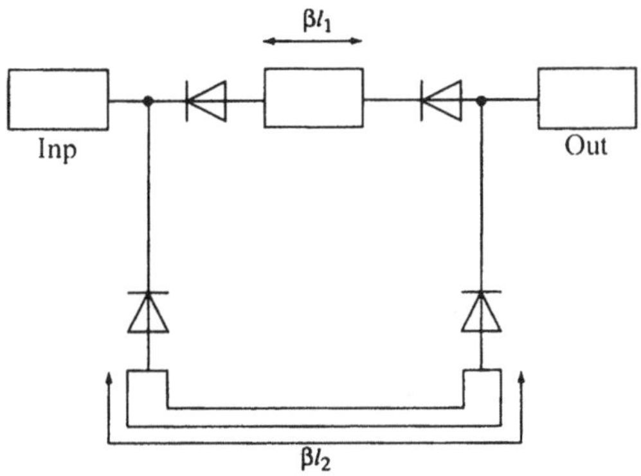

Fig. 11.42 A through-passage diode phase shifter on switched pieces of the transfer line

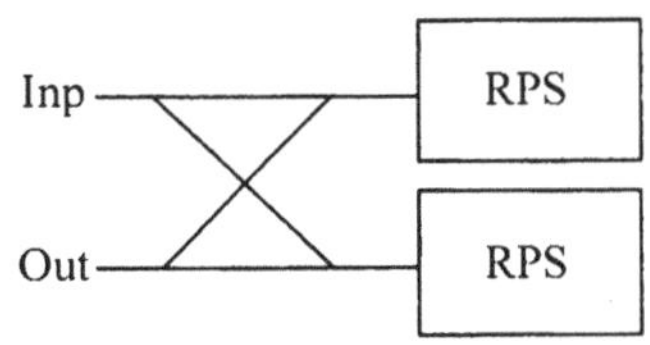

Fig. 11.43 Through-passage phase shifters of a bridge type

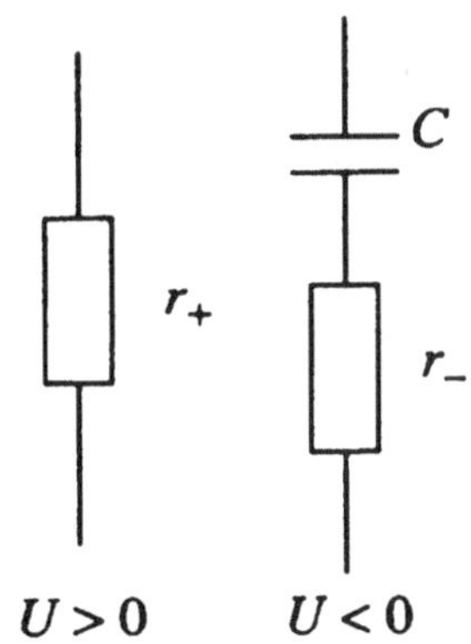

Fig. 11.44 Simplified equivalent circuits of the pin-diode ignoring the parameters of the case at direct r_+ and return r_- connections

The resistances r_+ and r_- are of the same order of magnitude (from tenths to few Ω), depending on the thickness of the diode's base (from 2 to 300 μm). The capacity C for various diodes can vary from 0.1 up to 3 pF. At change of the input voltage polarity the diode's resistance varies from a very low active value, close to short circuit, up to a high resistance of capacity character close to idling. Therefore, the diode included in the transfer line causes small losses of signal L_{los} at transmission (direct displacement) and great losses at blocking (return displacement), i.e., it operates similarly to a switch. The L_{los}–L_{bloc} ratio defines an important parameter K, named as the diode quality

$$K = \frac{\sqrt{L_{\text{bloc}}} - 1}{\sqrt{L_{\text{los}}} - 1}, \tag{11.7}$$

where $L_{\text{bloc}} = 1/\left|S_{21}^{3}\right|$ and $L_{\text{los}} = 1/\left|S_{21}^{\text{los}}\right|$ are the elements of the dispersion matrix of the switch at two states of the diode.

The frequency properties of the diode are estimated by the critical frequency f_{cr}, at which the diode's capacitance is numerically equal to the geometric mean of the active values r_+ and r_-

$$f_{\text{cr}} = \frac{1}{2\pi\, l\, \sqrt{r_+ r_-}}. \tag{11.8}$$

The diode quality K, describing the overall performance of the diode on some working frequency f, is related to the critical frequency with the expression:

$$K = \left(\frac{f_{\text{cr}}}{f}\right)^2. \tag{11.9}$$

Modern *pin*-diodes provide functioning of switches up to frequencies 200 ... 300 GHz.

Another important parameter of *pin*-diodes is the time of switching from one state to another. This time is limited by the value of accumulated charge Q_{acc} in the diode's base at a direct current I_0 traversing it and defined by the expression:

$$Q_{\text{acc}} = I_0 t, \tag{11.10}$$

where τ is the file time of carriers in the base.

The more Q_{acc}, the longer the time of transition from the open state to the closed one Q_{acc} increases with an increase in the thickness of the base. For modern thin-base diodes, Q_{acc} can amount to few ns, which provides the time of switching of few nC.

The power of switching devices is limited by the two parameters, namely: the maximum admissible power of dissipation on the diode (at direct current) and the break-down voltage of the *pin*-diode (at return displacement). The dissipated power can be from 0.5 up to 100 W, depending on the thickness of the base, the diode's volume, its design, and efficiency of its heat-conducting path. The break-down voltage is determined by the thickness of the base: the thicker the base, the higher the break-down voltage, but the lesser its speed of response (longer time of switching).

The requirements of a high response speed and a high power in *pin*-diodes are contradictory. With reduction of the thickness of the base, the lifetime and the accumulated charge of carriers decrease, the speed of response raises, but the break-down voltage of the base decreases. In Table 11.8, the parameters of some *pin*-diodes and in Table 11.9, those of some Russian ones [94] are presented.

The phase discrete 180° can be provided by a circuit on switched pieces of the transfer line. Such a phase shifter has been simulated on a computer by Microwave

Table 11.8 The parameters of some *pin*-diodes

Parameter	MA-47890	MA-47895	MA-47198	MA-47152	MA-47156
Thickness of base, μm	150	50	25	12	2
Diameter, mm	2.3	0.65	0.25	0.12	0.05
C, pF	3.0	0.7	0.2	0.1	0.1
r_+, Ω	0.2	0.4	0.9	1.0	1.0
r_-, Ω	0.2	0.6	2.0	4.0	3.0
f_{cr}, GHz	250	350	600	800	800
τ, μs	15.0	2.0	0.8	0.2	0.02
Direct current I_0, mA	250	50	25	25	25
Inverse voltage U_{inv}, V	200	50	25	10	10
Break-down voltage U_{br}, V	1,800	600	300	150	30
Thermal resistance, °C/W	1.5	7.0	15	30.0	40.0

Table 11.9 The parameters of some Russian *pin*-diodes

	Type of diode					
Parameter	KA 536	KA 528A-4	KA 537	KA 507	KA 536A-5	KA 543A-5
C, pF	3–4.5	1.8–2.4	3	0.8–1.2	0.08–0.16	0.12–0.19
f_{cr}, GHz	100	200	200	200	300	300
r_+, Ω	0,5	0.5	0.5	1.5	1.5	1.5
Q_{acc}, nC	500–2,500	900	1,000	200	150	0.5–3.0
Direct current I_0, mA	100	100	100	100	100	3
Inverse voltage U_{inv}, V	100	100	100	100	100	20
Break-down voltage U_{br}, V	1,000	1,000	600	300	300	100
Dissipation power $P_{dis.max}$, W	100	50	20	5	1	0.5

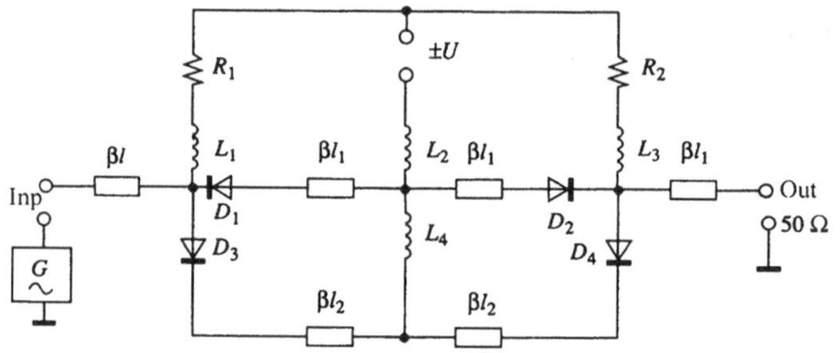

Fig. 11.45 The circuit of the phase shifter in microstrip design

Office 2002, *ver.* 5.51. The basic circuit of the phase shifter in microstrip design is shown in Fig. 11.45.

In this circuit, βl, βl_1, βl_2 are the electric lengths of pieces of a 50 Ω microstrip transfer lines. The lengths of pieces l_1 and l_2 are chosen so that

$$2\beta l_2 - 2\beta l_1 = \frac{\lambda}{2}. \tag{11.11}$$

The length βl can be any. The resistors R_1 and R_2 act as ballast resistances. The operating voltage source $\pm U$ switches the status of *pin*-diodes D_1–D_4. On the input of the circuit a high-frequency signal from the generator G is applied, on the output the signal is connected to a 50 Ω loading. At positive polarity of the operating pressure the diodes D_1 and D_2 are closed, and D_3 and D_4 are opened, and the signal passes from the input to the output of the circuit through the pieces of lines βl_2. At change of the voltage polarity, the diodes D_3 and D_4 are closed, and diodes D_1 and D_2 are open, and the signal goes through the pieces of the transfer lines βl_1. Thus, the output signal phase shifts by π.

At modeling, the *pin*-diodes were replaced in conformity with the circuits for direct and return connection shown in Fig. 11.44. The parameters of these circuits ($r_+ = 0.9\,\Omega; r_- = 2\,\Omega; C = 0.2\,\text{pF}$) are close to those of MA-47198 or KA536A-5 diodes. Specific circuits corresponding to different polarities of the operating voltage are shown in Fig. 11.46 – for a zero phase state and in Fig. 11.47 – for a π phase state.

At the chosen lengths of pieces l_1 and l_2 the exact value of phase shift $\Delta\varphi = \pi$ corresponded to a frequency of 3.85 GHz. The output signal voltage on this frequency for two states of the phase shifter is shown in Fig. 11.48.

At tuning out from the given frequency, the phase shift changed as shown in Fig. 11.49.

The phase jump at the central frequency from +180° down to −180° is explained by Microwave Office 2002 producing values of trigonometrical functions within the limits of $-\pi$ to $+\pi$ only. According to Fig. 11.49, the working frequency band of the phase shifter determined from $\Delta\varphi = 180° \pm 10\%$ is 7%.

In Fig. 11.50, the dependencies of the output–input power ratio which characterizes the losses in the phase shifter for $\varphi = 0$ and $\varphi = \pi$ are shown. The losses

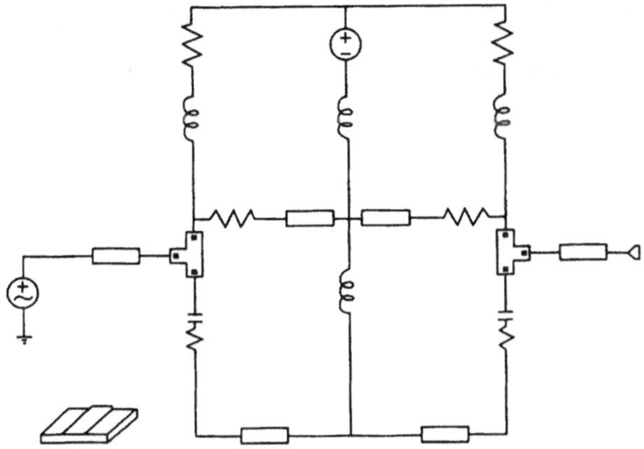

Fig. 11.46 The circuit corresponding to the operating voltage for a zero phase state

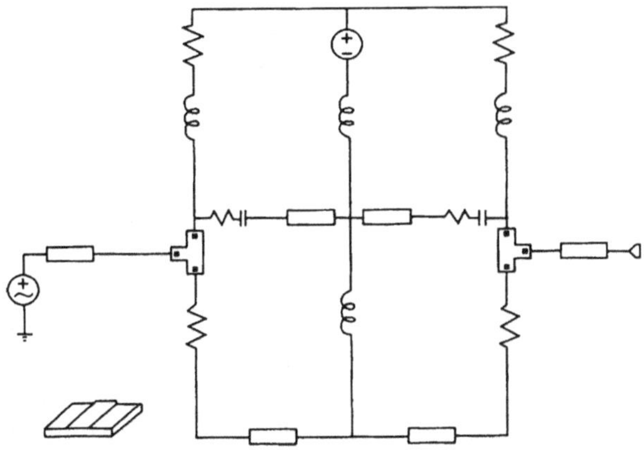

Fig. 11.47 The circuit corresponding to the operating voltage for a π phase state

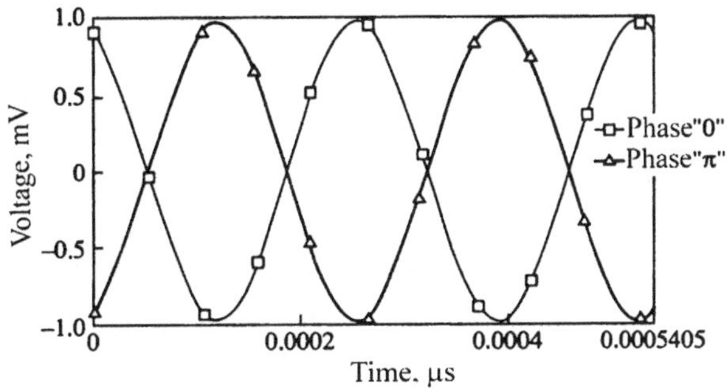

Fig. 11.48 The output signal voltage on a frequency 3.85 GHz for two states of the phase shifter

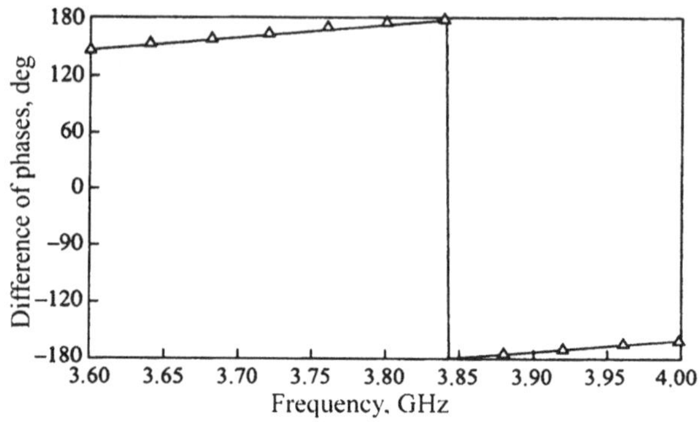

Fig. 11.49 The change of the phase shift at tuning out from the frequency 3.85 GHz

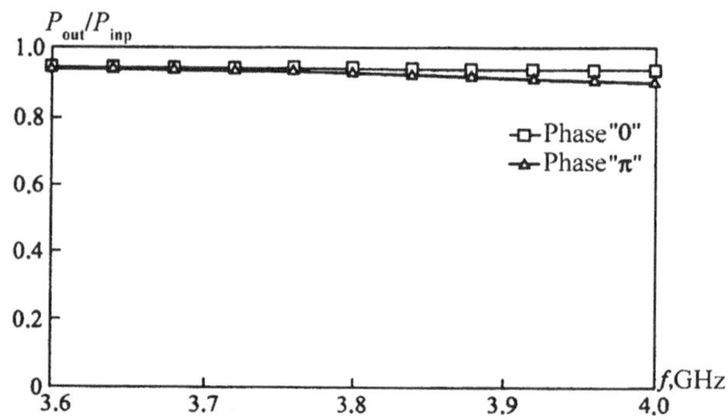

Fig. 11.50 The dependencies of the output–input power ratio which characterizes the losses in the phase shifter for $\varphi = 0$ and $\varphi = \pi$

on the central frequency are 0.34 dB, the maximal losses in the working frequency band do not exceed 0.45 dB.

Microwave Office 2002 could not determine the time of switching of the phase shifter from one state in another one, and so this parameter has been estimated approximately from two transients – transition of the diodes from their closed state into the opened one and vice versa – the latter being always longer as it needs removal of the accumulated charge from the base, which is defined by

$$Q_{acc} = I_{dir}\tau, \tag{11.12}$$

where I_{dir} is the direct current through the diode and τ is the lifetime of carriers in the base.

Reference books report one of the parameters Q_{acc} or τ.

At applying of the voltage U_{inv} the accumulated charge is extended from the base by the inverse current I_{inv} with its maximal value

$$I_{inv} = \frac{U_{inv}}{R'}, \tag{11.13}$$

where R' is the Ohmic resistance of the circuit at the initial moment of application of U_{inv} when the resistance of the base is still low enough.

An approximate estimation of the time of switching is

$$t_{sw} = \frac{Q_{acc}}{I_{inv}} = \frac{Q_{acc} \cdot R'}{U_{inv}}. \tag{11.14}$$

In the considered model, $U_{inv} = 3\,\text{V}$ and $R' = 100\,\Omega$. For a MA-47898 diode, the lifetime is $\tau = 0.8\,\mu\text{s}$, $I_{dir} = 25\,\text{mA}$, and $Q_{acc} = 20\,\text{nC}$; hence, the time of switching is $t_{sw} \cong 0.7\mu\text{s}$. For a KA543A-5 diode, $Q_{acc} = 3$ ns and $t_{sw} \cong 0.1\,\mu\text{s}$.

For reduction of the time of transition of *pin*-diodes from their open state into closed one it is necessary to increase the inverse current I_{inv}, that is to increase the voltage U_{inv}, as it is applied at designing of switching devices. The time of switching of modern low-power switching devices based on *pin*-diodes is few ns, and that of more powerful ones is few or tens μs.

Different states of *pin*-diodes in the circuits of modeling shown in Figs. 11.46 and 11.47 are provided by switching of the polarity of one of the power sources. In real designs for optimization of their direct current modes and maintenance of the necessary speed of switching, separate sources of direct current and return displacement are used as the voltage of these sources in an optimum variant should be different. These sources are switched by means of special switching circuits operated by logic 0 and 1 signals of a standard level. One of the possible elementary circuits of management of *pin*-diodes is shown in Fig. 11.51.

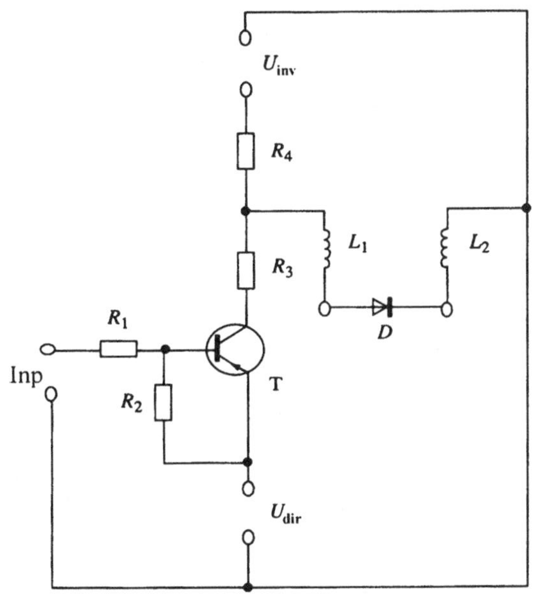

Fig. 11.51 The circuits of management of pin-diodes

Table 11.10 The parameters of multipurpose frequency synthesizers of a new generation based on HMT

Subrange	Equidistance (step of reorganization)	Spectral linewidth	RATE of frequency reorganization	Transient duration at frequency reorganization	Stability of frequency	Note
$30 \div 300$ MHz	Minimum 0.3 kHz (a theoretical estimation)	Not calculated	<10 ms per octave (a calculated value)	20–60 μs (a theoretical estimation)	$10^{-7} \div 10^{-9}$ depending on the type of the quartz basic generator and ambient temperature range	The VCM synthesizer has no clear advantages over digital synthesizers
$0.3 \div 3.0$ GHz	Minimum 0.3 kHz (a theoretical estimation)	1 kHz on a level 100 dB (theoretical calculation)	<10 ms per octave	2–6 μs	$10^{-7} \div 10^{-9}$ depending on the type of the quartz basic generator and ambient temperature range	Advantages of VCM synthesizers are shown most brightly
$3.0 \div 30$ GHz	Minimum 0.3 kHz (a theoretical estimation)	10 kHz on a level 100 dB (theoretical calculation)	<10 ms per octave	1–2 μs	$10^{-7} \div 10^{-9}$ depending on the type of the quartz basic generator and ambient temperature range	Advantages of VCM synthesizers are shown most brightly
$30 \div 100$ GHz	Minimum 0.3 MHz	Not calculated	<10 ms per octave	Not calculated	$10^{-7} \div 10^{-9}$ depending on the type of the quartz basic generator and ambient temperature range	Additional theoretical and experimental researches of transistors for this range are required

"0" and "1" signals are applied to the input of the circuit and change the transistor T into its open or close state. *in*-diode D is included in the circuit through the uncoupling inductances L_1 and L_2. At an opened transistor, a direct current traverses the diode D from the source U_{dir}; at a closed transistor, the inverse voltage U_{inv} is applied to the diode. Such a circuit of management provides optimum characteristics of the device.

Reflective phase shifters in a waveguide variant for the 8 and 3 mm ranges and the pulse power up to 5 kW are developed and experimentally investigated at a relative pulse duration of 1,000 with the time of switching 0.5–1.0 μs. At phase discrete values $\Delta\varphi \leq 60°$, the losses in the 8 mm range are below 1.2 dB, and in the 3 mm range below 2 dB. The working frequency band in the 8 mm range is about 10%, and in the 3 mm range it is of the order of 0.5% of the central frequency.

Thus, the circuits of phase shifters considered in this section allow implementation of PM PS in MCG. A specific circuitry realization of a phase shifter will be determined by the technical requirements to a frequency synthesizer: the frequency range, target power, etc.

11.11 Frequency Synthesizers on Generative Magnetotransistors

The key parameters which can be reached in multipurpose frequency synthesizers of a new generation based on HMT are presented in Table11.10.

Thus, the physical principles of functioning of a new type of multipurpose magnetoelectronic frequency synthesizers (MMFF) whose key element (elements) are electrically operated heteromagnetic generators are considered.

Chapter 12
Vector Sensors and Magnetometers with Heteromagnetic Interaction

12.1 Investigations of Properties of Double-Coil Coupling Elements

With the purpose of finding ways of effective interaction of a FMCR in the form of sphere with the conductors of double-coil elements, various schemes of MECE connection were calculated.

In Fig. 12.1, a scheme of MECE connection with grounded conductors (n_3 and n_4) through capacities $C = 100\,\text{pF}$ is presented.

In Fig. 12.2, there is a family of calculated AFC for several values of the capacities C, namely, 1 – 100 pF, 2 – 5 pF, 3 – 2 pF, 4 – 0.5 pF. With an increase in the capacity, microwave power transfer through MECE improves. The minimal losses of transfer are reached when $C > 5\,\text{pF}$.

In Fig. 12.3, a MECE with grounded conductors (n_3 and n_4) through the resistance R is presented.

Figure 12.4 presents calculated AFC for MECE with various connections of the conductors n_3 and n_4, namely, 1 – short circuit, 2 – a coordinated loading of 50 Ω, 3 – idling.

The maximal transfer losses are reached at short circuit. In the mode of a coordinated loading, the transfer losses worsen by 4 dB. In the mode of idling, there is no transfer of signal via MECE ($V_p = 0$) within the whole range of frequencies.

In Fig. 12.5, a scheme of MECE connection with grounded conductors n_3 and n_4 is presented.

In Fig. 12.6, calculated AFC for various factors of MECE coil filling of a FMCR in the form of sphere with a 0.7-mm diameter at several diameters of excitation coils D: 1–0.7 mm, 2–1.4 mm, 3–2.5 mm is shown.

The minimum level of transfer losses and the maximum decoupling level are reached at the maximal factor of ferrite coil filling – curve 1. By reduction of the factor of filling, the transfer losses increase and the decoupling in MECE decreases.

The dependences of the transfer losses of signal through MECE at the resonant frequency on the diameter of the conductors were investigated as well. At the diameter of the conductors within 10–600 μm, the transfer losses changed slightly. When choosing the diameters of MECE conductors, it is necessary to consider the wave resistance of the bringing lines for decreasing the level of input SWR.

A.A. Ignatiev and A.V. Lyashenko, *Heteromagnetic Microelectronics: Microsystems of Active Type*, DOI 10.1007/978-1-4419-6002-3_12,

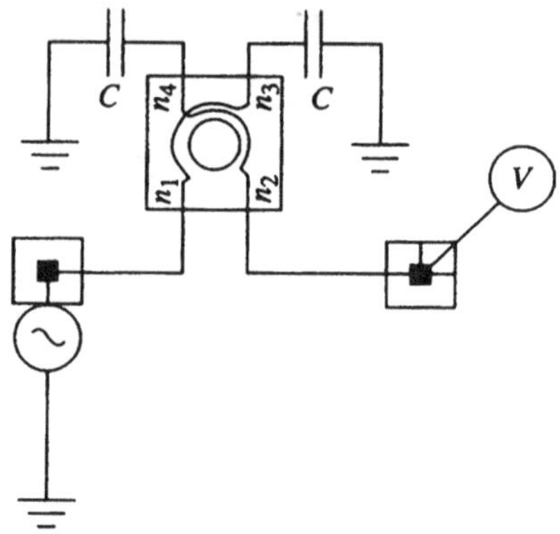

Fig. 12.1 A scheme of MECE connection with grounded conductors (n_3 and n_4) through capacities $C = 100\,\text{pF}$

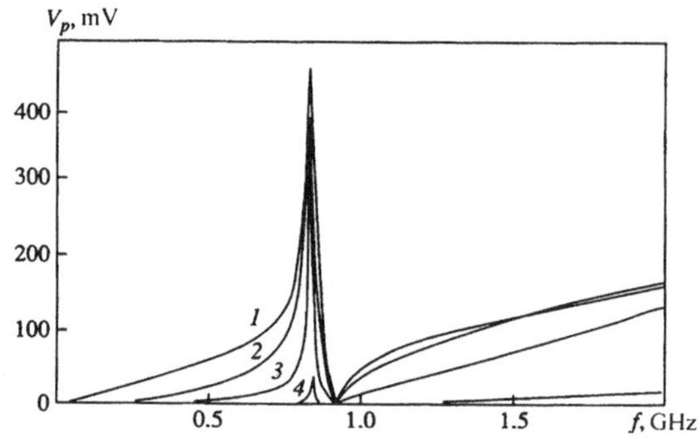

Fig. 12.2 A family of calculated AFC for several values of the capacities C, namely, 1 – 100 pF, 2 – 5 pF, 3 – 2 pF, and 4 – 0.5 pF

Fig. 12.3 A scheme of MECE connection with grounded conductors (n_3 and n_4) through the resistance R

Figure 12.7 presents experimental AFC of MECE on crossed-coupling coils with FMCR on the basis of a KG-12 sphere ($4\pi M_s = 140$ G) for several values of the central frequencies f and magnetic induction B : $1 - f = 656\,\text{MHz}$, $B = 234\,\text{G}$; $2 - f = 734\,\text{MHz}$, $B = 261\,\text{G}$; $3 - f = 820\,\text{MHz}$, $B = 296\,\text{G}$; $4 - f = 936\,\text{MHz}$, $B = 339\,\text{G}$; $5 - f = 1{,}049\,\text{MHz}$, $B = 380\,\text{G}$; $6 - f = 1{,}146\,\text{MHz}$, $B = 408\,\text{G}$.

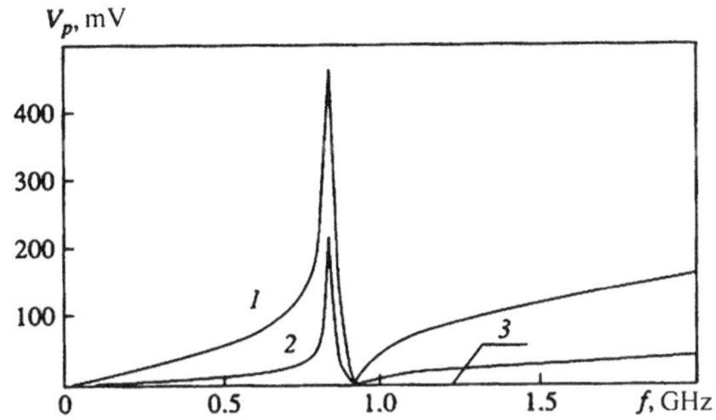

Fig. 12.4 Calculated AFC for MECE with various connections of the conductors n_3 and n_4, namely, 1 – short circuit, 2 – a coordinated loading of 50 Ω and 3 – idling

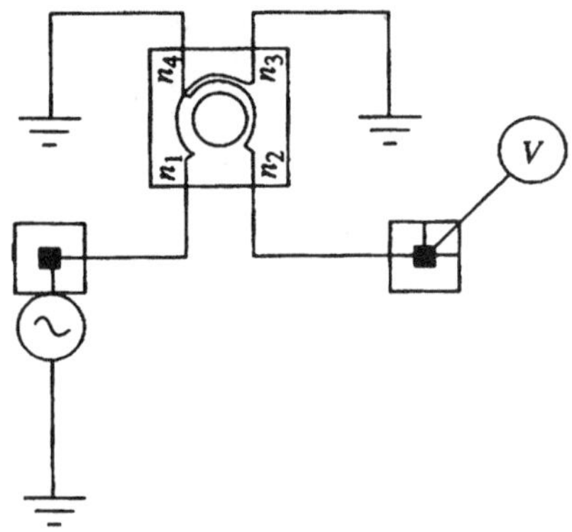

Fig. 12.5 A scheme of MECE connection with grounded conductors n_3 and n_4

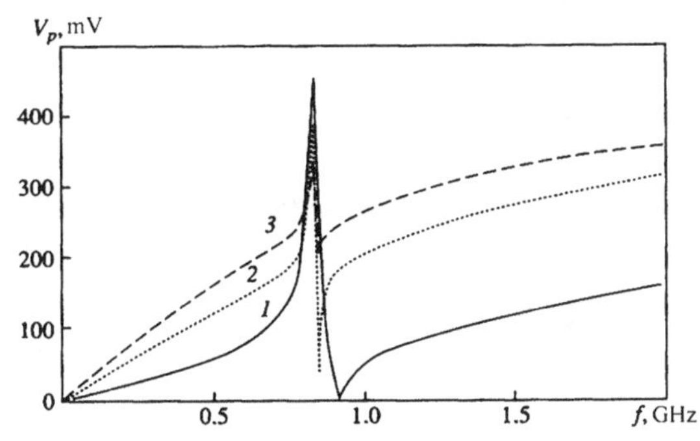

Fig. 12.6 Calculated AFC for various factors of MECE coil filling of a FMCR in the form of sphere with a 0.7-mm diameter at several diameters of excitation coils D: 1 – 0.7 mm, 2 – 1.4 mm, and 3 – 2.5 mm

Figure 12.8 presents calculated AFC of the above considered MECE for appropriate experimental values of bias field: 1 – f = 660 MHz, B = 234 G; 2 – f = 730 MHz, B = 261 G; 3 – f = 830 MHz, B = 296 G; 4 – f = 950 MHz, B = 339 G; 5 – f = 1,060 MHz, B = 380 G; 6 – f = 1,140 MHz, B = 408 G.

The results of our theoretical calculation agree with experimental data, which speaks for the correctness of the used MECE model with crossed-coupling coils and FMCR in the form of sphere.

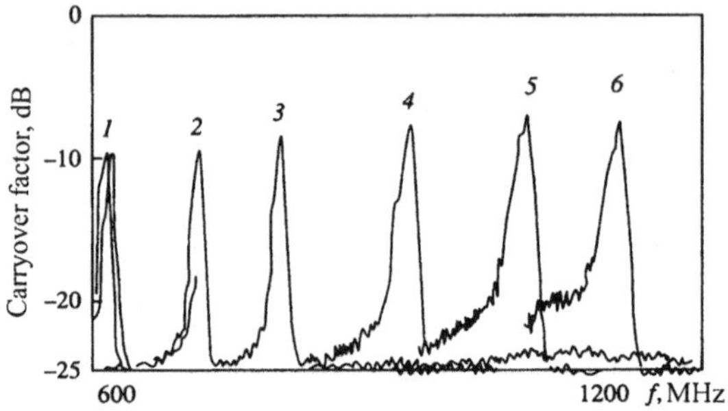

Fig. 12.7 Experimental AFC of MECE on crossed coupling coils with FMCR on the basis of a KG-12™ sphere ($4\pi M_s = 140\,\text{G}$) for several values of the central frequencies f and magnetic induction B: 1 – $f = 656\,\text{MHz}$, $B = 234\,\text{G}$; 2 – $f = 734\,\text{MHz}$, $B = 261\,\text{G}$; 3 – $f = 820\,\text{MHz}$, $B = 296\,\text{G}$; 4 – $f = 936\,\text{MHz}$, $B = 339\,\text{G}$; 5 – $f = 1{,}049\,\text{MHz}$, $B = 380\,\text{G}$; and 6 – $f = 1{,}146\,\text{MHz}$, $B = 408\,\text{G}$

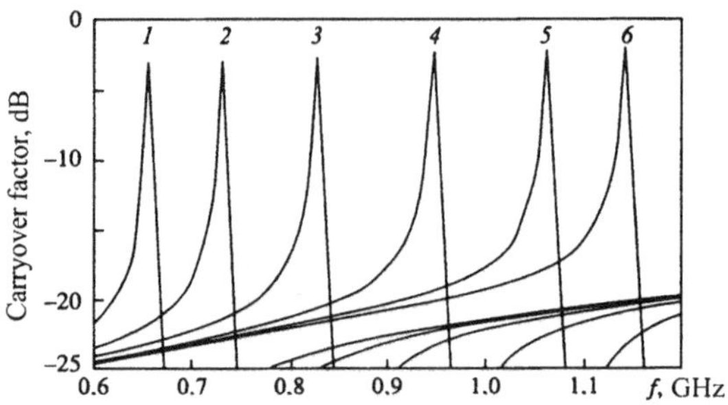

Fig. 12.8 Calculated AFC of MECE for appropriate experimental values of bias field: 1 – $f = 660\,\text{MHz}$, $B = 234\,\text{G}$; 2 – $f = 730\,\text{MHz}$, $B = 261\,\text{G}$; 3 – $f = 830\,\text{MHz}$, $B = 296\,\text{G}$; 4 – $f = 950\,\text{MHz}$, $B = 339\,\text{G}$; 5 – $f = 1{,}060\,\text{MHz}$, $B = 380\,\text{G}$; and 6 – $f = 1{,}140\,\text{MHz}$, $B = 408\,\text{G}$

12.2 Magnetosensitive Active Oscillator

In a magnetosensitive oscillator with a YIG sphere, the frequency of generation changes under the action of the magnetic induction vector B (field). Therefore, such a device can be used as a high-sensitivity vector sensor for $\overline{B}$ and its variable component. For the microwave range, circuits of oscillators with a common base on a bipolar transistor or with a common shutter on a field transistor, as a rule, are used. Thus, as the oscillatory system in the feedback circuit, MECE of this or that design with FMCR in the form of a sphere or a film is applied. The equivalent circuit of such an oscillator on a bipolar transistor is shown in Fig. 12.9.

The inductance L_1 included in the circuit of the base is an element of feedback. The loading R_{load} is connected through a matching circuit.

For realization of generation, the sum of the full conductivities of MECE with FMCR Y_r and of the transistor with an element of feedback Y_t should be equal to zero

$$Y_r + Y_t = 0. \tag{12.1}$$

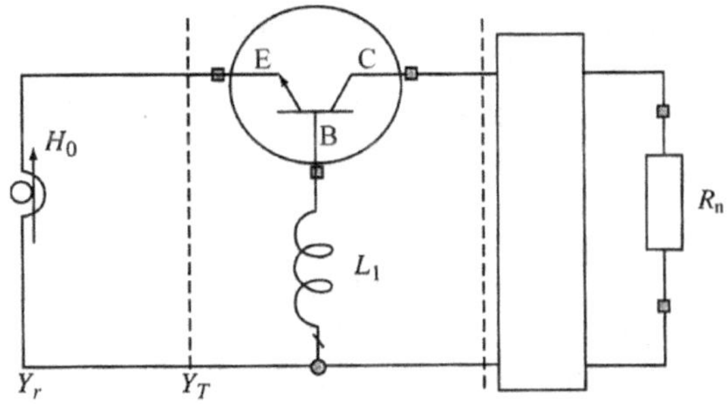

Fig. 12.9 The equivalent circuit of an oscillator on a bipolar transistor

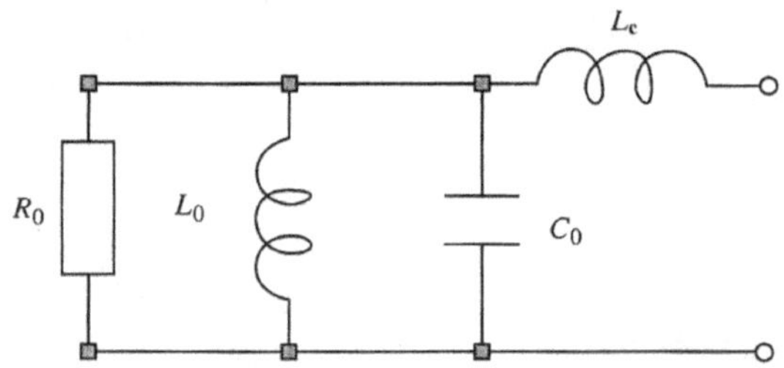

Fig. 12.10 The equivalent circuit of FMCR in the form of a parallel oscillatory contour connected in series with the inductance L_c formed by a coupling coil

Hence, both the sums of the active and reactive components of these conductivities must be equal to zero

$$G_r(\omega) + G_t(\omega, U_{hf}) = 0, \tag{12.2}$$

$$B_r(\omega) + B_t(\omega, U_{hf}) = 0, \tag{12.3}$$

where U_{hf} is the high-frequency component of the input voltage of the transistor.

Equation (12.2) defines the condition of self-excitation and the amplitude of established oscillations; Equation (12.3) does the frequency of oscillations.

To solve (12.2) and (12.3), it is necessary to have the equivalent circuits of FMCR and the transistor. The equivalent circuit of FMCR represents a parallel oscillatory contour connected in series with the inductance L_c formed by a coupling coil, as shown in Fig. 12.10, where L_0 and C_0 are the reactive components of the contour, R_0 its active component, and L_c is the inductance of the coupling coil with FMCR. The conductivity of this circuit is determined through the full resistance $Z_r = 1/Y_r$ by the following expression

$$Z_r = j\omega L_c + \frac{j\left(\omega \cdot \dfrac{\omega_0}{Q_{unl}}\right) R_0}{\omega_0^2 - \omega^2 + j\left(\omega \cdot \dfrac{\omega_0}{Q_{unl}}\right)}, \tag{12.4}$$

where $\omega = 2\pi f$ is the current circular frequency and $\omega_0 = 2\pi f_0$ is the resonant circular frequency.

The resonant frequency is

$$f_0 = \gamma H_0, \tag{12.5}$$

where H_0 is the external magnetic field and $\gamma = 2.8\,\mathrm{MHz/G}$ is the gyromagnetic electron ratio.

The active resistance R_0 in the circuit in Fig. 12.10 is defined by the expression

$$R_0 = \mu_0 V K^2 \omega_\mathrm{m} Q_\mathrm{unl}, \tag{12.6}$$

where $\mu_0 = 4\pi \times 10^{-7}\,\mathrm{H/m}$; $V = (1/6)\pi d^3$ is the volume of the YIG sphere; d its diameter; $K = 1/d_1$ the coupling factor; d_1 the diameter of the coupling coil covering the sphere; $\omega_\mathrm{m} = (2\pi\gamma \times 4\pi M_\mathrm{s})$ – the characteristic frequency; $4\pi M_\mathrm{s}$ the saturation magnetization; and Q_unl is the unloaded GB product of the YIG resonator determined by the resonant linewidth ΔH according to the formula:

$$Q_\mathrm{unl} = \frac{H_0 - \frac{1}{3}4\pi M_\mathrm{s}}{\Delta H}. \tag{12.7}$$

The inductance L_0 and capacity C_0 of the contour in Fig. 12.10 are defined by the expressions

$$L_0 = \frac{R_0}{\omega_0 Q_\mathrm{unl}}, \tag{12.8}$$

$$C_0 = \frac{1}{\omega_0^2 L_0}. \tag{12.9}$$

Thus, for calculation of the equivalent parameters of FMCR, it is necessary to know the diameter of the YIG sphere, its saturation magnetization, and the FMR linewidth. The inductance of the coupling coil L_c can be calculated approximately as

$$L_\mathrm{c} = \mu_0 b \left(\mathrm{ln}\frac{8b}{a} - \frac{7}{4} \right), \tag{12.10}$$

where b is the radius of the coil and a is the diameter of the wire.

For determination of the total conductivity of the oscillator, it is necessary to use an equivalent circuit of the transistor. There are some variants of such equivalent circuits of a varying degree of complexity. Some of them contain more than 30 elements (for example, Gummel–Poon's model). A simple enough equivalent circuit of the transistor reflecting its work on high frequencies is presented in Fig. 12.11, where r_e, r_b, and r_c are the resistance of the emitter, base, and collector, respectively; C_e, C_b, andC_c the capacities of the emitter, base, and collector, respectively; and C_eb is the capacity of the emitter–base junction,C_cb the capacity of the collector–base junction, and C_ec is the capacity of the emitter–collector junction.

For use of the equivalent circuits of the transistor, it is necessary to have data about all their elements. If these data are not known or known incompletely, it is necessary to make additional measurements.

For calculation of generating circuits by the method of S parameters, reference data of the transistor and MECE or parameters measured on special devices are necessary. Now there are computer programs to calculate the generating mode of

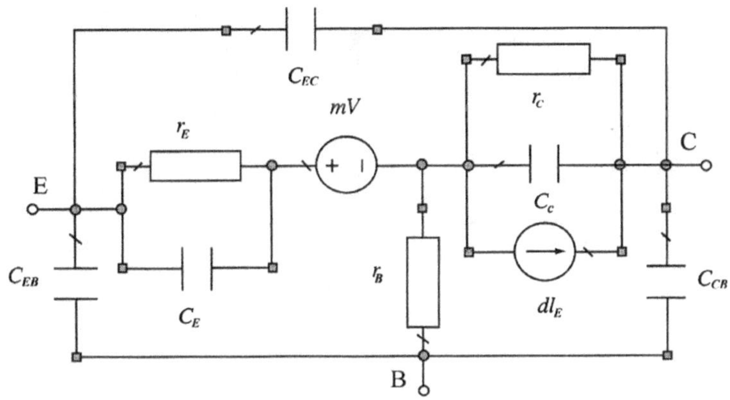

Fig. 12.11 The equivalent circuit of the transistor

Table 12.1 The equivalent parameters of the resonator in a range of frequencies 0.98–2.10 GHz

H_0, Oe	f_0, MHz	Q_{unl}	R_0, Ω	L_0, nH	C_0, pF
350	980	375	182.6	0.08	330
400	1,120	500	243.5	0.07	288
500	1,400	750	365.2	0.056	231
600	1,680	1,000	487	0.046	195
700	1,960	1,250	608	0.04	165
750	2,100	1,375	670	0.037	155

transistor circuits (Serenade 8.0, Microwave Office 2002, etc.). These programs contain reference data of a number of commercial transistors. For modeling of circuits on these transistors, their equivalent parameters are not needed as they are already in the program. As to the YIG resonator, Serenade 8.0 contains it as an element of circuits and Microwave Office 2002 lacks it, and it is necessary to enter the equivalent parameters of the YIG resonator calculated by (12.5)–(12.10).

As the YIG resonator, a ferrite sphere with a diameter of 0.6 mm with its saturation magnetization $4\pi M_s = 600$ G is taken and its resonant linewidth is 0.4 Oe. The coupling coil had a diameter of 0.7 mm and was made of a conductor with a diameter of 0.1 mm. By (12.5)–(12.10), the equivalent parameters of the resonator in a range of frequencies 0.98–2.10 GHz have been calculated. Table 12.1 summarizes the results of these calculations.

In Figs. 12.12 and 12.13, the equivalent circuits (a) and the results of calculation of the transfer factor of MECE with FMCR (b) for a field $H_0 = 500$ are shown for comparison.

The basic circuit of a vector magnetosensitive oscillator with FMCR is shown in Fig. 12.14.

For calculations, the value of inductance L_1 of the base circuit and the coordination elements on the output were selected. The criteria were generation in a preset

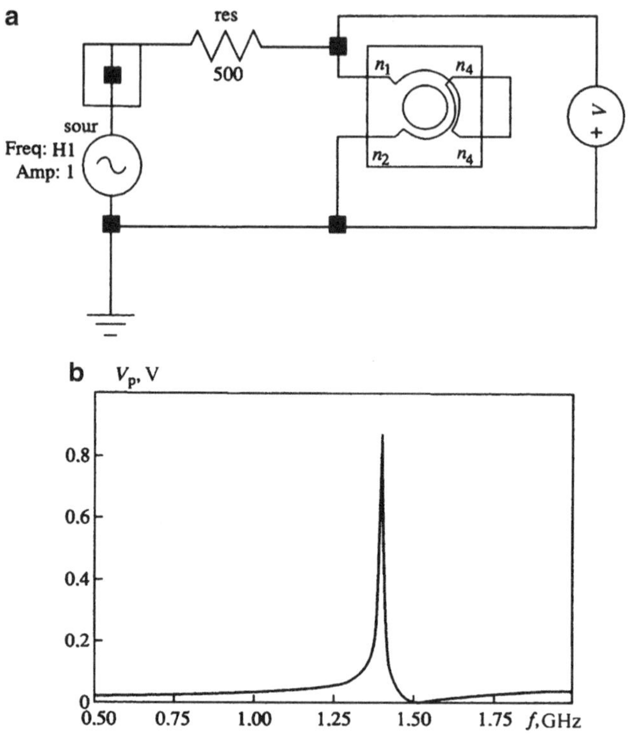

Fig. 12.12 The equivalent circuit (**a**) and the results of calculation of the transfer factor of MECE with FMCR (**b**) for a field $H_0 = 500$

range, the output power value of generation, and attenuation of the harmonic of a signal on the frequency f_1. Spectral characteristics of the oscillator in the range of frequencies of the basic signal f_0 are presented in Fig. 12.15.

In the middle part of the range, the output power was ~10 mW. In the lower part of the range, power reduction down to 6–8 mW due to reduction of the GB product of the YIG resonator was observed. In the upper part of the range, the output power also decreased due to the decrease in the amplification factor of the transistor. The nonuniformity of AFC in the set range did not exceed 3–4 dB.

In Fig. 12.16, the spectral components for the basic frequency ν_0 and harmonics ν_1 and ν_2 of the magnetosensitive oscillator are shown for one frequency within their range.

Attenuation of the harmonic ν_1 over the whole range of frequencies is not less than −18 dB and that of the harmonic ν_2 is not less than 36 dB.

Figure 12.17 shows the basic circuit of the magnetosensitive oscillator for a frequency $f_0 \approx 1.5$ GHz at $H_0 = 500$ Oe, simulated by Microwave Office 2002.

In Fig. 12.18, the spectrum of frequencies f_0, f_1, and f_2 of signal generation in the magnetosensitive oscillator is shown.

Comparison of Figs. 12.16 and 12.18 shows that the calculations done for the circuits in Figs. 12.14 and 12.17 yield close results.

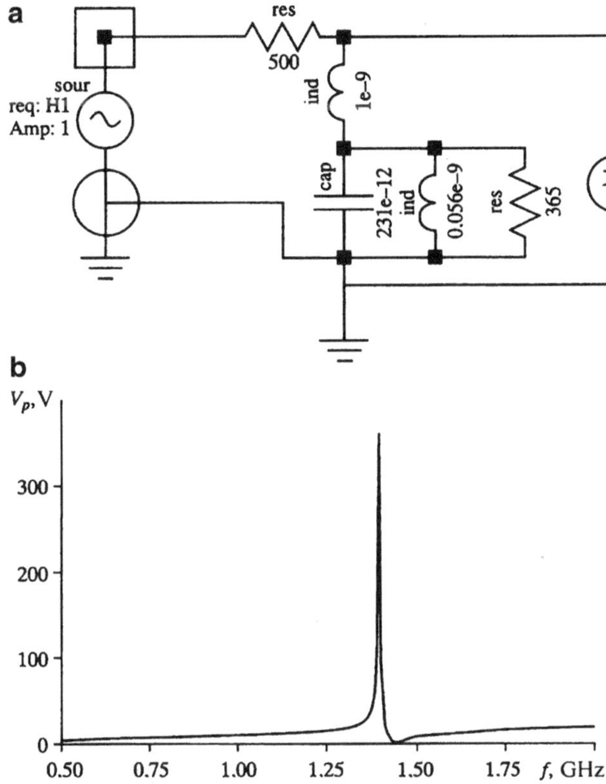

Fig. 12.13 The equivalent circuit (**a**) and the results of calculation of the transfer factor of MECE with FMCR (**b**) for a field $H_0 = 500$

12.3 Projection Element of Magnetosensitive Sensor

The frequency of oscillations is the most exact parameter to determine any physical quantity by means of indirect measurements. Changes of the frequency are related by means of various ways and devices with changes of the given physical quantity. This is caused by the existence of some frequency standards, comparison with which allows considering frequency measurements to be most exact. Proceeding from this, the radiofrequency method of determination of the power function of magnetic field vector (magnetic induction or magnetic field strength) is justified. In a magnetosensitive oscillator, the frequency of oscillations of output signal changes with changes of the magnetic field vector, and this change at small deviations of the latter should be suitable for subsequent transformation to the measured physical quantity in recalculation devices. For frequency resolution, resonant methods are used, allowing determining the frequency of oscillations with an accuracy comparable to quantum frequency standards. The GB product of the oscillatory system of the oscillator influences the accuracy of magnetic field determination at a given place.

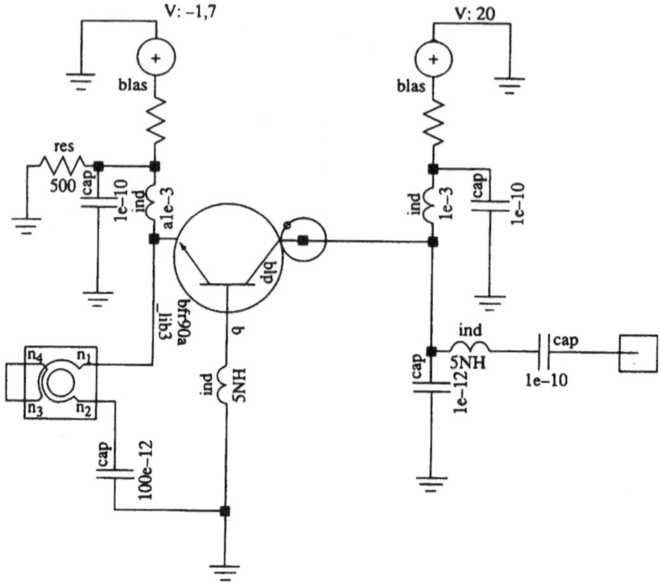

Fig. 12.14 The circuit of a vector magnetosensitive oscillator with FMCR

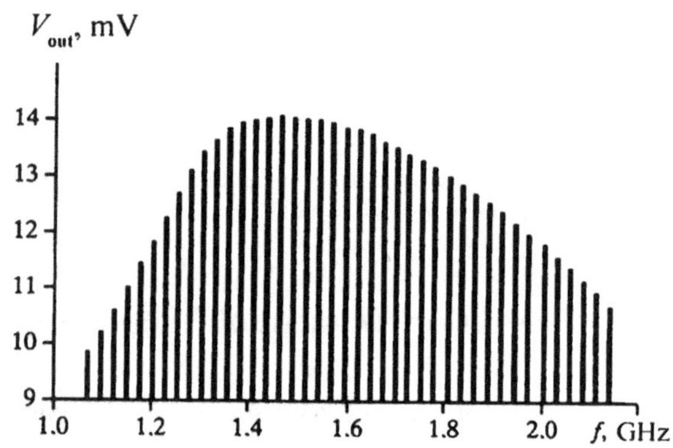

Fig. 12.15 Spectral characteristics of the oscillator in the range of frequencies of the basic signal f_0

The oscillator, whose frequency of oscillations is a “vector”-magnetodependent quantity, can be designed with use as the basic oscillatory system of the resonator on FMR in the microwave-range of lengths of waves where good quality can be very high (hundreds and thousand) at use of monocrystal ferrite for which with a high degree of accuracy simple enough mathematical dependences of the central frequency of a resonance on a magnetic field can be practically realized. So, for ferrite monocrystals in the form of spheres, the dependence $f_{res} = F(B)$ is linear

$$f_{res} = K \cdot B, \tag{12.11}$$

where K is a factor of proportionality.

Fig. 12.16 The spectral components for the basic frequency ν_0 and harmonics ν_1 and ν_2 of the magnetosensitive oscillator

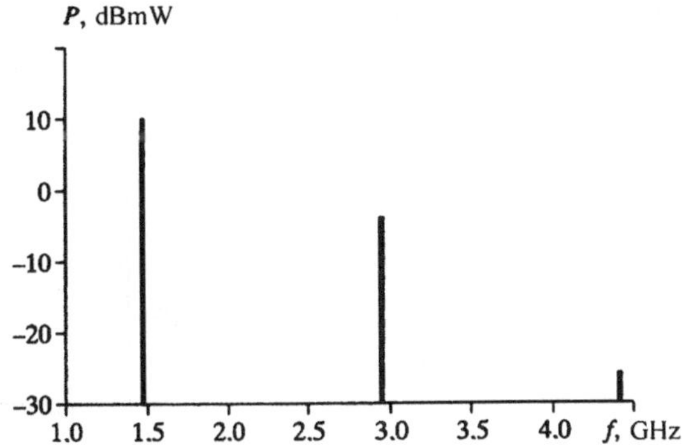

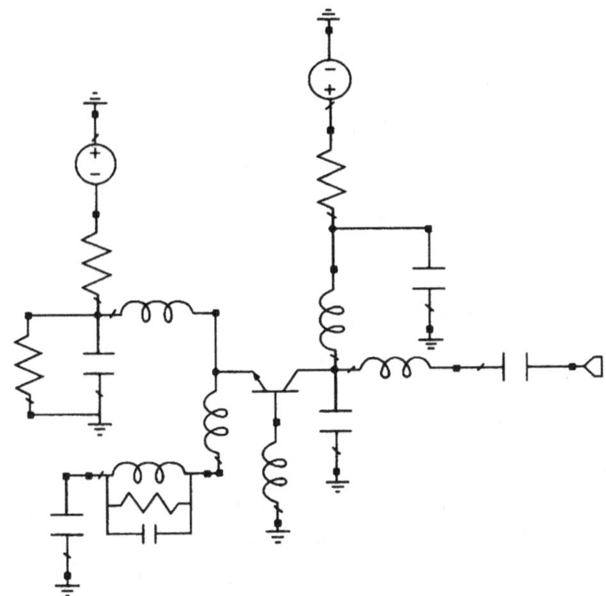

Fig. 12.17 The circuit of the magnetosensitive oscillator for a frequency $f_0 \approx 1.5\,\text{GHz}$ at $H_0 = 500\,\text{Oe}$, simulated by Microwave Office 2002

For example, for YIG spheres it is possible to assume

$$K = 2.8\,\text{MHz/G}, \tag{12.12}$$

in (12.11) as an initial point in calculations, while in (12.11) the value of f_{res} is given in MHz and the magnetic induction B in G.

Let us note that the numerical value of K depends on many factors, namely, the crystallographic structure of ferrite, MECE design, oscillation modes in the ferrite resonator, operating modes, etc. The value of K must be determined with a

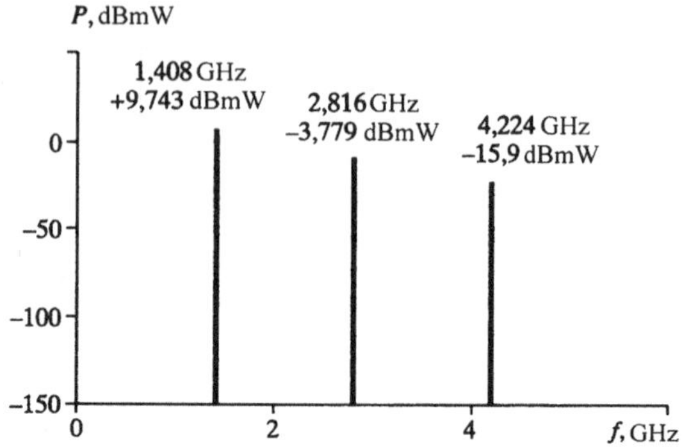

Fig. 12.18 The spectrum of frequencies f_0, f_1, and f_2 of signal generation in the magnetosensitive oscillator

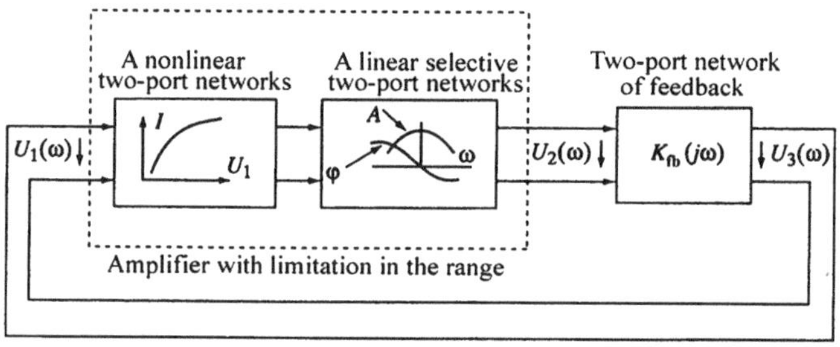

Fig. 12.19 The classical circuit of the oscillator

demanded accuracy in corresponding comparisons of f_{res} with the standard of magnetic induction B [1].

Equation (12.12) defines, in turn, the frequency sensitivity of the oscillator to magnetic field, which is 2.8 MHz/G or 2.8 Hz/nT under the condition of absolute frequency stability. For comparison, the frequency of oscillations in nuclear magnetic resonance with identical magnetic fields is $10^3 - 10^4$ lower than that in FMR.

For measurements of weak magnetic fields or their small changes in the microwave range of wavelength, frequency change of generation can be few fractions of percent of the basic frequency. Therefore, the design of such a sensor requires no additional elements providing continuous frequency change in a wide range (phase attack equalizers in the feedback circuits).

The aforesaid allows to use the classical circuit of the oscillator shown in Fig. 12.19.

The circuit contains three two-port networks, namely, a nonlinear one without inertia and two linear ones – selective one and that of feedback. Various variants of such a circuit are possible, in which advantages by CE, selectivity, output power, etc. can be realized.

As the nonlinear two-port network, any type of nonresonant microwave amplifier can be used. As the selective two-port network, a low-and-high-transmitting

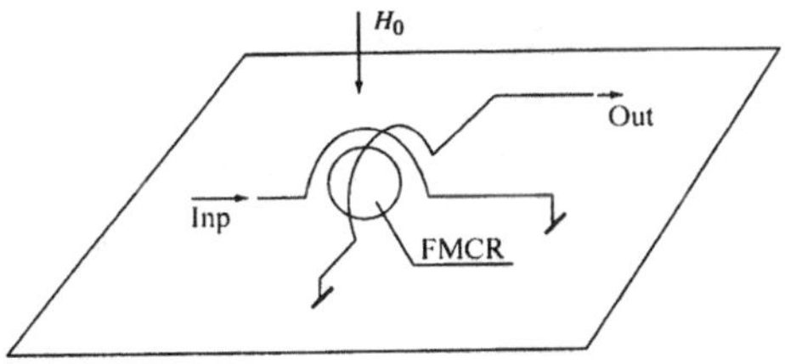

Fig. 12.20 A low-and-high-transmitting narrow-band filter – MECE on crossed coils with a YIG

narrow-band filter should be used – MECE on crossed coils with a YIG (Fig. 12.20) sphere or a film resonator.

The two-port network of feedback is a connecting transfer line which can be coaxial, strip, or waveguide one.

The collection of the oscillation energy from the magnetosensitive oscillator for subsequent processing can be one at any place of the connecting line. In microwave amplifiers, the input and output resistances differ; therefore, a matching element of the wave resistances is provided in the connecting line. Any change of the external magnetic field (a vector quantity) will change the frequency of FMR and shift the passband of the filter and the frequency generated by the device. For weak changes of the external magnetic field vector, the peak and phase conditions of generation practically do not differ from those provided in the field of a permanent magnet for the FMR filter with a long ($>10\Lambda$, Λ being the wavelength in the line) length of the feedback circuit. In a frequency range of 1–2 GHz, frequency change of the oscillator at changes in the external and internal magnetic fields in FMCR is 30–40 MHz and can be increased by reduction of the length of the feedback circuit. Changes of the power of oscillations in the oscillator appear insignificant until they reach the sensitivity threshold of the display device, that is, a frequency meter or its analogs.

Proceeding from the above, it is possible to offer the following basic requirements to design elements of a magnetosensitive vector oscillator for measuring weak magnetic inductions and their deviations:

- A working frequency band of the amplifier – not less than 200 MHz in a wavelength range of 20 – 30 cm with an amplification factor not less than 10 dBmW.
- In the absence of magnetic field, MECE should provide power transfer losses (the input–output decoupling) at a level not worse – 25–30 dB.
- Selection, with due account of the saturation magnetization $4\pi M_s$, of a YIG monocrystal as a small-diameter sphere or an epitaxial film; selection of the coil coupling element filling factor of the order of unity; for a film, its orientation should be chosen as well.
- Power transfer losses in MECE on the central frequency should be no higher than −5 dB.
- The HF power level of the oscillator for various electric modes of its work should provide operation of the registration device (usually not below than 1 mW).

12.4 Magnetosensitive One-Coordinate Sensor

Our breadboard model of the magnetometric sensor included:

- A broadband UHF amplifier for 300–1,200 MHz (of UHF-1 type, three-cascaded, on KT transistors of 3132 made at Tantal Corp^{TM}) with an amplification factor of 25 dBmW in the data-sheet mode ($U = 9$ W and $I = 28$ mA) in a frequency band of 600–1,000 MHz (Fig. 12.21).
- MECE contained a YIG monocrystal in the form of a KG-12 sphere with a diameter of 0.5 mm, located between two crossed (at 90°) strips of two short-circuited lines with their wave resistances 50 Ω (Fig. 12.22).

The HF power decoupling in the absence of magnetic field in MECE was not less than −30 dB. Frequency change in MECE in a range of 300–1,200 MHz was carried out by changes of the magnetic field from 50 up to 400 Oe (Fig. 12.23). The power transfer losses in MECE were $\Delta\nu_{-3\,\mathrm{dB}} \cong 2$–5 MHz. Calculated data agree well with our experimental results.

In Fig. 12.24, theoretical and experimental dependences of the resonant frequencies of FMCR transfer into MECE on crossed MSL depending on the field bias H_0 are shown. The insignificant distinction between the theoretical and experimental dependences when $H_0 < 250$ Oe is explained by the influence of the crystal anisotropy field (for KG-12, $H_A < 20$).

The feedback circuit in the considered design of the sensor was a coaxial connection of MECE with the input of the amplifier. While operating the device, a small change of the measured magnetic fields and, hence, a small frequency change of the

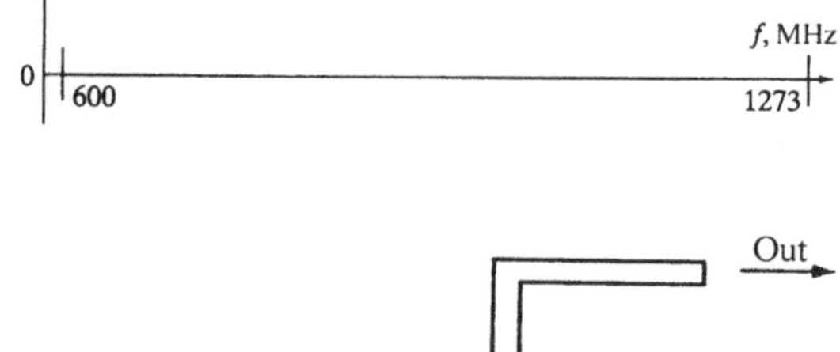

Fig. 12.21 The gain factor of a broadband UHF amplifier for 300–1,200 MHz

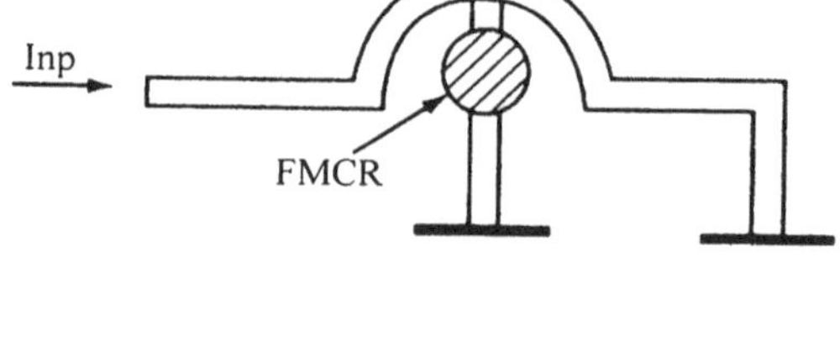

Fig. 12.22 MECE with a YIG sphere in crossed short-circuited striplines

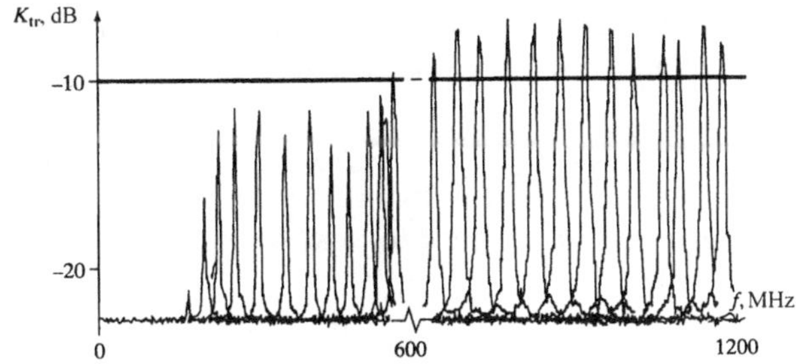

Fig. 12.23 Frequency change in MECE at changes of the magnetic field

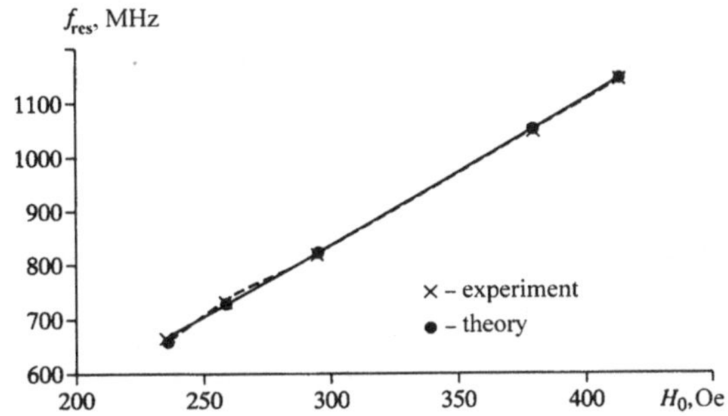

Fig. 12.24 Theoretical and experimental dependences of the resonant frequencies of FMCR transfer into MECE on crossed MSL depending on the field bias H_0

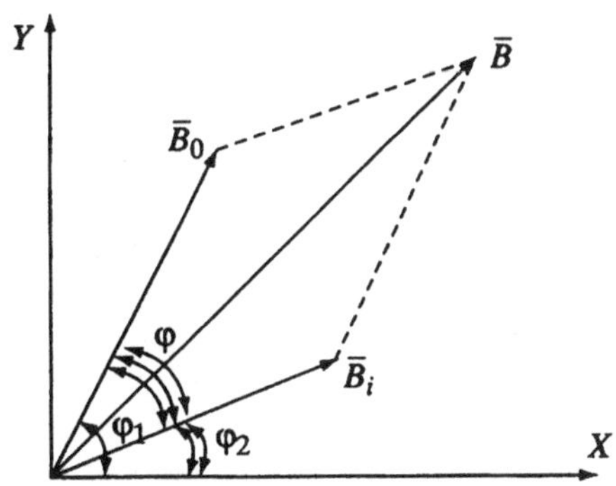

Fig. 12.25 The external magnetic induction $\overline{B}_0$ and magnetic induction of the sensor $\overline{B}_i$ form a resultant quantity $\overline{B} = \overline{B}_0 + \overline{B}_i$

oscillator were supposed, hence there was no necessity for selection of the feedback circuit length and placing of phase attack equalizers into it.

The external magnetic induction $\overline{B}_0$ and magnetic induction of the sensor $\overline{B}_i$ form a resultant quantity $\overline{B} = \overline{B}_0 + \overline{B}_i$ (Fig. 12.25).

The value of the resultant induction is

$$B = \left(B_0^2 + B_i^2 + 2\vec{B}_0\vec{B}_i\cos\varphi\right)^{\frac{1}{2}} \tag{12.13}$$

provided ferromagnetic resonance in MECE placed into the feedback circuit of the amplifier. At high GB products of the resonant curve of MECE, this is practically the central frequency of the passband. When the directions $\overline{B}_0$ and $\overline{B}_i$ coincided, the maximal frequency was generated:

$$f_{\max} = 2.8(B_0 + B_i). \tag{12.14}$$

The directions of vectors $\overline{B}_0$ and $\overline{B}_i$ being opposite, the minimal frequency was generated:

$$f_{\min} = 2.8(B_0 - B_i). \tag{12.15}$$

From (12.14) and (12.15), it follows that the measured value is

$$B_i = \frac{f_{\max} - f_{\min}}{5.6}. \tag{12.16}$$

Equations (12.12), (12.14)–(12.16) are valid when the field of anisotropy is much less than the internal magnetic field of the sensor on the frequency of FMR. If the field of anisotropy is a factor, it is necessary to modify the factor in (12.12), relating the frequency of FMR and magnetic induction. This modification can be carried out using an experimental curve $f = F(B)$ from which it is possible to evaluate the factor in the assumption of this dependence being linear. In view of this new constant, (12.16) becomes

$$B_i = \frac{f_{\max} - f_{\min}}{2K}, \tag{12.17}$$

where K is a factor found from the experimental dependence $f = F(B)$ at a constant position of the sensor in space.

As an example of detection of a weak magnetic field by means of a magnetometric sensor of the developed design, let us consider determination of the value and angle of inclination to a horizontal plane of Earth's magnetic field induction vector, which is known precisely enough and tabulated near the ground according to the World expedition on studying magnetism of the Earth 1985.

The sensor was fixed on a two-coordinate rotary measuring rack, allowing to be rotated in both horizontal and vertical planes by 360° relative to the axes passing through the center of the ferrite sphere. By rotation of the sensor with an arbitrary position in space, the frequency of oscillation of the oscillator changed. In Fig. 12.26, 1, 2, 3 – angle dependences of the frequency of generation angle $f(\theta)$ for three different orientations of the sensor in space are shown; 1.00 GHz, 1.01 GHz, and 1.02 GHz mean the frequencies of the oscillator.

The maximal and minimal frequency values corresponded to the orientation of magnetic induction $\overline{B}_i$ of the sensor along the magnetic meridian. Alignment of the directions of vector projections gives the highest frequency, and their antiparallelism does the lowest frequency of oscillations of the oscillator. Having located the sensor along the magnetic meridian direction and rotating it in a vertical plane, we obtain a diagram represented in Fig. 12.27.

Maximal and minimal values of the frequencies of the oscillator were observed as well. With respect to the horizontal plane, the straight line connecting the points with $f_{\max}$ and $f_{\min}$ and passing through the center of the diagram defined the direction

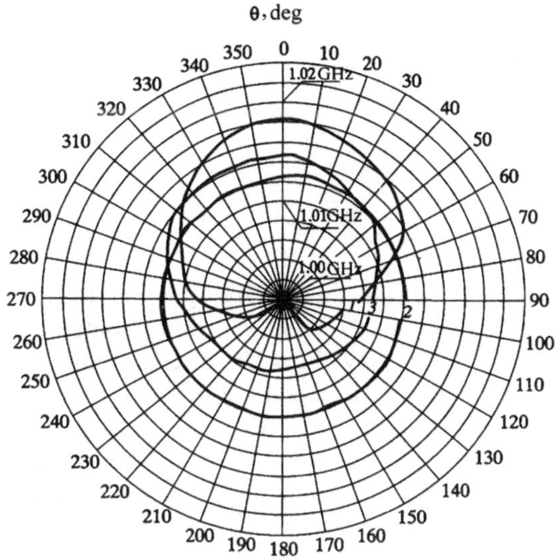

Fig. 12.26 Angle dependences of the frequency of generation angle $f(\theta)$ for three different orientations of the sensor in space

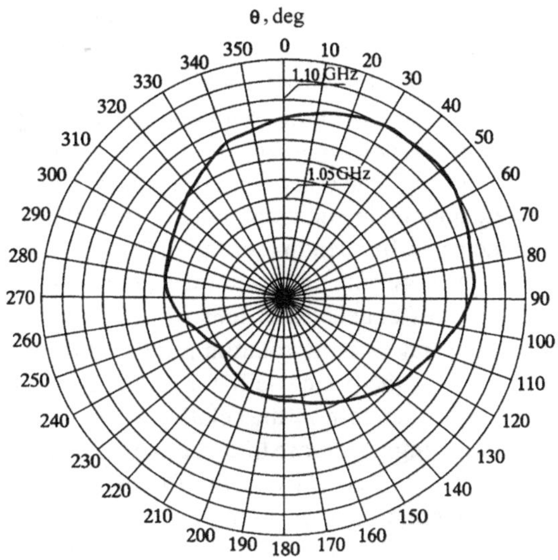

Fig. 12.27 Diagram at rotation of the sensor along the magnetic meridian

of Earth's total vector of magnetic induction, and the absolute value of the vector $\left|\overline{B}_{\text{Earth}}\right|$ was determined from (12.17). By rotations of the sensor in both horizontal and vertical planes, the frequency of oscillations of response signal was determined by the total magnetic induction acting on the YIG monocrystal. Therefore, it is impossible to identify the measured frequency at sensor rotation in a horizontal plane at its arbitrary arrangement with the projection of the magnetic induction vector onto the horizontal plane and to relate the angle of inclination of the magnetic induction vector of the sensor to the horizontal plane with the projection of Earth's magnetic

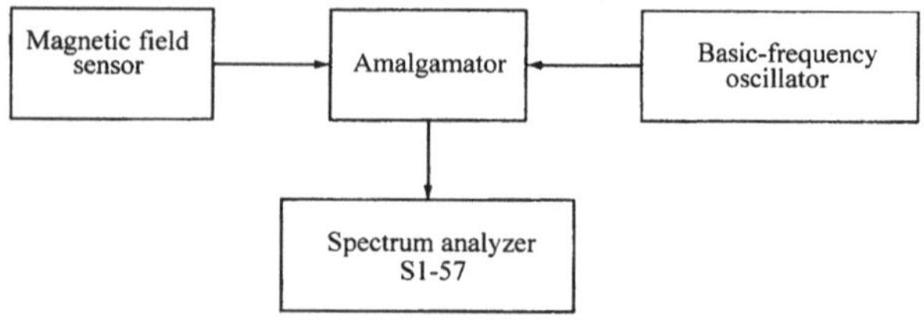

Fig. 12.28 The flowchart of measurements on an intermediate (difference) frequency

induction onto this plane. It is possible to liken the designed sensor to a compass with its free axes of rotation in horizontal and vertical planes. In its unperturbed state, it will always have the direction aligned with the direction of the magnetic field vector at a given point of the country. Unlike such a "bidimensional" compass, the designed sensor allows determination of the total vector of Earth's magnetic field. Knowing the angle of inclination of the vector of Earth's magnetic field, it is possible to determine the horizontal (along the magnetic meridian) and vertical components of this vector.

The presented circular diagrams of Earth's magnetic field are drawn directly for the frequency of the sensor's oscillator. As the maximal frequency change at rotation of the sensor in the horizontal and vertical planes (in a plane perpendicular to the horizontal one, located in the direction of the highest change of frequency) does not exceed few MHz, which corresponds to tenth fractions of percent of the average frequency of the oscillator, it seems expedient to raise the relative deviation of frequency, passing to an operating mode on an intermediate frequency.

In Fig. 12.28, the flowchart of measurements on an intermediate (difference) frequency obtained in an amalgamator from the oscillator of the magnetic field sensor and the basic-frequency oscillator (G4-76A) is presented.

On the spectrum analyzer, measurements similar to those described above were realized on frequencies of tens of MHz. As the frequency changes of the oscillator of the sensor at its rotation remained the same, as it was noted above, and the frequency of the basic oscillator remained constant, an essential (by two orders of magnitude) increase in the relative deviation of frequency occurred that allows to correspondingly increase the accuracy of measurements of frequency leaving (under the condition of stability of the frequencies of the basic oscillator and sensor). In operating conditions on an intermediate frequency, it is also possible to provide an increase of the relative deviation with operation of the oscillator of the sensor on harmonics. In this case, the absolute leaving of frequency at rotation of the sensor is multiple to the number of the harmonic.

In Fig. 12.29, polar diagrams of changes of the difference frequencies are shown when the oscillator of the sensor operates on the harmonics: $1 - \nu_1, 2 - \nu_2, 3 - \nu_3$.

So, for a difference frequency of 40–60 MHz on the third harmonic of the oscillator of the sensor, the distinction of frequencies along the direction of their maximal change (along Earth's magnetic field) is 6 MHz; that with a 6-digit frequency measuring instrument allows estimation of frequency leaving of few Hz (certainly, with an absolute stability of the frequencies of the basic oscillator and the sensor).

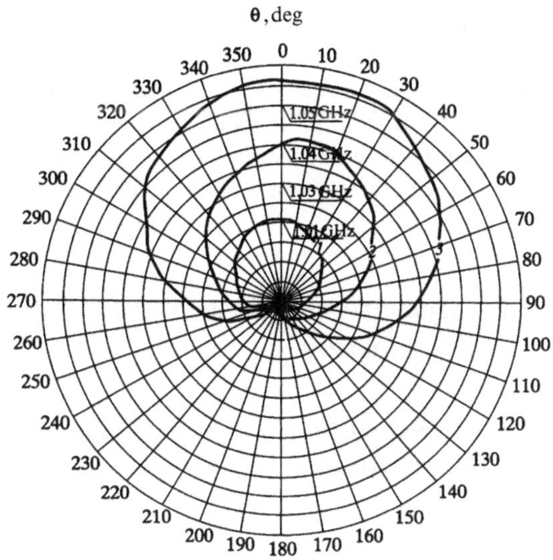

Fig. 12.29 Polar diagrams of changes of the difference frequencies, when the oscillator of the sensor operates on the harmonics: $1 - \nu_1$, $2 - \nu_2$, and $3 - \nu_3$

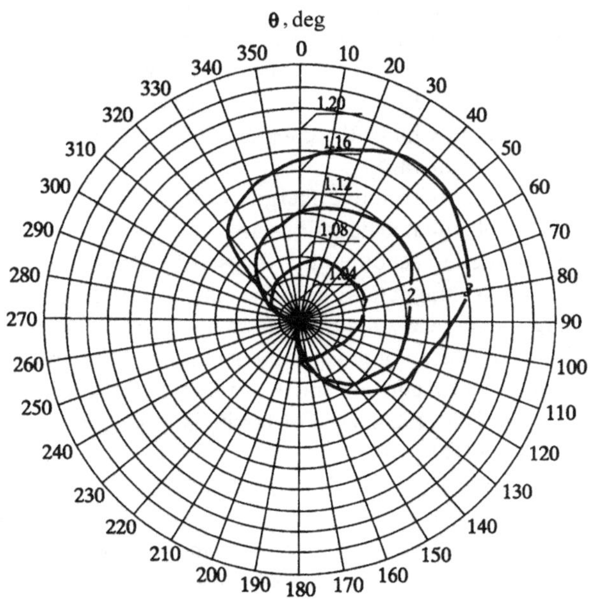

Fig. 12.30 The circle diagram of magnetic induction changes, which was registrable by sensor

A similar estimation of frequency leaving allows measuring a weak magnetic field of hundredth fractions of nT ($\Delta f = 1$ Hz corresponds to a change of $\Delta B = 0.035$ nT at FMR).

From the diagrams in Figs. 12.26, 12.27, 12.29, and 12.30, the value of the magnetic induction vector of Earth's field is determined as $(0.4 - 0.6) \times 10^5$ nT. For 51°32′19″ latitude North and 46°00′27″ longitude East, this value is 51, 197 μT.

The angle of inclination of the vector B_3 to the horizontal surface at the point of measurement is 67°6′. According to Figs. 12.27 and 12.30, this angle, at measurements in a vertical plane in the direction of the greatest change of frequency, lies within 65–70°. Measurements were made in laboratory conditions, that is, in a building with ferroconcrete overhead covers. Therefore, it is possible to consider the results of comparison with reference data to be quite satisfactory.

12.5 Measurement Procedures of Ferrite Microresonator Parameters

12.5.1 Determination of Equilibrium Orientation of Magnetization for Cubic Ferrite Monocrystals

To explore magnetic oscillations in ferromagnetics, the crystal structure must be oriented in a magnetic field. The field may be more or less homogeneous. The equilibrium directions of magnetization orientation are generally defined by the minimum magnetic energy of the crystal [17, 18] (the sum of the energy of the ferromagnetic crystal in the external field, the crystallographic anisotropy energy, the internal energy, and, in the presence of elastic external stresses, the extra magnetoelastic and elastic energies). For the considered FMCR in the form of spherical samples, there is no shape anisotropy.

Account of the above energies of the crystal in a magnetic field generally leads to rather sophisticated equations and, as experiment shows, in many cases it is enough to be limited to account of the crystallographic anisotropy energy for spherical samples of ferromagnetics only. The equilibrium directions with a minimum anisotropy energy define easy magnetization axes (AEM) of the crystal, and the directions in which the energy of anisotropy is maximal do hard magnetization axes (AUM).

AEM and AUM are most simply defined in monoaxial crystals. In an external magnetic field, the crystal of ferrite is guided along AEM, and AUM lay in a plane perpendicular to AEM.

For cubic crystals [17–19] with account of the first constant of anisotropy $K_1 > 0$ only, AEM will be $\langle 100 \rangle$, that is any of the equivalent directions [100], [010], [001], and AUM will be $\langle 111 \rangle$, that is, the diagonal axes of the cube. When $K_1 < 0$, axis $\langle 100 \rangle$ will be easy and $\langle 100 \rangle$ will be hard. Involving the second constant of anisotropy ($K_2 \neq 0$), the hard and easy axes become the same, if $|K_2| < \frac{g}{H}\,|K_1|$.

Homogeneous magnetization, characteristic of ellipsoid samples of ferromagnetic crystals, is not always their equilibrium state. In the absence of an external magnetic field or when this field is weak, the one-domain state disappears and a multidomain one is formed. This process now has no finished experimentally proven theory. The existing models of the multidomain state are not always adequate to experimental data and can be used at this stage for qualitative comparison with experimental results only. Therefore, we consider the one-domain state of the ferrite sample, which is realized in saturation modes. It is accepted [17] that the saturated mode of ferromagnetics is provided under external magnetic fields with

$$H_0 > \frac{4\pi M_s}{3}, \tag{12.18}$$

where M_s is the saturation magnetization.

For cubic crystals in the mode of saturation, the magnetization $\overline{M}$ and the internal magnetic field $\overline{H}_{e_0}$ coincide, and the one-domain mode is realized. Under these conditions in plane {110}, the conditions of ferromagnetic resonance, in view of the anisotropy constants K_1 and K_2, have been obtained [17] in the form of

$$\begin{aligned}\left[\frac{\omega}{\gamma}\right]^2_{(110)} &= \left[H_{e0z} + H_{A_1}\cdot\left(-\frac{3}{8}2\cos 2\theta_0 + \frac{3}{8}\cos 4\theta_0\right)\right.\\ &\quad \left. + H_{A_2}\times\left(\frac{5}{8}+\frac{5}{8}\cos 2\theta_0 + \frac{9}{8}\cos 4\theta_0\right)\cdot\sin^2\theta_0\right]\\ &\quad \times\left[H_{e0z} + H_{A_1}\cdot\left(\frac{1}{2}\cos 2\theta_0 + \frac{3}{2}\cos 4\theta_0\right)\right.\\ &\quad \left. + H_{A_2}\left(-\frac{7}{16}+\frac{3}{16}\cos 2\theta_0\right)\cdot\sin^2\theta_0\right].\end{aligned} \tag{12.19}$$

In Fig. 12.31, coordinate axes in the sphere of a cubic ferromagnetic are shown.

Of greatest interest is the case when $\varphi_0 = \pi/4$ and lay in plane {100}. Equation (12.19) is written for this practically important case.

At collinearity of $\overline{M}_0$ and $\overline{H}_{e0}$, for special cases of coincidence of $\overline{M}_0$ (and consequently $\overline{H}_{e0}$) with the axes of symmetry of the crystal, (12.19) looks like:

For direction ⟨100⟩ at $\theta_0 = 0$

$$\frac{\omega}{\gamma} = H_{e_0} + 2H_{A_1}; \tag{12.20}$$

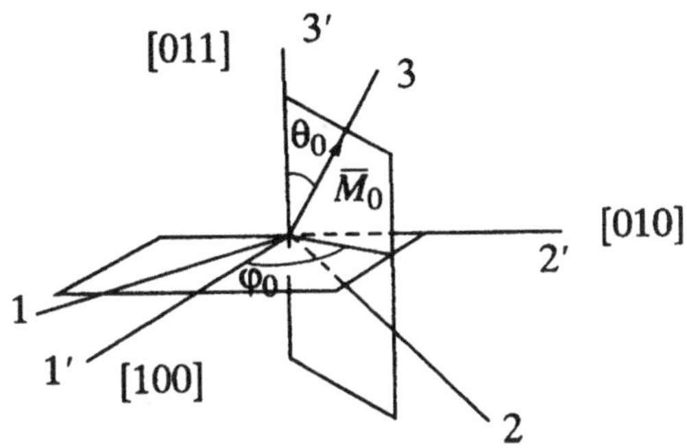

Fig. 12.31 Coordinate axes in the sphere of a cubic ferromagnetic

For direction $\langle 110 \rangle$ at $\theta_0 = 54° \, 44'$

$$\frac{\omega}{\gamma} = H_{e0} - \frac{4}{3} H_{A_1} - \frac{4}{9} H_{A_2}; \tag{12.21}$$

for direction $\langle 110 \rangle$ at $\theta_0 = 90°$

$$\frac{\omega}{\gamma} = \left[(H_{e0} - 2H_{A_1}) \left(H_{e0} + H_{A_1} + \frac{1}{2} H_{A_2} \right) \right]^{\frac{1}{2}}. \tag{12.22}$$

As will be shown below, (12.20)–(12.22) well enough describe the resonant frequencies at small and greater constants of anisotropy in the cubic structures of ferromagnetics.

In the case of ferrite sphere orientation along axis [110], the use of (12.20)–(12.22) assumes that the change of the angle θ at its rotation around an axis perpendicular to plane $\{110\}$ will cause (at ω_0 = const) the necessity of a change of the external magnetic field corresponding to the ferromagnetic resonance on the given frequency. By coincidence of any of the three basic axes of symmetry of the crystal with the direction of the external magnetic field, the changes have an extremum. From these values of magnetic field by (12.20)–(12.22), it is possible to find the constants of anisotropy K_1 and K_2. At H_0 = const and rotation of the oriented sphere around an axis perpendicular to plane $\{110\}$, a change of the ferromagnetic resonance frequency was observed at coincidence of the easy, hard, and intermediate axes of the crystal with the direction of the external magnetic field. In this case, it is necessary to expect that in the AEM, AUM, and $\langle 110 \rangle$ directions, the maximal, minimal, and intermediate frequency of resonance will be found, respectively. Thus, it is necessary to remember that sign inversion of the first constant of anisotropy changes the position AEM and AUM. Changes of the FMR frequency at rotation of a spherical sample of ferrite monocrystal in a constant magnetic field underlie the technique of determination of its equilibrium orientation and constants of anisotropy described below.

12.5.2 Determination of Equilibrium Orientation of Magnetization of Spheric Specimen

Our experimental module for FMR observations is shown in Fig. 12.32.

The width of the strip is 500μm and the thickness of the polycor substrate is 500 μm. The ferrite sphere is located near the short-circuited ends of the strip lines, where the HF magnetic field is maximal. Several variants of the arrangement of the sphere are presented in Fig. 12.32: 1, 3 – on the strips and 2 – between them. Depending on the sphere arrangements, either absorption or transfer of power in the mode of FMR was observed on the output. The distance between the short-circuited ends of the input and output strip lines was selected so as to reach a necessary

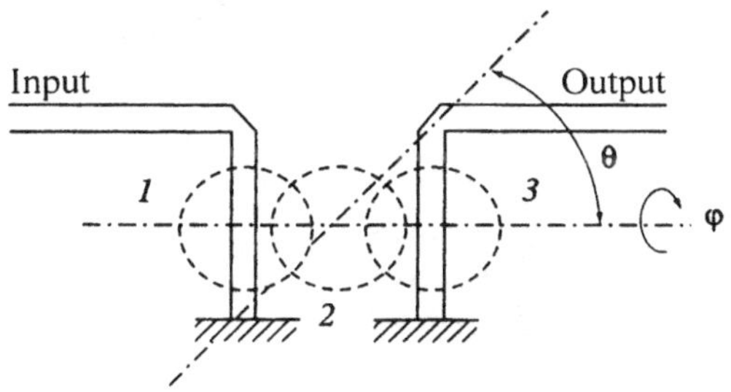

Fig. 12.32 The arrangement of the ferrite sphere in experiments: 1, 3 – on the strips; 2 – between strips

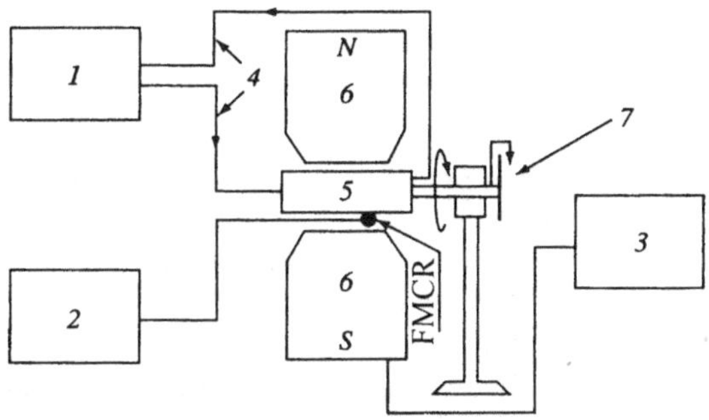

Fig. 12.33 The flowchart of a measuring device for observation of the effects of absorption and transfer of HF power in the mode of FMCR

level of power reemission in the output element. This distance was one order of magnitude with the diameter of the spherical sample (0.4–0.6 mm). By such a close arrangement of the input and output elements of the transfer line, reemission of the HF field takes place. Its level was about −30 dB of the level of the acting HF power. The use of the transfer line made of two strips with significant reemission has been dictated by the conditions of experiment in a wide band of frequencies. So, in particular, in a range of frequencies of 2 – 4 GHz it was higher absorption of HF power at FMR, and in a range of 4 – 8 GHz transfer was more effective. On the plane of the strips, a grid with angular coordinates from 0 to 360° was placed for orientation of the sphere relative to the axis of rotation.

In Fig. 12.33, the flowchart of a measuring device for observation of the effects of absorption and transfer of HF power in the mode of FMCR is presented: 1 – a measuring instrument of SWR_e and attenuation Ya2R-75, Ya2R-70; 2 – a measuring instrument of magnetic induction Sh1-8; 3 – a power unit of an electromagnet B5-50; 4 – input and output bringing devices, including ferrite gates E8-14, attenuators, a detector head, and HF coaxial cables; 5 – a module to be examined with a ferrite sphere; 6 – an electromagnet OEM-1; and 7 – a device to measure the angle between the direction $\overline{H}_0$ and the axis of easy magnetization of the ferrite sphere.

The method of determination of the equilibrium orientation of the sphere (plane {110} perpendicular to the axis of rotation of the module) consisted in that in a strong ($H_0 \cong 2{,}000$ Oe) magnetic field the ferrite sphere is oriented along its easy axis (irrespective of the sign of K_1). At this position, the holder of the sample was rigidly fixed on a small plane, which allows orienting the sphere at the corresponding angle to the axis of rotation on the strip according to the angular coordinate grid.

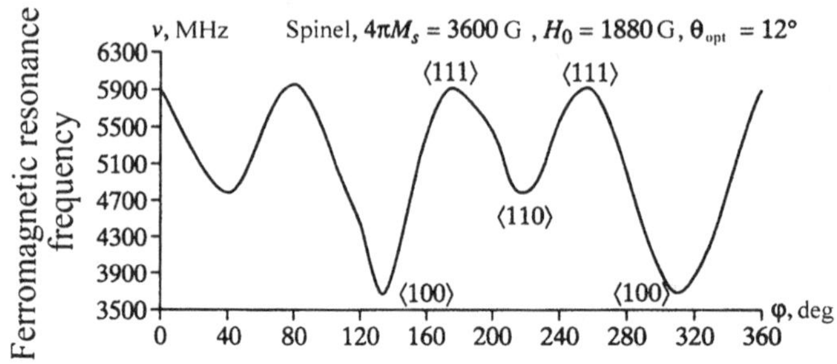

Fig. 12.34 Dependence of the FMR frequency ν on the angle of rotation φ for spinel

The position of the easy axis in the initial state of the module (perpendicular to the field direction $\overline{H}_0$) did not vary. Having fixed the spherical sample at a certain angle to the axis of rotation of the module, the module was rotated in a magnetic field ($H_0 = \text{const}$) by means of the rotary device. The screen of the measuring instrument of SWR_e and attenuation showed changes of the FMR frequency at absorption of the reradiated power and at transfer of HF power to the output element.

In Fig. 12.34, dependence of the FMR frequency ν on the angle of rotation φ for spinel (a cubic crystal) with a saturation magnetization $4\pi M_\text{s} = 3{,}600$ G for the optimum position of the spherical sample relative to the axis of rotation of the rotary device is shown.

Apparently, for various angles φ, the difference between the maximal and minimal frequencies is maximal when plane {110} locates the sphere perpendicularly to the axis of rotation of the rotary device.

From the maximal and minimal frequencies corresponding to AEM and AUM, by (12.20)–(12.22), it is possible to determine the constants of anisotropy K_1 and K_2.

Calculation by (12.20)–(12.22) has confirmed [17, 19] the values of constants K_1 and K_2 for Mn–Zn ferrite, and

$$\frac{K_1}{M_\text{s}} = H_{A_1} = -239\,\text{Oe},$$
$$\frac{K_2}{M_\text{s}} = H_{A_2} = -54\,\text{Oe}.$$

In our calculations, the following data were used: $H_0 = 1{,}800$ Oe = const, $\nu_{\max} = 6{,}000$ MHz, and $\nu_{\min} = 3{,}700$ MHz.

The obtained values have the same order, as the values of constants K_1 and K_2 of anisotropy for spinels like [19, Table 5.7]:

$$\text{NiFe}_2\text{O}_4 : 4\pi M_\text{s} = 3{,}120\,\text{G};\; H_{A_1} = \frac{K_1}{M} = -244\,\text{Oe};\; H_{A_2} = \frac{K_2}{M} = -40\,\text{Oe},$$

$$\text{Li}_{0.5}\text{Fe}_{2.5}\text{O}_4 : 4\pi M_\text{s} = 3{,}500\,\text{G};\; H_{A_1} = \frac{K_1}{M} = -288\,\text{Oe};\; H_{A_2} = \frac{K_2}{M} = -46\,\text{Oe}.$$

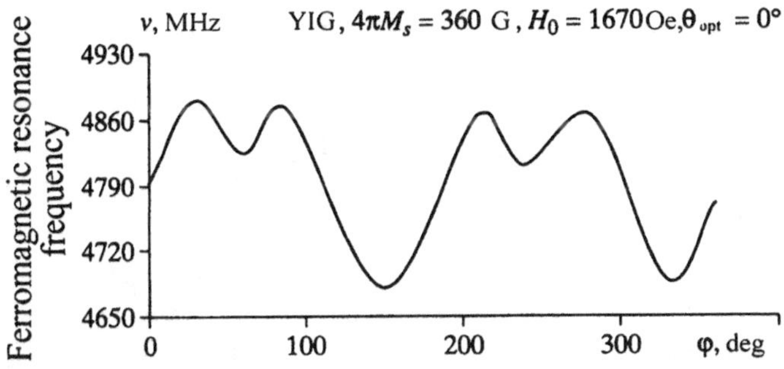

Fig. 12.35 A similar dependence of the FMR frequency on the angle of rotation φ relative to the axis of the module for various angles of the YIG sphere position with a saturation magnetization of $4\pi M_s = 360\,\mathrm{G}$

In Fig. 12.35, a similar dependence of the FMR frequency on the angle of rotation φ relative to the axis of the module for various angles of the YIG sphere position with a saturation magnetization of $4\pi M_s = 360\,\mathrm{G}$ is shown.

The plane [110] corresponds to the maximal change of frequency at rotation of the module around its axis in a constant homogeneous magnetic field ($H_0 =$ 1,670 Oe). The calculated constant of anisotropy K_1 corresponds to the value resulted [19, Table 5.2] for $Y_3Ga_{1.0}I_{pos}Fe_{3.6}O_{12}$.

Equations (12.20)–(12.22) allow calculation of both the constants of anisotropy (the fields of anisotropy H_{A_1} and H_{A_2}) and the g factor (or the value of γ related with it) at a preset external magnetic field N_{e_0}.

For the corresponding axes and angles θ_{unl}, let us write the expressions relating the resonant frequencies and fields H_{e_0}, H_{A_1}, H_{A_2}:

$$\text{(a) } [100], \theta_{\mathrm{unl}} = 0°, \ \frac{\omega_1}{\gamma} = H_{e_0} + 2H_{A_1}; \tag{12.23}$$

$$\text{(b) } [111], \theta_{\mathrm{unl}} = 54°44', \ \frac{\omega_2}{\gamma} = H_{e_0} - \frac{4}{3}H_{A_1} - \frac{4}{9}H_{A_2}; \tag{12.24}$$

$$\text{(c) } [110], \theta_{\mathrm{unl}} = 90°, \ \frac{\omega_3}{\gamma} = \left[\left(H_{e_0} - 2H_{A_1}\right) \times \left(H_{e_0} + H_{A_1} + \frac{1}{2}H_{A_2}\right)\right]^{\frac{1}{2}}. \tag{12.25}$$

From (12.23), we express H_{A_1}

$$H_{A_1} = \frac{\frac{\omega_1}{\gamma} - H_0}{2}. \tag{12.26}$$

In (12.22) we substitute (12.24) and find H_{A_2}

$$H_{A_1} = \frac{9}{4}\left[H_0 - \frac{\omega_2}{\gamma} - \frac{4}{3}\left(\frac{\frac{\omega_1}{\gamma} - H_0}{2}\right)\right]. \tag{12.27}$$

From (12.25) after transformations we derive an equation for γ

$$\gamma^2 - \gamma \underbrace{\left[\frac{2.875\omega_1 + 2.25\omega_2}{4.75H_0}\right]}_{A} + \omega_1 \underbrace{\left[\frac{0.25\omega_1 + 1.125\omega_2}{4.75H^2}\right]}_{B} - \omega_3^2 = 0. \tag{12.28}$$

From experiment, we have the values: $\nu_1 = 3{,}700$ MHz, $\nu_2 = 6{,}000$ MHz, $\nu_3 = 4{,}820$ MHz at $H_0 = 1{,}800$ Oe, after substitution of which (in view of $\omega_{1,2,3} = 2\pi\nu_{1,2,3}$) into (12.28) we get a value $\gamma = 2.7$ which differs from the value $\gamma = 2.8$ by $\approx 3.6\%$.

Considering that for spinel $\gamma = 2.8$, calculate H_{A_1} and H_{A_2}. From (12.26) we have $H_{A_1} = -239$ Oe and from (12.27) we have $H_{A_2} = -56.5$ Oe. From the plot in Fig 12.34 and from the difference of frequencies $(\nu_2 - \nu_1)$, a value of $H_{A_1} \approx -247$ Oe can be found. Then, from (12.27), we get $H_{A_2} = -31$ Oe. The order of magnitude of the obtained values H_{A_1} and H_{A_2} corresponds to the manufacturer's data.

12.6 Experimental Investigation of Parameters of a Vector Magnetoelectronic Sensor

In the approximation, when the induction is $\overline{B} \ll \overline{B}_0$, the frequency of generation of the sensor is determined by

$$f = \gamma\left(B_0 + \frac{\overline{B}\,\overline{B}_0}{B_0}\right), \tag{12.29}$$

where $\gamma = 28$ Hz/nT.

As $\overline{B} \cdot \overline{B}_0 = B \cdot B_0 \cdot \cos\alpha$, where α is the angle between the vectors $\overline{B}$ and $\overline{B}_0$, then, rotating the sensor in three mutually perpendicular planes, it is possible to find the components of the measured magnetic field B_x, B_y, and B_z from the obtained frequencies of generation.

The investigated sensor is assembled under the circuit of a two-cascade amplifier on field transistors, the input and output of which are connected through MECE (Fig. 12.36).

The electronic part of the sensor fastens on a polycoric substrate of a thickness of 0.5 mm and sizes of 7.5×9.5 mm, which is soldered to a fixing metal plate of a

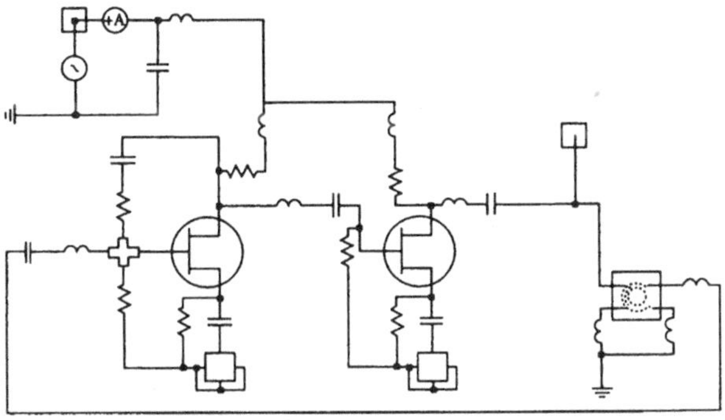

Fig. 12.36 The investigated sensor is assembled under the circuit of a two-cascade amplifier on field transistors, the input and output of which are connected through MECE

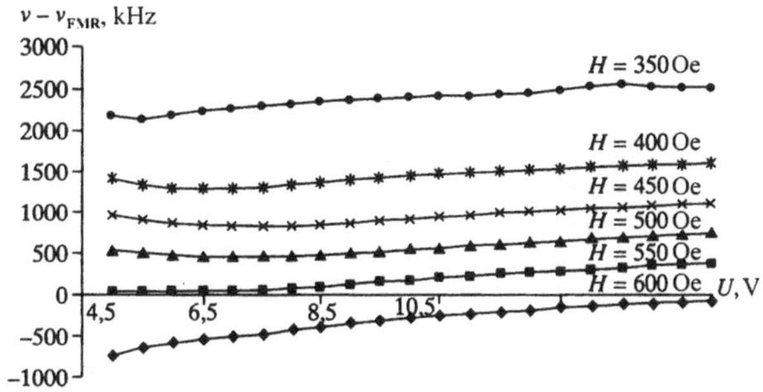

Fig. 12.37 Calculated dependences of the difference of frequencies (the frequency of generation – ν and the FMR frequency – ν_{FMR}) on the power source voltage U for several bias fields (from 350 to 600 Oe with a step of 50 Oe, 1 Oe = 10^5 nT)

thickness of 0.5 mm and sizes of 17 × 9.5 mm. The own volume of the electronic plate is 0.035 mm^3. The fixing plate is fixed inside a brass shielding case with sizes 20 × 21 × 10.5 mm. A constant magnet with a diameter of 5 mm and a thickness of 2.5 mm, which can be displaced on a groove in the axial direction, is built in the case. While adjusting the sensor, the magnet is moved so as to provide a steady mode of generation.

Theoretical calculations have shown that the frequency of generation of a similar circuit does not coincide with the FMR frequency though it is close to it. Calculated dependences of the difference of frequencies (the frequency of generation – ν and the FMR frequency – ν_{FMR}) on the power source voltage U for several bias fields (from 350 to 600 Oe with a step of 50 Oe, 1 Oe = 10^5 nT) are shown in Fig. 12.37.

It is seen that the coincidence of the frequency of generation and that of FMR can be reached only at one value of the magnetic field and in a narrow range of changes of the power supply voltage of the transistors. All this leads to the dependence of the frequency of generation of the sensor on magnetic field being no longer strictly linear, and the differential steepness of frequency change $\gamma = dF/dB$ becomes a function of magnetic field. The difference of the frequency of generation of the oscillator with FMCR from the FMR frequency can be due to the presence of reactive components of the resistance of the transistor and the coupling element.

All the sensor parameters can be divided into three groups, namely:

– Magnetosensitive ones, describing transformation of the parameters of measured magnetic induction to the frequency of response signal, errors of determination of the value, and direction of the vector of measured magnetic induction, etc.

- Electric ones, describing the electric operating mode of the sensor
- Operational ones, defining the conditions of sensor operation

According to this, the magnetometric sensor can be characterized by the following system of parameters:

1. Magnetosensitive parameters:
   - $\gamma = \frac{\Delta f}{\Delta B}$ is the sensitivity of the sensor to Earth's magnetic field in three mutual perpendicular planes, Hz/nT
   - $\gamma_{\text{ang}} = \frac{\Delta f}{\Delta \varphi}$ is the angular sensitivity of the sensor to the direction of rotation relative to the vector of Earth's magnetic field in a horizontal and two vertical mutual perpendicular planes, kHz/deg
   - The maximal relative regular error of induction measurements of a constant magnetic field, %
   - The maximal root-mean-square random error of magnetic induction measurement of a constant magnetic field, nT
2. Electric parameters:
   - f is the average working frequency of the oscillator of the sensor, kHz
   - U is the power supply voltage of the sensor, V
   - I is the current consumed by the sensor, mA
3. Operational parameters:
   - m is the weight of the converting element of the sensor (without the screen), g
   - M is the weight of the sensor in a shielding case, g
   - V_{el} is the volume of the converting element of the sensor, cm^3
   - V_{g} is the volume of the sensor in a shielding case, cm^3
   - $T_{\min}$ and $T_{\max}$ are the limiting admissible values of ambient temperature, °C

The basic experimental characteristics of breadboard sensor models are presented in Figs. 12.38– 12.43.

In Fig. 12.38, static characteristics of the sensor are shown: a is the dependence of current of the sensor I and b is the dependence of power consumption P on the voltage U on the sensor.

From these curves, it is obvious that the static VAC of transistors is close to linearity at low voltages ($U \leq 2$ V), and when $U > 4$ V, the VAC is saturated.

A demerit of the sensor is eventual changes of the frequency of generation because of gradual and nonuniform warming up of the transistors and other elements of the circuit. In Fig. 12.39, the time dependences recorded simultaneously are shown at a fixed position of the sensor in space of change: a– of frequency df; b – of current I. The change of temperature of the transistor estimated from the change of current obviously occurs during a long enough interval (more than 40 min). The frequency of generation of the sensor after the sharp recession (which should be expected at a rise in temperature) starts to raise in 15 min and then goes down. This speaks, most likely, for nonuniform warming up of various elements of the sensor.

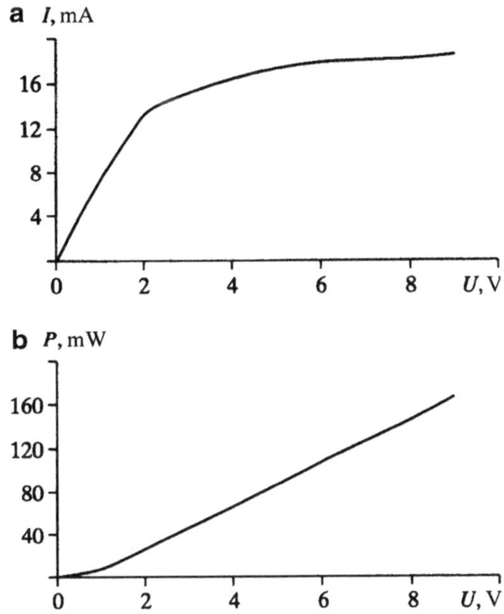

Fig. 12.38 Static characteristics of the sensor are shown: (**a**) the dependence of current of the sensor I; (**b**) the dependence of power consumption P on the voltage U on the sensor

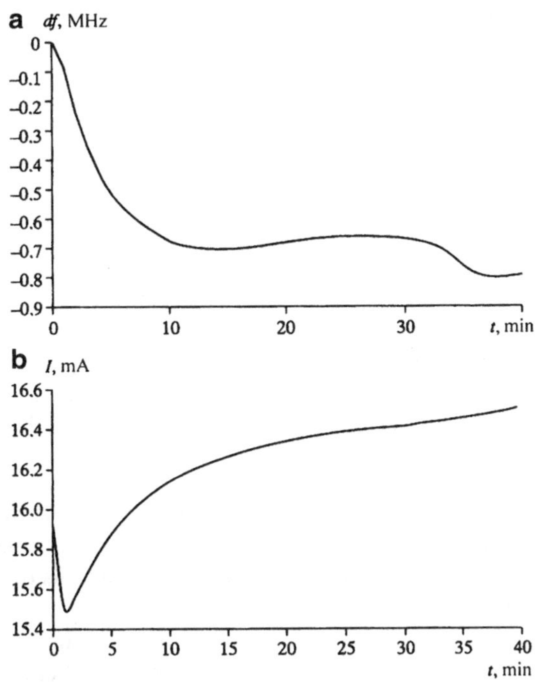

Fig. 12.39 The time dependences recorded simultaneously are shown at a fixed position of the sensor in space of change: (**a**) of frequency $\mathrm{d}f$; (**b**) of current I

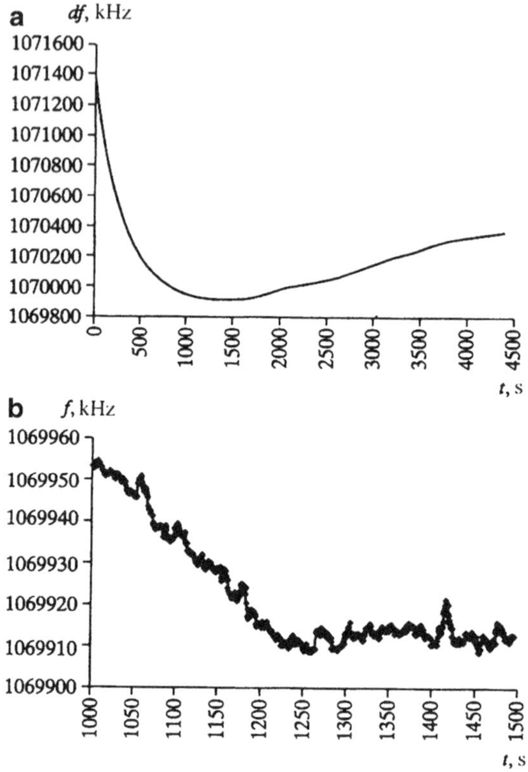

Fig. 12.40 Dependences of changes of the frequency of generation of the sensor df on time t in the case of isolation of the case from the environment with a heat-insulating material for several intervals of time are presented: (**a**) 0–5,000 s; (**b**) 0–1,500 s

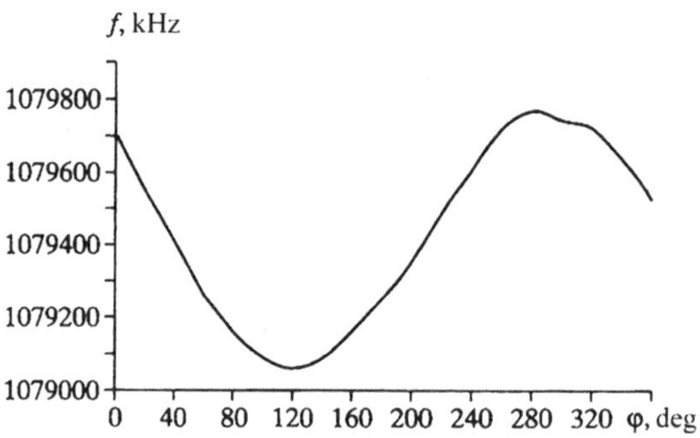

Fig. 12.41 The corresponding curves for one of our breadboard sensor models at its rotation in the horizontal plane XOY

In Fig. 12.40, dependences of changes of the frequency of generation of the sensor df on time t in the case of isolation of the case from the environment with a heat-insulating material for several intervals of time are presented: (a) (0 – 5,000) s; (b) (0 – 1,500) s.

It is seen that the general character of the time dependence of the frequency of generation after the sensor turns on has not changed. This confirms the conclusion that the smooth growth and smooth fluctuations of frequency in time are mainly due to nonstationary thermal processes occurring in the sensor.

Fig. 12.42 The corresponding curves for one of our breadboard sensor models at its rotation in the vertical plane *X0Z*

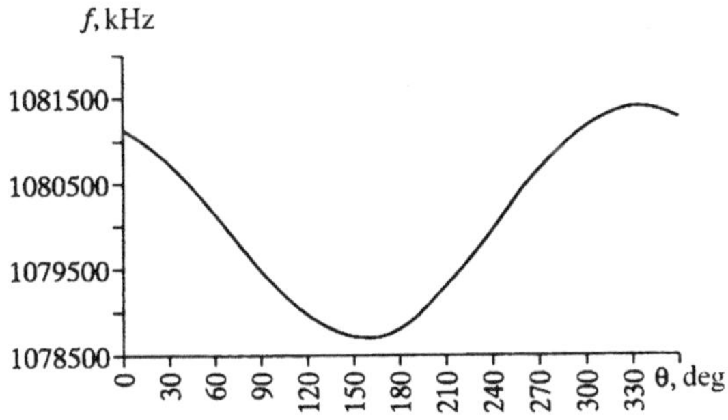

Fig. 12.43 The corresponding curves for one of our breadboard sensor models at its rotation in the vertical plane *Y0Z*

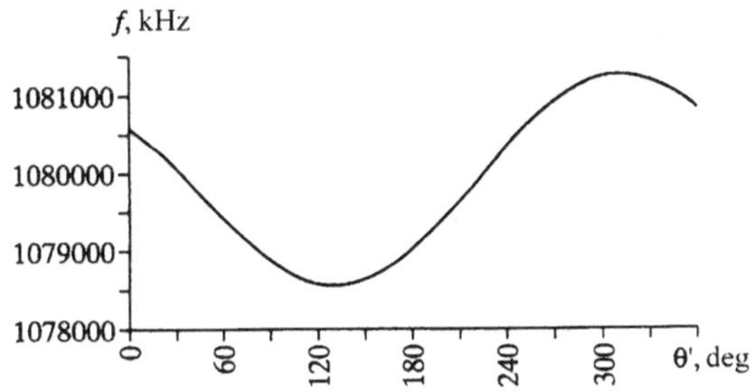

In Fig. 12.40b, a part of the general time dependence of the frequency of generation of the sensor is scaled up. It is seen that the fluctuation component of frequency does not exceed several kHz. This means that the relative short-term stability of the frequency of generation of the sensor can be brought to 10^{-6} by using heat stabilization or heat compensation.

One of the basic characteristics of the magnetic field sensor is its angular sensitivity, i.e., the change of the frequency of generation when the sensor rotates relative to the vector of Earth's magnetic field [84–86]. Figures 12.41–12.43 show the corresponding curves for one of our breadboard sensor models at its rotation in three mutual perpendicular planes (in Fig. 12.41 – in the horizontal plane $X0Y$, in Fig. 12.42 – in the vertical plane $X0Z$, in Fig. 12.43 – in the vertical plane $Y0Z$). In the horizontal plane, the maximal angular sensitivity is 7 ± 0.13 kHz/deg, and in the vertical planes, it is -20 ± 0.5 kHz/deg.

This high value of the angular sensitivity at the points of its maximal steepness prompts that the direction of the corresponding projection of the vector of Earth's magnetic field is most exactly defined not by maximum frequency where the angular sensitivity is equal to zero, but by the points of the maximal angular sensitivity. The angular distance between these points is (precisely) 180°, and these points are displaced (precisely) by 90° to the left and right from the maximum. The corresponding data processing has shown that the accuracy of determination of the direction of the vector of Earth's magnetic field induction in this case rises to ±2.5°, whereas determination of this direction by maximum and minimum frequencies is ±5°.

The calculated sensitivities of the sensor to magnetic induction changes according to Figs. 12.41 – 12.43 are:

$$\gamma_{x0y} = (24.4 \pm 2.5)\,\mathrm{Hz/nT} = (2.44 \pm 0.25)\,\mathrm{MHz/mT};$$
$$\gamma_{x0z} = (29.9 \pm 2.6)\,\mathrm{Hz/nT} = (2.99 \pm 0.26)\,\mathrm{MHz/mT};$$
$$\gamma_{y0z} = (31.8 \pm 2.7)\,\mathrm{Hz/nT} = (3.18 \pm 0.27)\,\mathrm{MHz/mT}.$$

Apparently, the experimentally obtained sensitivities differ from their theoretical value ($\gamma = 28\,\mathrm{Hz/nT}$) and are different in different planes. Special experiments have shown that it is most likely due to the presence of ferromagnetic details in the sensor design, in particular, a nickel substrate at the crystal of the amplifier and a nickel covering at the energy output socket, etc.

A possible application of the magnetometric sensor is detection of ferromagnetic bodies by their distortion of Earth's magnetic field.

To check such an opportunity, dependences of the frequency of generation of the sensor on the angle of its rotation in a vertical plane for several distances from the sensor to the object (an iron sheet with sizes 710 × 610 mm and a thickness of 1.5 mm) were obtained. The corresponding dependences are presented in Fig. 12.44.

It is obvious that the maximal change of frequency in the presence of a ferromagnetic object is observed at the angles corresponding to either a maximum or minimum of frequency.

Some results of determination of the registration range of response signals from ferromagnetic bodies and objects of various weights and sizes are presented in Table 12.2.

One can see that the heavier the ferromagnetic object and the larger its surface area, the wider the range of registration of response signals.

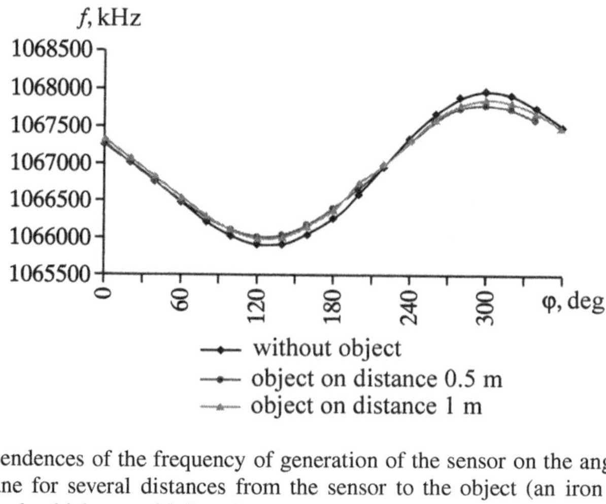

Fig. 12.44 Dependences of the frequency of generation of the sensor on the angle of its rotation in a vertical plane for several distances from the sensor to the object (an iron sheet with sizes 710 × 610 mm and a thickness of 1.5 mm)

Table 12.2 Some results of determination of the registration range of response signals from ferromagnetic bodies and objects of various weights and sizes

Body/object	Mass, kg	Sizes, mm	Range of registration, cm
Sphere	0.055	Ø25.5	10
Sphere	0.285	Ø43	15
Sphere	3.780	Ø98.7	35
Cylinder	1.000	Ø62 × 42	20
Cylinder	3.000	Ø90 × 60	30
sheet (from its flat side)		710 × 610 × 1.5	150
sheet (from its edge)	5.500	710 × 610 × 1.5	120

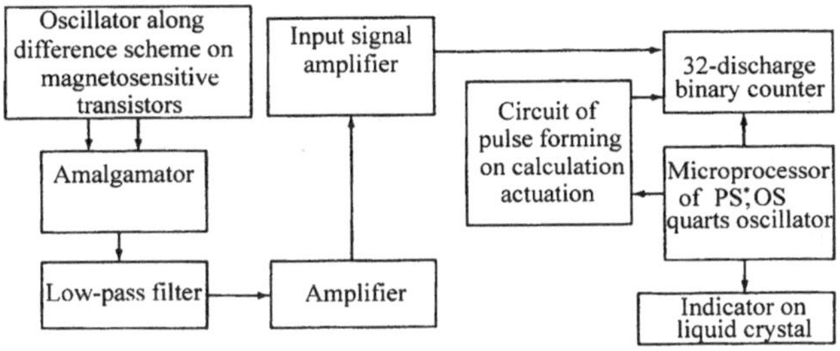

Fig. 12.45 The flowchart of a promising magnetometer

A major characteristic of a magnetic field sensor is the measurement error of the value and direction of the field vector, which can be presented as the sum of regular and random errors. In our case, regular errors are caused by the following major factors:

- Inaccuracy of determination of the field direction of a constant magnet in the field of FMCR and its uncontrollable heterogeneity
- Inaccuracy of orientation of the easy magnetization axis of FMCR in relation to the field of a constant magnet
- Difference of the frequency of generation of the sensor from the FMR frequency because of the presence of reactive elements in the sensor's circuit and of reactivities of the transistor, depending on the operating mode of the sensor
- Long-term instability of the sensor frequency due to, first of all, changes of its temperature and nonsimultaneity of frequency readouts at various orientations of the sensor in space
- Inaccuracy (error) of determination of the differential sensitivity of the sensor to changes of the value of an external magnetic field ($K = \Delta f / \Delta B$)
- Presence of magnetic hysteresis

By calibration of the parameters of the designed sensor using indications of a standard MPF-3MG magnetometer, the regular error did not exceed 3.5%.

Another part of the offered magnetometric sensor is a block (soon – a microcircuit) of processing of response signal by the principle of a digital frequency meter and a spectrum analyzer of the frequency response from a magnetoelectronic converter.

In Fig. 12.45, the flowchart of a promising magnetometer is shown.

12.7 Determination of Earth's Magnetic Field Vector by a Heteromagnetic Sensor

The components of the magnetic induction vector $\overline{B}$ in the YIG sphere of the heteromagnetic sensor are related to the geophysical coordinates (Fig. 12.46) as follows:

$$\begin{aligned} B_x &= B\cos\theta\ \cos\varphi, \\ B_y &= B\cos\theta\ \sin\varphi, \\ B_z &= B\sin\theta, \end{aligned} \tag{12.30}$$

where θ is the angle between the vector $\overline{B}$ and horizontal plane $X0Y$ and φ is the angle between the projection of $\overline{B}$ onto the plane $X0Y$ and axis X.

The value of the resultant induction $\overline{B}$ from addition of the magnetic induction of FMCR $\overline{B}_0(B_{0x}, B_{0y}, B_{0z})$ and Earth's induction $\overline{B}_{\text{Earth}}(B_{\text{Earth},x}, B_{\text{Earth},y}, B_{\text{Earth},z})$ will be

$$B = \sqrt{(B_{0x} + B_{\text{Earth},x})^2 + (B_{0y} + B_{\text{Earth},y})^2 + (B_{0z} + B_{\text{Earth},z})^2}, \tag{12.31}$$

The frequency of generation of the sensor is $f = \gamma B$, where $\gamma = 28\,\text{Hz/nT}$. The value of γ appears a little bit less than its theoretical value, and each copy of the sensor should be adjusted experimentally.

Provided that $B_{\text{Earth}} \ll B_0$, from (12.31) we approximately get

$$\frac{f}{\gamma} = B_0 + \frac{\overline{B}_0\overline{B}_{\text{Earth}}}{B_0} \tag{12.32}$$

and, in view of (12.30),

$$\frac{f}{\gamma} = B_0 + (B_{\text{Earth},x}\cos\theta\cos\varphi + B_{\text{Earth},y}\cos\theta\sin\varphi + B_{\text{Earth},z}\sin\theta). \tag{12.33}$$

For determination of the components B_{Earth} , the heteromagnetic sensor is rotated in three mutually perpendicular planes and in each of them the maximal and minimal

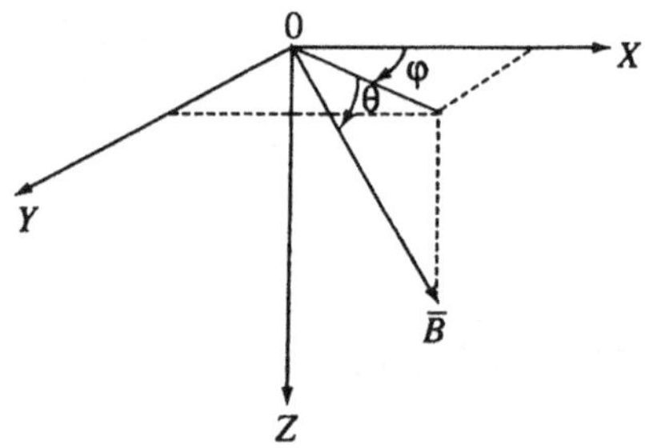

Fig. 12.46 The components of the magnetic induction vector $\overline{B}$ in the YIG sphere of the heteromagnetic sensor are related to the geophysical coordinates

values of frequency are determined. In such a way, the error in determination of the direction B_{Earth} appears high as the angular sensitivity is minimal at a maximum and minimum of the frequency diagram of the sensor. To increase the accuracy of measurements of $\overline{B}_{\text{Earth}}$, a method of registration of frequencies at turning of the sensor by discrete angles (0°, 90°, 180°, and 270°) has been proposed.

To measure $B_{\text{Earth},x}$, it is necessary to position the sensor so that the vector $\overline{B}_0$ is directed along the axis X that gives $\sin\theta_1 = 0$ and $\sin\varphi_1 = 0$. For this purpose, at initial calibration of the made sensor its case should bear the direction of the field $\overline{B}$. Then, for the frequency f_1, we have

$$f_1 = \gamma \cdot B_0 + \gamma \cdot B_{\text{Earth},x} = f_0 + \gamma \cdot B_{\text{Earth},x}. \tag{12.34}$$

Then the sensor is turned in a horizontal plane by the angle $\varphi_2 = 180°$ at $\theta_1 = 0°$. For f_2 we get:

$$f_2 = f_0 - \gamma \cdot B_{\text{Earth},x}. \tag{12.35}$$

The difference between the values of frequencies f_1 and f_2, referred to 2γ, gives the value of $B_{\text{Earth},x}$ – the component of Earth's magnetic induction

$$B_{\text{Earth},x} = \frac{f_1 - f_2}{2\gamma}. \tag{12.36}$$

To measure $B_{\text{Earth},x}$ (the component of Earth's field), it is necessary to direct the vector $\overline{B}_0$ of the sensor along the axis Y, that gives $\sin\theta_1 = 0$, $\sin\varphi_3 = 1$, $\cos\varphi_3 = 0$. For this purpose, the sensor is turned from its initial position by 90° in a horizontal plane clockwise, looking at the sensor from above. The measured value of frequency f_3 for such a position of the sensor is

$$f_3 = f_0 + \gamma \cdot B_{\text{Earth},y}. \tag{12.37}$$

Then the sensor is turned in a horizontal plane by 180° that gives values $\sin\theta_1 = 0$ and $\sin\varphi_4 = -1$. In this case, the frequency is

$$f_4 = f_0 - \gamma \cdot B_{\text{Earth},y}, \tag{12.38}$$

$$B_{\text{Earth},y} = \frac{f_3 - f_4}{2\gamma}. \tag{12.39}$$

For determination of $B_{\text{Earth},z}$, it is necessary to direct the vector $\overline{B}_0$ along the axis Z that gives $\cos\theta_1 = 0$ and $\sin\theta_1 = 1$. For this purpose, the sensor should be turned by 90° in a vertical plane. The angle φ can be chosen arbitrarily. The value of frequency f_5 will be

$$f_5 = f_0 + \gamma \cdot B_{\text{Earth},z}. \tag{12.40}$$

After the turn of the sensor in a vertical plane by 180°, the value of frequency is

$$f_6 = f_0 - \gamma \cdot B_{\text{Earth},z}, \tag{12.41}$$

Table 12.3 The frequencies measurements f for the various angles φ and θ

φ, deg	0	180	90	270	0	180
θ, deg	0	0	0	0	90	270
f, MHz	1,079.70	1,079.25	1,079.13	1,079.75	1,081.10	1,078.80

Then

$$B_{\text{Earth},z} = \frac{f_5 - f_6}{2\gamma}. \tag{12.42}$$

Thus, for determination of all the components $\overline{B}_{\text{Earth}}(B_{\text{Earth},x}, B_{\text{Earth},y}, B_{\text{Earth},z})$ at a given point of the FMCR position, six measurements of frequencies $f_1 \div f_6$ at six different positions of the sensor are enough.

By such a technique, measurements of the frequencies of response signals from the sensor have been made with their results in Table 12.3.

The values of the components of Earth's magnetic induction vector calculated by (12.36), (12.39), and (12.42) are:

$$\begin{aligned} B_{\text{Earth},x} &= 8.04 \times 10^3\,\text{nT}, \\ B_{\text{Earth},y} &= -11.07 \times 10^3\,\text{nT}, \\ B_{\text{Earth},z} &= 41.07 \times 10^3\,\text{nT}. \end{aligned}$$

The horizontal component of Earth's magnetic induction is:

$$B_{\text{hor}} = \sqrt{B^2_{\text{Earth},x} + B^2_{\text{Earth},y}} = 13.68 \times 10^3\,\text{nT}, \tag{12.43}$$

and the modulus is

$$B_{\text{Earth}} = \sqrt{B^2_{\text{hor}} + B^2_{\text{Earth},z}} = 43.3 \times 10^3\,\text{nT}. \tag{12.44}$$

The obtained values agree with the results of measurements of $\overline{B}_{\text{Earth}}$ at rotation of the sensor in three mutually perpendicular planes with findings of maxima and minima of the corresponding frequencies.

In the case of an undefined direction of $\overline{B}_{\text{Earth}}$, a technique is proposed in which some initial values are set: $\varphi = \varphi_0$ and the angle $\theta = \theta_0$. If

$$\frac{f}{\gamma} = B_0 + B_{\text{Earth},x} \cos\theta \cos\varphi + B_{\text{Earth},y} \cos\theta \sin\varphi + B_{\text{Earth},z} \sin\theta, \tag{12.45}$$

then at angles $\varphi_1 = \varphi_0; \varphi_2 = \varphi_0 + (\pi/2); \varphi_{\text{Earth}} = \varphi_0 + \pi; \varphi_0 + (3/2)\pi$, and a constant angle $\theta = \theta_0$, at rotation of the sensor with a step of $\pi/2$ in a horizontal plane, we obtain four corresponding values of the frequency of generation $f_1(\varphi_1)$, $f_2(\varphi_2)$, $f_3(\varphi_3)$, and $f_4(\varphi_4)$.

If from the angle $\theta = \theta_0$, the sensor is turned in a vertical plane by $\pi/2$ $(\theta = \theta_0 + (\pi/2))$ and then is rotated again in a horizontal plane with a step $(\pi/2)(\varphi_1' = \varphi_0, \varphi_2' = \varphi_0 + (\pi/2), \varphi_3' = \varphi_0 + \pi, \varphi_4' = \varphi_0 + (3/2)\pi)$, then the values of frequencies will be $f_5(\varphi_1')$, $f_6(\varphi_2')$, $f_7(\varphi_3')$, $f_8(\varphi_4')$, and then from (12.45) we get

$$\begin{aligned}
\frac{f_5}{\gamma} &= B_0 - B_{\text{Earth},x} \sin\theta_0 \cos\varphi_0 - B_{\text{Earth},y} \sin\theta_0 \sin\varphi_0 + B_{\text{Earth},z} \cos\theta_0,\\
\frac{f_6}{\gamma} &= B_0 + B_{\text{Earth},x} \sin\theta_0 \sin\varphi_0 - B_{\text{Earth},y} \sin\theta_0 \cos\varphi_0 + B_{\text{Earth},z} \cos\theta_0,\\
\frac{f_7}{\gamma} &= B_0 + B_{\text{Earth},x} \sin\theta_0 \cos\varphi_0 + B_{\text{Earth},y} \sin\theta_0 \sin\varphi_0 + B_{\text{Earth},z} \cos\theta_0,\\
\frac{f_8}{\gamma} &= B_0 - B_{\text{Earth},x} \sin\theta_0 \sin\varphi_0 + B_{\text{Earth},y} \sin\theta_0 \cos\varphi_0 + B_{\text{Earth},z} \cos\theta_0.
\end{aligned} \quad (12.46)$$

At subsequent turns in a vertical plane by the angles $(\pi/2)$ $(\theta = \theta_1 + \pi, \theta = \theta_1 + (3/2)\pi)$, we obtain extra eight values of frequency $f_9 \div f_{12}$ and $f_{10} \div f_{16}$, respectively. Solving this set of eight equations, we find the values of the components of Earth's magnetic field induction vector $B_{\text{Earth},x}$, $B_{\text{Earth},y}$, $B_{\text{Ear}th,z}$ and also $\sin\varphi_0, \cos\varphi_0, \sin\theta_0, \cos\theta_0$ in the geophysical system of coordinates.

If we continue turning the sensor in a vertical plane with a step $\pi/2$ and rotating it again in a horizontal plane with the same step $\pi/2$, we get a similar set of eight equations for other values of frequency. The solution of such a set can be used for verification of the sensor.

12.8 Algorithms and Circuit Engineering Solutions for Investigations of Frequency Signal Responses from a Heteromagnetic Sensor

With the purpose of measuring and recording of informative parameters of the output signal from a heteromagnetic sensor, a device on the basis of a Ch3-63/1 frequency meter, a computer interface block, and a personal computer (Fig. 12.47) has been made.

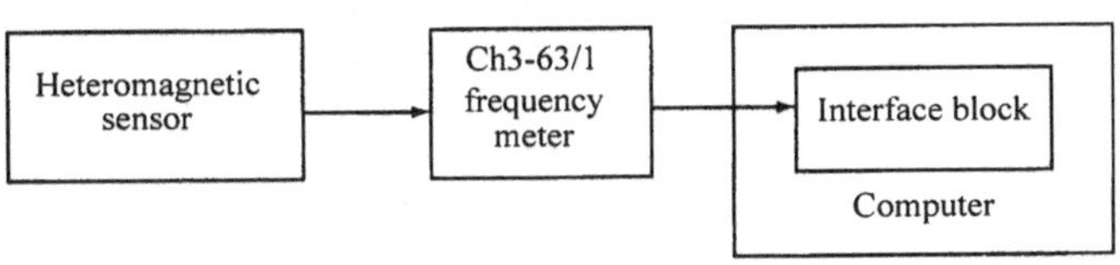

Fig. 12.47 A device on the basis of a Ch3-63/1 frequency meter, a computer interface block, and a personal computer

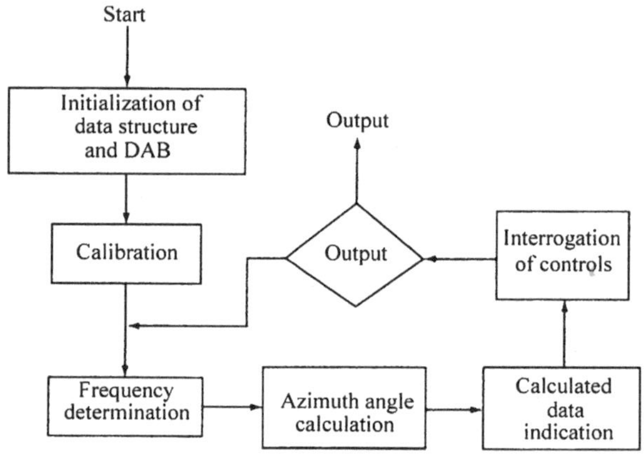

Fig. 12.48 The algorithm for determination of the direction of the sensor to the northern magnetic pole

Our algorithms of the software [87] allow:

- To make calibration of the heteromagnetic sensor
- To record the dependence of the frequency of response signals of the sensor on the angle between the axis of the sensor and the magnetic meridian or azimuth
- To record the angle between the direction to the northern magnetic pole and the axis of the sensor
- To register the long-term and short-term drift of the frequency of response signals
- To register the 6root-mean-square fluctuations of the frequency of response signals

The algorithm for determination of the direction of the sensor to the northern magnetic pole requires carrying out of its preliminary calibration and is shown in Fig. 12.48.

Calibration of the sensor consists in its continuous rotation in the plane of measurements by an angle exceeding 360°. The computer fixes the maximal and minimal values of the frequencies of response signals from the sensor.

The angular dependence of the frequency of response signals of the sensor is discrete. The value of step should be a multiple of 360°. The algorithm of determination of the angular dependence of the frequency of response signals is shown in Fig. 12.49.

The algorithm of recording of the angular dependence of frequency of the sensor is realized in two appendices in [87]: “Diagrams” and “sevender” (“seven”). The “Diagram” appendix realizes the algorithm shown in Fig. 12.49 and outputs a report of the dependences in the form of a text file (protocol.txt) containing angles and the corresponding frequencies of response signals. Processing of the report file and presentation of the results graphically is made by the “sevender” appendix for dependence in an azimuthal plane and by the “seven” appendix for meridian planes. Sample reports of the angular dependences of the frequency of the sensor at rotation in azimuthal and meridional planes are shown in Figs. 12.50 and 12.51.

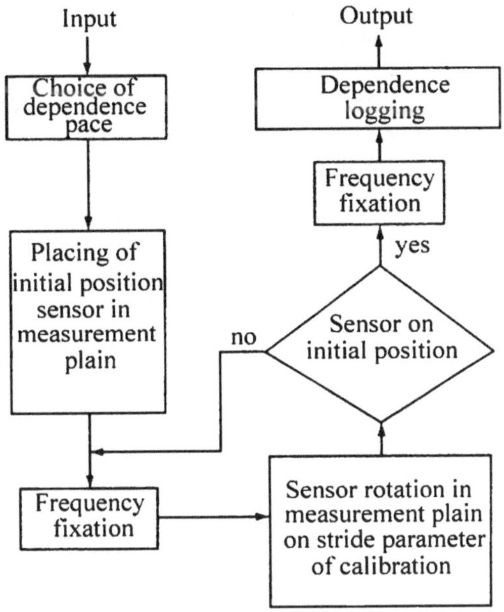

Fig. 12.49 The algorithm of determination of the angular dependence of the frequency of response signals

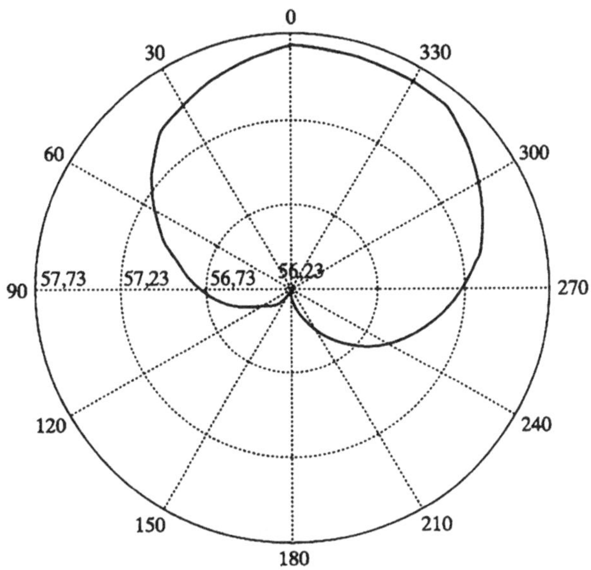

Fig. 12.50 The angular dependences of the frequency of the sensor at rotation in azimuthal planes

Heteromagnetic sensors, whose flowcharts shown in Figs. 12.52 and 12.53, were tested.

In these sensors, a resonant amplifier of 10-cm range with an output power below 5 mW is used.

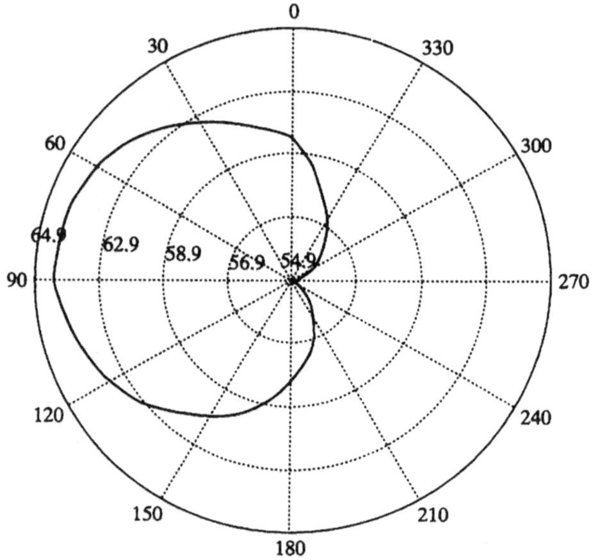

Fig. 12.51 The angular dependences of the frequency of the sensor at rotation in meridional planes

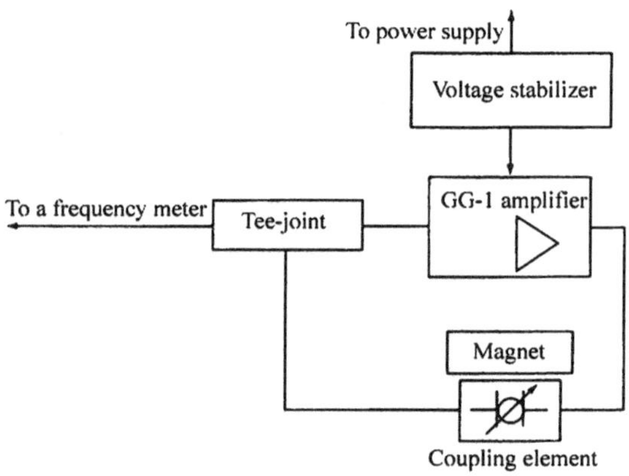

Fig. 12.52 The flowcharts of heteromagnetic sensors

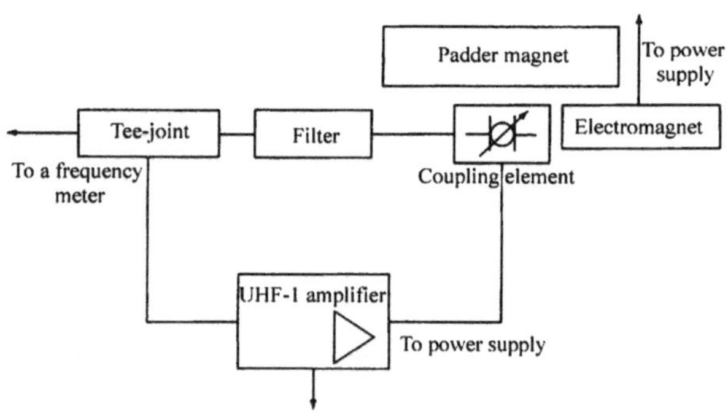

Fig. 12.53 The flowcharts of heteromagnetic sensors

The transfer line is a system of two short-circuited strip elements on a polycoric substrate. The total length of the external feedback circuit, including the transfer line with ferrite, was 15 cm, which provided the presence of the basic frequency $\nu_0 = 1{,}250\,\text{MHz}$ and a harmonic $\nu_2 = 2{,}500\,\text{MHz}$ in the output spectrum of signals.

To increase the relative sensitivity of the sensor to weak changes of magnetic induction, measurements were made on a difference frequency with the use of a heterodyne and an amalgamator from which the intermediate (difference) frequency was recorded to the display device. Changes of the intermediate frequency as a function of magnetic induction corresponded to the law of change of the frequency of the sensor.

In Fig. 12.54, a magnetometer on the basis of a heteromagnetic sensor is shown: 1 – a sensor; 2 – a frequency meter; and 3 – a computer. In Fig. 12.55, a D-1

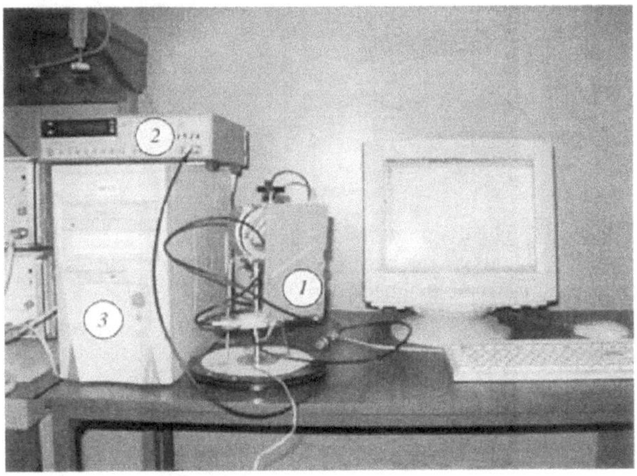

Fig. 12.54 A magnetometer on the basis of a heteromagnetic sensor: 1 – a sensor; 2 – a frequency meter; and 3 – a computer

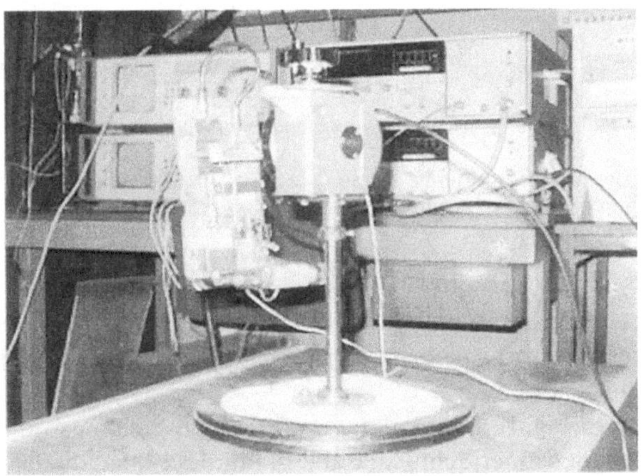

Fig. 12.55 A D-1 sensor assembled under the circuit in Fig. 12.52 (fixed on a two-coordinate rotary mechanism)

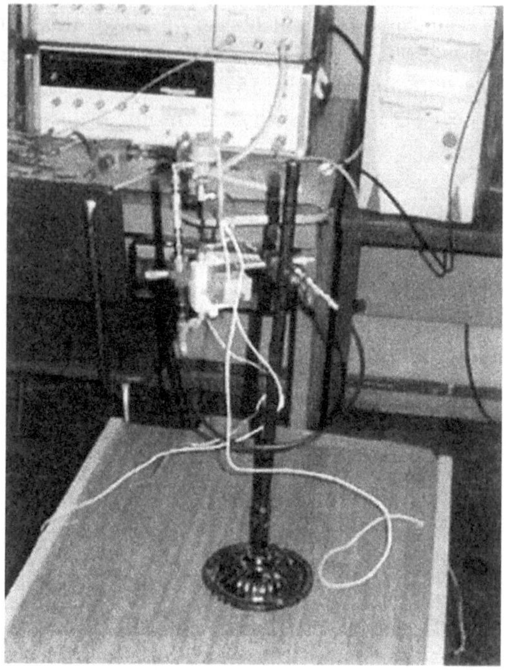

Fig. 12.56 A D-2 sensor assembled under the circuit in Fig. 12.53

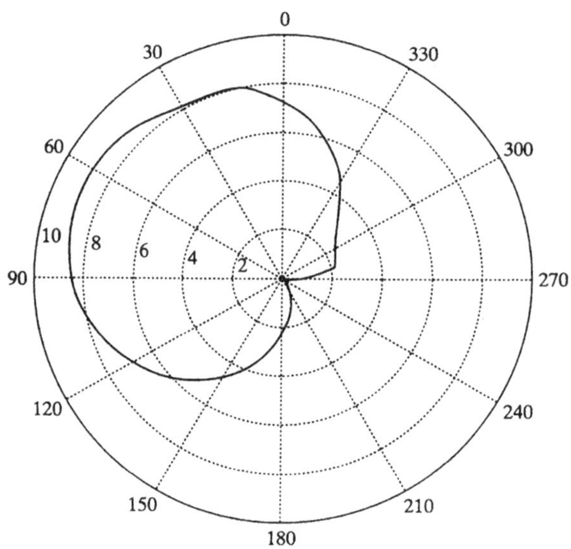

Fig. 12.57 The diagram of response signals on the basic frequency ν_0 in the magnetic meridian plane

sensor assembled under the circuit in Fig. 12.52 (fixed on a two-coordinate rotary mechanism) is shown. Figure 12.56 presents a D-2 sensor assembled under the circuit in Fig. 12.53.

In Fig. 12.57 the diagram of response signals on the basic frequency ν_0 in the magnetic meridian plane and in Fig. 12.58 that on the harmonic component ν_1 in an azimuthal plane are shown.

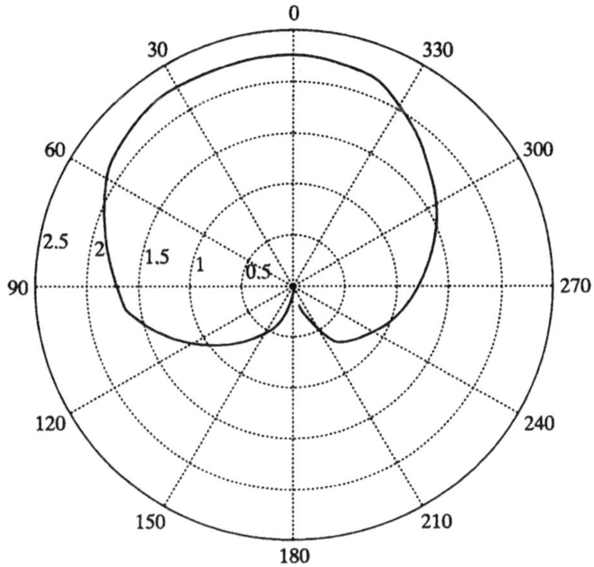

Fig. 12.58 The diagram of response signals on the harmonic component ν_1 in an azimuthal plane

With our heteromagnetic sensors, the following parameters were obtained:

- The absolute sensitivity for $\nu_0 = 1{,}200$ MHz (the basic frequency) – 16.6 kHz/deg and for $\nu_3 = 3{,}600$ MHz (the third harmonic) – 50 kHz/deg
- The relative sensitivity for the basic frequency and its harmonics was 1.38×10^{-5} 1/deg and for $\nu_{\text{int.f}} = 50$ MHz on the third harmonic – 10^{-3} 1/deg
- The maximal deviation of frequency of the third harmonic in the magnetic meridian plane was 9 MHz, that is, $H_3 = 5 \times 10^4$ nT, that corresponds to the reference value $H_3 = 51{,}197$ nT for Saratov City (Russian Federation)
- By rotation in the magnetic meridian plane (Fig. 12.57) by the maximum of frequency deviation, the inclination of Earth's magnetic induction vector $\theta_3 = 60°$ was determined, which also corresponds to the reference value $\theta = 67°$ for Saratov City
- By rotation in an azimuthal plane (Fig. 12.58), the direction to the magnetic pole was determined.

Chapter 13
Low-Noise Amplifiers on Magnetotransistors Below 40 GHz

The results of theoretical researches of the physical principles of functioning and ways of design of low-noise amplifiers on the basis of transistor-magnetic structures, which can be promising for satellite communication systems with a raised density of information channels, are considered. Directions of noise decrease (the noise factor below 3.5 dB), and management of the selective properties in magnetotransistor amplifiers with a dynamic range not less than 70 dB, with a minimum level of signal −156 dBW, and with a saturation signal above −70 dBW is investigated.

CAD methods and means necessary for calculation of a similar sort of amplifiers are selected.

13.1 Power Level and Dynamic Range. Choice of a Linear Transistor Model for Calculation

The input cascades of amplifiers of satellite communication systems should be sensitive to the levels of thermal noise power P_0 and possess an expanded linear dynamic range of the order of magnitude of 70 dB. According to Naiquist's theorem, these noises are at a level $P_0 = NkT_0\Delta f$, where $k = 1.38 \times 10^{-23}$ J/K is the Boltzmann constant, T_0 the temperature of free space (near space), Δf the band of frequencies, and N is a number. For overlapping from the VHF up to the near part of the EHF range, these are power levels of 140–156 dBW or 110–126 dBW. With the purpose to check the CAD capabilities regarding the minimum power levels and to estimate the dynamic range for low-signal linear modes of the one-cascade transistor, a field transistor PTSHG-1000 (a field transistor with a Schottky barrier and a shutter length of 1,000 μm) has been chosen. A nonlinear model *Materka FET* was used. The circuit of the amplifier is presented in Fig. 13.1.

On a frequency of 5 GHz, the input of the amplifier was subjected to signals whose power varied from −200 dBmW up to +70 dBmW with a step of 1 dBmW. The dependence of the transfer factor of such an amplifier on the input power level was calculated (Fig. 13.2).

One can see that on change of the input power level from −170 dBmW up to −116 dBmW, the transfer factor of the amplifier drops with a slope

A.A. Ignatiev and A.V. Lyashenko, *Heteromagnetic Microelectronics: Microsystems of Active Type*, DOI 10.1007/978-1-4419-6002-3_13,

Fig. 13.1 The circuit of the amplifier on the base of a nonlinear model Materka FET

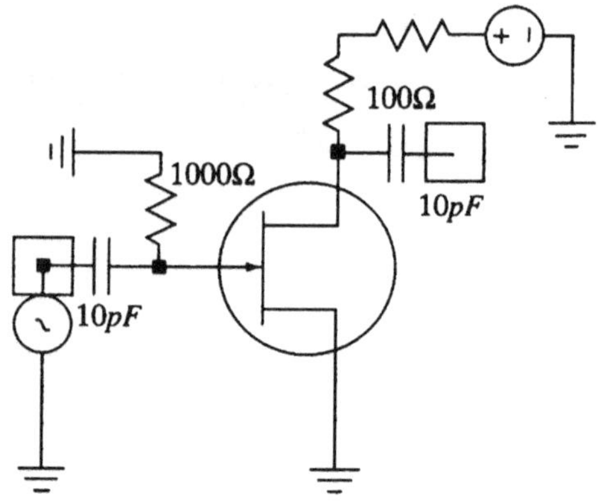

Fig. 13.2 The dependence of the transfer factor of an amplifier on the input power level

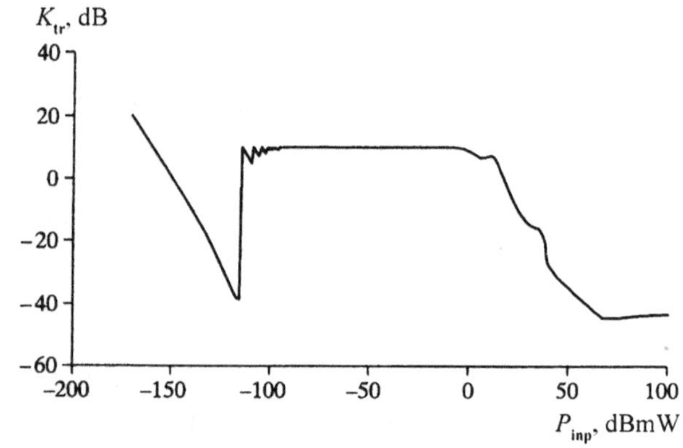

Fig. 13.3 The circuit of usable MECE with a short-circuited, grounded, loop-like conductors

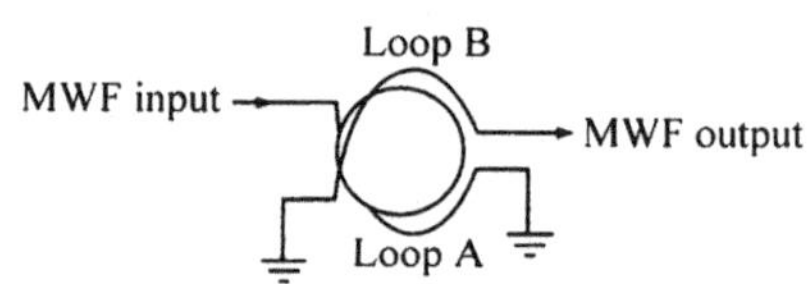

−1.1 dB/dBmW. At a power level of the order of magnitude −115 dBmW, the transfer factor of the amplifier sharply increases, and within the limits of power changes from −115 up to −95 dBmW, the transfer factor of the amplifier has a constant value and does not depend on the level of input power. Thus, CAD can be used for theoretical calculations of low-noise amplifiers. We shall use further a model of the transistor in the form of a two-port network described by the S parameters of scattering and noise parameters.

The low-noise magnetoamplifier consists of two basic parts, namely, an active element – a transistor and a magnetoelectronic coupling element. MECE is connected to the microwave path, with short-circuited, grounded, loop-like conductors (Fig. 13.3).

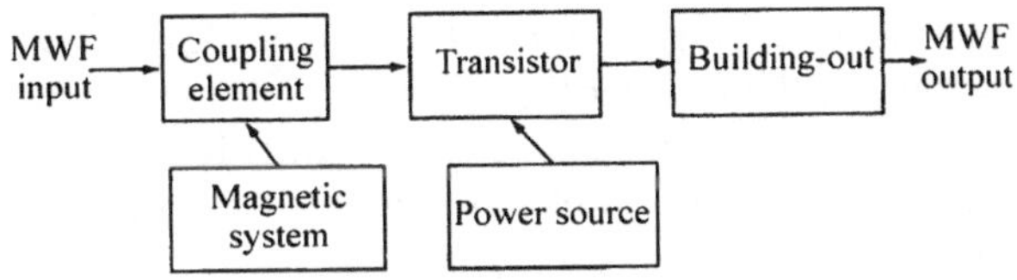

Fig. 13.4 The flowchart of the magnetoamplifier

The YIG sphere is located in the external magnetic field created by a constant magnet or an electromagnet. The coupling element has its resonant transfer characteristic, whose central frequency is changed by the magnetic field.

The flowchart of the magnetoamplifier is shown in Fig. 13.4. The coupling element is located on the input of the transistor and implements two functions, namely, coordination of the input of the transistor with the microwave path and change of the central frequency of the amplifier.

Selectivity is reached due to the resonant transfer characteristic of MECE, which has its GB product of ~100. The central frequency linearly depends on the external bias magnetic field. Variation of the value of the external magnetic field provides frequency-selective amplification of input signals in the passband of the amplifier.

By calculations, the frequency transfer characteristics and the dependences of the noise factor of magnetoamplifiers on changes of the external magnetic field were estimated.

13.2 Choice and Substantiation of Coupling Element for a Frequency Band Below 40 GHz

Calculations involve the following parameters of YIG spheres:

- ds – The diameter of the sphere
- msat – Saturation magnetization
- dh – The resonant linewidth

The parameters of a coupling element are as follows:

- h – The value of the external magnetic field
- dla – The diameter of loop A
- dlb – The diameter of loop B
- dwa – The diameter of loop A 's wire
- dwb – The diameter of loop B's wire
- aa – The angle at which loop A wraps the sphere
- ab – The angle at which loop B wraps the sphere
- aab – The angle between loops A and B

For a range of 0.3–0.4 GHz, a YIG sphere of the 8KG type was chosen. The coupling element had the following parameters: $h = 115$ Oe, ds $= 0.8$ mm, msat $= 90$ G, dh $= 0.5$ Oe, dla $= 0.81$ mm, dwa $= 0.02$ mm, dlb $= 0.81$ mm, dwb $= 0.02$ mm, aa $= 360°$, ab $= 360°$, and aab $= 90°$. The magnetic field h varied from 115 up to 130 Oe with a step of 5 Oe. Figure 13.5 presents the dependence of the noise factors K_n (hereinafter the upper curves) and the transfer factor $K_{car.f}$ (hereinafter the lower curves) in the magnetoamplifier for different magnetic fields. The transfer factor on the resonant frequency was ~ -3.5 dB ± 0.5 dB, and the noise factor was 3.0 dB $\pm$ 0.2 dB.

For a range of 3–4 GHz, a YIG sphere of the 65KG type was chosen. For the coupling element, the parameters were set as follows: h = 1,215 Oe, ds = 0.4 mm, msat = 820 G, dh = 0.3 Oe, dla = 0.41 mm, dwa = 0.02 mm, dlb = 0.41 mm, dwb = 0.02 mm, aa = 180°, ab = 180°, and aab = 90°. The magnetic field h varied from 122 up to 1,385 Oe with a step of 20 Oe. In Fig. 13.6 , the dependences of the noise factors and transfer factors for different magnetic fields are presented. On the resonant frequency, the transfer factor was ~ -0.1 dB, and the noise factor was of the order of magnitude of 0.1 dB.

For a range of 7.0–8.0 GHz, a YIG sphere of the 65KG type was selected. For the coupling element, the parameters were set as follows: h = 2,580 Oe, ds = 0.4 mm, msat = 820 G, dh = 0.3 Oe, dla = 0.41 mm, dwa = 0.02 mm, dlb = 0.41 mm, dwb = 0.02 mm, aa = 180°, ab = 180°, and aab = 90°. The magnetic field h varied from 2,580 up to 2,760 Oe with a step of 20 Oe. In Fig. 13.7, the dependences of the factors of noise and transfer for different magnetic fields are presented. On the resonant frequency, the transfer factor was ~ -0.1 dB, and the noise factor was $\sim$0.1 dB.

For a range of 20–21 GHz, a YIG sphere of the 140KG type was chosen. For the coupling element, the parameters were set as follows: h = 7,100 Oe, ds = 0.4 mm,

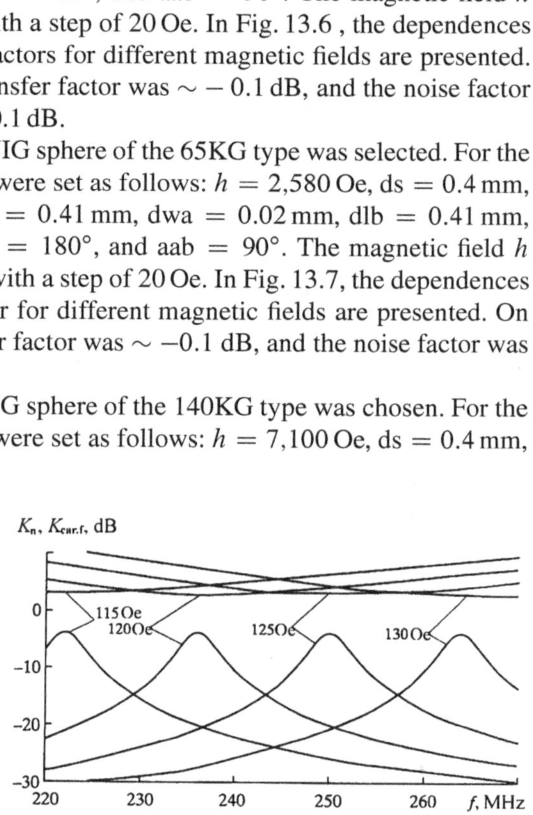

Fig. 13.5 The dependence of the noise factors K_n (hereinafter the *upper curves*) and the transfer factor $K_{car.f}$ (hereinafter the *lower curves*) in the magnetoamplifier for different magnetic fields

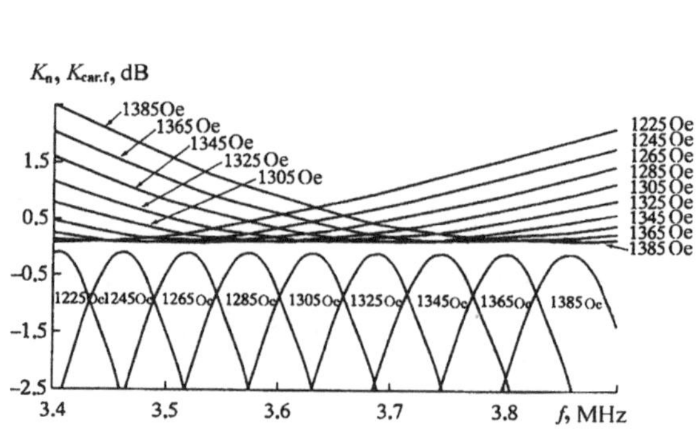

Fig. 13.6 The dependences of the noise factors and transfer factors for different magnetic fields

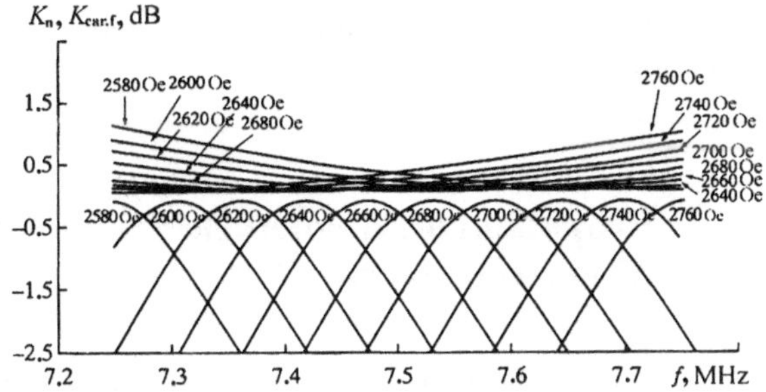

Fig. 13.7 The dependences of the factors of noise and transfer for different magnetic fields

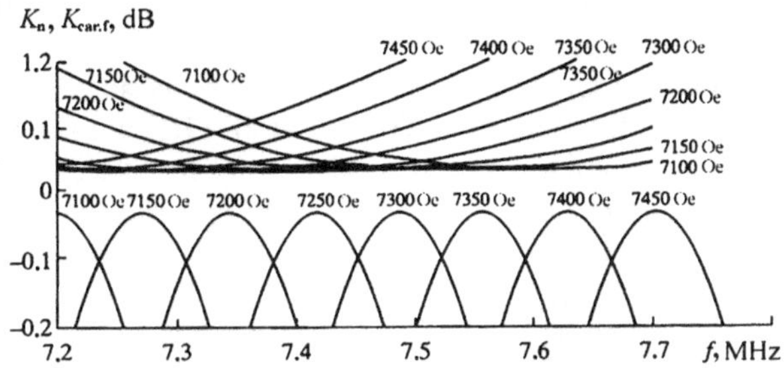

Fig. 13.8 The dependences of the factors of noise and transfer for different magnetic fields

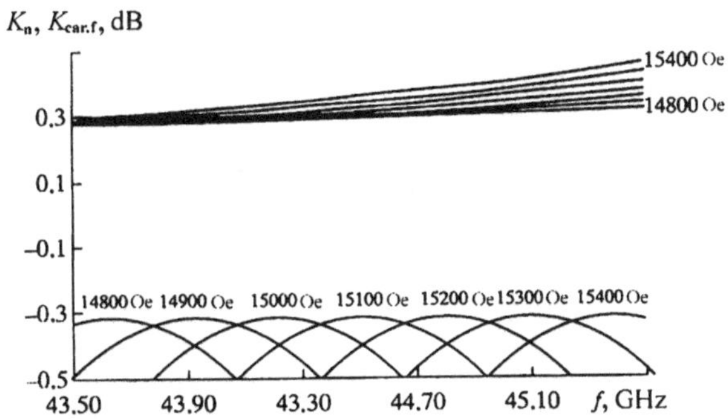

Fig. 13.9 The dependences of the factors of noise and transfer of the magnetoamplifier for several magnetic fields

msat = 1,750 G, dh = 0.2 Oe, dla = 0.41 mm, dwa = 0.02 mm, dlb = 0.41 mm, dwb = 0.02 mm, aa = 180°, ab = 180°, and aab = 90°. The magnetic field h varied from 7,100 up to 7,450 Oe with a step of 50 Oe. In Fig. 13.8, the dependences of the factors of noise and transfer for different magnetic fields are presented. The transfer factor on the resonant frequency of the magnetoamplifier was ~ -0.1 dB, and the noise factor of noise was ~0.1 dB.

For a range of 43.5–45.5 GHz, a YIG sphere of the 8HB type (doped barium hexaferrite) was chosen. For the coupling element, the parameters were set as follows: $h = 14,800$ Oe, ds = 0.4 mm, msat = 4,520 G, dh = 35 Oe, dla = 0.41 mm, dwa = 0.02mm, dlb = 0.41 mm, dwb = 0.02 mm, aa = 180°, ab = 180°, and aab = 90°. The magnetic field h varied from 14,800 up to 15,400 Oe with a step of 100 Oe. In Fig. 13.9, the dependences of the factors of noise and transfer of the magnetoamplifier for several magnetic fields are presented.

The transfer factor on the resonant frequency in all the cases was ∼ − 0.3 dB. The noise factor was ∼0.3 dB.

13.3 Projection of Magnetoelectronic One-Stage Amplifier on Magnetotransistor

The subsequent calculations involve NEC's field transistor *N*24200 *A*, for which parameters of dispersion and noise parameters up to 30 GHz that are set were known.

For a range of 0.3–0.4 Hz, the scheme of the amplifier shown in Fig. 13.10 was used.

In Figs. 13.11 and 13.12 , the transfer characteristic of the amplifier and the noise factor in a frequency band of 0.32–0.37 GHz are shown, respectively.

In the magnetoamplifiers, MECE for the corresponding range and an additional dividing capacitor were placed at the input instead of an inductance – see the circuit in Fig. 13.13.

In Fig. 13.14, the dependences of the factors of amplification (a) and noise (b) in the selective-type magnetoamplifier resulted from change of the external magnetic field in MECE from 115 to 130 Oe with a step of 5 Oe.

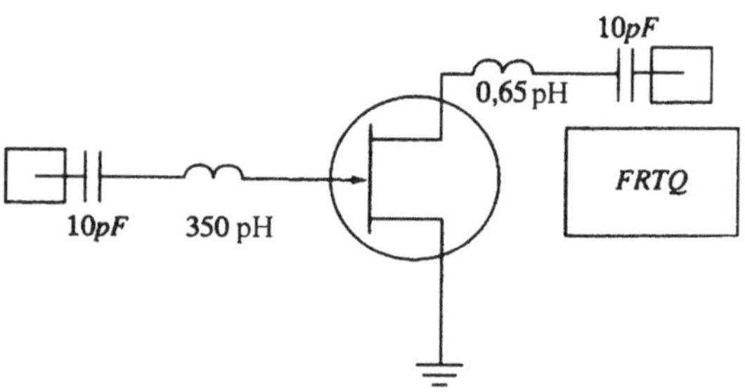

Fig. 13.10 The scheme of the amplifier for a range of 0.3–0.4 Hz

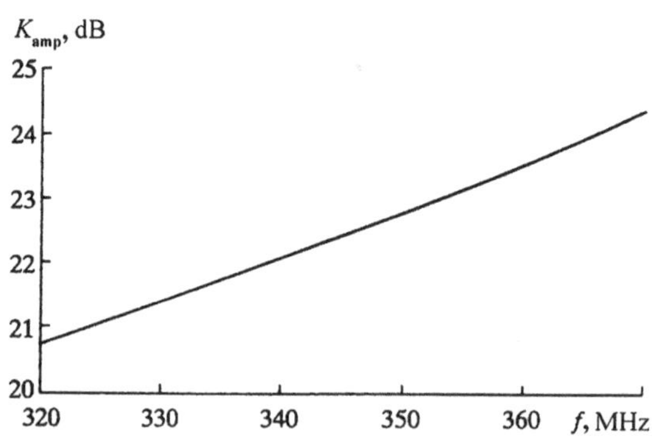

Fig. 13.11 The transfer characteristic of the amplifier

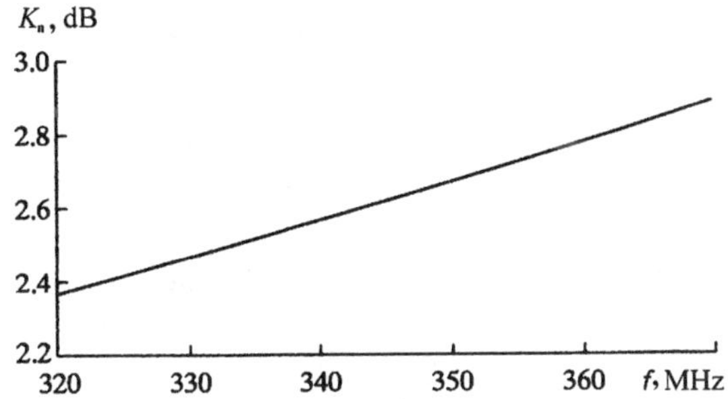

Fig. 13.12 The noise factor of the amplifier in a frequency band of 0.32–0.37 GHz

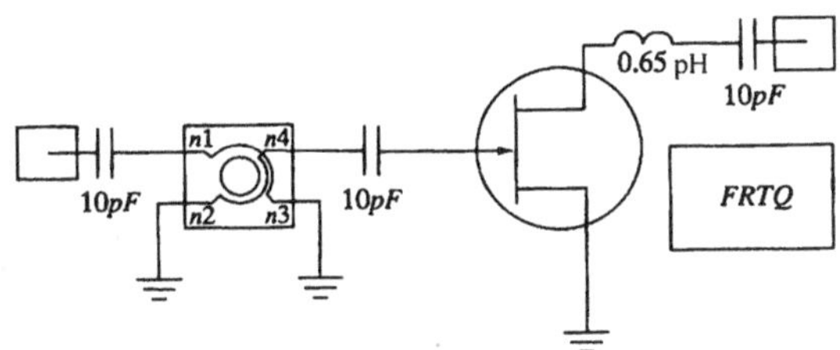

Fig. 13.13 The magnetoamplifiers with MECE for the corresponding range and an additional dividing capacitor

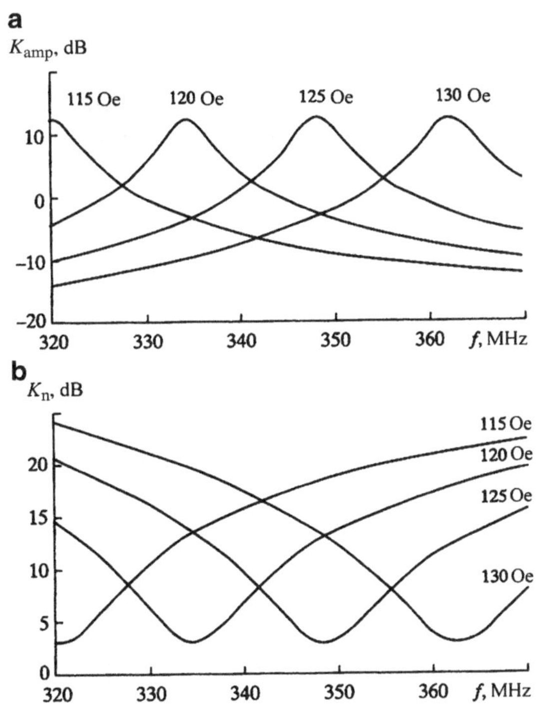

Fig. 13.14 The dependences of the factors of amplification (**a**) and noise (**b**) in the selective-type magnetoamplifier on change of the external magnetic field in MECE from 115 to 130 Oe with a step of 5 Oe

From Fig. 13.14b, it is seen that the noise factor of the magnetoamplifier reaches 3.0 dB with MECE, and in the common operating mode of the transistor, it was 2.36–2.88 dB (see Fig. 13.12).

An amplifier for a range of 3–4 GHz is shown in Fig. 13.15.

In Fig. 13.16, the transfer characteristic of the transistor amplifier (a) and its noise factor (b) are shown.

The equivalent circuit of a magnetoamplifier for a range of frequencies (3.4–3.9 GHz) is shown in Fig. 13.17. MECE for the corresponding range and an additional dividing capacitor were placed at the input of the amplifier.

In Fig. 13.18, the dependences of K_{amp} (a) and K_n (b) in the selective-type magnetoamplifier resulted from change of the external magnetic field in MECE from 1,225 to 1,375 Oe with a step of 50 Oe.

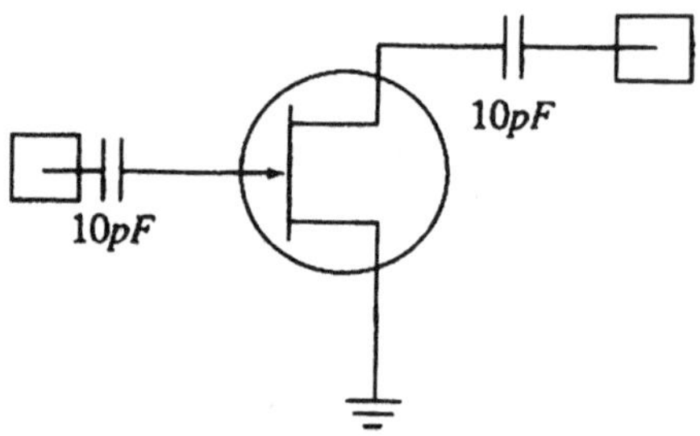

Fig. 13.15 An amplifier for a range of 3–4 GHz

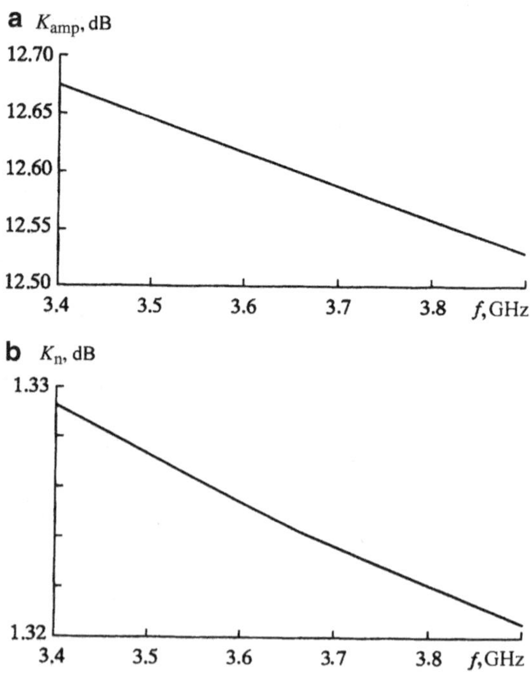

Fig. 13.16 The transfer characteristic of the transistor amplifier (**a**) and its noise factor (**b**)

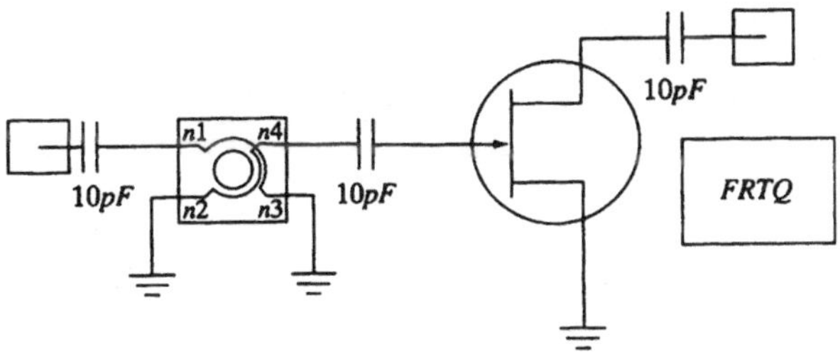

Fig. 13.17 The equivalent circuit of a magnetoamplifier for a range of frequencies 3.4–3.9 GHz

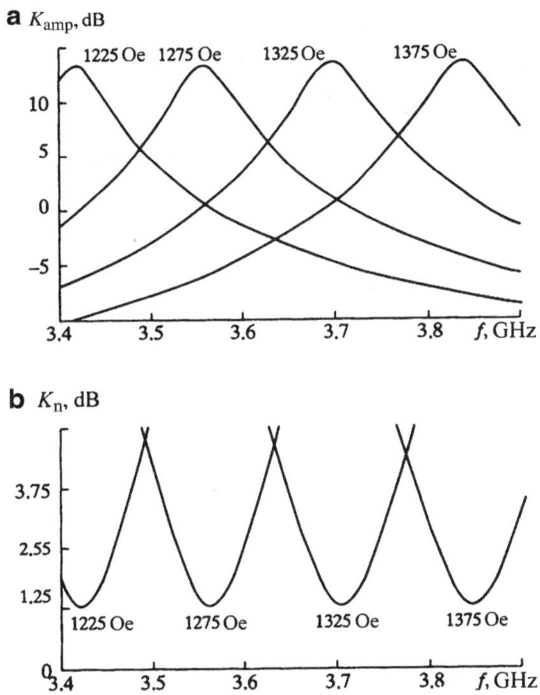

Fig. 13.18 The dependences of K_{amp} (**a**) and K_n (**b**) in the selective-type magnetoamplifier on change of the external magnetic field in MECE from 1,225 to 1,375 Oe with a step of 50 Oe

From Fig. 13.18b, it is obvious that the noise factor in the magnetoamplifier was 1.25 dB with MECE, which is lower in comparison with the amplification mode of the transistor (1.33 dB – see Fig. 13.16).

A field-transistor amplifier for a range of 7–8 GHz is shown in Fig. 13.19.

In Fig. 13.20, the transfer characteristic of the transistor amplifier (a) and its noise factor (b) are shown.

In the magnetoamplifier, MECE for the corresponding range and an additional dividing capacitor were placed at the input of the field transistor (Fig. 13.21).

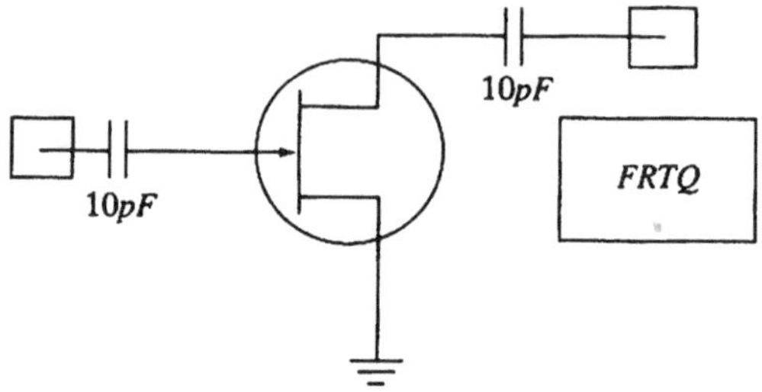

Fig. 13.19 A field-transistor amplifier for a range of 7–8 GHz

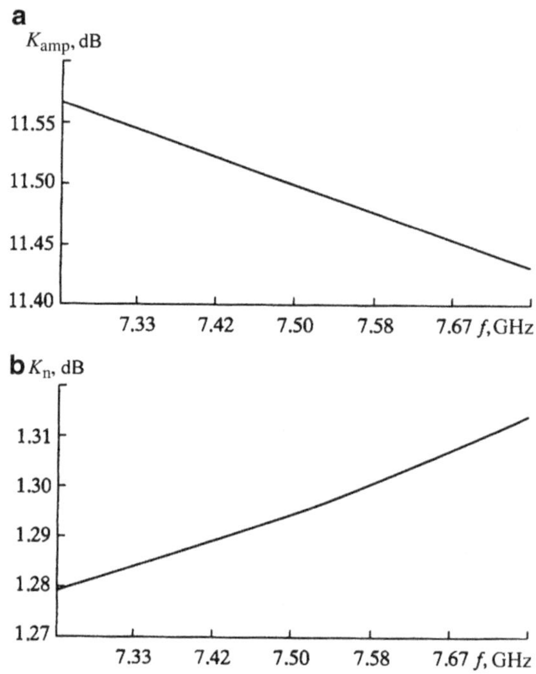

Fig. 13.20 The transfer characteristic of the transistor amplifier (**a**) and its noise factor (**b**)

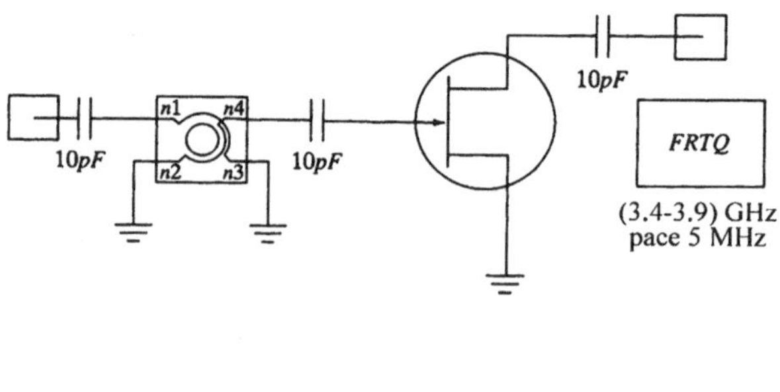

Fig. 13.21 The magnetoamplifier with MECE at the input of the field transistor with additional dividing capacitor

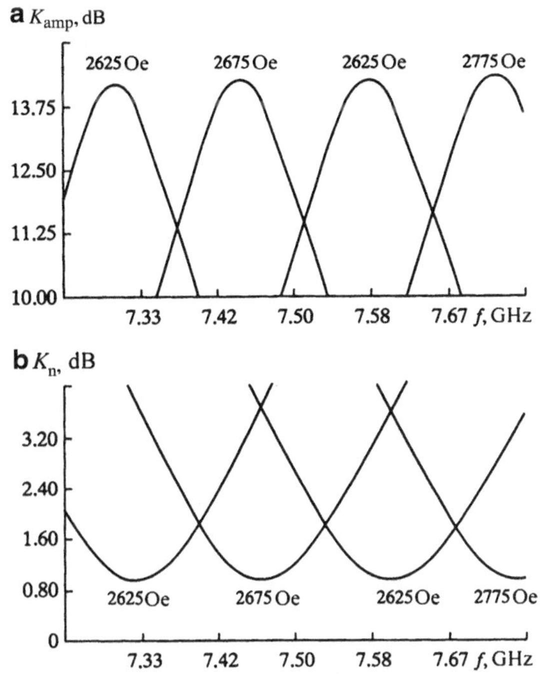

Fig. 13.22 The dependences of K_{amp} (**a**) and K_n (**b**) in the magnetoamplifier of selective type on change of the external magnetic field in MECE from 2,625 to 2,775 Oe with a step of 50 Oe

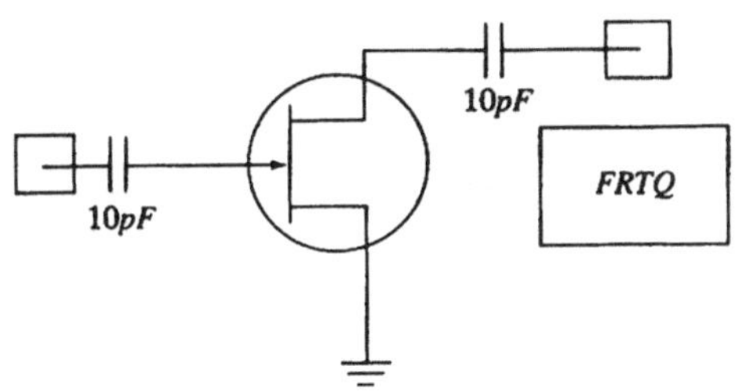

Fig. 13.23 An amplifier for a range of 20–21 GHz

In Fig. 13.22, the dependences of (a) K_{amp} and (b) K_n in the magnetoamplifier of selective type resulted from change of the external magnetic field in MECE from 2,625 to 2,775 Oe with a step of 50 Oe.

From Fig. 13.22b, it is seen that the noise factor in the magnetoamplifier was 1.0 dB with MECE, which is lower than K_n in the transistor amplifier (1.3 dB – Fig. 13.20).

An amplifier for a range of 20–21 GHz is shown in Fig. 13.23.

In Fig. 13.24, the transfer characteristic (a) and the noise factor (b) of the transistor amplifier are shown.

The circuit of a magnetic amplifier with MECE at the input of the transistor is presented in Fig. 13.25 – it differs from that in Fig. 13.21 by different parameters of the block *FRTQ* only.

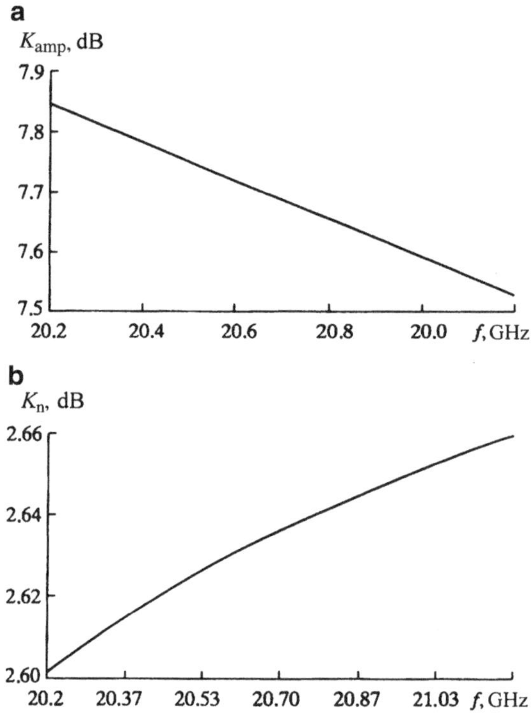

Fig. 13.24 The transfer characteristic (**a**) and the noise factor (**b**) of the transistor amplifier

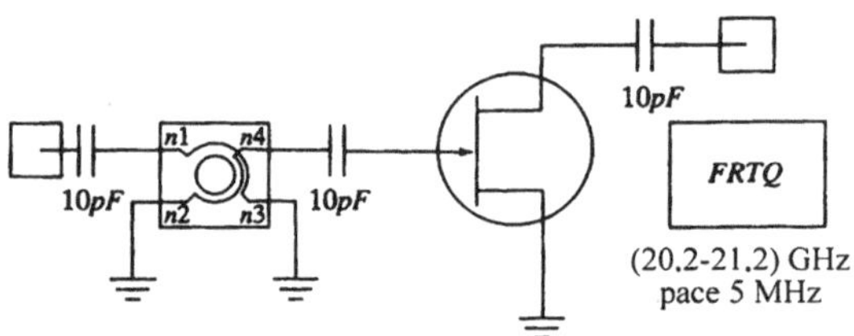

Fig. 13.25 The circuit of a magnetic amplifier with MECE at the input of the transistor

In Fig. 13.26, the dependences of (a) K_{amp} and (b) K_n in the magnetoamplifier resulted from change of the external magnetic field in MECE from 6,950 to 7,250 Oe with a step of 100 Oe.

From Fig. 13.26b, it is obvious that the noise factor in the magnetoamplifier was 1.9 dB, which is lower than K_n in the transistor amplifier (2.6 dB).

A transistor amplifier for a range of 45–46 GHz is shown in Fig. 13.27.

In Fig. 13.28, the transfer characteristic of such a transistor amplifier (a) and its factor of noise (b) are shown.

The circuit of a magnetoamplifier with MECE at the input of the transistor is shown in Fig. 13.29.

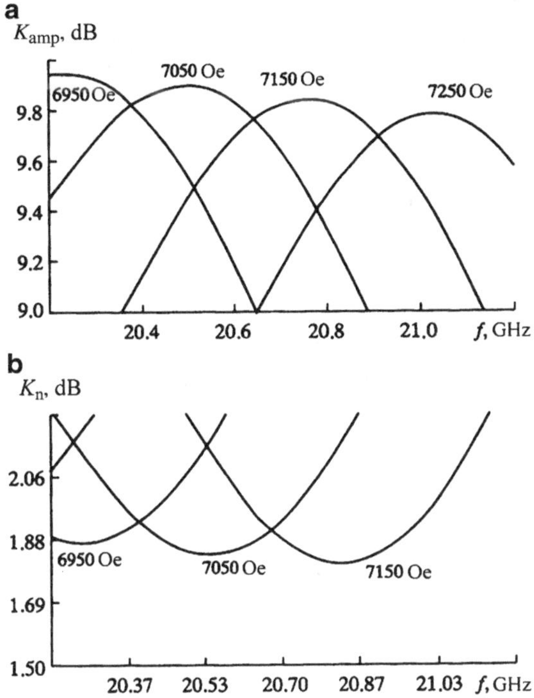

Fig. 13.26 The dependences of (**a**) K_{amp} and (**b**) K_n in the magnetoamplifier resulted from change of the external magnetic field in MECE from 6,950 to 7,250 Oe with a step of 100 Oe

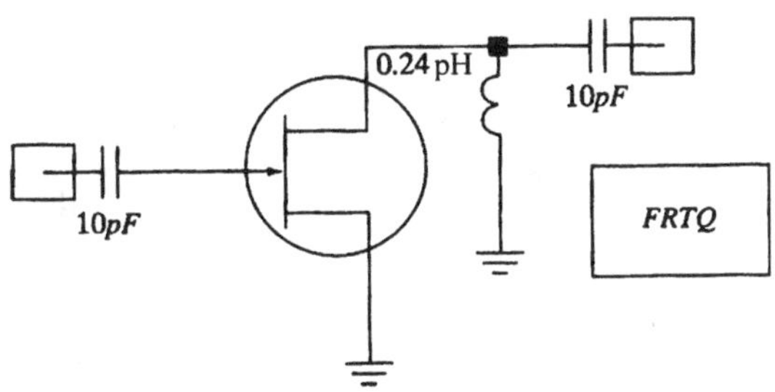

Fig. 13.27 A transistor amplifier for a range of 45–46 GHz

In Fig. 13.30, the dependences of (a) K_{amp} and (b) K_n in the magnetoamplifier resulted from change of the external magnetic field in MECE from 14,500 to 15,100 Oe with a step 200 Oe.

By changing the external magnetic field of the coupling element from 14,500 to 15,100 Oe with a step 200 Oe, we obtain a series of curves for the factor of transfer of the magnetoelectronic amplifier (Fig. 13.38) and its factor of noise (Fig. 13.39).

From Fig. 13.30b, it is seen that the factor of noise of the magnetoamplifier was made 2.5 dB, which is lower than K_n in the mode of amplification of the transistor (2.7 dB).

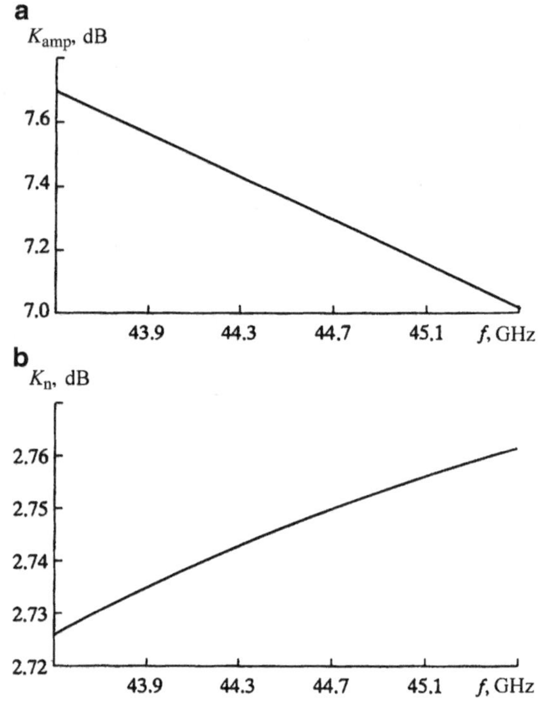

Fig. 13.28 The transfer characteristic of such a transistor amplifier (**a**) and its factor of noise (**b**)

Fig. 13.29 The circuit of a magnetoamplifier with MECE at the input of the transistor

From the figures, it follows that in the amplifier with MECE the factor of noise on 0.5–0.7 dB is below in comparison with the base transistor amplifier. In a range of 0.3–0.4 GHz, there is no gain in the factor of noise, but selective frequency-change amplification is provided, and the factor of noise is less than 3.5 dB.

In Table 13.1, the values of K_{amp} of the used single-cascade transistor amplifiers, their factors of noise K_n, the factors of noise of the magnetoamplifiers designed on these transistors K_{nm}, and the difference $\Delta K_n = K_{nm} - K_n$ for ranges of frequencies of 0.3–0.4 GHz; 3.0–4.0 GHz; 7.0–8.0 GHz; 20.0–21.0 GHz; and 43.0–46.0 GHz are shown.

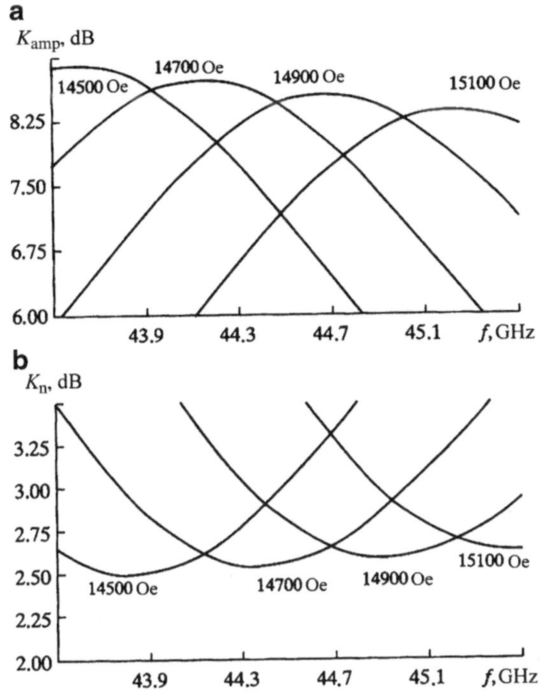

Fig. 13.30 The dependences of K_{amp} (**a**) and K_n (**b**) in the magnetoamplifier on change of the external magnetic field in MECE from 14,500 to 15,100 Oe with a step of 200 Oe

Table 13.1 The values of K_{amp} of the used single-cascade transistor amplifiers, their factors of noise K_n, the factors of noise of the magnetoamplifiers designed on these transistors K_{nm}, and the difference $\Delta K_n = K_{nm} - K_n$

Range of frequencies, GHz	0.3–0.4	3.0–4.0	7.0–8.0	20.0–21.0	43.0–46.0
K_{amp}, dB	20–25	13–12.5	11.6–11.4	8.0–7.5	7.8–7.0
K_{nm}, dB	3.0	1.25	1.0	1.9	2.5
K_n, dB	2.5	1.3	1.3	2.6	2.7
ΔK_n, dB	+0.5	−0.05	−0.3	−0.7	−0.2

Chapter 14
Magnetotransistors and Their Technologies

Bipolar and field magnetic transistors of LPL and HPL in a range of frequencies below 1,000 GHz in various modes: amplification, generation, and transformation of signals are considered. The properties and parameters of various kinds of microstrip coupling elements of high (few Ws) power levels are analyzed. Multiparametric vector sensors of mechanical displacements, statistical, and dynamic quantities are considered.

14.1 Magneto-FET of High Power Level in Intense and Generator Modes

Our researches have shown that the maximum level of power which can be passed through a coupling element depends not only on the factor of filling of this coupling element of FMCR, but also on the threshold power level [17–19]. By the use of two and more FMCRs in the coupling element, its threshold power increases, which ensures the functioning of HMT at high power levels.

The coupling elements of a high power level contain two coupling elements with FMCR connected in parallel (Fig. 14.1) for frequency ranges: (a) 0.2–1.0 GHz and (b) 1.0–5.0 GHz. By the use of the summation principle with a greater number of FMCR (three, four, and more), the level of working power will be higher.

Figure 14.2 shows an HPL MECE with a high factor of filling of FMCR of the coupling loop, which has:

- Losses of transfer of 5 dB in the FMR mode
- A range of frequencies from 0.2 to 18.0 GHz.

Low-power field magnetic transistors with Schottky's barrier and a 300−μm shutter for output power levels below 100 mW have been developed for research. All the FEMT had the basic circuit as shown in Fig. 14.3.

Figure 14.4 shows the topology of low-power FEMT for ranges of frequencies: (a) 0.3–2 GHz; (b) 1–4 GHz; (c) 0.3–2.2 GHz; and (d) 2.5–15.0 GHz. The size of FEMT's crystal did not exceed $2.5 \times 2.5\,\mathrm{mm}^2$.

A.A. Ignatiev and A.V. Lyashenko, *Heteromagnetic Microelectronics: Microsystems of Active Type*, DOI 10.1007/978-1-4419-6002-3_14,

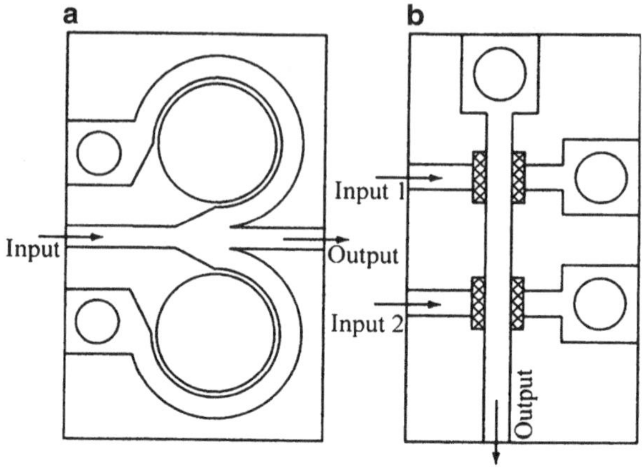

Fig. 14.1 The coupling elements of a high power level for frequency ranges: (**a**) 0.2–1.0 GHz, (**b**) 1.0–5.0 GHz

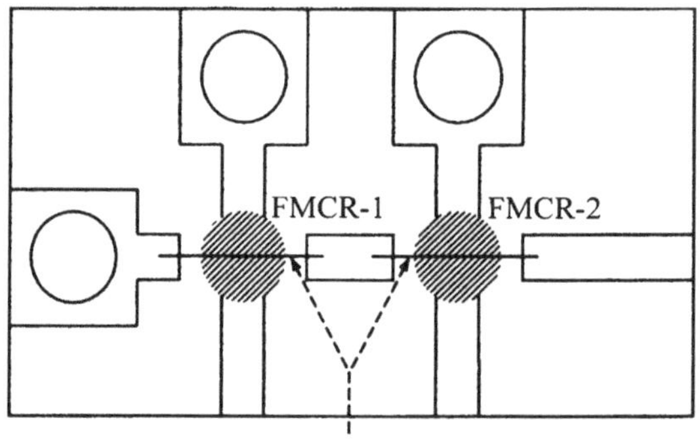

Fig. 14.2 An HPL MECE with a high factor of filling of FMCR of the coupling loop

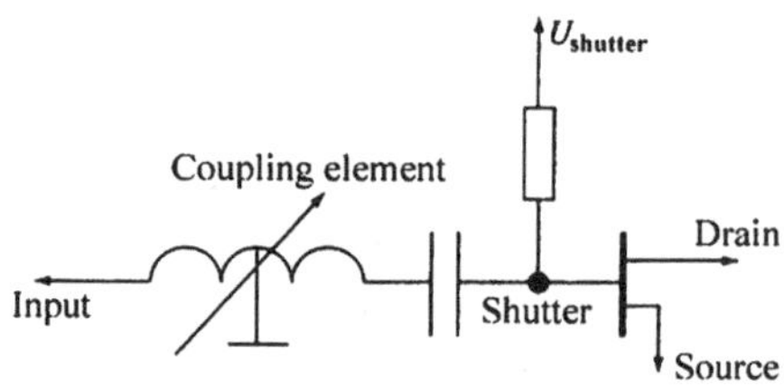

Fig. 14.3 The basic circuit of FEMT

For realization of frequency-selective amplification mode, the coupling element with FMCR was placed on the input of the field transistor shutter through a dividing capacitor. The working frequency of such a FEMT was set by an external magnetic field.

Figure 14.5 represents the AFC in a range of frequencies from 0.3 to 2.0 GHz at change of the external magnetic induction from 14 to 70 mT of FEMT, whose design is presented in Fig. 14.4a.

The level of noise was −55 dB (the lower border of the dynamic range of the used panoramic measuring instrument), the symbol "*K*" designates the level of calibration, and figures from 1 to 10 are the numbers of the resonances of the AFC of the amplifier. The amplification factor of FEMT was 1.8–6.0 dB.

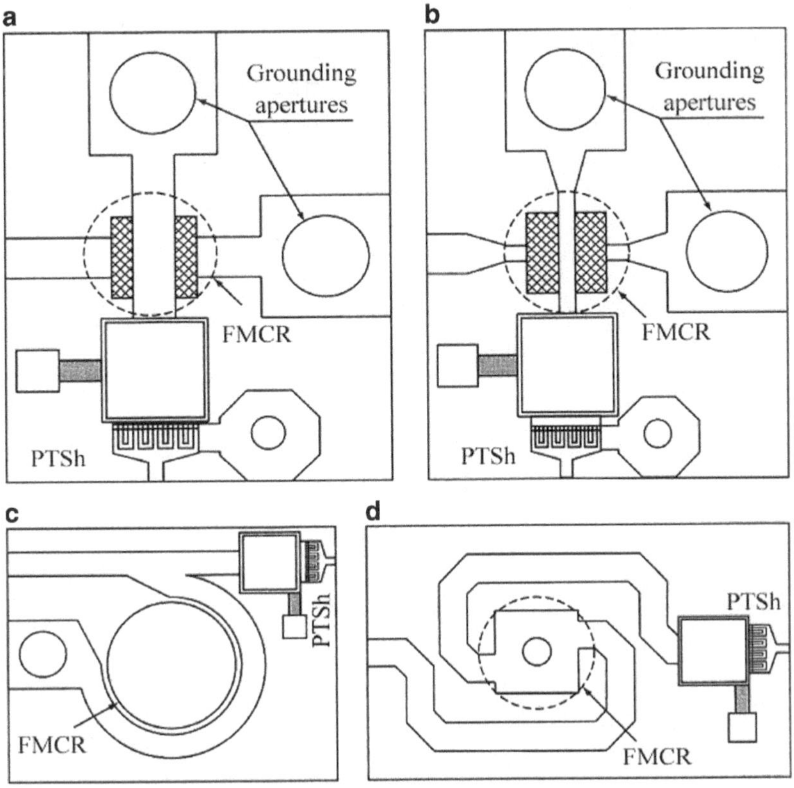

Fig. 14.4 The topology of low-power FEMT for ranges of frequencies: (**a**) 0.3–2 GHz; (**b**) 1–4 GHz; (**c**) 0.3–2.2 GHz; (**d**) 2.5–15.0 GHz

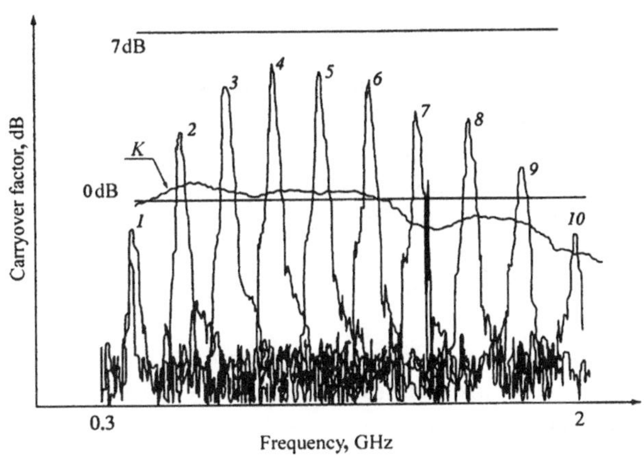

Fig. 14.5 The AFC in a range of frequencies from 0.3 to 2.0 GHz at change of the external magnetic induction from 14 to 70 mT of FEMT

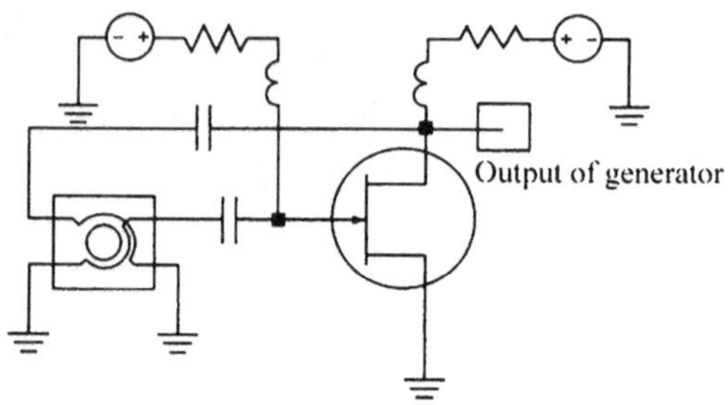

Fig. 14.6 The circuit of the generator on FEMT with MECE on crossed coils

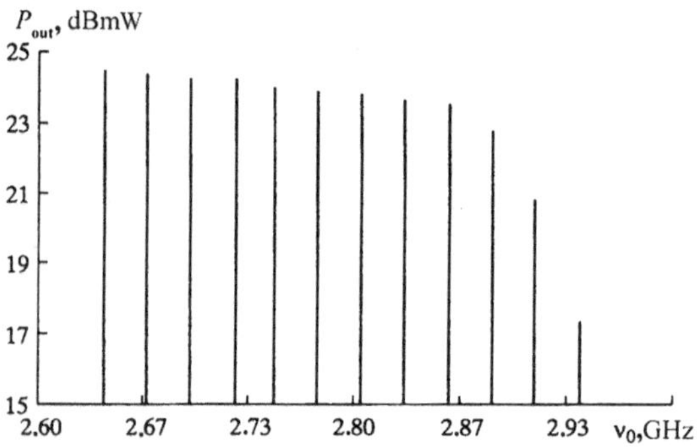

Fig. 14.7 The calculated dependences of output power on the frequency of generation

Our experimental researches have shown that FEMT has selective amplification, whose frequency is governed by the magnetic field.

For the generating operating mode, the coupling element in FEMT was placed between the shutter and the drain of the transistor and provided a positive feedback on the frequency of FMCR.

In Fig. 14.6, the circuit of the generator on FEMT with MECE on crossed coils is presented.

The external magnetic field applied to FMCR changed from 96 to 107 mT with a step of 1 mT. The calculated dependences of output power on the frequency of generation are presented in Fig. 14.7. It is obvious that with frequency growth the signal power falls from 24.6 dBmW on a frequency of 2.64 GHz at $B_0 = 96$ mT down to 17.4 dBmW on a frequency of 2.97 GHz at $B_0 = 107$ mT.

In experimental research, the output power of the generator of the order of 1 mW was attained, whose frequency was linearly changed in a range from 1.0 to 1.7 GHz at change of the magnetic induction from 36 to 61 mT. The steepness of frequency change on the value of magnetic induction was 28 MHz/mT. The spectrum of signal on the basic frequency was a noisy grid of frequencies with an equidistance of 200 kHz.

Alongside with the basic frequency of the generator, four noise-like spectral components were observed, whose central frequencies were close to the frequencies of the basic signal harmonics. At a basic frequency $\nu_0^{NS} = 1.6$ GHz, the spectral components were observed on frequencies $\nu_1^{NS} = 3.3$ GHz; $\nu_2^{NS} = 5.0$ GHz; $\nu_3^{NS} = 6.7$ GHz; and $\nu_4^{NS} = 8.3$ GHz.

In Fig. 14.8, FEMTs with various MECE made by the GaAs technology are shown.

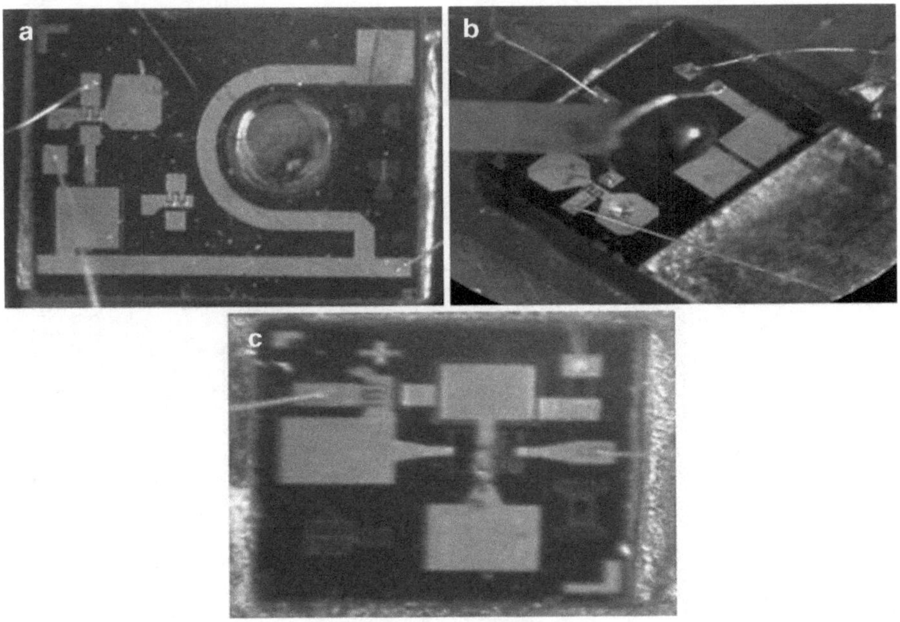

Fig. 14.8 FEMTs with various MECE made by the GaAs technology: (**a**) ansiform, (**b**) turns, (**c**) turns

14.2 Bipolar Magnetotransistors in Intense Mode on High Power Level in UHF Range

Below are the results of our experimental researches of BMT of various designs in an amplification operating mode at a high power level. In Fig. 14.9, BT-1 in the commercial KT-83 case is presented, which was used as a test one in the adjustment of the equivalent parameters.

In Fig. 14.10, BMT-2 with a coupling element on asymmetrical crossed MSL with FMCR in the circuit of the transistor base is shown.

In Fig. 14.11, BMT-3 with MECE on the principle of power division with a half-coil of an asymmetrical MSL for coupling with FMCR is shown.

Oscillograms of the AFC of the amplifier on BMT-3, recorded on a panoramic measuring instrument Ya2R-74, are presented in Figs. 14.12–14.15.

In Fig. 14.12, for the bipolar transistor KT-83, the following are presented in a range of frequencies 400–620 MHz: a calibration curve K of a panoramic meter on a level of 0 dB; the amplification factor $K_{amp.0}$ of bipolar transistor (BT); and the amplification factor of BMT for magnetic fields B_{01}, B_{02}, B_{03}, B_{04}. The maximal amplification factor on a frequency of 550 MHz was $K_{amp} = -8$ dBmW.

In Fig. 14.13, similar dependences in a range of frequencies 550–1,100 MHz are presented. The maximal amplification factor of BMT on a frequency of 600 MHz was $K_{amp} = -7$ dB.

Fig. 14.9 BT-1 in the commercial KT-83

Fig. 14.10 BMT-2 with a coupling element on asymmetrical crossed MSL with FMCR in the circuit of the transistor base

In Figs. 14.14 and 14.15, dependences of $K_{amp.0}$ and K_{amp} in ranges of frequencies 0–700 MHz and 550–1,300 MHz, respectively, are presented. The maximal amplification factor of BMT on a frequency of 600 MHz was $K_{amp} = 0.5$ dB.

On BT-1 (Fig. 14.9), families of the static output characteristics of the bipolar transistor (Fig. 14.16) were studied, which were used for calculation of the parameters of Gummel–Poon's nonlinear model of bipolar transistor and modeling of the parameters of the transistor with CAD (*SPICE* model).

Fig. 14.11 BMT-3 with MECE on the principle of power division with a half-coil of an asymmetrical MSL for coupling with FMCR

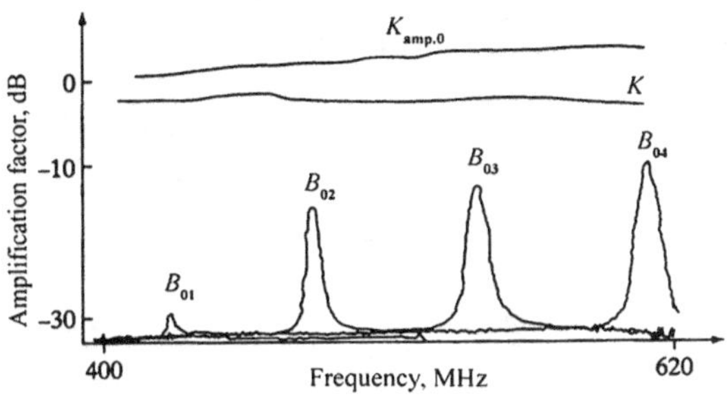

Fig. 14.12 The following for the bipolar transistor KT-83 are presented in a range of frequencies 400–620 MHz: a calibration curve K of a panoramic meter on a level of 0 dB; the amplification factor $K_{amp.0}$ of bipolar transistor (BT); and the amplification factor of BMT for magnetic fields B_{01}, B_{02}, B_{03}, B_{04}

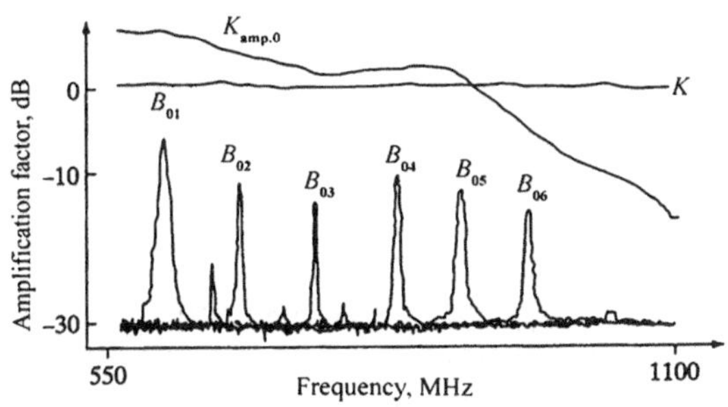

Fig. 14.13 The dependences K_{amp}, $K_{amp.0}$ in a range of frequencies 550–1,100 MHz for a magnetic field H_{01}, H_{02}, H_{03}, H_{04}

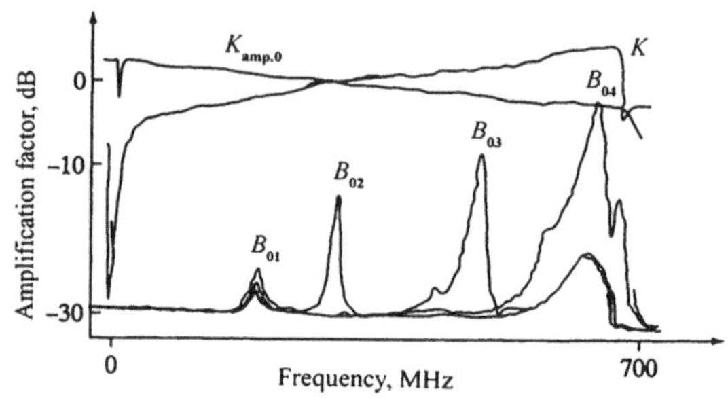

Fig. 14.14 The dependences of $K_{amp.0}$ and K_{amp} in ranges of frequencies 0–700 MHz

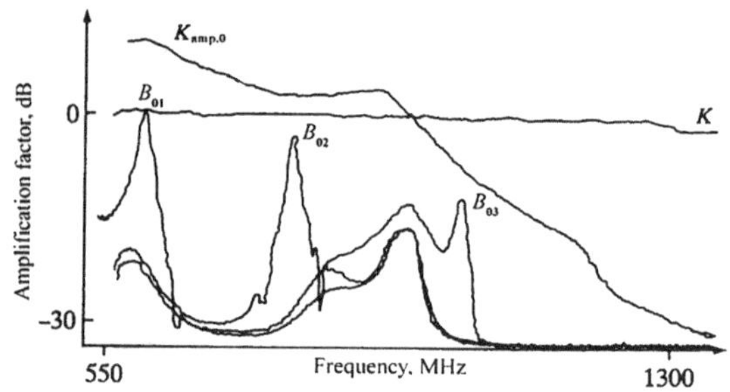

Fig. 14.15 The dependences of $K_{amp.0}$ and K_{amp} in ranges of frequencies 550–1,300 MHz

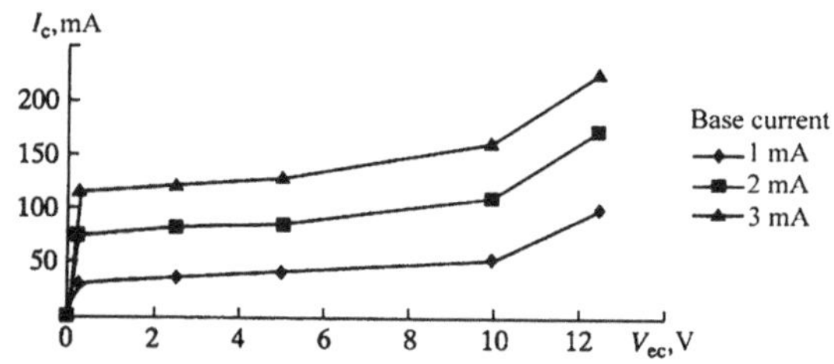

Fig. 14.16 The families of the static output characteristics of the bipolar transistor

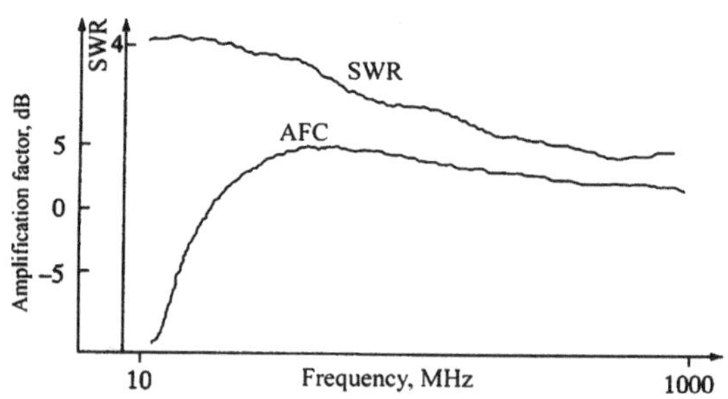

Fig. 14.17 Oscillograms of amplification and SWR in a range of frequencies 10–1,000 MHz for BT-1 in various operating modes

The parameters of AFC and SWR of the input of a test BT-1amplifier were investigated on a panoramic measuring instrument Ya2R-74. Oscillograms of amplification and SWR in a range of frequencies 10–1,000 MHz for BT-1 in various operating modes are presented in Figs. 14.17–14.20.

In Fig. 14.17, AFC and SWR are shown. The maximal amplification factor on a frequency of 320 MHz was $K_{amp} = 4.2$ dB at $SWR_{out} \approx 3.74$; the passband laid in a range of frequencies 155–930 MHz.

Fig. 14.18 Oscillograms of amplification and SWR in a range of frequencies 10–1,000 MHz for BT-1 in various operating modes

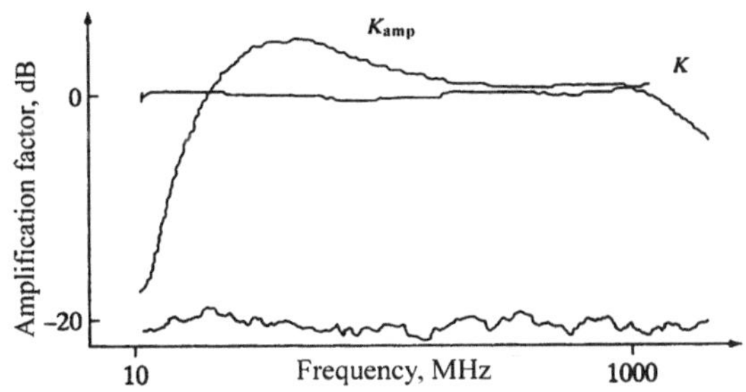

Fig. 14.19 Oscillograms of amplification and SWR in a range of frequencies 10–1,000 MHz for BT-1 in various operating modes

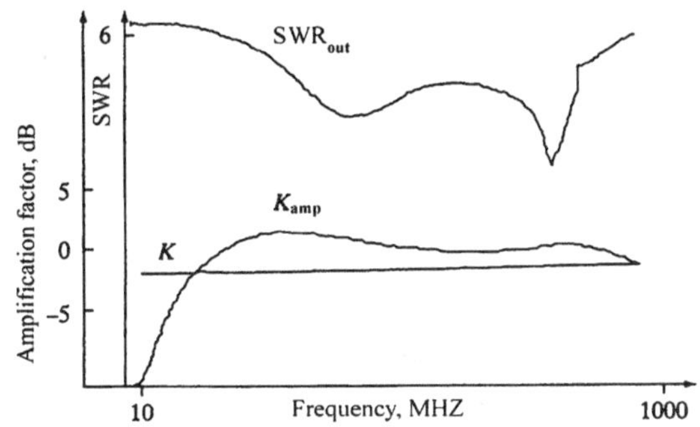

Fig. 14.20 Oscillograms of amplification and SWR in a range of frequencies 10–1,000 MHz for BT-1 in various operating modes

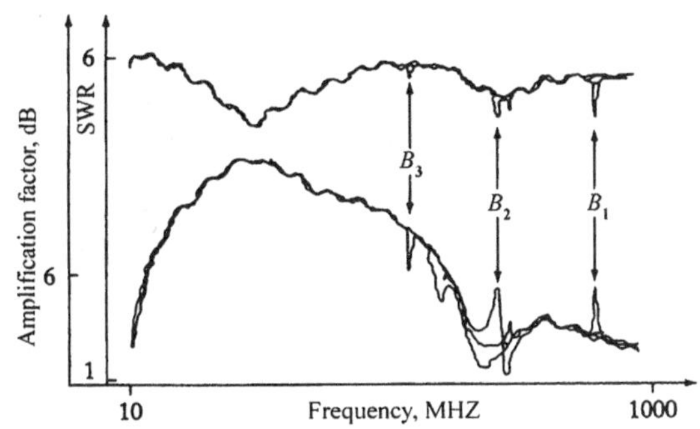

In Fig. 14.18, the level of calibration K of the measuring instrument, the amplification factor, and the level of noise of the panoramic measuring instrument ($-56.6\,\mathrm{d}B$) are presented. The test amplifier had an unoptimal values of tuning capacitors, namely, $C_4 = 4$–$13\,\mathrm{pF}$ and $C_{5,9} = 1$–$3\,\mathrm{pF}$. The maximal amplification factor on a frequency of 260 MHz is $K_{amp} = 5.4\,\mathrm{dB}$; the passband laid in a range of frequencies 15–850 MHz.

In Fig. 14.19, the level of calibration of the measuring instrument, the amplification factor, and SWR_{out} are presented. The test amplifier had the optimum value of tuning capacitor $C_4 = 4$–$13\,\mathrm{pF}$ and $C_{5,9} = 1$–$3\,\mathrm{pF}$. The maximal amplification factor on a frequency of 300 MHz is $K_{amp} = 6.1\,\mathrm{dB}$; the amplification frequency band was 130–1,000 MHz.

In Fig. 14.20, the characteristics of BMT-2 (see Fig. 14.11, a YIG KG-35 sphere) in an amplification mode are presented: the amplification factor and SWR_{out} at

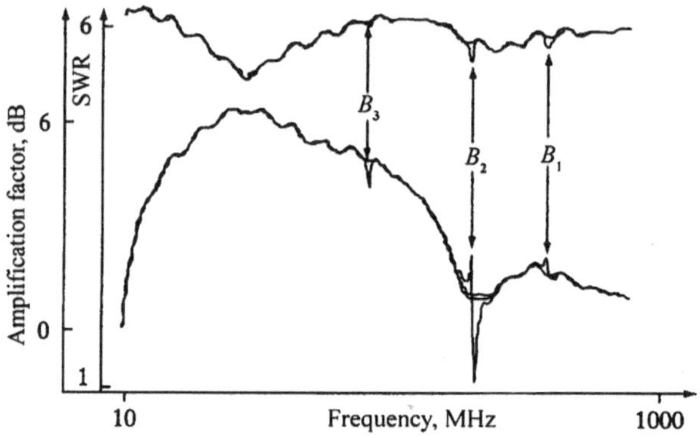

Fig. 14.21 The dependences of K_{amp} and SWR for BMT-2 with FMCR on the basis of a YIG KG-12 sphere are presented at several values of the magnetic field

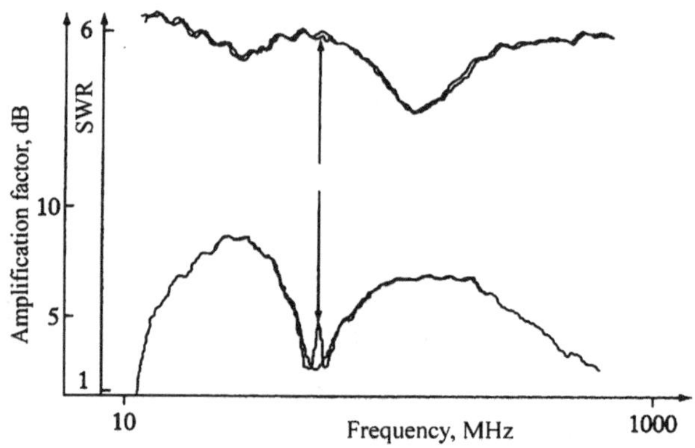

Fig. 14.22 The dependences of K_{amp} and SWR for BMT-2 with a YIG KG-35 sphere at several values of magnetic field

magnetic fields B_1, B_2, and B_3. The maximal amplification factor on a frequency of 300 MHz is $K_{amp} = 6.3$ dB at $SWR_{out} \approx 2.8$; the amplification frequency band is 100–510 MHz. The level of output power varied within 0.7–0.9 W.

In Fig. 14.21, similar dependences for BMT-2 with FMCR on the basis of a YIG KG-12 sphere are presented at several values of the magnetic field. The maximal amplification factor on a frequency of 240 MHz is $K_{amp} = 6$ dB; the amplification frequency band is 100–510 MHz. The level of output capacity varied within 0.7–0.9 W.

In Fig. 14.22, dependences for BMT-2 with a YIG KG-35 sphere are shown at several values of magnetic field. The amplification factor is $K_{amp} = 5.5$ dB at the magnetic field B_1; the amplification frequency band is 100–300 MHz.

Similar dependences were observed for BMT-3 in a range of frequencies 300–600 MHz at several magnetic fields B_1, B_2, and B_3 with FMCR on the basis of YIG spheres KG-35 and KG-12 – Figs. 14.23 and 14.24, respectively. The maximal

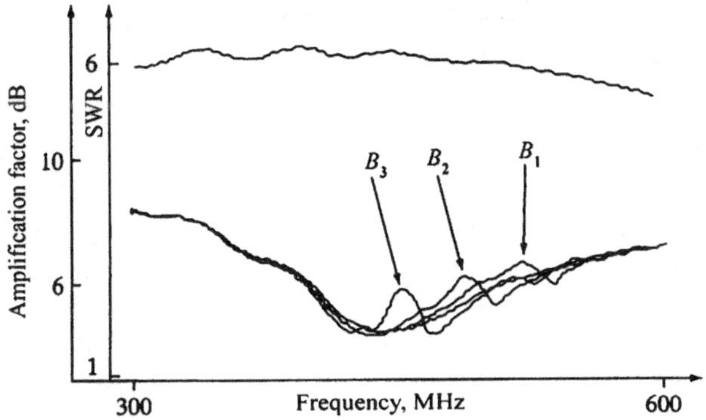

Fig. 14.23 The dependences of K_{amp} and SWR on frequency 240 MHz for the field B_3 at a level of output power 0.8–1.2 W

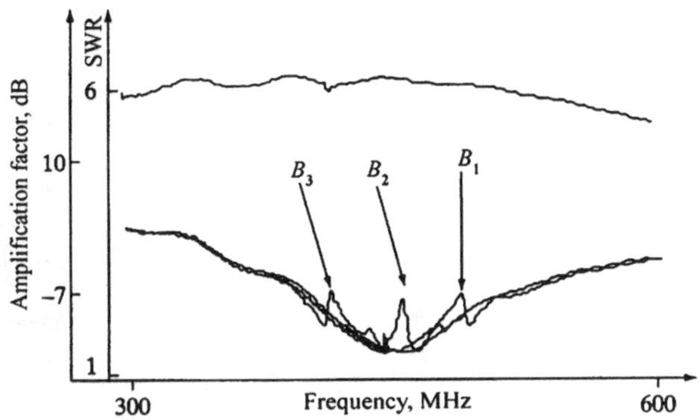

Fig. 14.24 The dependences of K_{amp} and SWR on frequency 240 MHz for the field B_2 at a level of output power 0.8–1.2 W

amplification factor on a frequency of 240 MHz is $K_{amp} = 6.1$ dB for the field B_3 (Fig. 14.23) and $K_{amp} = 6.6$ dB for the field B_2 (Fig. 14.24), at a level of output power 0.8–1.2 W.

Experimental studies have shown that BMT-2 and BMT-3 (Figs. 14.10 and 14.11) had an increase of the amplification factor within 100–500 MHz and a sharp fall within 600–900 MHz (Figs. 14.20–14.24). The best interaction with FMCR was observed in the range of the fall of the amplification factor (from 600 to 900 MHz).

A demerit of the breadboard models BMT-2 and BMT-2 was the high wave resistance of the grounding circuit (through the capacitor and case of the transistor) that did not allow obtaining a good interaction with FMCR in a wide range of frequencies. The drop of the amplification factor above 600 MHz is explained by the mismatch introduced by the 100-pF capacitor in the coupling element.

In the amplification band of the investigated BMT, the coupling element practically did not distort the transfer factor of the amplifier that spoke of a low decoupling level of the chosen MECE types.

14.3 Experimental Investigation of Bipolar Magnetotransistors Based on KT9175A Crystals

In Fig. 14.25, a photo of our laboratory breadboard model of bipolar transistor on the basis of KT9175A crystals is presented. In Fig. 14.26, BMT with MECE on asymmetrical crossed MSL is shown. In Fig. 14.27, FMCR in the form of a YIG sphere in the coupling element is shown.

In Figs. 14.28–14.31, oscillograms of the transfer factor of BMT of various types in a range of frequencies from 400 to 1,400 MHz with FMCR on the basis of ferrite KG-15 are presented: at a magnetic field $H_0 = 209\,\text{Oe}$ on a frequency

Fig. 14.25 A photo of our laboratory breadboard model of bipolar transistor on the basis of KT9175A crystals

Fig. 14.26 BMT with MECE on asymmetrical crossed MSL

Fig. 14.27 FMCR in the form of a YIG sphere in the coupling element

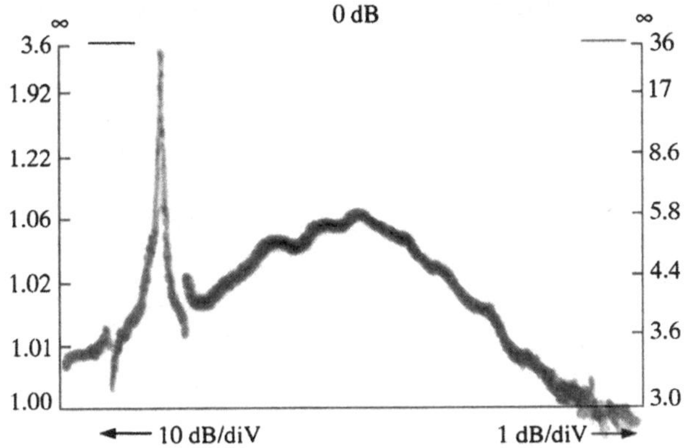

Fig. 14.28 The oscillogram of the transfer factor of BMT on the basis of ferrite KG-15 at a magnetic field $H_0 = 209\,\text{Oe}$ on a frequency $\nu = 585\,\text{MHz}$

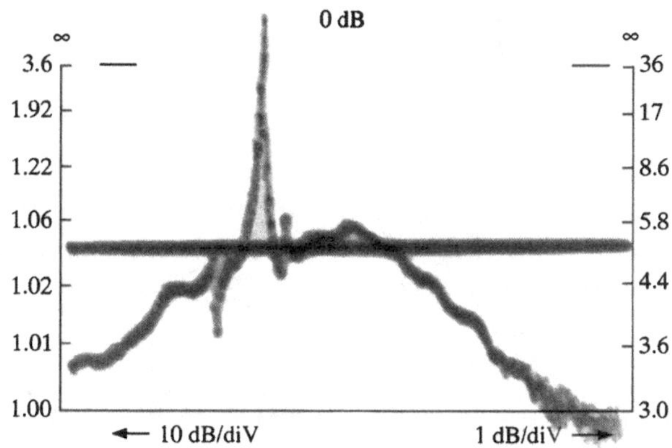

Fig. 14.29 The oscillogram of the transfer factor of BMT on the basis of ferrite KG-15 without magnetic field

$\nu = 585\,\text{MHz}$ $K_{\text{amp}} = -6.7\,\text{dB}$ and without magnetic field $K_{\text{amp}} = -28\,\text{dB}$ (Fig. 14.28); at a magnetic field $H_0 = 261$ Oe on a frequency $\nu = 730\,\text{MHz}$ $K_{\text{amp}} = -2.4\,\text{dB}$ and without magnetic field $K_{\text{amp}} = -22\,\text{dB}$ (Fig. 14.29); at a magnetic field $H_0 = 310\,\text{Oe}$ on a frequency $\nu = 987\,\text{MHz}$ $K_{\text{amp}} = -4.5\,\text{dB}$ and without

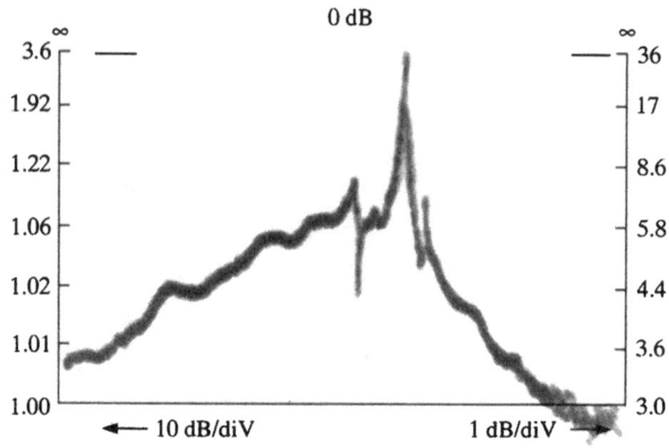

Fig. 14.30 The oscillogram of the transfer factor of BMT on a frequency $\nu = 987\,\text{MHz}$ without magnetic field

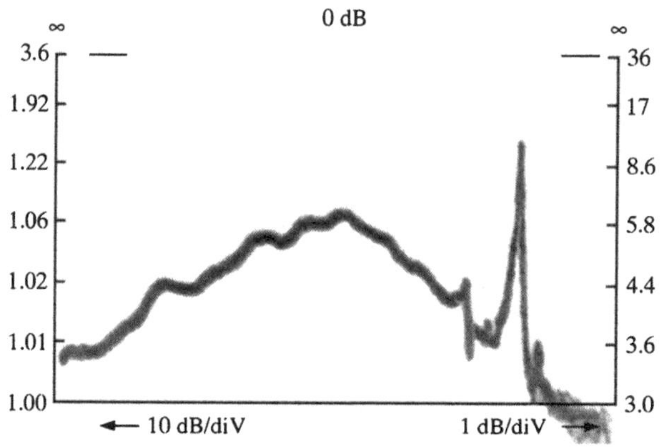

Fig. 14.31 The oscillogram of the transfer factor of BMT on a frequency $\nu = 1{,}214\,\text{MHz}$

magnetic field $K_{\text{amp}} = -17\,\text{dB}$ (Fig. 14.30); at $H_0 = 355\,\text{Oe}$ on a frequency $\nu = 1214\,\text{MHz}$ $K_{\text{amp}} = -13.6\,\text{dB}$ and without magnetic field $K_{\text{amp}} = -32\,\text{dB}$ (Fig. 14.31). The passband is $\Delta\nu_{-3\,\text{dB}} \leq 10\,\text{MHz}$.

14.4 Magneto-FET in EHF Range in Boost Regime

The development trends of modern active elements are associated with broadening of the range of working frequencies below 100 GHz and above. However, at designing devices on the basis of field transistors in the EHF range, difficulties arise with design of an adequate mathematical model of HEMT structures with a raised mobility of carriers.

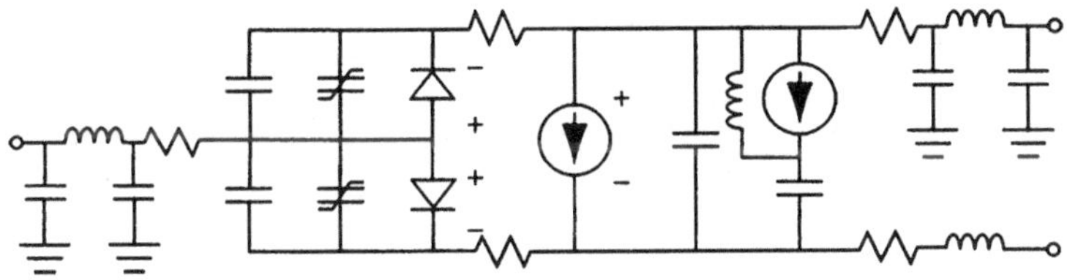

Fig. 14.32 Chalmers' model for the HEMT structure in a nonlinear operating mode

The comprehensive model considering the features of the HEMT structure in a nonlinear operating mode is Chalmers' model (Fig. 14.32), whoseparameters are shown in Table 14.1.

These parameters determine:

- The current of the generator

$$I_\mathrm{d}(V_\mathrm{g}, V_\mathrm{d}) = \mathrm{IPK}(1 + \tan(\psi)\tanh(\alpha V_\mathrm{d})(1 + \mathrm{LAMBDA} \cdot V_\mathrm{d}), \tag{14.1}$$

where $\psi = P_{1\mathrm{m}}(V_\mathrm{g} - V_\mathrm{pk}) + P3(V_\mathrm{g} - V_\mathrm{pk})^2 + P3(V_\mathrm{g} - V_\mathrm{pk})^3 + \cdots,$

$$\begin{aligned}\alpha &= \mathrm{ALPHR} + \mathrm{ALPHS}(1 + \tan(\psi)),\\ P_{1\mathrm{m}} &= \mathrm{PI}(1 + \mathrm{BI}/\cosh^2(B2 \cdot V_\mathrm{d}),\\ V_\mathrm{pk} &= \mathrm{VPKO} + (\mathrm{VPKS} - \mathrm{VPKO})\tanh(\mathrm{ALPHS} \cdot V_\mathrm{d})\end{aligned}$$

- The phenomena of breakdown

$$\begin{aligned}I_\mathrm{d}(V_\mathrm{g}, V_\mathrm{d}) &= \mathrm{IPK}\tanh(\psi)\tanh(\alpha V_\mathrm{d})(1 + \mathrm{LAMBDA} \cdot V_\mathrm{d}\\ &\quad + \mathrm{LAMSB} \cdot e^{V_\mathrm{dg} - \mathrm{VT}}), \qquad (14.2)\\ V_\mathrm{pk} &= \mathrm{VPKO} + (\mathrm{VPKS} - \mathrm{VPKO})\tanh(\mathrm{ALPHA} \cdot V_\mathrm{d})\\ &\quad - \mathrm{VSB2} \cdot (V_\mathrm{dg} - VTR)^2; \qquad (14.3)\end{aligned}$$

- The shutter current

$$I_\mathrm{g} = \frac{\partial Q_\mathrm{g}}{\partial V_\mathrm{g}}\frac{\mathrm{d}V_\mathrm{g}}{\mathrm{d}t} + \frac{\partial Q_\mathrm{g}}{\partial V_\mathrm{gd}}\frac{\mathrm{d}V_\mathrm{gd}}{\mathrm{d}t} \tag{14.4}$$

- The shutter–source capacity

$$\begin{aligned}C_\mathrm{gs1}(V_\mathrm{g}, V_\mathrm{d}) &= \mathrm{ADIV} \cdot \mathrm{CGSO}[1 + \tanh(P20 + P21 \cdot V_\mathrm{d})]\\ &\quad \times (1.0 + \tanh(P10 + P11 \cdot V_\mathrm{g})),\\ \tilde{N}_\mathrm{gs2}(V_\mathrm{g}, V_\mathrm{d}) &= [(1 - \mathrm{ADIV}) \cdot \mathrm{CGSO}(1 + \tanh(P20 + P21 \cdot V_\mathrm{d})]\\ &\quad \times [1.0 + \tanh(P110 + P111 \cdot V_\mathrm{g})],\\ C_\mathrm{gs}(V_\mathrm{g}, V_\mathrm{d}) &= \frac{\partial Q_\mathrm{g}}{\partial V_\mathrm{g}} = C_\mathrm{gs1} + C_\mathrm{gs2} \qquad (14.5)\end{aligned}$$

Table 14.1 The parameters of Chalmers' model

Parameter	Description	Unit
IPK	Current at maximal amplification	A
$P1, P2, P3\ldots$	Polynomial factors I_d	
$B1, B2$	Parameters P_{1m}	
VPKS	Shutter voltage at the maximal amplification in the saturated state	V
VPK0	Shutter voltage at the maximal amplification at V_{ds}	V
ALPHR	Parameter of the point of VAC inflection of drain	
ALPHS	Parameter of the point of VAC inflection of drain in the saturated state	
LAMBDA	Parameter of drain–source resistance	
LAMSB	Parameter of superficial breakdown	
VTR	Threshold breakdown voltage	V
VSB2	Parameter of superficial breakdown	
TAU	Shutter–drain delay time	s
CGS0	Shutter–source capacity	F
CGSP	Linear component of C_{gs}	F
PC10, PC11, PC20, PC21	Polynomial factor of shutter–source capacity	
PC110, PC111	Polynomial factor for modeling of the maximal C_{gs}	
ADIV	Parameter for modeling of the maximal C_{gs}	
PC20, PC21	Polynomial factor of shutter–source capacity	
CGD0	Shutter–drain capacity	F
CGDP	Linear component of C_{gd}	F
PC30, PC31, PC40, PC41	Polynomial factor of shutter–drain capacity	
CDS0	Constant shutter–source capacity	F
CDSW	Variable shutter–source capacity	F
CPG	Parasitic shutter capacity	F
CPD	Parasitic drain capacity	F
ISG	Shutter diode current	A
NG	Diode parameter	Ω
RG	Shutter resistance	Ω
RS	Source resistance	Ω
RI	Internal resistance	Ω
RD	Drain resistance	Ω
RGD	Shutter–drain resistance	Ω
RCW	Shutter–source resistance on microwaves	Ω
CRF	Capacity for definition of R_{ds} on microwaves	F
LS	Source inductance	H
LG	Shutter inductance	H
LD	Drain inductance	H

– The shutter–drain capacity

$$C_{gd}(V_g, V_d) = \frac{\partial Q_g}{\partial V_{gd}} = \text{CGDO}[1 + \tanh(P30 + P31 \cdot V_d)] \times (1.0 + \tanh(P40 + P41 \cdot V_{gd})) \tag{14.6}$$

– The current of the generator

$$I_{\mathrm{Rd}}(V_{\mathrm{g}}V_{\mathrm{d}}) = \frac{(1 + \tanh(\psi))}{\mathrm{RCW}} V_{\mathrm{d}} \tag{14.7}$$

However, the practical use of Chalmers' model is connected with the necessity of carrying out a big number of special measurements for determination of equivalent parameters. In the EHF range, a specialized vector circuit analyzer *PNA* N 5250 of *Agilent Technologies Corp*® should be applied. In this connection, as a base element for calculation of FEMT in the EHF range, broadband amplifiers UA1S65LM (USA, Centellax, Inc.) and D01MH (USA, OMMIC) are chosen.

Calculation was done by a simplified empirical nonlinear model based on the use of the basic experimental characteristics of UA1S65LM amplifiers (the working range of frequencies from 1 to 65 GHz, the output power 40 mW) and D01MH (a working range from 50 to 75 GHz).

For modeling of UA1S65LM, the scattering parameters given in Table 14.2 were used.

In Figs. 14.33 and 14.34, the appearance and transfer factor of the amplifier UA1S65LM are presented.

The offered model allows doing calculations in modern CADs (MWO, Serenade, etc.). A nonlinear amplifier model with a strip filter of lower frequencies

Table 14.2 The scattering parameters for modeling of UA1S65LM

Parameter	Frequency range	Value
Gain factor $S21$, dB	1–26	$18 \div 21$
	26–45	$15 \div 17$
Input reflectivity factor $S11$, dB	1–26	$-(12 \div 10)$
Reflectivity factor from output $S11$, dB	26–45	$-(10 \div 8)$
Maximal output power P, dBmW	1–26	$-(15 \div 12)$
	26–45	$-(12 \div 9)$
	1–45	$16 \div 40$

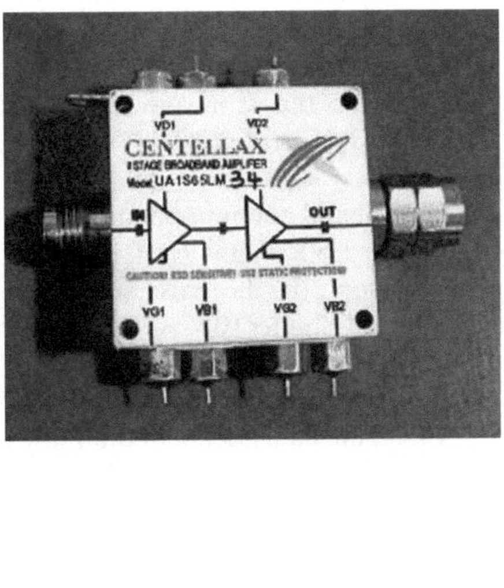

Fig. 14.33 The appearance of the amplifier UA1S65LM

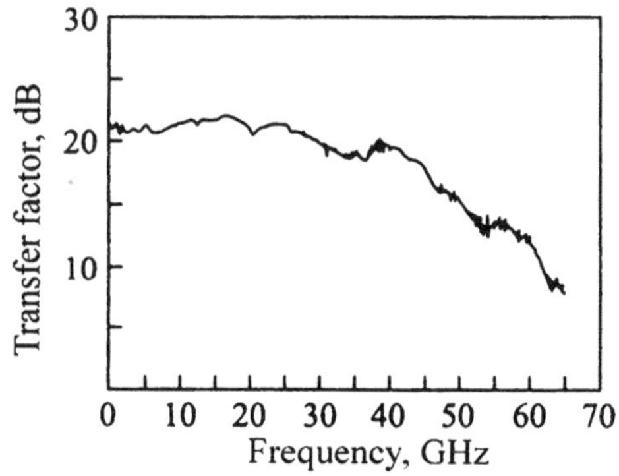

Fig. 14.34 The appearance of transfer factor of the amplifier UA1S65LM

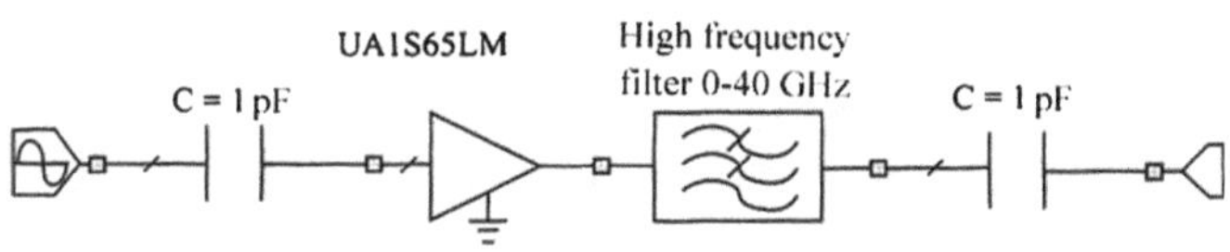

Fig. 14.35 The equivalent circuit of the amplifier UA1S65LM and the parameters of its elements

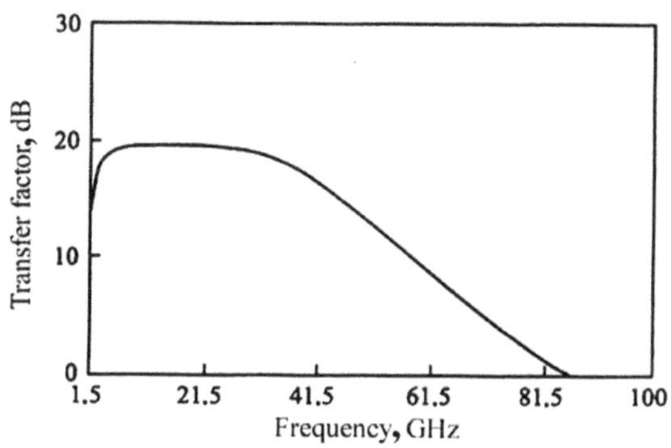

Fig. 14.36 The transfer factor of the model of amplifier UA1S65LM

on its output was used. In Fig. 14.35, the equivalent circuit of the amplifier UA1S65LM and the parameters of its elements are shown, and in Fig. 14.36, the transfer factor of this model is presented.

A plot of the input-power dependence of output power for the amplifier UA1S65LM is presented in Fig. 14.37 and that of its models is presented in Fig. 14.38.

A broadband low-noise amplifier on the basis of the transistors D01MH with its CHIP sizes $1.5 \times 3\,\mathrm{mm}^2$ (Fig. 14.39) was simulated. In Fig. 14.40, (a) the topology of conductors; (b) the cross section (the *HEMT* technology); and (c) the appearance of the T-shutter ($0.2 \times 100\,\mu\mathrm{m}$) of the AlGaN–GaN *HEMT* transistor on a SiC substrate are presented.

A model of the linear amplifier with a strip filter on its output was used. In Fig. 14.41, the equivalent circuit of the amplifier on the basis of D01MH transistors

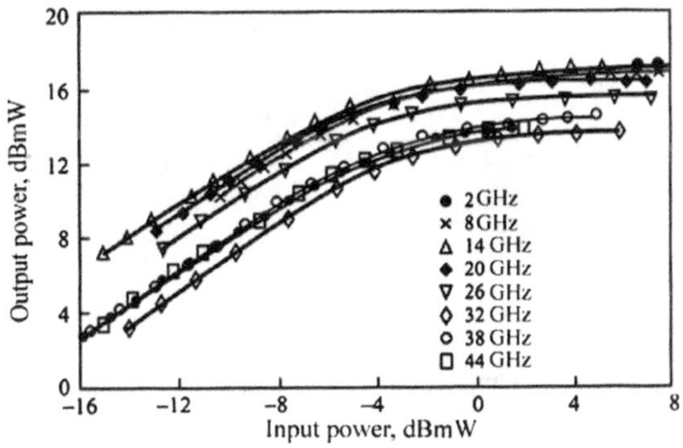

Fig. 14.37 The input-power dependence of output power for the amplifier UA1S65LM

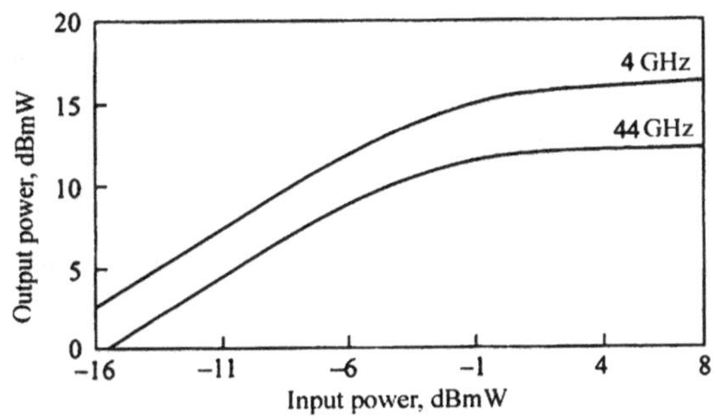

Fig. 14.38 The input-power dependence of output power of models of the amplifier UA1S65LM

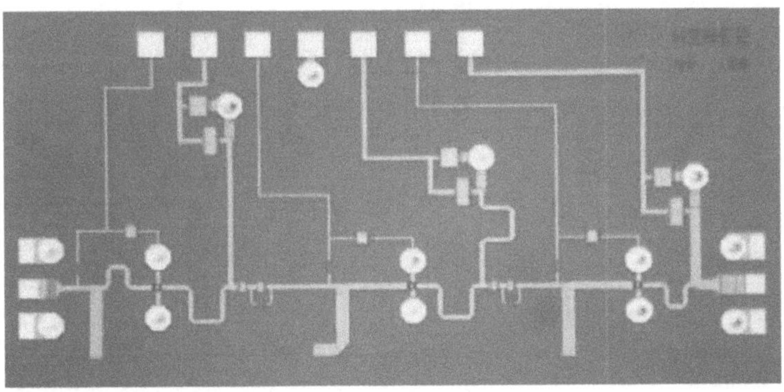

Fig. 14.39 A broadband low-noise amplifier on the basis of the transistors D01MH with its CHIP sizes $1.5 \times 3\,mm^2$

and the parameters of its elements are presented. The transfer factors of the amplifier and its model are presented in Figs. 14.42 and 14.43, respectively.

In the investigated ranges of frequency and power, the results of calculations of the transistor coincide with its experimental characteristics with an error of 10%. The obtained results allow developing active FEMT for the EHF range in the mode of big signals.

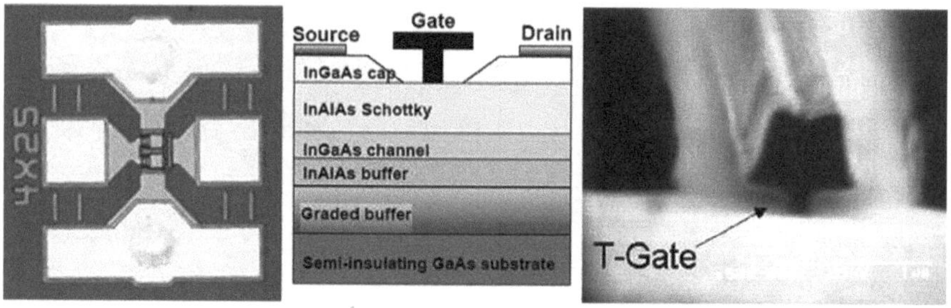

Fig. 14.40 (**a**) The topology of conductors; (**b**) the cross-section (the HEMT technology); (**c**) the appearance of the T-shutter ($0.2 \times 100\,\mu m$) of the AlGaN-GaN HEMT transistor on a SiC substrate

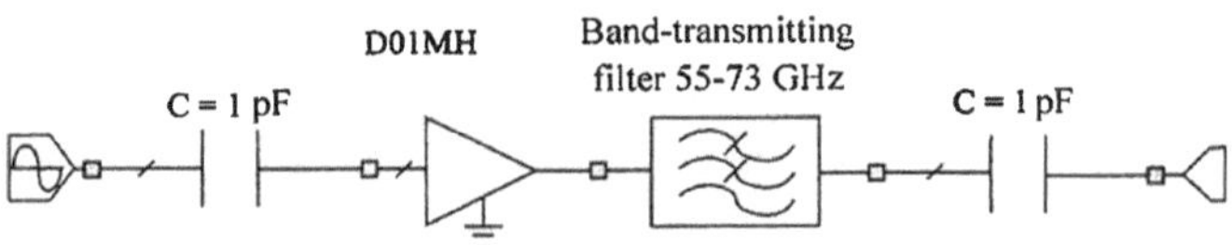

Fig. 14.41 The equivalent circuit of the amplifier on the basis of D01MH transistors with parameters of its elements

Fig. 14.42 The transfer factors of the amplifier

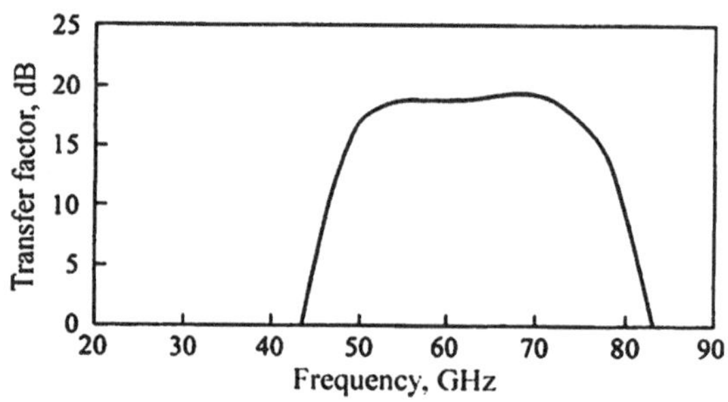

Fig. 14.43 The transfer factors of the model of amplifier

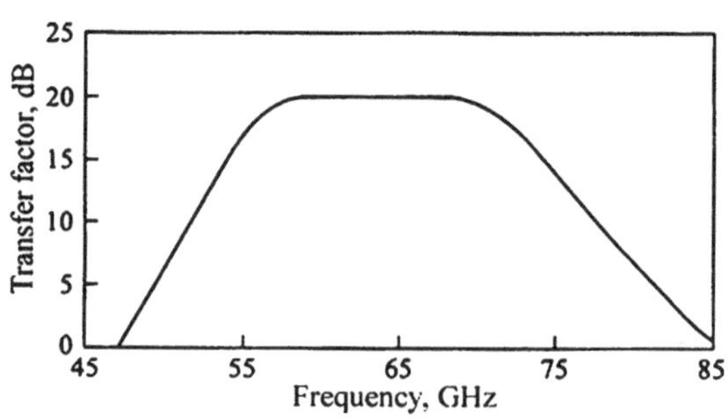

14.5 FET and Bipolar Magnetotransistor in Microwave Frequency Range of High Power Level

14.5.1 Magneto-FET of High Power Level

Field magnetic transistors of a high power level (up to 2 W) were designed on the basis of the designs of powerful MECE (Sect. 14.1). The basic circuit of FEMT connection is presented in Fig. 14.44. The resistance of 100 Ω sets the working point of the transistor. The 12-pF capacitors on input and output are dividing. The drain of the transistor is powered with a positive voltage through a resistance. On the shutter, a negative voltage of displacement is applied.

In Fig. 14.45, the vtopology of various types of powerful FEMT on pin transistors with Schottky's barrier PTSh-5000 (Almas-Phasotron Corp.) with a maximal output power of 2–2.5 W for the following ranges of frequencies are shown: (a) 0.2–1.5 GHz; (b) 1.0–5.0 GHz; and (c) 0.2–7.0 GHz.

The appearance of a variant of powerful FEMT is presented in Fig. 14.46.

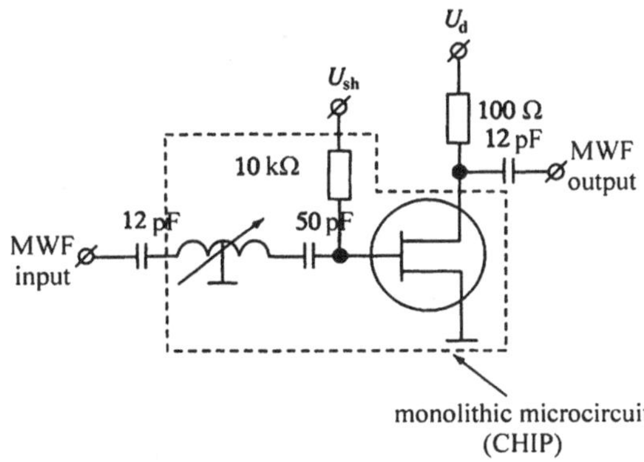

Fig. 14.44 The circuit of FEMT connection of a high power level

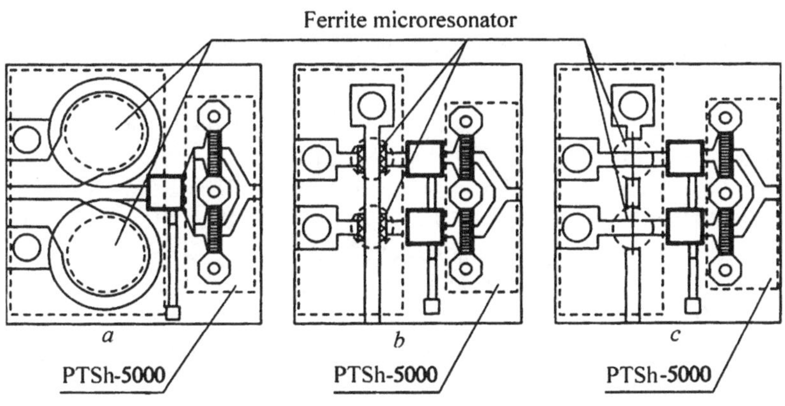

Fig. 14.45 The vtopology of various types of powerful FEMT on pin transistors with Schottky's barrier PTSh-5000 (Almas-Phasotron Corp.) with a maximal output power of 2–2.5 W for the following ranges of frequencies: (**a**) 0.2–1.5 GHz; (**b**) 1.0–5.0 GHz; (**c**) 0.2–7.0 GHz

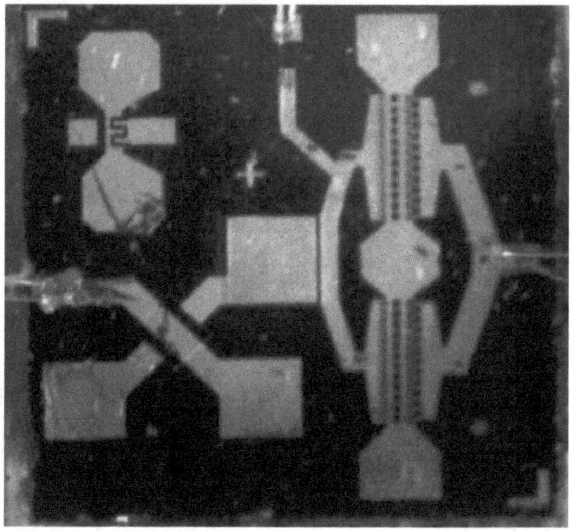

Fig. 14.46 The appearance of a variant of powerful FEMT

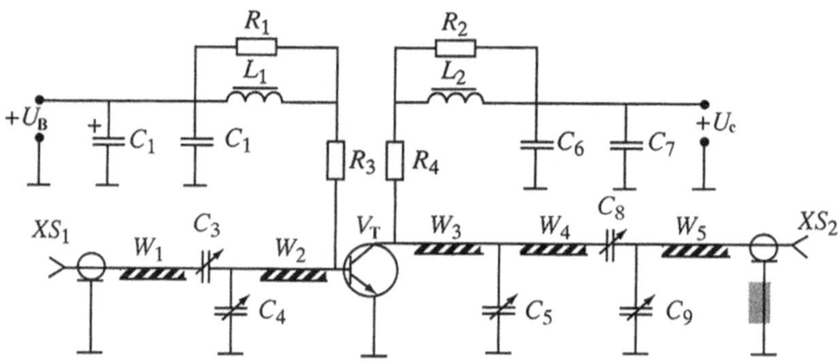

Fig. 14.47 The circuit for investigation of parameters of a powerful BMT

14.5.2 *Bipolar Magnetotransistors of High Power Level*

Let us consider the results of designing a powerful BMT on the basis of KT9175B crystals in the KT-83 case (2 W, 900 MHz, NIIET Corp.).

A circuit has been developed for research in Fig. 14.47, here C_3, C_4, C_5, C_8, C_9: KT4-25b, 100 W, 1/5 pF; C_2, C_6: K10-17-3-H50, 0.68 μF ± 20%; C_1, C_7: K50-35, 100 W, 10 μF; L_1, L_4: a high-frequency choke DM-3-1 ± 20%; R_1, R_2: C2-33H-0,5I, 11 Ω ± 10%; R_3, R_4: 100 Ω; asymmetrical strip lines on FAF-4D-1,0: W_1, $W_5 = (3 \pm 0.1)$ μm; $l = (20 \pm 1)$ mm; $W_2 = (3 \pm 0.1)$ mm; $l = (13 \pm 1)$ mm; $W_3 = (3 \pm 0.1)$ mm; $l = (33 \pm 1)$ mm; $W_4 = (3 \pm 0.1)$; $l = (5 \pm 1)$ mm; XS_1 and XS_2 – a coaxial-strip transition.

In Fig. 14.48, (a) a drawing of the assembly plate for the bipolar transistor KT9175; (b) a plate; and (c) the arrangement of the pieces on the plate are shown.

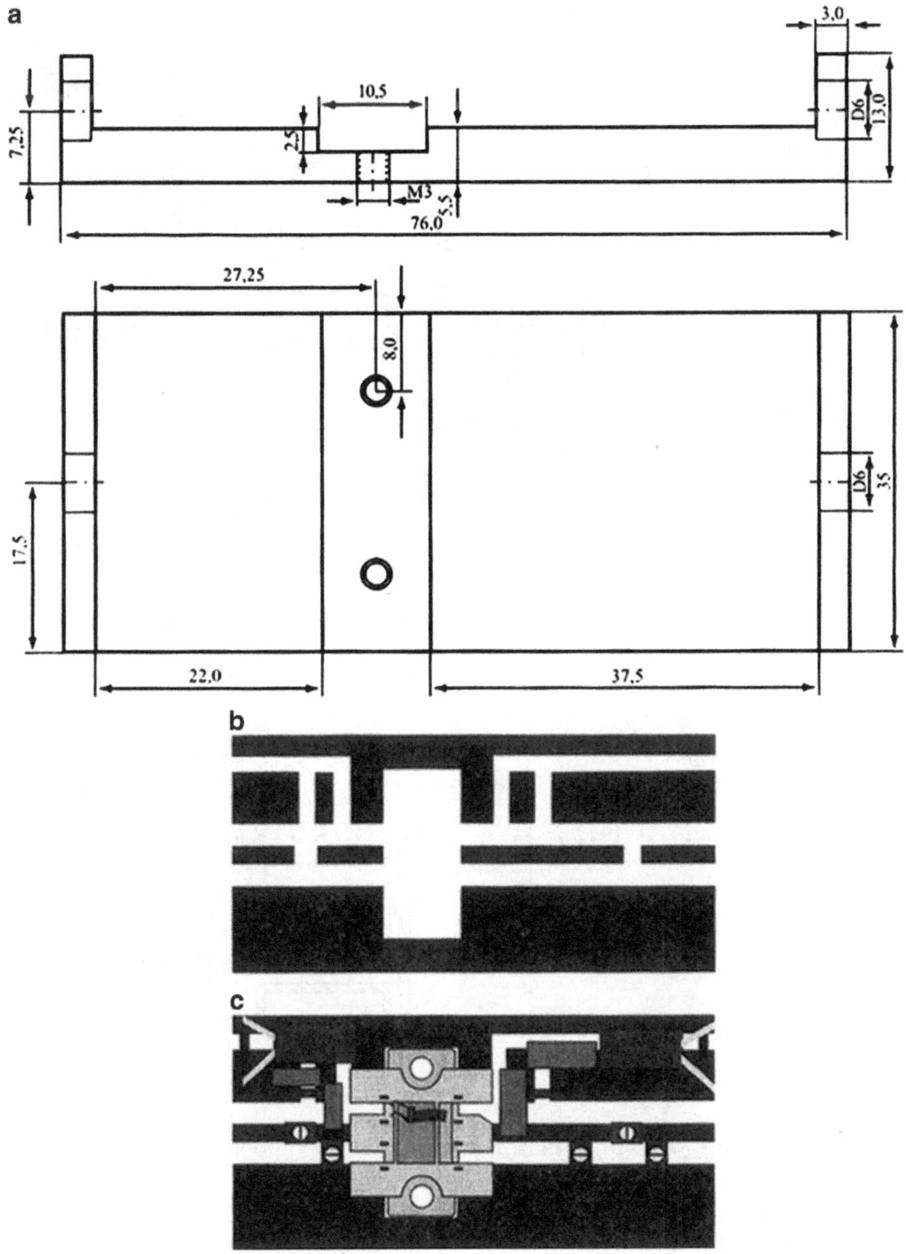

Fig. 14.48 (**a**) A drawing of the assembly plate for the bipolar transistor KT9175; (**b**) a plate; (**c**) the arrangement of the pieces on the plate

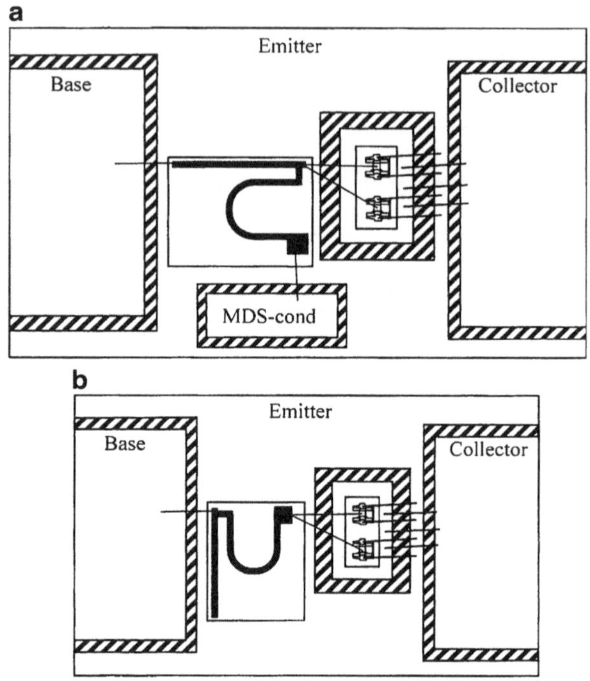

Fig. 14.49 A BMT with MECE in the form of a semiloop asymmetrical MSL connected to the transistor's base circuit in several ways: (**a**) earthing to MDS-condenser, (**b**) connection to collector

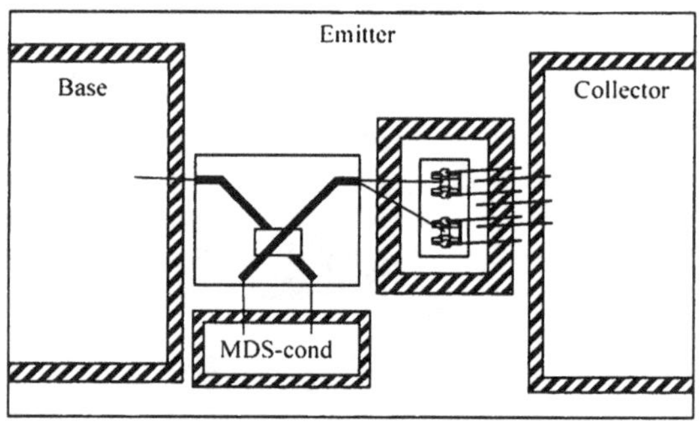

Fig. 14.50 A BMT with MECE on crossed asymmetrical MSL

In Fig. 14.49, a BMT with MECE in the form of a semiloop asymmetrical MSL connected to the transistor's base circuit in several ways is shown. In Fig. 14.50, a BMT with MECE on crossed asymmetrical MSL is shown.

Experimental researches have shown that the influence of the case KT-83 limits the working range of frequencies of BMT and reduces the transfer losses by 6 dB. Therefore, such designs were not used for development of magnetocontrolled amplifiers and generators.

Let us consider the results of our development of a powerful BMT on the basis of KT9175A crystals in an unpackaged enclosure on a polycorial basis. Sketches of the corresponding plates are presented in Figs. 14.51–14.53. In these designs, shortened

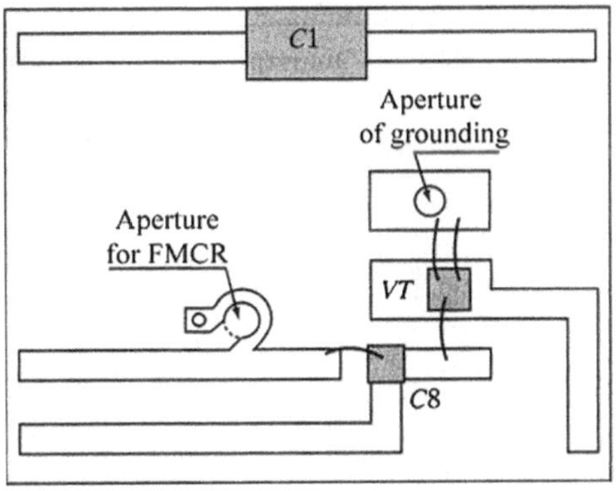

Fig. 14.51 The topology of powerful BMT on the basis of KT9175A crystals

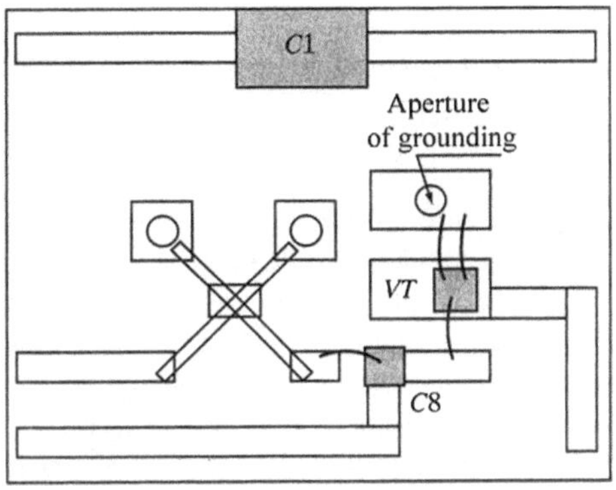

Fig. 14.52 The topology of powerful BMT on the basis of KT9175A crystals

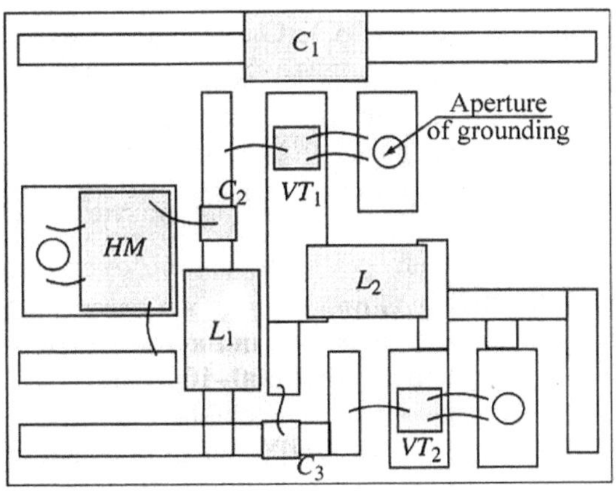

Fig. 14.53 The topology of powerful BMT on the basis of KT9175A crystals

grounding lines of MECE were used. In the design shown in Fig. 14.51, a through aperture was made at the center of the microstrip coupling loop in the polycoric plate for placement of a ferrite sphere to improve the efficiency of interaction.

In Fig. 14.54, a breadboard model of BMT for research is shown.

Fig. 14.54 A breadboard model of BMT for research

14.6 Ferrite Semiconductor Structures in Regime of Oscillation Conversion in a Frequency Band 100–1,000 GHz

Ferrite Semiconductor Structures (FSS) in a mode of transformation of harmonic components with FMCR in a range of frequencies below 1,000 GHz were simulated.

As the prototype at modelin g, the transistor amplifier developed by the California University (USA) was chosen: the working range of frequencies 140–220 GHz, $K_{amp} = 8.5$ dB on a frequency of 195 GHz. This amplifier is designed by the transferring substrate technology on HBT transistors (heterojunction InAlAs/InGaAs). The structure of the *HBT* transistor is shown in Fig. 14.55.

As the bearing basis of the amplifier, a crystal GaAs serves on which a gold grounding coating is sputtered through an adhesive (In/Pb/Ag) layer. Between the earthing, covering, and microstrip lines, a layer of benzocyclobuthene with a thickness of 5 μm and permeability $\varepsilon_r = 2.7$ is deposited. The appearance of the crystal of the amplifier with sizes $1.66 \times 0.59\,\text{mm}^2$ is presented in Fig. 14.56.

In modeling this amplifier in CAD MWO-2002, linear *S* parameters (Fig. 14.57) and the input-power dependence of the maximal output power (Fig. 14.58) were used.

A MECE with FMCR on the basis of an epitaxial BHF film (barium hexaferrite H35, the field of anisotropy $H_A = 35$ kOe, $4\pi M_s = 1,400$ G, $\Delta H = 3.5$ kOe) of the size $0.3 \times 0.3\,\text{mm}^2$ was introduced into the model of HMG to connect the two asymmetrical microstrip lines with a width of 10 μm (Fig. 14.59). The substrate of MECE is made of benzocyclobuthene ($\varepsilon_r = 2.7$) with a thickness of 5 μm.

The model of FMCR represented a parallel *RLC* contour, whose parameters were defined by the properties of the ferrite film and the external magnetic field by the following equations:

$$L = \frac{R}{\omega_0 Q_s} \text{ (H)},$$

$$C = \frac{1}{\omega_0^2 L} \text{ (F)},$$

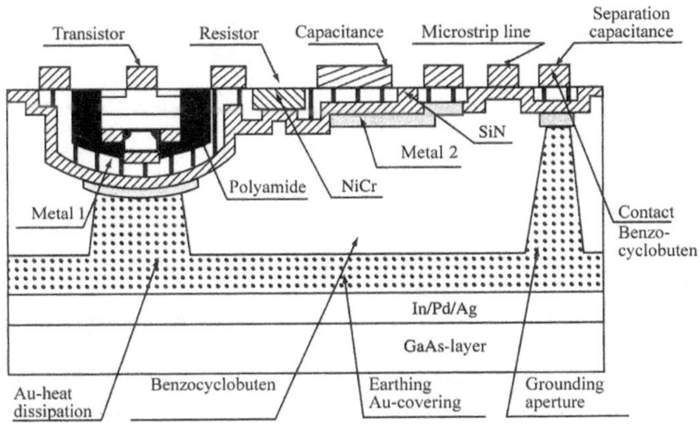

Fig. 14.55 The structure of the HBT transistor

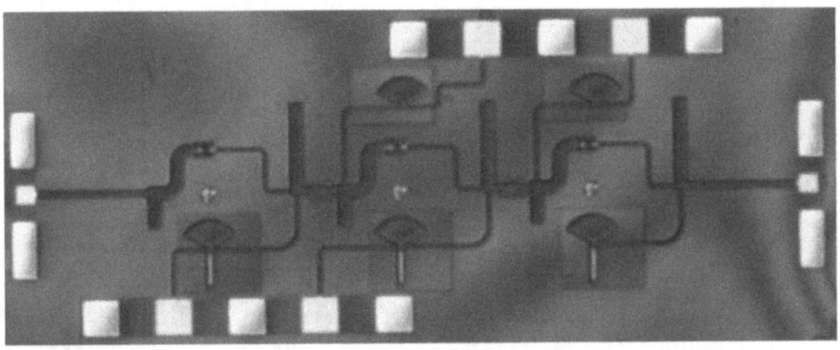

Fig. 14.56 The appearance of the crystal of the amplifier

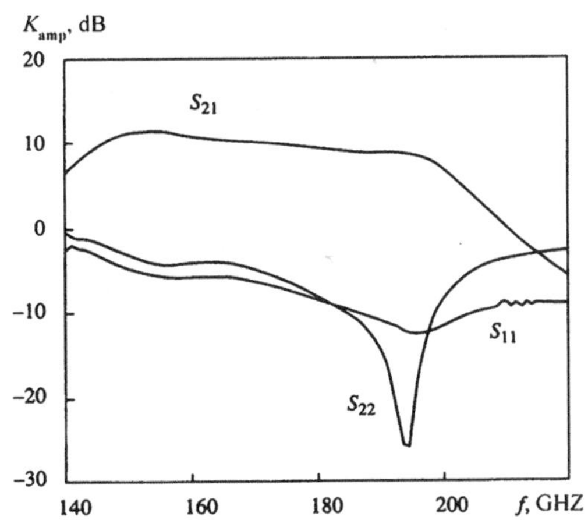

Fig. 14.57 The linear S parameters

Fig. 14.58 The input-power dependence of the maximal output power

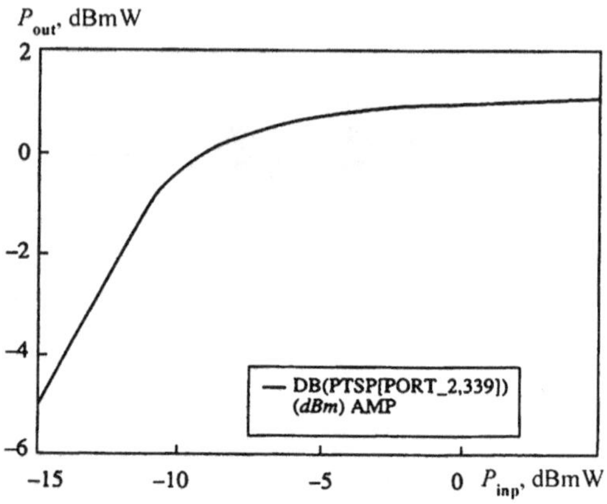

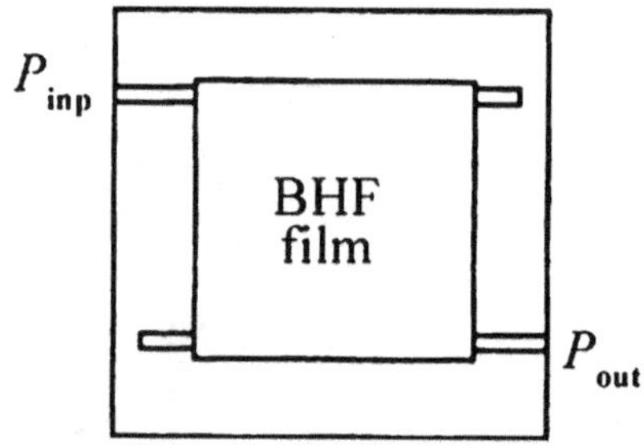

Fig. 14.59 A MECE with FMCR on the basis of an epitaxial BHF film (barium hexaferrite H35, the field of anisotropy $H_A = 35\,\mathrm{kOe}$, $4\pi M_s = 1{,}400\,\mathrm{G}$, $\Delta H = 3.5\,\mathrm{kOe}$) of the size $0.3 \times 0.3\,\mathrm{mm}^2$ was introduced into the model of HMG, to connect the two asymmetrical microstrip lines with a width of $10\,\mu\mathrm{m}$

$$R = K(h) \cdot Q_s\,(\Omega),$$

$$Q_s = \frac{(H_0 + H_A) - \frac{1}{3}4\pi M_s}{\Delta H} \quad \text{unloaded GB product,}$$

$$\omega_0 = 2\pi\gamma(H_0 + H_A) \quad \text{ferromagnetic resonance frequency, and}$$

$K(h)$ is the coupling factor of the ferrite film with microstrips.

In Fig. 14.60, theoretical AFC of the amplifier – 1 and that of FMCR at several magnetic fields: 2 – 21 kOe, 3 – 27 kOe, and 4 – 30 kOe are presented.

The basic circuit of HMG is presented in Fig. 14.61. The conditions of generation are satisfied in a range of frequencies 150–200 GHz at changes of the magnetic field within 20–32 kOe. On the output of the generator, there is a frequency multiplier that allows isolation of the harmonic $n = 5$ with losses of transformation (−20 dB). The nonlinearity of the amplifier is adjusted so that the amplitude of the harmonic $n = 5$ was maximal.

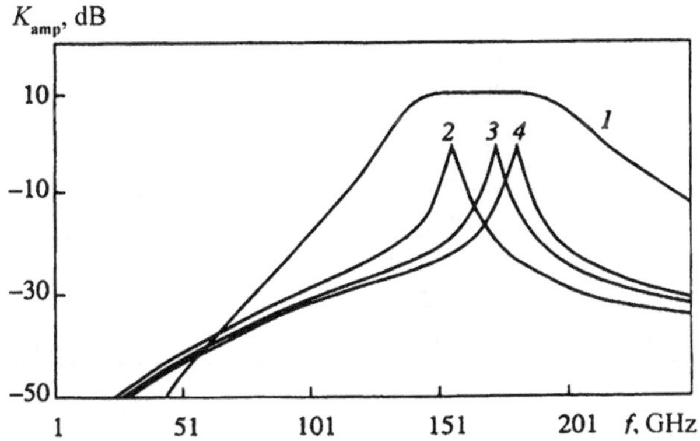

Fig. 14.60 Theoretical AFC of the amplifier – 1 and that of FMCR at several magnetic fields: 2 – 21 kOe, 3 – 27 kOe, 4 – 30 kOe

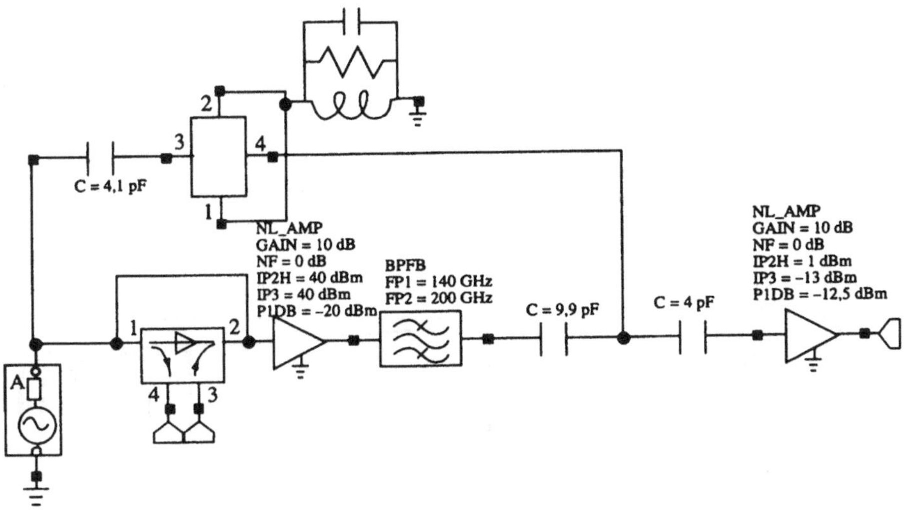

Fig. 14.61 The basic circuit of HMG

In Table 14.3, the results of our numerical experiment for signals on the basic frequency ν_0 and ν_1, ν_2, ν_3, ν_4 are presented at three values of the magnetic field.

The output path for a range of frequencies below 1,000 GHz is simulated in the form of a microstrip line on a benzocyclobuthene substrate ($\varepsilon_r = 2.7$) with a thickness of 5 μm. The capabilities of MWO-2002 CAD for modeling of structures on frequencies below 1,000 GHz with greater sizes of the calculation field have been investigated. In Fig. 14.62, the instability of calculations is presented at a field size of 200×400 μm and in Fig. 14.63 it is presented for a working field of 400×800 μm. In these cases, the size of the point was 2×2 μm.

Table 14.3 The results of our numerical experiment for signals on the basic frequency ν_0 and $\nu_1, \nu_2, \nu_3, \nu_4$ at three values of the magnetic field

Magnetic field, kOe	Frequency ν_0 (GHz) [Amplitude, dBmW]	1 harmonic. frequency ν_1 (GHz) [Amplitude, dBmW]	2 harmonic. frequency ν_2 (GHz) [Amplitude, dBmW]	3 harmonic. frequency ν_3 (GHz) [Amplitude, dBmW]	4 harmonic. frequency ν_4 (GHz) [Amplitude, dBmW]
21	158 [−20.6]	316 [−42.9]	474 [−31.3]	632 [−44.9]	790 [−38.3]
27	173 [−20.6]	346 [−42.3]	519 [−31.7]	692 [−44.5]	865 [−39.3]
30	180 [−20.6]	360 [−43.2]	540 [−31.5]	720 [−45.0]	900 [−38.1]

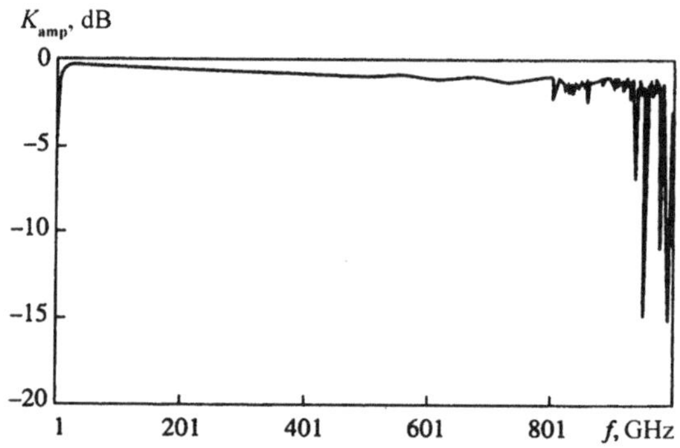

Fig. 14.62 The instability of calculations is presented at a field size of 200 × 400 μm

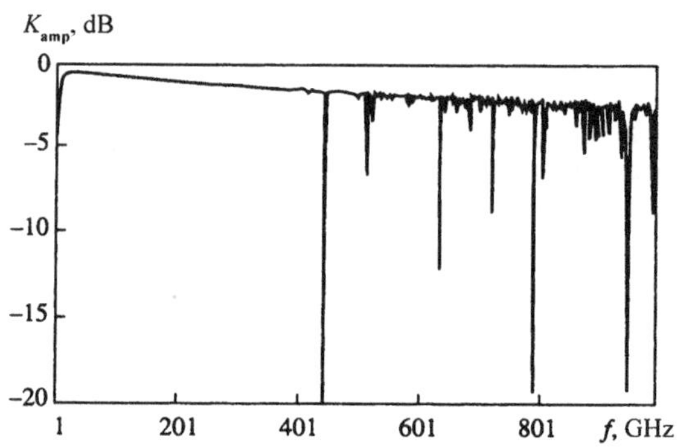

Fig. 14.63 The instability of calculations is presented at a field size of 200 × 400 μm for a working field of 400 × 800 μm

For increase of the level of output power of HMG in a mode of frequency transformation, an amplifier with a working range of frequencies below 1,000 GHz, an amplification factor of 10 dB, and the maximal output power of 1 mW was simulated. For increase of the amplification factor and the level of output power, a two-cascade circuit with power summation in the second cascade (Fig. 14.64) has been used.

In such an amplifier on the harmonic with $n = 5$, the power of signal has increased from 0.3 μW (−37.1 dBmW) to 4.4 mW (6.4 dBmW).

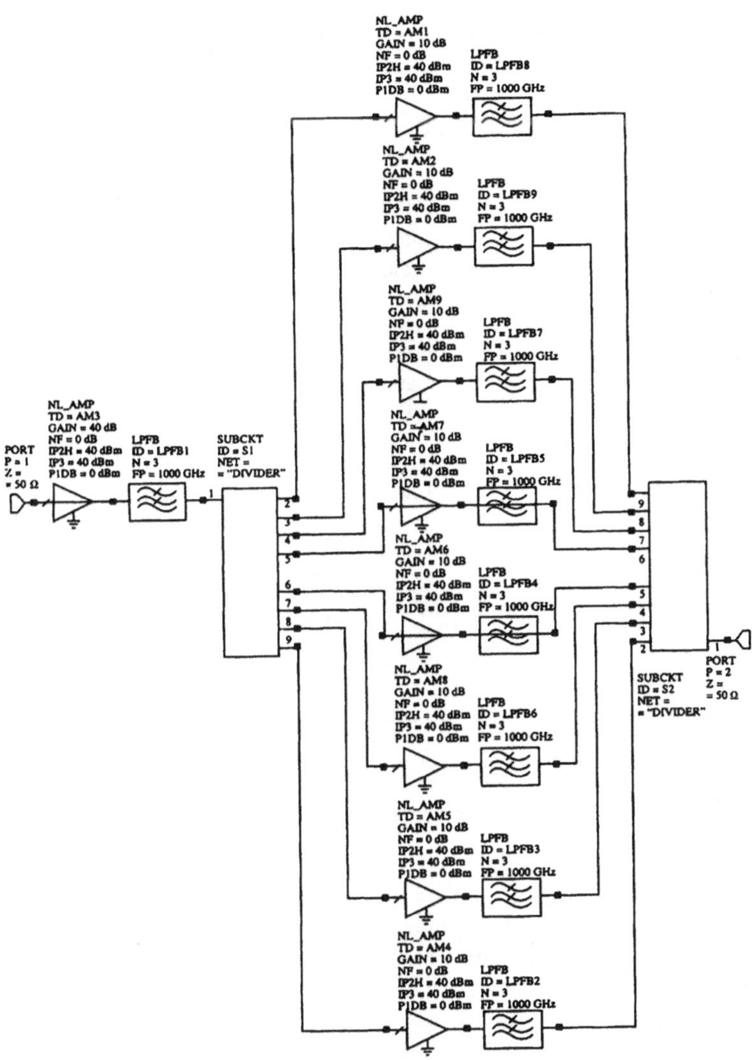

Fig. 14.64 A two-cascade circuit with power summation in the second cascade

14.7 Manufacturing Methods

14.7.1 FET Parameters

In Table 14.4, parameters of several field microwave transistors on GaAs with Schottky's gate (*MESFET*) for coupling with MECE by power level and working frequency range in various types and sorts of FEMT are presented.

Table 14.4 Parameters of several field microwave transistors on GaAs with Schottky's gate

FT type	f, GHz	K_{amp}, dB	P_{out} on 1 dB compression, mW	I_{op}, mA	$U_{drain-source}$, V	$U_{test.\ drain-source}$, V
PT-300 (pin)	18	7.0	100	35	6–7	14
A foreign analog	12	7.5	140	35	6–7	14
PT-300-TC 1201	12	7.5	140	40	7	–
PT-600 (pin)	18	7.0	200	70	6–7	14
A foreign analog	12	7.0	250	70		
PT-600-MGF1601B	12	7.5	160	70	6	–
PT-900 (pin)	12	7.0	350	120	6–7	14
PT-1200 (pin)	12	7.0	500	180	6–7	14
A foreign analog	12	7.0	450	130	9	–
PT-1200-ATF-46101	3	5–8	120	350	7–9	14
PT-2500 (pin with metallized apertures for grounding sources)	9	5–8	1,000	350	7–9	14
PT-3000 (pin with metallized apertures for grounding sources)	3	5–8	1,500	400	7–9	14
PT-5000 (pin with metallized apertures for grounding sources)	3	5–8	2,000	700	7–9	14

Our parametrical series of FT (see Table 14.4) covers a range of working frequencies up to 18 GHz with an output power up to 0.2 W and a range of frequencies up to 3 GHz with an output power from 1.2 W to 2 W (Almas-Phasotron Corp.).

The design of these FTs and their sizes are optimal for construction of circuits of power division and summation. The gate and gravity lines of FT together with their active area can be components of MECE.

Thus, FTs are integrated with MECE on one crystal. To integrate powerful FTs with MECE, it is necessary to reduce the overall dimensions of the latter ones as they do not allow to improve heat removal due to thinning of FEMT down to $30 \div 50\,\mu$m. It is necessary to introduce coupling circuits between MECE and powerful FTs as well.

Further substantial improvement of the parameters of the field transistors intended for integration with MECE is connected with development and manufacturing of pseudoamorphous FT with a high electron mobility (*PHEMT*), having a specific output power of 1 W/mm and with *GAN* transistors having a specific output power of few W/mm.

14.7.2 Technological Peculiarities of Manufacturing of GaAs FET

The design and know-how elements of FT and MC intended for the use in circuits with planar MECE to increase the power level are described.

Simple technological solutions of increase of the specific output power and breakdown voltage of FT are offered. Pin FTs with a shutter width of 300 μm for frequencies below 18 GHz and 5,000 μm for obtaining an output power of more than 2 W on frequencies below 3 GHz have been designed and made. A monolithic amplifier for a range of frequencies 1.5–4.5 GHz with a high amplification factor, low SWRH on input and output, a low current consumed has been designed and made.

FT topologies with various shutter widths are presented in Fig. 14.65: (a) 300 μm and (b) 5,000 μm. The structure of the channel of the developed FT is shown in Fig. 14.66.

As is obvious, the shutter is displaced toward the source. The double recess of the shutter allows keeping a high steepness of FT at sufficient breakdown voltages. The proper choice of the orientation of the masostructure forms flat walls of the channel's recess.

For maintenance of the high interelement isolation and decrease in the leakage currents, the active areas of FT were formed by double mesa-etching and four-stage proton bombardment. In the manufacturing of FT, $n^+ - (n_a - n_a^+)$ structures of gallium arsenide were applied to get a specific output power of more than 0.5 W/mm at CE > 40%, for increase of breakdown voltages up to 20 V.

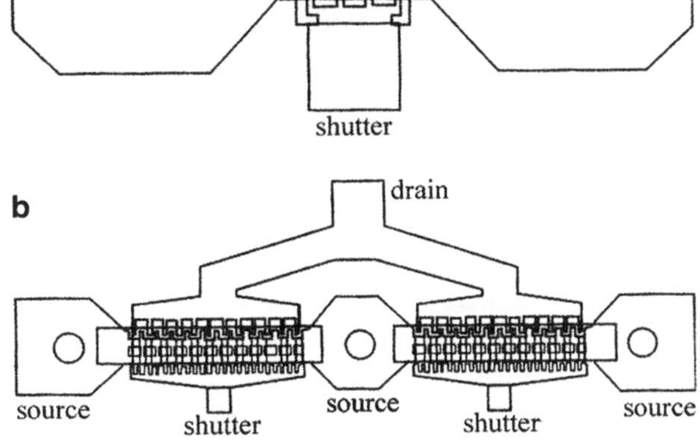

Fig. 14.65 FT topologies with various widths of the shutter: (**a**) 300 μm; (**b**) 5, 000 μm

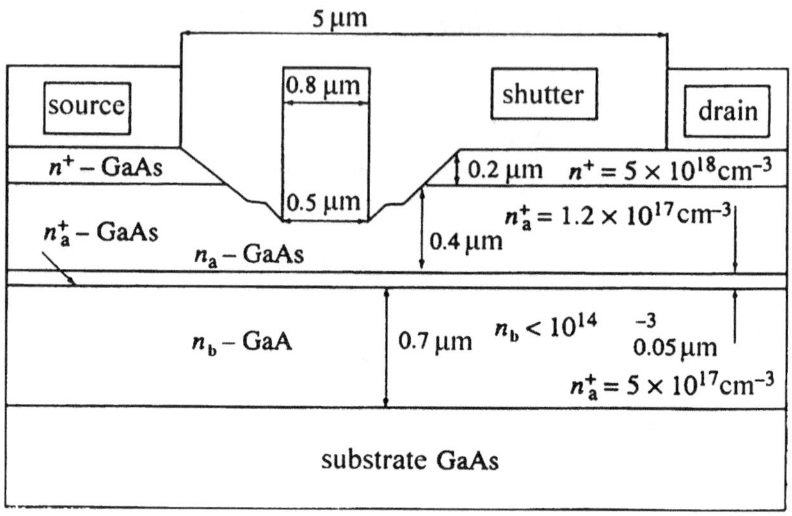

Fig. 14.66 The structure of the channel of the developed FT

The fabrication stages FT are as follows:

1. Formation of the active areas of FT and semiconductor resistors and etching of the mesa-structures with subsequent proton bombardment through a photoresist mask
2. Low-temperature deposition of a SiO_2 film serving (in further) as a masking layer at formation of FT electrodes
3. Formation of ohmic FT contacts, source grounding decks via metallized apertures, and contact decks of the semiconductor resistors
4. Formation of a gate mask, etching of a recess for the shutter, evaporation of gate metallization, formation of the bottom facings of the MDM capacitors and first-level metallization, and film *Ti* resistors
5. Passivation of the active area of FT, formation of displacement resistors, and deposition of a dielectric for the MDM capacitors
6. Restriction of the dielectric of the MDM capacitors and deposition and heat treatment of the underbridge dielectric
7. Formation of second-level metallization: PLG, spraying, galvanic escalating, and etching
8. Plate refinement, formation of through metallized apertures, splitting of the plate into crystals, and grading and parameter measurements

The results of measurements of the FT parameters in a mode of bilateral coordination are presented in Table 14.5.

By the measured VAC of low-level *S* parameters of the made FT with a shutter width of 300 μm, an equivalent circuit of the transistor (Fig. 14.67) has been designed, on the basis of which a monolithic amplifier in a range of frequencies from 1.5 to 4.5 GHz has been calculated ($C_{gs} = 0.35$ pF, $C_{gd} = 0.04$ pF, $C_{ds} = 0.1$ pF, $R_i = 8\ \Omega$, $R_g = 2.5\ \Omega$, $R_s = 3\ \Omega$, $R_{ds} = 3\ \Omega$).

Table 14.5 The results of measurements of the FT parameters in a mode of bilateral coordination

FT type	ν, GHz	V_{drain}, V	I_{op}, mA	K_{amp}, dB	P_{out}, mW
FT-300	12	7	40	9	160
FT-300	18	7	40	8	122
FT-5000	1	8	610	9	2,500
FT-5000	3	8	610	8	2,200

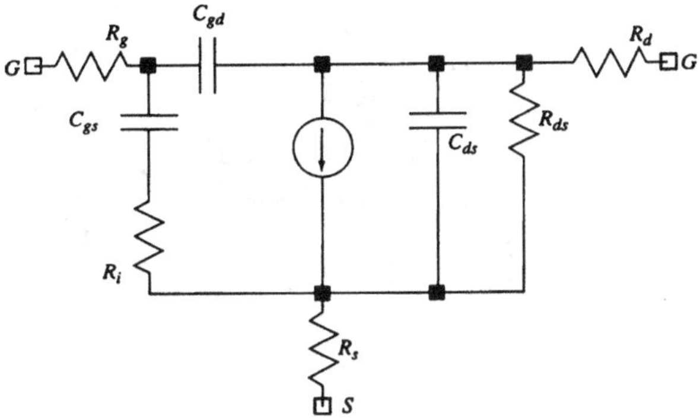

Fig. 14.67 The ready-built equivalent circuit of the transistor by the measured VAC of low-level S parameters

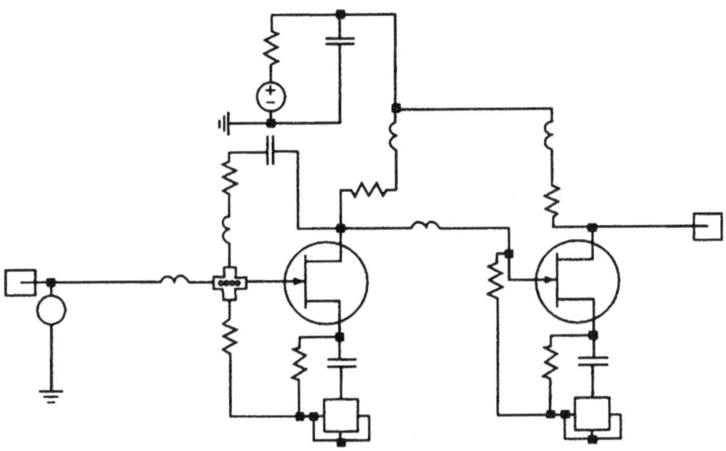

Fig. 14.68 The electric circuit of the FT model

The parameters of the FT model were investigated in modes: V_{ds} $5 \div 7$ V, $V_{gs} \approx -1.7$ V. The amplifier was computed with the use of CAD. In Fig. 14.68 the electric circuit and in Fig. 14.69 the topology of the developed monolithic amplifier are presented.

For maintenance of amplification not less than 12 dB in a range of frequencies from 1.5 to 4.5 GHz, a two-cascade amplifier circuit with a negative feedback has been chosen. The width of the FT shutter of the input and output cascade was 1.0 and 0.8 mm, respectively.

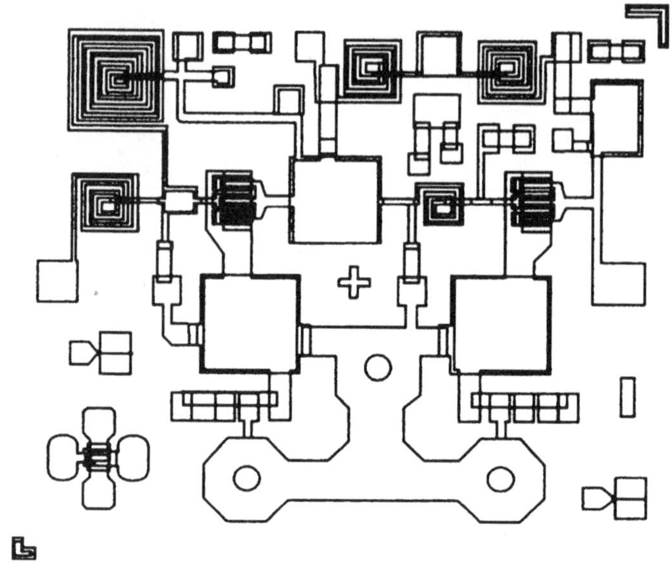

Fig. 14.69 The topology of the developed monolithic amplifier

Fig. 14.70 Characteristics of factor of strengthening

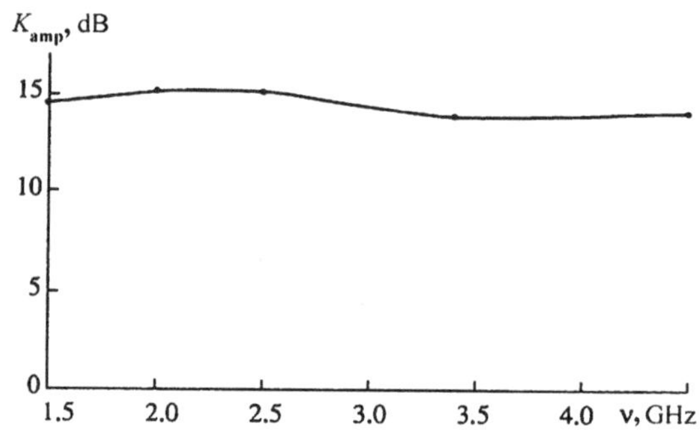

In Fig. 14.70 characteristics of factor of strengthening and in Fig. 14.71 $SWR_{e,inp}$ and $SWR_{e,out}$ samples MC in a range of frequencies from 1.5 up to 4.5 GHz are presented.

The output power of the monolithic amplifier is not less than 25 mW at a current consumed from 45 to 55 mA. The nonuniformity of the amplification factor in a range of frequencies from 1 to 4.5 GHz did not exceed ±0.8 dB.

The reached parameters and design and technological solutions allow one to use FT and MC in a combination with microstrip MECE for design of multipurpose magnetodevices intended for amplification, generation, and transformation of signals.

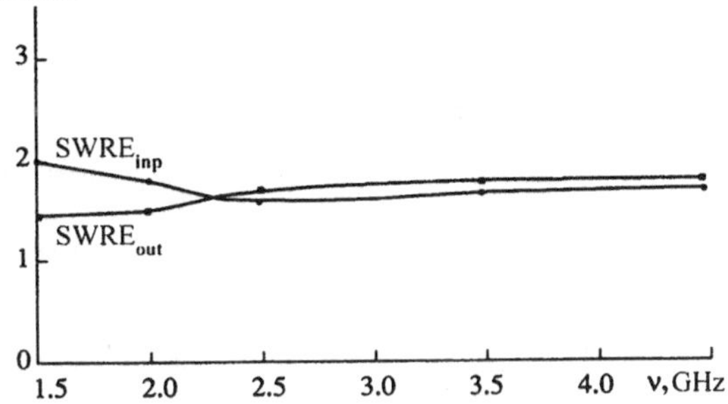

Fig. 14.71 $SWR_{e,inp}$ and $SWR_{e,out}$ samples MC in a range of frequencies from 1.5 up to 4.5 GHz

14.8 Manufacturing Methods of an Integral Magnetosemiconductor Device

The known types and structures of ferrites allow covering ranges from the radiowave one on ferro- and ferrimagnetics (yttrium-iron garnets $Y_3Fe_5O_5$, spinels $MgAl_2O_3$, barium hexaferrite) up to the optical one on antiferromagnetics ($\alpha \cdot Fe_2O_3$, hematite, $FeBO_3$, iron borate, $NiCO_3$ nickel carbonate). Nonlinear effects are determined by the level of threshold high-frequency power (of the order of magnitude 0.5–1 mW on a frequency of 1 GHz), and they influence the modes of parametrical multiplication, division, and frequency modulation in FMCR. Besides, FMCR can be made of various materials with various magnetic parameters. New kinds of structures – epitaxial film, including multilayered ones with preset cross gradients of magnetic parameters (saturation magnetizations), reduction of the thickness of the resonator down to several μm (5–50 μm), including multilayered (up to seven layers) ones – allow the sizes of FMCR to be reduced considerably, in comparison with spheres [1].

Below we consider some variants of the manufacturing techniques of monolithic-type HMG – generating magnetosemiconductor devices, CHIPS [23].

In Fig. 14.72, the topology of n–p–n BMT is shown: 1 – a terminal of the substrate of p-type carriers; 2 – a terminal of the collector; 3 – a terminal of the base; 4 – a terminal of the emitter; 5 – an additional electrode of coupling with the base through FMCR; 6 – an additional electrode of coup;ing with the emitter through FMCR; 7 – an additional electrode of coupling with the collector through FMCR; 8 – FMCR between the collector and the additional electrode of the collector; 9 – FMCR between the base and the additional electrode of the base; 10 – FMCR between the emitter and base; 11 – FMCR between the emitter and the additional electrode of the emitter; 12 – FMCR between the emitter and collector; 13 – FMCR between the base and collector; 14 – a layer of SiO_2; 15 – substrates with p-conductivity; 16 – an epitaxial layer with an n-area; 17 – p-area of the base; 18 – n^+-area of the emitter; 19 – n^+-area of the collector.

The magnetization fields $\overline{B}_1, \overline{B}_2, \overline{B}_3, \overline{B}_4, \overline{B}_5, \overline{B}_6$ are applied to the corresponding FMCR in the magnetotransistor. On the sections A–A and B–B (see Fig. 14.72), the transistor's electrodes – 1, 2, 3, 4; additional electrodes – 5, 6, 7; FMCR – 8, 9, 10, 11, 12, 13; a layer of SiO_2 – 14 are shown.

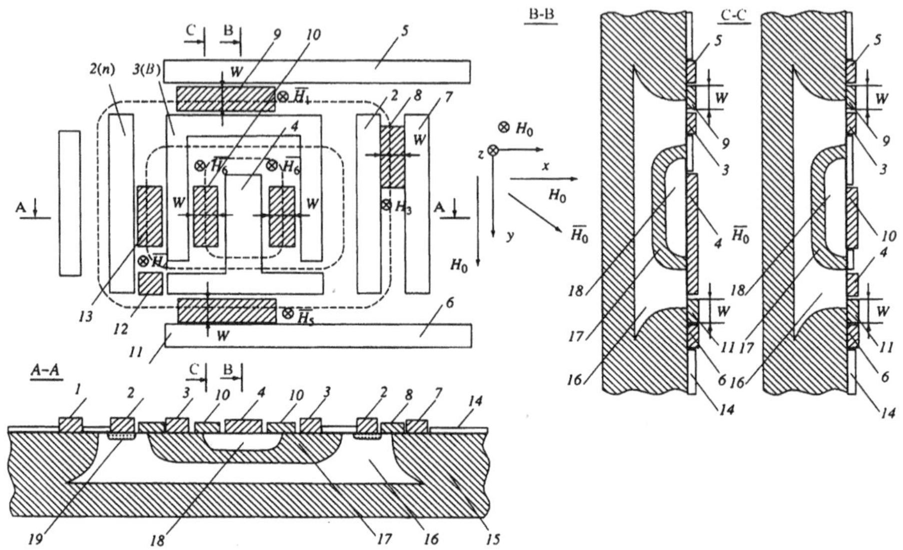

Fig. 14.72 The topology of n–p–n BMT

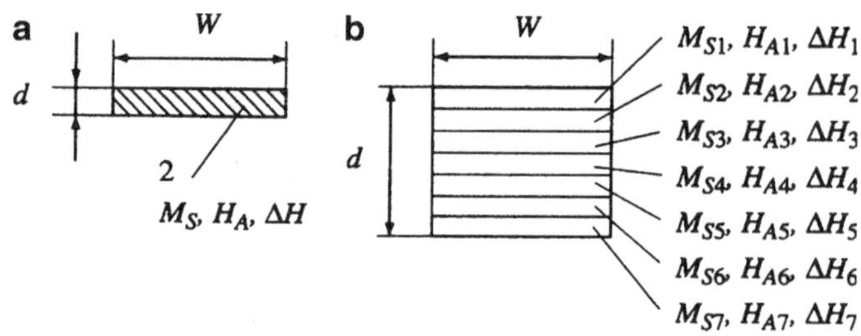

Fig. 14.73 Sections of film FMCRs – magnetic dielectrics are shown: (**a**) a single-layered one with constant parameters over the thickness of the ferrite layer (film) (saturation magnetization M_s, the field of anisotropy H_A, the linewidth of ferromagnetic resonance ΔH); (**b**) a seven-layer one with the magnetizations M_{s1}, M_{s2}, M_{s3}, M_{s4}, M_{s5}, M_{s6}, M_{s7}, the fields of anisotropy B_{A1}, B_{A2}, B_{A3}, B_{A4}, B_{A5}, B_{A6}, B_{A7}, the linewidth of ferromagnetic resonance ΔH_1, ΔH_2, ΔH_3, ΔH_4, ΔH_5,ΔH_6, ΔH_7

In Fig. 14.73, sections of film FMCRs – magnetic dielectrics are shown: (a) a single-layered one with constant parameters over the thickness of the ferrite layer (film) (saturation magnetization M_s, the field of anisotropy H_A, the linewidth of ferromagnetic resonance ΔH); (b) a seven-layer one with the magnetizations M_{s1}, M_{s2}, M_{s3}, M_{s4}, M_{s5}, M_{s6}, M_{s7}, the fields of anisotropy B_{A1}, B_{A2}, B_{A3}, B_{A4}, B_{A5}, B_{A6}, B_{A7}, and the linewidth of ferromagnetic resonance $\Delta H_1, \Delta H_2, \Delta H_3, \Delta H_4, \Delta H_5, \Delta H_6, \Delta H_7$.

The width of a ferrite resonator W is not less than the thickness of domains; at a thickness of the ferrite layer d, the size must be $W \geq 10^{-2}\sqrt{d}$. In the interelectrode gap of the semiconductor device, one or several FMCRs with their width W and length L is placed. FMCR can have the forms: a rectangular with the sizes $d \times W \times L$; an ellipsoid with the semiaxis sizes $a \times b \times c$, and $a = d/2, b = W/2, c = L/2$; a spheroid (sphere) with a diameter $D \geq W$. FMCR can have their magnetic parameters over thickness, width, and length: (a) homogeneous and (b) changing according to some law, in particular, to be multilayered with cross gradients of their parameters $\nabla_x M_s, \nabla_x H_A, \nabla_x \Delta H_A$. FMCR should be placed at the maximum of high-frequency magnetic fields $\tilde{\tilde{h}}$ in the semiconductor device. If the semiconductor device contains two or more FMCR, they can have identical or different magnetic parameters and be:

- In one magnetization field
- In different magnetization fields
- In identical or different static operating modes (domain, transitive, or one-domain ones), which are determined by magnetic parameters, the sizes of FMCR, as well as the value of magnetization field
- In identical or different dynamic states, namely, linear ($\tilde{h} \ll \underset{\sim}{h}_{\text{threshold}}, \tilde{h} \ll \underset{\sim}{m}$), transitive or quasilinear ($\tilde{h} \approx \tilde{h}_{\text{threshold}}, \tilde{h} \approx \tilde{m}$), nonlinear ($\tilde{h} \gg \underset{\sim}{h}_{\text{threshold}}, \tilde{h} \gg \tilde{m}$) ones.

The magnetization field, along with its constant component H_0, can have a varying component h_{mod}, so the total field magnetization is $\overline{H}_0 + \tilde{h}_{\text{mod}} e^{j\omega_{\text{mod}} t}$, where ω_{mod} is the frequency of modulation. For high-quality monocrystal ferrites like YIG, the relaxation time of transients is 10^{-9} s that corresponds to the upper frequencies of modulation ($\omega_{mod}/2 \cong 1$ GHz).

Let us note that FMCR can also have the appearance of a sphere that is placed in the area of the maximal sagging of the high-frequency magnetic fields in the semiconductor device, has a diameter $D \gg W$, and is in the domain, transitive, or one-domain modes and in one of the possible states – linear, transitive, and nonlinear ones. One or several FMCRs can operate without any magnetization field – in the mode of autoresonance on one of the characteristic frequencies, defined by the magnetic parameters – magnetization, the anisotropy field, the sizes and shape of the sample, and its demagnetization factors.

For a semiconductor device designed in the form of layers with various conductivities, including epitaxial technologies and dopation, FMCR can be placed in the areas of the maximum high-frequency magnetic fields in the interelectrode areas or between the basic and additional electrodes.

The basic fabrication stages of a multipurpose epitaxial bipolar magnetotransistor with FMCR built in one of the layers in the field of localization of high-frequency magnetic fields in the interelectrode "emitter–base" areas are presented in Fig. 14.74. The fabrication stages of the bipolar transistor [88] are supplemented new ones, providing formation of film FMCRs.

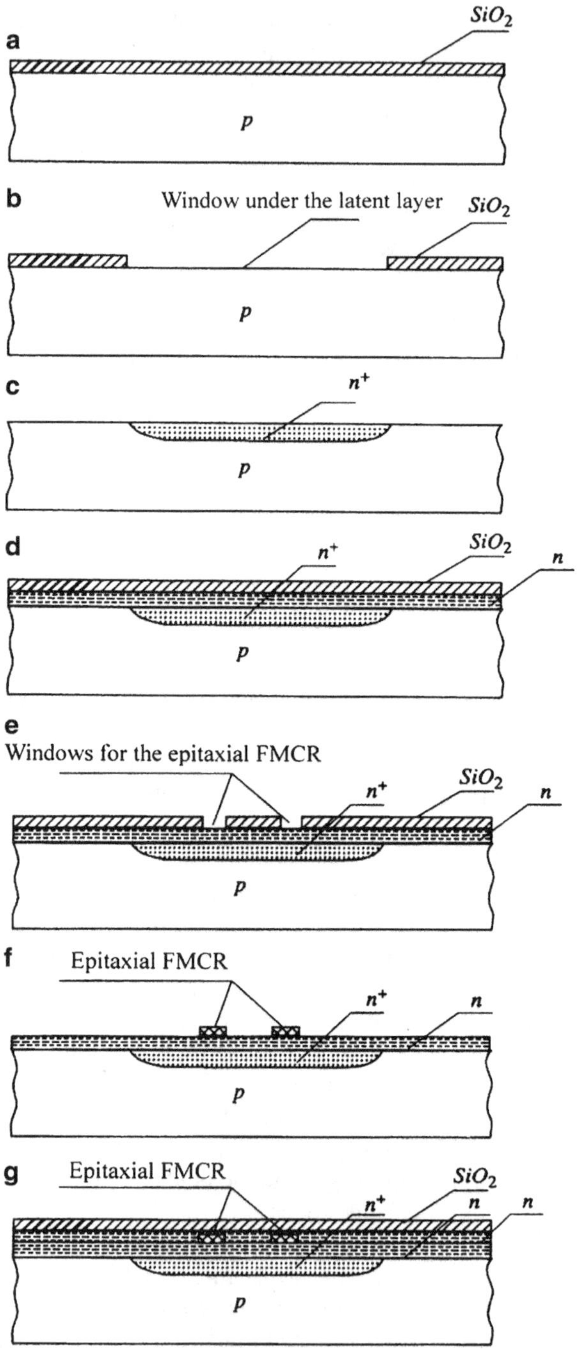

Fig. 14.74 Fabrication stages of a multipurpose epitaxial bipolar magnetotransistor with FMCR built in one of the layers in the field of localization of high-frequency magnetic fields in the inter-electrode "emitter–base" areas

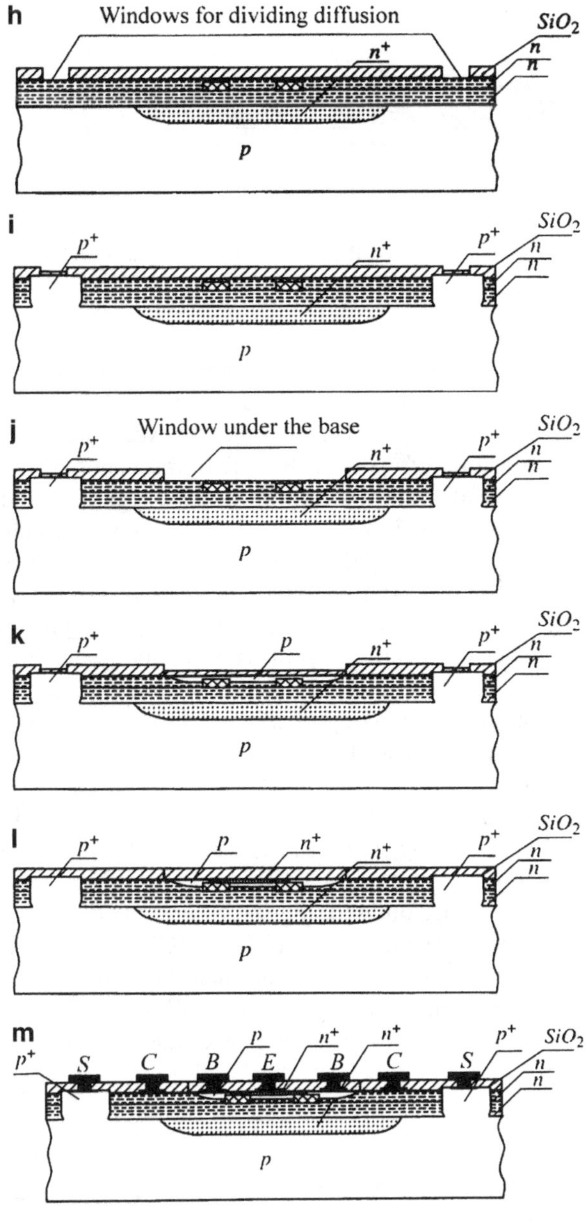

Fig. 14.74 (continued)

In Fig. 14.74, the following are shown: (a) preparation of a plate with p-conductivity and its oxidizing with a SiO_2 layer; (b) photolithography for opening windows under the latent layer; (c) diffusion of the donor impurity n^+ and removal of the SiO_2 layer; (d) deposition of the epitaxial layer with n-conductivity

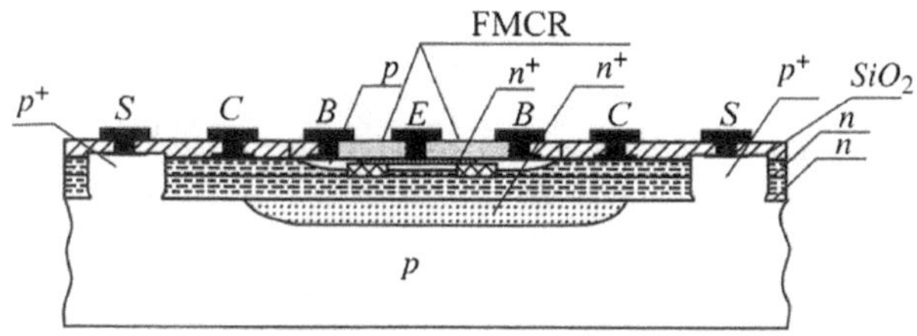

Fig. 14.75 The placing of epitaxial FMCR between the ohmic contacts of a bipolar magnetotransistor

and oxidation with a SiO_2 layer; (e) photolithography for opening windows under the epitaxial FMCR, placed inside the layer with n-conductivity in the field of localization of high-frequency magnetic fields between the emitter and base in the transistor; (f) deposition of epitaxial FMCR in the windows on the layer with n-conductivity and removal of the protective SiO_2 layer; (g) deposition of an epitaxial layer with n-conductivity and oxidation with a SiO_2 layer; (h) photolithography for opening windows for dividing diffusion; (i) diffusion of the p-type impurity and oxidation with SiO_2; (j) photolithography for opening windows under the base; (k) diffusion of the p-type impurity, oxidation with SiO_2, and photolithography for opening windows under the emitter; (l) diffusion of the donor impurity, removal of the SiO_2 oxide, and deposition of pure oxide; (m) photolithography for opening windows under the contacts, spraying of a metal film, photolithography to form contact decks of the emitter E, base B, collector C, and connecting conductors with the substrate S.

By placing epitaxial FMCR between the ohmic contacts of a bipolar magnetotransistor (Fig. 14.75), the basic stages a, b, c, d (Fig. 14.78) are kept, the stages e, f, g, that is, formation of epitaxial FMCRs between the emitter and base inside the layer of the transistor, are omitted, the stages h, i, j, k, l, m are carried out, followed by the stages of formation of epitaxial FMCRs on the place between the future electrodes of the emitter and base, and then the stage n follows – PLG for opening windows for contacts, spraying of a metal film, and PLG for formation of contact decks of the emitter E, base B, collector C, and the connecting conductors with the substrate S.

14.9 Multivariate Vector Sensors of Mechanical Dynamic Quantities

The autogenerating circuit on FSS, a heteromagnetic sensor with MPMS placed on an elastic suspension bracket, allows registering deviations of the frequency response of the generator to the displacement vector along the axis $0X$ of easy magnetization $\overline{X}$ for the time Δt (Fig. 14.76).

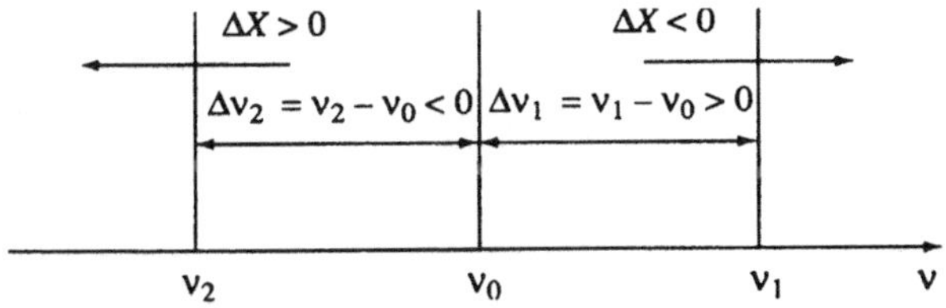

Fig. 14.76 The registration of deviations of the frequency response of the generator to the displacement vector along the axis $0X$ of easy magnetization $\overline{X}$ for the time Δt

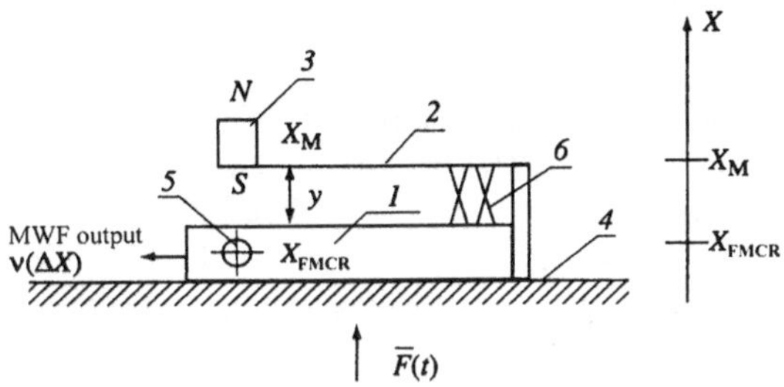

Fig. 14.77 1 – HMS; 2 – an elastic suspension bracket; 3 – the mobile part of the magnetic system; 4 – a platform subjected to force influence $\overline{F}(t)$; 5 – a FMCR; 6 – a damper

If the own frequency of MPMS oscillations is low $\nu_0 \approx (1-10)$ Hz, the external short-term pulse or periodic influence of force $F(t)$ with a frequency $\nu_F > \nu_0$ onto the platform on which HMS is fixed leads to the change in distance along the axis $0X$ between MPMS and FMCR, which is the displacement vector, $\Delta\overline{X}$, with $\Delta\overline{X} = X_{\mathrm{M}} - X_{\mathrm{FMCR}}$, where X_{M} is the coordinate of the center of MPMS and X_{FMCR} is the coordinate of the center of FMCR. At reduction of $\Delta\overline{X}$, the frequency of HMS response will increase and $\Delta\nu_1 = (\nu_1 - \nu_2) > 0$, $\Delta X < 0$, and at increase of $\Delta\overline{X}$, the frequency of HMS response will decrease and $\Delta\,\nu_2 = (\nu_2 - \nu_1) < 0$, $\Delta X < 0$.

In Fig. 14.77, the following are shown: 1 – HMS; 2 – an elastic suspension bracket; 3 – the mobile part of the magnetic system; 4 – a platform subjected to force influence $\overline{F}(t)$; 5 – a FMCR; and 6 – a damper.

At reduction of $\Delta X(\Delta X < 0)$, the magnetic field acting on FMCR increases, and at increase of $\Delta X(\Delta X > 0)$, it decreases. For the case of small displacements of MPMS, the own frequency of HMS oscillations and the magnetic induction acting on FMCR are related as ν_0 (MHz) $= 2.8 \times B_0$ (G). Then, the deviation of frequency is ν_0 (MHz) $= 2.8 \times \Delta B_0$, where ΔB_0 is the value of magnetic induction deviation.

MPMS's magnetic induction is strongly nonuniform. Experimental researches have allowed resolving linear sites in a wide range of magnetic induction, and, hence, of the frequency of generation. For example, for a cylindrical washer made of samarium–cobalt alloy of diameter 8 mm and of height $h = 4$ mm, a linear site at a 1-mm change of MPMS's position was realized at $\Delta B_0 = 60$ G that corresponded

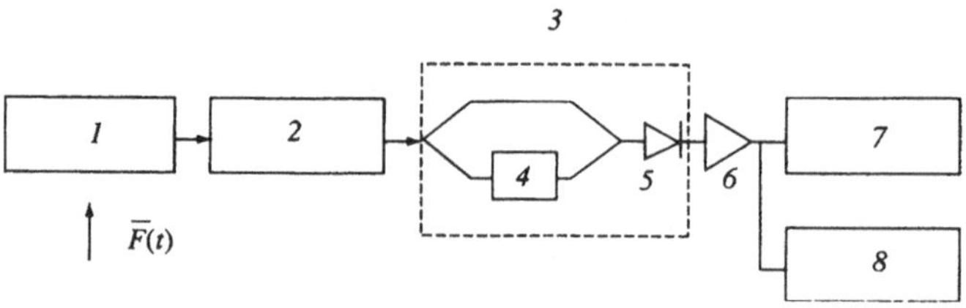

Fig. 14.78 The flowchart of the sensor with a phase detector is shown: 1 – a HMS of displacements; 2 – an amplifier; 3 – a phase detector; 4 – a phase-shifting circuit; 5 – a peak detector; 6 – an operational amplifier with $K_{amp} = 20\,dB$; 7 – an oscillograph; 8 – a millivoltmeter

to the change of frequency of generation by 160 MHz. As the frequency instability of the generator was at a level of 10 kHz, the sensor fixed force influences with an amplitude of 6×10^{-8} m (60 nm).

Transformation of the frequency change $\Delta\nu$ of HMS to the change of voltage ΔU requires introduction of a frequency or phase detector into the circuit of measurements.

In Fig. 14.78, the flowchart of the sensor with a phase detector is shown: 1 – a HMS of displacements; 2 – an amplifier; 3 – a phase detector; 4 – a phase-shifting circuit; 5 – a peak detector; 6 – an operational amplifier with $K_{amp} = 20\,dB$; 7 – an oscillograph; and 8 – a millivoltmeter.

The circuit in Fig. 14.78 at a HF power of 200–300 μW allows detection of, for example, oscillations of an internal windowpane in a massive building from the cars passing at a distance of 30 m, with a deviation of frequency $|\Delta\nu| \cong 2.5$ MHz and an amplitude of voltage change on the voltmeter (0.8–1 V).

The sensitivity of the sensor at a short-term force impact by speed and pressure was estimated as follows. A cup was placed on the case of the sensor, into which a body of a certain mass made of an inelastic material fell from various heights. Thus, an absolutely inelastic impact on the case of the sensor was realized. With the known area of cup–body contact, the weights of the sensor and magnet and the changes of speed and pressure upon the sensor were calculated. The sensitivity to relative speed changes is $S_\upsilon = \Delta V/\Delta\upsilon = 66\,(\mathrm{V/(m/s)})$, and the sensitivity to pressure changes is $S_\mathrm{P} = \Delta V/\Delta P = 800\,\mu\mathrm{V/Pa}$.

The characteristics of the commercial sensors applied in geophysics are [89] as follows:

- A geophone – $S_\upsilon = 28\,\mu\mathrm{V/Pa}$
- A hydrophone – $S_\mathrm{P} = 120\,\mu\mathrm{V/Pa}$

Registration of the displacement vector $\overline{X}$ for the time interval Δt allows determination of all the basic mechanical and dynamic quantities along the axis $0X$. For registration of the radius vector increment $\bar{r}(\Delta\overline{X}, \Delta\overline{Y}, \Delta\overline{Z})$, a three-dimensional sensor registering the projections $\Delta\overline{X}$, $\Delta\overline{Y}$, $\Delta\overline{Z}$ for the time intervals Δt_x, Δt_y, Δt_z is necessary [90]. It is convenient to have $\Delta t_x \approx \Delta t_y \approx \Delta t_z \approx \Delta t$. In the general case (Fig. 14.79), the trajectory of movement of the vector is $\bar{r}(t) = \bar{r}_0(t) \cdot r(t)$, then the full velocity vector is

Fig. 14.79 The trajectory of movement of the vector $\bar{r}(t)$

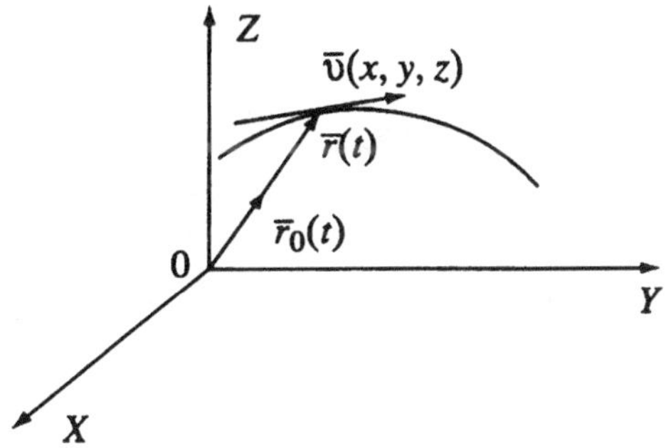

$$\bar{\upsilon}(t) = r\frac{\Delta\bar{r}_0}{\Delta t} + \bar{r}_0\frac{\Delta r}{\Delta t}, \quad (14.8)$$

where $\bar{r}_0$ is the unit vector directed toward the point of moving and r is the radius vector directed toward the point of observation.

The first component in (14.8) gives the change of velocity by direction, and the second does its change by value.

For registration of the full velocity vector $\bar{\upsilon}(t)$ according to (14.8), it is necessary to set $\bar{r}_0$ and r and to determine $\Delta r/\Delta t$ by sensors. At short intervals of time (mathematically, $\Delta t \to 0$ is required, in practice, Δt is a physically short time), the moving of HMS in space in comparison with the interval of time of supervision over object with which the sensor $-\Delta t_{\text{sup}}$ is connected, that is, $\Delta t \ll \Delta t_{\text{sup}}\Delta$.

Algorithmically, in (14.8) the first component $r(\mathrm{d}\bar{r}/\mathrm{d}t)$ and the second component $\bar{r}_0(\mathrm{d}r/\mathrm{d}t)$ can be determined.

The full acceleration vector $\bar{a}$ according to [90] is defined by

$$\bar{a} = \frac{\mathrm{d}\bar{\tau}}{\mathrm{d}t}\upsilon(t) + \bar{\tau}\frac{\mathrm{d}\upsilon}{t} = \frac{\upsilon}{R^2}\bar{n} + \bar{\tau}\frac{\mathrm{d}\upsilon}{\mathrm{d}\bar{\tau}}, \quad (14.9)$$

where $\bar{\tau}$ is the unit vector of the tangent to the trajectory, which coincides with $\bar{\upsilon}$ for accelerated movement ($\bar{a} > 0, \bar{\upsilon} \uparrow\uparrow \bar{\tau}$) and is oppositely directed for retarded movement – ($\bar{a} < 0, \bar{\upsilon} \downarrow\uparrow \bar{\tau}$), and $\bar{n}$ is the unit vector of the normal directed along the radius of curvature $\bar{R}$ to the center of the circle ($\bar{R} \uparrow\downarrow \bar{n}$).

The values $\bar{\upsilon}$, $\overline{R}$, $\bar{\tau}$, and $\mathrm{d}\upsilon/\mathrm{d}\bar{\tau}$ require their determination and design of an algorithm (14.9) of finding the full acceleration vector.

In Fig. 14.80, trajectories of movement are shown: (a) uniformly accelerated one and (b) uniformly retarded one.

Further, the following can be determined:

- With the use of (14.8) – the momentum is

$$\overline{P} = m\overline{V} \quad (14.10)$$

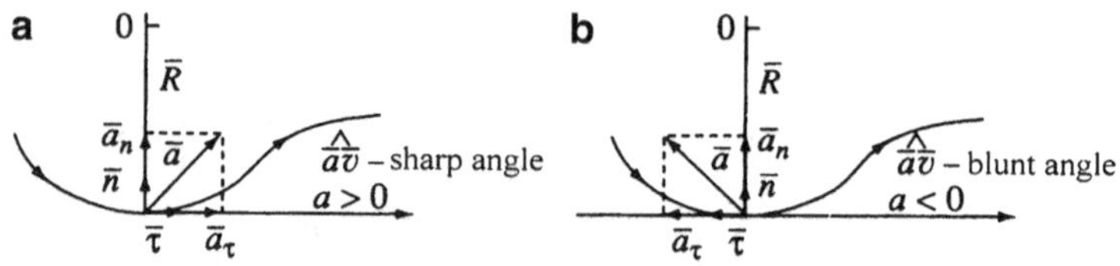

Fig. 14.80 The trajectories of movement: (**a**) uniformly accelerated one; (**b**) uniformly retarded one

– With the use of (14.9) – the force is

$$\overline{F} = m\overline{a} \tag{14.11}$$

Or from Newton's second law, through the momentum $\bar{P}(t)$ we have

$$\overline{F} = \frac{\Delta \overline{P}}{\Delta t}. \tag{14.12}$$

Then the momentum is

$$P = \int_0^t F(t)\mathrm{d}t. \tag{14.13}$$

The moment of force, as is known [101, 102], is defined by

$$\overline{N} = r \cdot F \cdot \sin(\overline{r}, \overline{F}). \tag{14.14}$$

Substituting (14.11) into (14.14), we get $N = r \cdot m \cdot a \cdot \sin(\overline{r}, \overline{F})$, the three of vectors $\overline{r}, \overline{F}, \overline{N}$ is right hand.

The moment of momentum is

$$L = r \cdot P \cdot \sin(\overline{r}, \overline{P}). \tag{14.15}$$

Substituting (14.10) into (14.15), we obtain

$$L = r \cdot m \cdot V \cdot \sin(\overline{r}, \overline{P}), \tag{14.16}$$

the three of vectors $\overline{r}, \overline{\upsilon}, \overline{L}$ is right hand. From the equation of rotary movement

$$\frac{\mathrm{d}\overline{L}}{\mathrm{d}t} = \overline{N}_{\mathrm{ext}}. \tag{14.17}$$

With the use of (14.15), the moment of external forces can be found $\overline{N}_{\mathrm{ext}} = \Delta\overline{L}/\Delta t$.

Let the system represented in Fig. 14.81 move with a velocity $\overline{\upsilon}_0$, relative velocity $\overline{\upsilon}'$, and rotate with an angular speed $\overline{\omega}$ around an axis passing through the origin of coordinates of the mobile system $0'$.

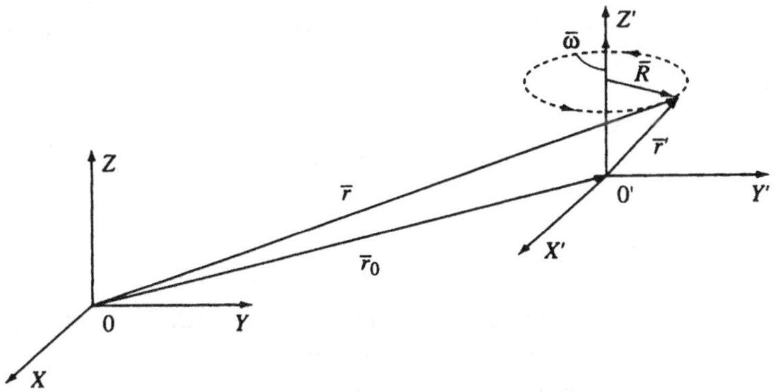

Fig. 14.81 The system $X'Y'Z'$ moves with a velocity $\overline{\upsilon}_0$, relative velocity $\overline{\upsilon}'$ and rotate with an angular speed $\overline{\omega}$ around an axis passing through the origin of coordinates of the mobile system $0'$

For the case of classical mechanics $\upsilon \ll c$, where c is the speed of light, and $\mathrm{d}t = \mathrm{d}t'$, we find for the vector of displacement

$$\Delta\overline{r} = \Delta\overline{r}_0 + \Delta\overline{r}' + [\overline{\omega r}']\Delta t, \tag{14.18}$$

where the primed quantities are referred to the mobile (primed) $X'Y'Z'$ system of coordinates.

From (14.18) for speeds $\upsilon \ll c$ and also $\Delta t = \Delta t'$, we have

$$\overline{\upsilon} = \overline{\upsilon}_0 + \overline{\upsilon}' + [\overline{\omega r}']. \tag{14.19}$$

If $\omega = \text{const}$, then from (14.19) we derive

$$\overline{a}_{\text{iner}} = \overline{a}_0 + \overline{a}_n + 2 \cdot [\overline{\omega\upsilon}'] + [\overline{\omega}[\overline{\omega r}']]. \tag{14.20}$$

From (14.20) for $\bar{\omega} = \text{const}$ we find the force of inertia

$$\overline{F}_0 = -m \cdot (\overline{a}_{\text{iner}} - \overline{a}_{\text{noniner}}) = -m \cdot \overline{a}_0 + 2 \cdot m \cdot [\overline{\upsilon}'\overline{\omega}] + m \cdot [\overline{\omega}[\overline{\omega r}']], \tag{14.21}$$

where the first component $m\overline{a}_0$ is the force of inertia at progressive motion; the second component $2 \cdot m \cdot [\overline{\upsilon}'\overline{\omega}]$ is the Coriolis' force; and the third component $m \cdot [\overline{\omega}[\overline{\omega r}']]$ is the centrifugal force of inertia.

Thus, HMS allows determination of mechanical dynamic parameters and characteristics of movements. Such sensors can be promising in seismography, in geological exploration, and in systems of protection against unauthorized access; they are more informative in determination of the characteristics of external mechanical influence in comparison with seismic sensors of induction type as the latter ones react to the speed of magnetic flux change and not to magnetic induction changes. In our case, the rate of change $\overline{B}$ can be very small, but during a long time it leads to an appreciable change of frequency. In comparison with capacitor sensors, the described device provides considerably larger changes of frequency that essentially increases the dynamic range of measurements of the characteristics of mechanical force influence.

The basic advantage of such sensors consists in the application of one element base, including HMS and microcircuits of processing of the signal responses from them.

14.10 Multivariate Vector Sensors of Electromagnetic and Mechanical Physical Quantities for New Generations of Metrical, Checking, and Tested Microsystems, Including Intellectual Ones

One of the independent directions of heteromagnetic microelectronics are multiple parametric sensors and converters (MPS and MP) of electromagnetic and mechanical dynamic quantities and microsystems of active type, including intellectual ones.

The distinctive features of MPS and MT are as follows:

- A uniform element base
- Uniform technologies and means of designing
- A raised informativity
- Small weights and dimensions
- A high technical CE for raised power levels
- A low cost
- Raised accuracy and sensitivity
- The vector character of measured quantities, including their deviations
- Digital or volt output
- An opportunity of design of microsystems with various I.Q. from two-state key coding to distinction of image portraits
- Contactlessness
- Passivity (no requirements to illumination)
- A mode of background or specialized illumination

MPS and MT are based on various types of magnetosensitive generators – CHIPS in whose feedback circuits volumetric (a sphere) or film FMCR are connected. The frequency or volt response signal bears information on the measured vector electromagnetic or mechanical dynamic quantity, including their modulation characteristics in a range of frequencies below GHz.

The most essential advantages of MPS and MT are as follows:

1. They satisfy the criteria of critical technologies (their performance is improved by several orders of magnitude; the weights, dimensions, and cost decrease; the reliability raises; the ecological loads decrease; new generations of tiny and, further, supertiny devices are designed to form completed signals within the CHP limits).

2. The following is common:
   - The element base
   - The technologies
   - The means of designing, including CAD
   - The means of control and measurements
   - Tests on EEF and special factors
   - Processing of response signals
   - Controllers and microprocessors
3. Multiparametric, multipurpose interactions with a high efficiency (a high technical CE) in the CHIP; the microsystem with multichannel electric management of the kinds of spectra; and the quality of signals by both the semiconductor and magnetic subsystems, including dynamic ways of HF power management.
4. The high stability to mechanical and thermal influences and special factors (radiations of various kinds).
5. The record ratios: "power/weight," "spectral density of power/weight," "multifunctionality/weight."
6. The presence of a radiofrequency friend-or-foe channel with a high level of protection.
7. The possibility of introduction of MPS, MT, and CHIPs directly into the microprocessor for creation of intellectual microsystems of navigation, control, guide, operation, information reading, etc.
8. The widest interspecific, interbranch application in microelectronics, microsystem engineering, instrument making, aviation and space industries, geophysics, biomedicine, means of control over great people streams, search means for electric cables, communications, illegal branching in pipelines, defectoscopy, etc.

Below are listed on what the heteromagnetic sensor reacts and the basic attributes of image (object) recognition:

1. The vector of Earth's magnetic induction $\overline{B}_{\text{Earth}}$ and the modulation component $\tilde{b}$ in the form of a multichannel frequency response of "frequencies or voltage, as a vector," whose value depends on the orientation the FMCR sensor in space. It is the first attribute of image (object) recognition.
2. Close to ferroobjects (bodies), the magnetic induction line are bent and the heavier the ferrobody is, the stronger their perturbation. It is the second attribute of image (object) recognition.
3. The characteristic modulation frequency $\nu_{\text{mod}} = 50\,\text{Hz}, 60\,\text{Hz}$, or another one by which $\tilde{b}$ changes, including a complex spectrum code. It is the third attribute of image (object) recognition.
4. The characteristic frequency (frequencies) of vibration(s) of rotation(s) of the moving object $\nu_{\text{vib}}, \omega_{\text{rot}}$. It is the fourth attribute of image (object) recognition.
5. If the object moves along the sensor or the sensor does along the object, the AFC of the frequency response (its amplitude, shape) and the spectrum, including its fine structure, bear helpful information. It is the fifth attribute of image (object) recognition.

6. The spectra of electromagnetic radiation of objects in an active mode (the characteristic "portraits" of the spectra of location stations, enterprises, power complexes, power substations). It is the sixth attribute of image (object) recognition.
7. The spectra of electromagnetic radiation reflected from objects (static or moving ones) under natural illumination by television signals, *FM* communication at reception, and direct processing by a heteromagnetic detector (passive multiparametric location of an object by the magnetic component). It is the seventh attribute of image (object) recognition.
8. For increase of reliability, it is necessary to use a feature set of image (object) recognition.

The key parameters of vector heteromagnetic magnetic-induction–frequency converters are presented in Table 14.6.

The parameters of new multipurpose vector heteromagnetic sensors of mechanical dynamic quantities are presented in Table 14.7.

Table 14.6 The key parameters of vector heteromagnetic magnetic-induction–frequency converters

No.	Parameter	Theoretical estimation	Experimental value	Advantages
1	Absolute additive sensitivity to magnetic field induction	1 nT and less	10–100 nT	Dynamic range >60 dB, stability to magnetic pulses up to 1 T and higher
2	Sensitivity to magnetic field induction	28–56 Hz/nT	28–56 Hz/nT	The "magnetic induction–frequency" converter, the target signal in digital form. No additional delays and errors at digitization
3	Absolute error of magnetic induction measurements	1%	3%	Calibration by frequency
4	Time of transformation	Below 1 ns	1 ms	A short time of transformation in comparison with induction and magnetoresistive sensors
5	Absolute additive sensitivity to angular coordinate measurements	0.3″ and less	0.65′	
6	Sensitivity to angular coordinate changes	10–20 kHz/deg	10–20 kHz/deg	Target signal in digital form
7	Absolute error of angular coordinates	0.5°	265°	Calibration by frequency deviation

Table 14.7 The parameters of new multipurpose vector heteromagnetic sensors of mechanical dynamic quantities

No.	Parameter	Theoretical estimation	Experimental value	Advantages
1	Sensitivity to force influence (MHz/mN)		40	
2	Sensitivity to frequency changes by voltage:	The maximal sensitivity 1 pT		
	• On the PM-FM-PM transformation channel (mV/MHz)		2.5–3.0	
	• On the channel with a direct current amplifier (mV/MHz)		250	
3	Vector of $0X$ displacement $\Delta\bar{x}$ of the mobile part of the sensor (m)	$1 \times 10^{-10} \div 1.7 \times 10^{-6}$	5×10^{-6}	Common element base Common technologies
4	Linear $0X$-axis velocity vector for a time interval $\Delta t = 10^{-6}$ from displacement $\Delta x = 5 \times 10^{-6}$ m (m/s)	$5 \times 10^{-4} \div 1.7$	5	Common means of designing Common test base
5[a]	Linear 0X-axis acceleration vector for a time interval $\Delta t = 10^{-6}$ from displacement $\Delta x = 5 \times 10^{-6}$ m(m/s^2) (g)	$1 \times 10^{2} \div 1.7 \times 10^{-2} (10 \times g) \div (1.7 \times 10^{-1} \times g)$	$5 \times 10^{-2} (5 \times 10^{-3} \times g)$	Target signal in digital and/or analog forms
6[a]	Angular velocity vector (at $r = 10^{-2}$ m) $\Delta\bar{\varphi}/\Delta t$ (rad/s)	$1 \times 10^{-2} \div 1.7 \times 10^{2}$	$5 \times 10^{2} (50 \times g)$	
7[a]	Angular acceleration vector (at $r = 10^{-2}$ m) $\Delta^2\bar{\varphi}/\Delta t^2$ (rad/s^2)	$1 \times 10^{-2} \div 1.7 \times 10^{2}$	$5 \times 10^{8} (5 \times 10^{7} \times g)$	
8[a]	Vector of speed change by value $\frac{\Delta \bar{r}_0}{\Delta t} r$ (at $\Delta \bar{r}_0 = 2.5 \times 10^{-6}$ m, $\Delta t = 10^{-6}$ m, $r = 10^{-1}$) (m/s)	2.5×10^{-1}	2.5×10^{-1}	
9[a]	Vector of velocity change by direction $\bar{r}_0 \frac{\Delta r}{\Delta t}$ (at $r = 5 \times 10^{-6}$ m, $\Delta t = 10^{-6}$ m, $r = 10^{-3}$) (m/s)	5×10^{-3}	5×10^{-3}	
10[a]	Normal acceleration $a_n = v^2/r$ (at $r = 10^{-1}$ m) (m/s^2)(g)	$1 \times 10^{-7} \div 3 \times 10^{1}$	$250 (25 \times g)$	

(continued)

Table 14.7 (continued)

No.	Parameter	Theoretical estimation	Experimental value	Advantages
11[a]	Tangential acceleration $a_\tau = \Delta\upsilon/\Delta t$ (at $\Delta\upsilon = 2.5\,\text{m/s}$, $\Delta t = 0.5 \times 10^{-6}\,\text{s}$) $(\text{m/s}^2)(g)$	$2 \times 10^2 \div 0.6$	$5 \times 10^6 (5 \times 10^5 \times g)$	
12[a]	Force (at $m = 5 \times 10^3\,\text{kg}$, $a_\tau = 5 \times 10^6\,\text{m/s}^2, a_n = 250\,\text{m/s}^2$) (N)			
	• Tangential	$1 - 3 \times 10^{-3}$	25×10^3	
	• Normal	5×10^{-10}	1.25	
		1.5×10^{-1}		
13[a]	Pressure (Pa):			
	• Normal (at $F_n = 1.25\,\text{N}, S = 1 \times 10^{-6}\,\text{m}^2$)	$2 \times 10^{-5} \div 1 \times 10^{-3}$	1.25×10^6	
	• Tangential (at $F_\tau = 2.5 \times 10^{-3}\,\text{N}, S = 1 \times 10^{-6}\,\text{m}^2$)	$1 \times 10^{-14} \div 5 \times 10^{-2}$	25×10^3	
14[a]	Centrifugal force (at $m = 5 \times 10^3\,\text{kg}, \omega = 5 \times 10^2\,\text{rad/s}, r = 10^{-2}\,\text{m}$) (N)	$5 \times 10^{-9} \div 1.5$	12.5	
15[a]	Coriolis' force (at $m = 5 \times 10^3\,\text{kg}$, $\upsilon = 5\,\text{m/s}$, $\omega = 5 \times 10^2\,\text{rad/s}$) (N)	$10^{-8} \div 3$	25	
16[a]	Moment of force (at $r = 10^{-2}\,\text{m}$, $F_\tau = 25 \times 10^{-3}\,\text{N}$) (N m)	$10^{-8} \div 3$	25×10^{-5}	
17[a]	Moment of momentum (at $r = 10^{-2}\,\text{m}$, $P_1 = 2.5 \times 10^{-1}\,\text{kg m/s}$) (kg m/s^2)	2.5×10^{-3}	12.5×10^{-6}	
18[a]	Momentum P_1 of speed change by value (at $\frac{\bar{\Delta}\bar{r}_0}{\Delta t} r = 2.5 \times 10^{-1}\,\text{m/s}$, $m = 5 \times 10^{-3}\,\text{kg}$) (kg m/s)	2.5×10^{-1}	12.5×10^{-4}	
19[a]	Momentum P_2 of velocity change by direction (at $\frac{\bar{\Delta}\bar{r}_0}{\Delta t} r = 5 \times 10^{-3}\,\text{m/s}$, $m = 5 \times 10^{-3}\,\text{kg}$) (kg m/s)	25×10^{-6}	25×10^{-6}	
20	Power voltage (V)	3	7	
21	Consumed current (mA)	10	60	

[a] Parameters is estimate for experimental data of subsection 14.9.

The parameters in 5–19 of Table 14.7 have been determined by calculation from the experimental data in 1–4 in view of variations of Young's modulus within the limits of (5×10^4)–$(3 \times 10^{11})\mathrm{H/m^2}$ for various materials – rubber, composites, steel, and bronze [91].

The parameters of programmable heteromagnetic generators of complex signals are presented in Table 14.8.

The influence of external factors on the normal operation of heteromagnetic radiocomponents and microsystems is shown in Table 14.9.

Table 14.8 The parameters of programmable heteromagnetic generators of complex signals

No.	Parameter	Theoretical estimation	Experimental value	Advantages
1	Range of working frequencies	From 0.5 to 1,000 GHz (overlapped by several letters)	From 0.5 to 37.5 GHz	
2	Shape of signal	Spectral pure, quasiharmonic Equidistant frequency spectra Noisy	Spectral pure, quasiharmonic Equidistant frequency spectra Noisy	Use of uniform CAD, means of measurements, element base
3	Target power	Below 400 W per pulse	10 W in a continuous mode	
4	CE	60%	50%	

Table 14.9 The influence of external factors on the normal operation of heteromagnetic radiocomponents and microsystems

EEF name	Requirements	Results of theoretical estimations in statics	Results of tests
Sine wave vibration	Range of frequencies from 1 to 5,000 Hz, amplitude of acceleration $40 \times g$	Brazed fastening of active crystal: break-off force in normal direction defined by acceleration $4 \times 10^6 \times g$	Corresponds
Mechanical impact of single action	Peak shock acceleration $(>3,000 \times g)$ at duration shorter 2 ms	Break-off force in tangential direction defined by acceleration $3,200 \times g$ (solder thickness of 1 μm)	Corresponds
Mechanical impact of repeated action	Peak shock acceleration $150 \times g$ at duration 2 ms	Fastening of ferrite microresonator: break-off force in normal direction, defined by acceleration $17 \times 10^4 \times g$	Corresponds

(continued)

Table 14.9 (continued)

EEF name	Requirements	Results of theoretical estimations in statics	Results of tests
Linear acceleration	Value of acceleration (>500 × *g*)	Deformation of the technological case: walls' thickness 4–5 mm; natural mechanic frequency not less than 2 × 10^3 kHz At a sharp change of external pressure, bending effort no more than 2 MPa Beam distributing with a wire diameter of 0.02–0.05 mm: break-off force defined by acceleration 8 × 10^5 × *g*; break-off force at right angle defined by acceleration 3 × 10^5 × *g*	Corresponds
A raised temperature of environment	Maximal value of operation (+125°C)	Theoretical value of shift factor of the central generated frequency −0.6 MHz/deg wrt room temperature	
A lowered temperature of environment	Minimal value of operation −60°C Minimal value of transportation and storage −60°C	A sharp decrease in temperature factor of central frequency shift is possible due to application of magnetic thermocompensators in a range from −40 to +80°C. For sensors of differential type, thermal shift of generated frequency are caused only by technological frequency leavings	For a range from −60 to +125°C application of a thermocompensator, changes of magnetic induction, program account of temperature shift of frequency are necessary at processing of indications
Fast change of ambient temperature	From maximal value at operation up to minimal value at transportation and storage	No influence on workability of systems at observance of coordination condition of TEC of interfaced elements	Corresponds
A raised air humidity	Relative humidity of 100% at +35°C	No influence on workability of systems at observance of condition of hermetic sealing of the case	Experimental check is expedient for a each product as a whole
A lowered air humidity	Dew point at −40°C		
A lowered atmospheric pressure	Value at operation (less than 5 mm hg)	No influence on workability of systems at observance of condition of hermetic sealing of the case	
A raised atmospheric pressure	Value at operation of 2,207 mm hg		

(continued)

Table 14.9 (continued)

EEF name	Requirements	Results of theoretical estimations in statics	Results of tests
Pressure change	Range and speed of change – on demand	No influence on workability of systems at observance of condition of hermetic sealing of the case	
Atmospheric precipitation: falling and condensed		No influence on workability of systems at observance of condition of hermetic sealing of the case	
Dust: static and dynamic		No influence on workability of systems at observance of condition of hermetic sealing of the case	
Sunlight	Upper value of integrated flux density at operation 1, 120 W/m^2	Influence on a device with a tight case only by thermal factor. A reflecting covering of the case is needed (polishing of the metal case)	
	Upper value of UV radiation flux density at operation 68 W/m^2		
Biological factors, aggressive, and testing media		Defined by design of the case, conditions of hermetic sealing, sheetings, etc.	
Neutron radiation	On demand	Steady to various densities	Corresponds

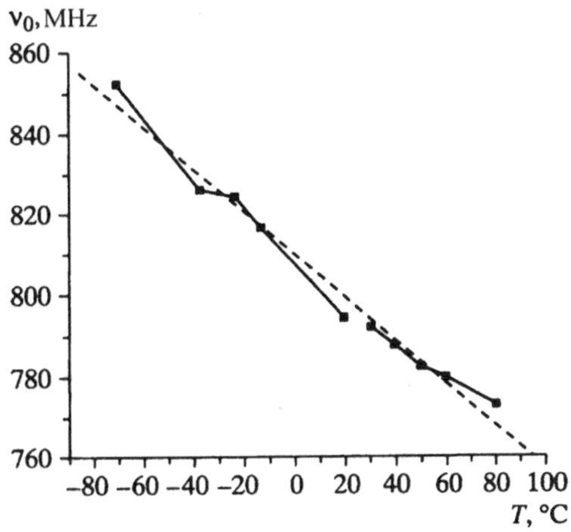

Fig. 14.82 The experimental dependence of the central frequency on ambient temperature of HMG on the basis of PTSh with a ferrite KG-30 microresonator

In Fig. 14.82, the experimental dependence of the central frequency on ambient temperature of HMG on the basis of PTSh with a ferrite KG-30 microresonator is shown.

The high linearity of the temperature shift of generated frequency is confirmed. The average value of the frequency shift factor is $-0.53\,\text{MHz/K}$.

Chapter 15
Nonlinear Effects in Magnetotransistors and Their Elements

Theoretical and experimental studies of the behavior of magnetodielectrics in a mode of ferromagnetic resonance, including nonlinear one, are full enough presented in [38, 92, 93]. On the basis of these researches, microwave devices working in various ranges of wavelengths have been designed. These are terminators of HF power, noise blankers, band-elimination and low-and-high-pass filters, phase shifters, and generating devices operated by a magnetic field. The last ones are typically nonlinear systems with generation of regular signals, generation of frequency-equidistant and noise-like signals with various band of frequencies, and generation of the harmonics of a regular signal, mixture frequencies, etc.

15.1 Peculiarities of Nonlinear Processes in Ferromagnetics

Three kinds of nonlinear modes (processes) depending on the level of applied HF power are distinguished [17]. They are caused by the first-order and second-order nonlinearity of the magnetization variable.

Consider a mode of coincidence of the frequencies of the basic and additional resonances. At homogeneous precession of magnetic moments, energy is transferred to separate groups of spin waves. This process is most intensive when the HF magnetic field achieves its critical value h_{cr} at which the precession ceases to increase and there comes saturation of the basic resonance, and the HF magnetic permeability decreases [38, 92, 93]. This is the so-called first-order nonlinearity effect at which spin waves with a frequency $f_0/2$ are excited in FMCR, where f_0 is the FMR frequency.

Another treatment of this effect is parametrical excitation of spin waves by the pumping frequency f_0. The mode of coincidence of the basic and additional resonances has a very low saturation threshold – few μW for monocrystals. The additional resonance in this mode coincides with the basic one, being a nonlinear (parametrical) effect of the first order, and characterizes coupling between spin waves and the amplitude of homogeneous precession [38]. For spherical samples of YIG monocrystals, the range of frequencies in this mode is

A.A. Ignatiev and A.V. Lyashenko, *Heteromagnetic Microelectronics: Microsystems of Active Type*, DOI 10.1007/978-1-4419-6002-3_15,

$$\gamma\frac{1}{3}4\pi M_s < f_0 < \gamma\frac{2}{3}4\pi M_s, \tag{15.1}$$

where M_s is the saturation magnetization of ferrite (mT) and $\gamma = 28\,\text{MHz/mT}$ is the hydromagnetic ratio ($1\,\text{G} = 10^5\,\text{nT}$).

Consider a mode of saturation of the basic resonance named as a nonlinear effect of the second order. The amplitude of spin waves is proportional to the squared amplitude of homogeneous precession of the magnetization vector [38]. The precession of magnetization is connected with degenerate spin waves on the frequency $f_{\text{s.w.}} = f_0$. In the mode of frequency coincidence of the basic and additional resonances (the first-order mode), this condition is realized on the frequency $f_{\text{s.w.}} = f_0/2$, where $f_{\text{s.w.}}$ is the frequency of wave excitation. The considered mode takes place at the frequency

$$f_0 > 28 \times \frac{2}{3}4\pi M_s. \tag{15.2}$$

In this case, the threshold level of saturation by HF power is almost four orders of magnitude higher than in the previous modes (being few or tens mW).

At further increase in the HF power when condition (15.2) is satisfied, restriction by the additional resonance is observed when $B < B_0$. In this case, the influence on homogeneous precession is considerably smaller than in the modes of the first and second orders, and restriction by the FMR frequency (f_0) will come at a HF power of tens or hundreds Watts. This mode, as well as that considered above, is a nonlinear mode of the second order. The condition of occurrence of an additional resonance is

$$f_0 < f_{\text{s.w.}} + \frac{1}{3} \times 28\pi M_s, \tag{15.3}$$

where $f_{\text{s.w.}} = f_0/2$ and f_0 is the FMR frequency.

Thus, the proper choice of FMR frequency, saturation magnetization of ferrite M_s allow realization of various modes of the saturation level by HF power in HMG.

15.2 Peculiarities of Ferromagnetic Resonance in Structures with First-order Nonlinearity

According to [38], the volume-average magnetic moment of a ferromagnetic under parametrical excitation changes. The changes of the magnetic moment vector by value and direction are proportional to the changes of the linewidth of absorption and the pumping field amplitude. The state of the ferromagnetic behind the threshold of excitation depends on the relaxation mechanism of magnetic moment, that is, the times of relaxation T_1 and T_2 (in the form of Bloh–Blombergen's dissipative term [38]), where T_1 is the time of "longitudinal" relaxation (that of the magnetization component parallel to the constant magnetic field) and T_2 is the time of "cross" relaxation (that of the magnetization component perpendicular to the constant magnetic field). In the modes when the HF magnetic field exceeds its threshold value, there are periodic oscillations of the magnetization vector of the sample as a whole (collective oscillations) with a frequency $\Omega \ll f_0$ (self-modulation of oscillations on the FMR frequency [105]). In the single-wave approximation, the frequency of these slow oscillations of the magnetization vector was calculated:

$$\Omega = \frac{2}{T_2}\sqrt{\frac{\left(\dfrac{h}{h_{\text{thr}}}\right)^2 - 1}{1 + 2\dfrac{T_1}{T_2}}}. \tag{15.4}$$

From (15.4), it follows that the frequency is determined by the value of the HF magnetic field excess above its threshold value and by the times of relaxation T_1 and T_2, depending on temperature, surface processing, uniformity of the structure, and other factors [38, 105].

The instability of nonlinear ferromagnetic resonance is characterized by a spectrum of frequencies (satellites) relative to the basic line of parametrical excitation of spin waves on the frequency $f_0/2$ at tuning-outs of $n\Omega$ from it ($n = \pm 1, 2, \ldots$). In this connection, on the FMR frequency f_0, the occurrence of a frequency modulation spectrum should be expected. As the frequency of excitation for a spin wave is $f_{\text{s.w.}} = f_0/2$, satellites should be expected at distances $(2n \times f_0)$ from the basic line of the FMR frequency.

15.3 Experimental Observations of Nonlinear Ferromagnetic Resonance

A two-cascade (double-link) FMCR of a low-and-high-passing type with samples of a YIG monocrystal of a diameter 0.4 mm, $4\pi M_s = 175$ mT was under study (Fig. 15.1).

Near the FMCR frequency, side frequencies (satellites) appeared, the distance between which W increased with the HF power is shown in Fig. 15.2: two spheres, $4\pi M_s = 175$mT, $f = 3{,}040$ MHz.

Nonlinear FMR was observed experimentally up to a frequency of 3,450 MHz, which confirmed the presence of the upper border of parametrical excitation of spin

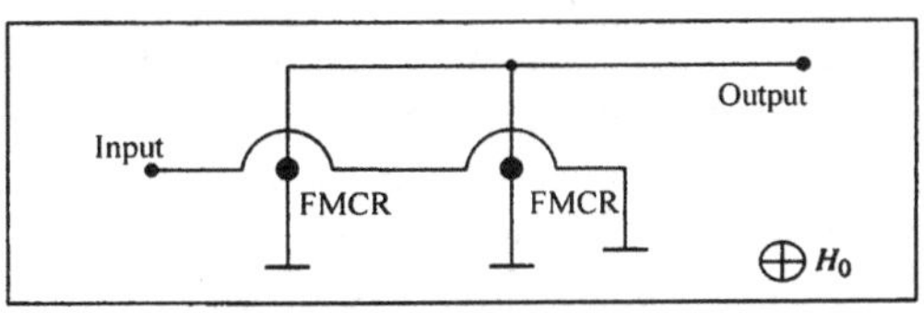

Fig. 15.1 A two-cascade (double-link) FMCR of a low-and-high-passing type with samples of a YIG monocrystal of a diameter 0.4 mm, $4\pi M_s = 175$ mT

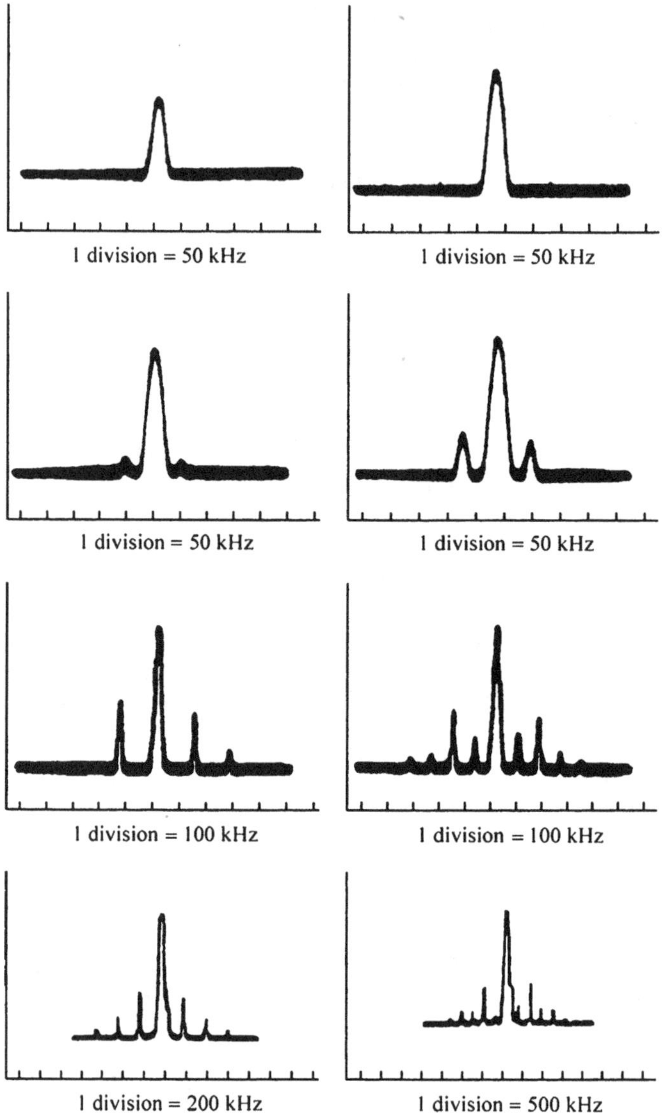

Fig. 15.2 Near the FMCR frequency, side frequencies (satellites) appeared, the distance between which W increased with the HF power

waves for a spherical YIG monocrystal near the frequency of 3,300 MHz ($4\pi M_s =$ 175 mT), calculated in [38].

Self-modulation at FMR took place up to a level of input HF power of 3–4 mW. With increase in power, the side frequencies vanished and the mode of restriction was realized at intensive excitation of spin waves. It is possible to assume that at higher values of input HF power, self-modulation disappears because of thermal fluctuations of the crystal lattice owing to transfer the power of spin waves to it.

In a mode of frequency mixture of the setting generator f_0 and the frequency f_g from the additional generator near the frequency $|f_0 - f_g|$, a spectrum of frequencies $n\Omega$ (Fig. 15.3) was observed: for the case of two spheres, $4\pi M_s = 175\,\mathrm{mT}$, $f = 3{,}040\,\mathrm{MHz}$.

It is obvious that the frequency of self-modulation according to (15.4) is within 100–400 kHz at FMR on a frequency of 3,040 MHz.

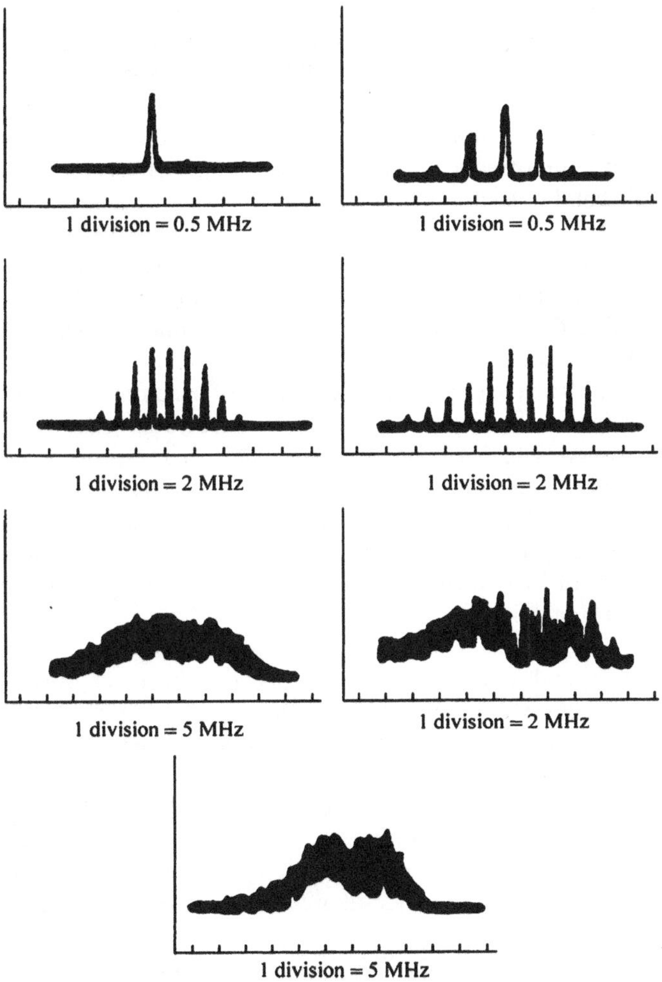

Fig. 15.3 In a mode of frequency mixture of the setting generator f_0 and the frequency f_g from the additional generator near the frequency $|f_0 - f_g|$, a spectrum of frequencies $n\Omega$ was observed

15.4 Generation of Signals in Regime of Nonlinear Ferromagnetic Resonance

Reference [26] provides experimental data on transition to chaos in a parametrical system with a nonlinear ferrite resonator in the external feedback circuit of the amplifier. A similar system can be a HF oscillator, and the amplitude and phase conditions satisfied can provide excitation of stationary oscillations in the frequency band of the low-and-high-passing filter on FMR.

For a spherical YIG monocrystal, the range of frequencies of parametrical excitation of spin waves by (15.1) is within 1,630–3,260 MHz. Our experimental breadboard model of HMG consisted of an amplifier on a range of frequencies 3,000 MHz in whose feedback circuit a FMCR with two spherical samples of YIG monocrystal with a diameter of 0.4 mm has been included. At achievement of the threshold power on the output of FMCR, stationary oscillations in the generator turned to periodic multifrequency oscillations. At further increase in the generator's power, the band of each spectral component broadened and merged to form a noise-type signal in a band of frequencies $f_0 \pm n\Omega$, where Ω is the frequency of collective oscillations (the frequency of self-modulation) of magnetization [26, 93].

In Fig. 15.4, spectra of oscillations are shown at increase in the HMG power. Note that a similar development of the spectra of oscillations was found at some mismatch from the basic FMR frequency.

On the basic FMR frequency, a different order of spectrum transformation took place, namely: first, a noise-type signal was formed; then multifrequency oscillations were formed, which became periodic ones; then a noise-type signal was formed; and again periodic multifrequency oscillations were formed.

Such a dynamics of the spectra could be explained by the fact that the generator power was changed by the power source voltage of the transistor amplifier, that is, the working point of VAC changed and transition from the periodic to stochastic spectra of output signal occurred.

Frequency change of the generator by a magnetic field has found the absence of multifrequency and stochasticity of the spectra above the frequency $f_0 =$ 3,400 MHz, which confirmed our concept about the first-order nonlinearity at FMR in the mode of coincidence of the basic and additional resonances [38, 93].

Above the frequency $f_0 = 3{,}400$ MHz, a nonlinear second-order mechanism took place, at which the threshold power level was four orders of magnitude higher than that at the first-order nonlinearity [38], and excitation of spin waves almost did not influence the basic FMR. The spectral structure of the generated signal had no satellites, and the frequency of generation was determined by the usual condition of FMR.

By the use of a band-passing, FMCR with spherical YIG samples with a smaller saturation magnetization (a YIG sphere: $4\pi M_s = 81.6$ mT) in the feedback circuit of parametrical instability of NFMR was displaced toward lower frequencies (761 MHz $< f_0 <$ 1,522 MHz).

As is seen from Fig. 15.4, the character of transition from a monochromatic signal to a multifrequency one and then to a noise signal remained (see Fig. 15.3), but

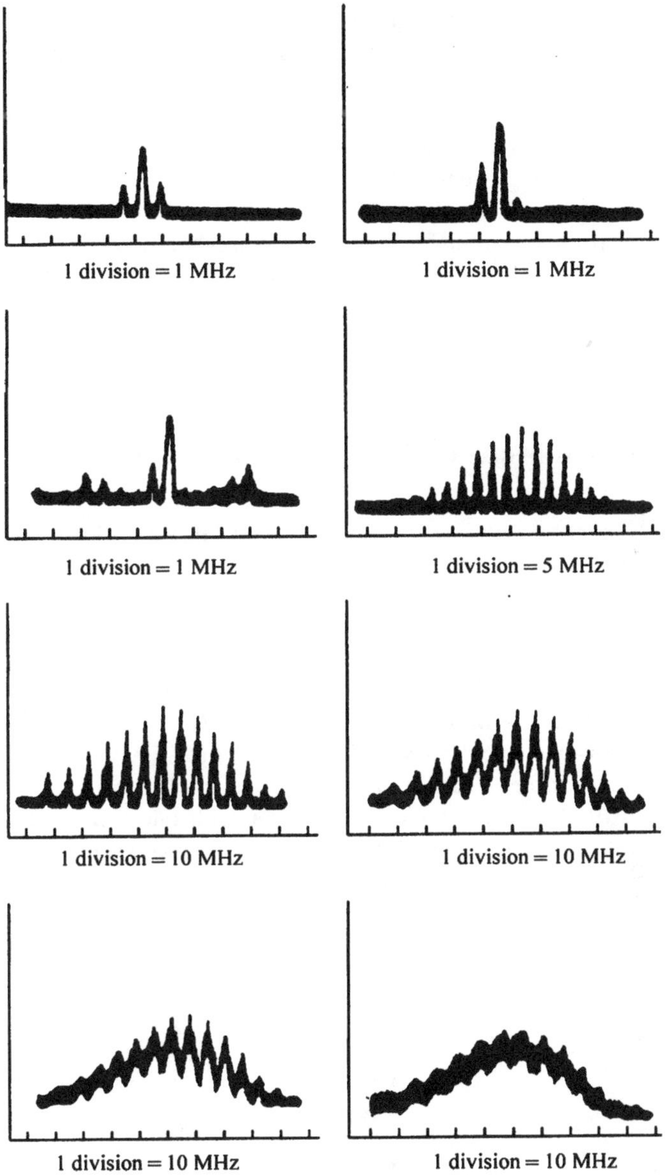

Fig. 15.4 Spectra of oscillations at increase in the HMG power

the frequency range between the satellites became much wider. In [38, 105], this was associated with a lot of quickly relaxing ions of nonmagnetic impurities in the samples with a smaller saturation magnetization in comparison with pure YIG monocrystals that led to reduction of the relaxation times T_1 and T_2 and, according to(15.4), to an increase in the frequency Ω of self-modulation.

A similar dynamics of the spectra in HMG with YIG samples with even a smaller saturation magnetization (KG-35, $4\pi M_s$) was observed on frequencies 440 and 660 MHz.

The effect of parametric instability was observed below the frequency $(2/3)\gamma 4\pi M_s$ above which generation was monochromatic with no satellites. As it was supposed, in the presence of self-modulation, the spectrum of regular signals took place. At some levels of HF power and fine tuning of the field bias, the regular spectrum was transformed into a noise-like one near the basic FMR frequency. These processes were of bifurcational character that spoke of the multiwave mechanism of parametrical excitation of spin waves and confirmed theoretical estimations [93].

15.5 Saturation Mode of Principal Resonance

For studying the possibilities to use ferromagnetics at raised levels of HF power, YIG spheres doped with rare-earth elements with a saturation magnetization $4\pi M_s = 80$ mT in the mode of the basic resonance saturation were investigated. In FMCR with a YIG monocrystal ($4\pi M_s = 175$ mT) in the mode of coincidence of the basic and additional resonances [38], the threshold power and the corresponding level of restriction were 0.1 mW. In the mode of the basic resonance restriction on frequencies below 3 GHz for a sample with $4\pi M_s = 80$ mT, the level of HF-power saturation was above 40 mW (Fig. 15.5). In FMCR with two YIG samples, the level of saturation reached 80 mW and the losses of transfer in the filter vary from −3 to −4 dB (Fig. 15.6, where 1 is a filter with one sphere, 2 a filter with two spheres KG-65, $f = 3{,}000$ MHz). The output signal spectrum contained no satellites and was determined by the condition of the basic FMR. Studies of NFMR in the mode of the basic resonance restriction under the ring circuit were made on a generator consisting of an amplifier at a 1-W power level with an amplification

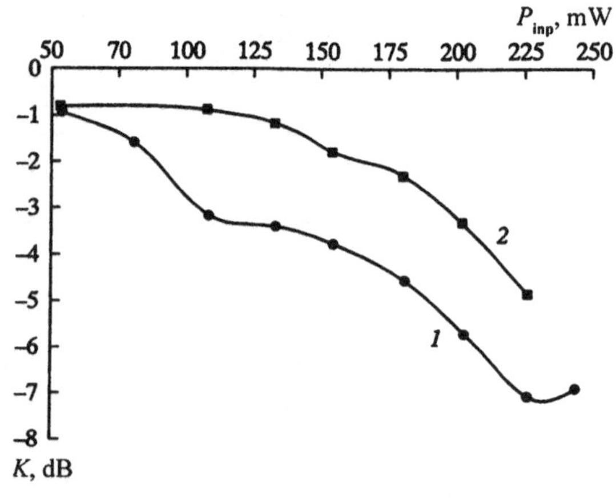

Fig. 15.5 In the mode of the basic resonance restriction on frequencies below 3 GHz for a sample with $4\pi M_s$=80 mT, the level of HF-power saturation was above 40 mW

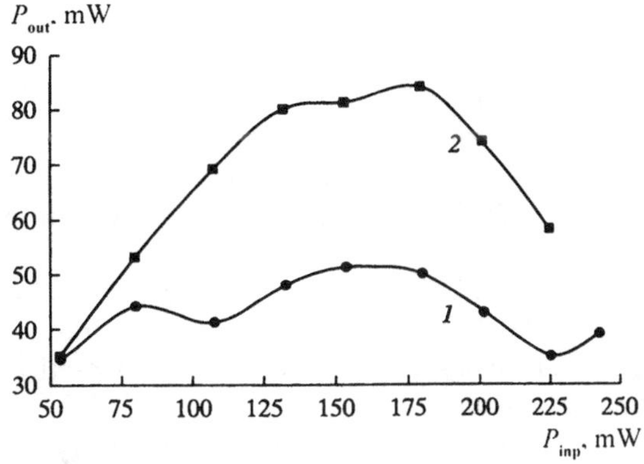

Fig. 15.6 In FMCR with two YIG samples, the level of saturation reached 80 mW, the losses of transfer in the filter vary from −3 to −4 dB (1 is a filter with one sphere, 2 a filter with two spheres KG-65, $f = 3{,}000$ MHz)

factor of 10 dB in a range of frequencies near 7 GHz and MECE with two YIG spheres. The output power of HMG was 250 mW and was changed by a magnetic field in a band of frequencies from 200 to 300 MHz. The output signal spectrum was determined, as was supposed, by the conditions of basic FMR.

Our experimental researches showed the following:

1. In various ranges of frequencies, the first-order nonlinearity (intense excitation of spin waves on the FMR half-frequency) leads, at achievement of the level of threshold power, to the instability of FMR and to excitation, along with the basic frequency, of satellites distant from each other by a frequency $\Omega \ll f_0$ (collective oscillations of magnetization), which at further increase of the HF pumping power, turn to acyclic, stochastic fluctuations. In this process, restriction by power at a level not exceeding 1 mW takes place.
2. In the mode of the basic resonance saturation, when the FMR frequency $f_0 > (2/3)\gamma 4\pi M_s$, a second-order NFMR is realized, at which spin waves are excited on the basic frequency, and the threshold mode comes at a HF power by 3–4 orders of magnitude greater than at a first-order nonlinearity. It allows controlling HF power from 100 to 200 mW in HMG.
3. In HMG, the nonlinearities of the first and second orders on FMR allow realization of various operating modes (monochromatic, periodic multifrequency, and noise-type ones) in one device.

15.6 Power Limiting in FMCR

At a level of input HF power above h_{cr}, the nonlinear properties of the magnetization variable manifest themselves in FMCR and the energy of homogeneous precession of spin dipoles passes to a certain group of spin waves and into thermal fluctuations of the crystal lattice. The magnetic permeability of the sample starts to decrease, and the basic resonance is saturated [17, 38]. This saturation of the basic resonance due to an unstable growth of spin waves occurs at an amplitude of the HF magnetic field considerably lower than the field value at which the magnetization precession angle should reach 90°.

At other values of the FMR frequencies and other values of h_{cr}, an additional resonant absorption of HF power by the ferrite occurs. The additional resonance occurs when $H_{add} \ll H_{cr}$. The basic and additional resonances can coincide under certain conditions.

In Fig. 15.7, dependences of the output pulse power on the pulse (peak) HF power on the FMCR input for the basic types of restrictions [38] are shown: (a) at coincidence of the basic and additional resonances; (b) at saturation of the basic resonance; and (c) at the additional resonance.

For a spherical sample of a YIG monocrystal in the mode of coincidence of the basic and additional resonances, the threshold power is of the order of several μW. At saturation of the basic resonance, the level of power restriction is of the order of 1 mW. At operation in the mode of an additional resonance, the level of restriction reaches 10–100 W.

The distinction in the levels of threshold power at the specified types of restrictions speaks that the phenomenon of additional resonance, should it occur at the value of external magnetic field H_{add} or coincides with the basic resonance, is a nonlinear effect of the first order. The relation between spin waves and homogeneous precessions is proportional to its amplitude.

At saturation of the basic resonance, a nonlinearity of "second order" takes place. In this case, the amplitude of spin waves is proportional to the squared amplitude of precession. At nonlinear effects of the first order, the relation with spin waves is carried out with a frequency twice lower than the frequency of homogeneous precession. At nonlinear effects of the second order, the precession of magnetization is connected with spin waves on the FMR frequency.

At coincidence of the resonances, restriction by power in FMCR comes at rather low power levels (Fig. 15.7a).

At saturation of the basic resonance, the probability of energy transfer to spin waves is much lower. The interaction of homogeneous precession and the varying field is rather effective than on the FMR frequency, and the level of threshold power is considerably higher (Fig. 15.7b).

By the use of the additional resonance H_{add}, the external magnetic field is lower than the value H_{res} at FMR. In this case, the efficiency of the influence of the external varying field on homogeneous precession is much weaker than in the first two cases of restriction. For achievement of higher precession angles at which restriction arises, much higher levels of output power than in the first two modes of restriction (Fig. 15.7c) are required.

The realization of the above modes of restriction in FMCR can be reached by the proper choice of: 1 – the range of working frequencies;nd 2 – the saturation magnetization of the ferrite sample $4\pi M_s$.

In [38], ranges of working frequencies of FMCR power restrictions are given at: coincidence of resonances; saturation of the basic resonance; and an additional resonance in the YIG monocrystals with $4\pi M_s = 180\,\text{mT}$ of various geometries,

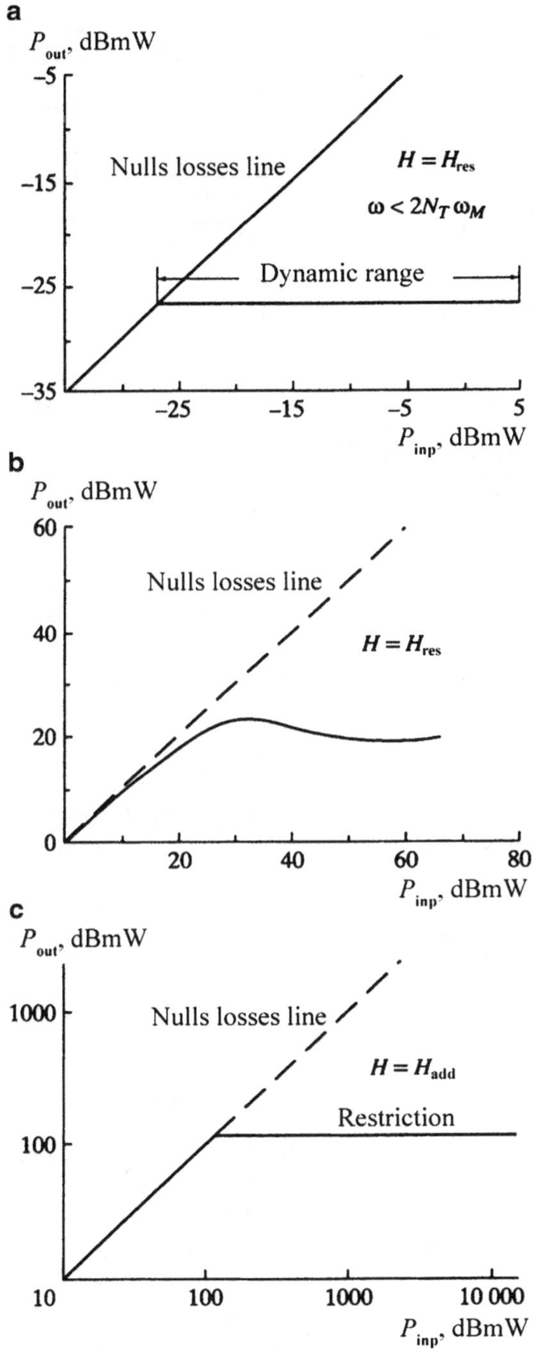

Fig. 15.7 The restriction by power in FMCR at (**a**) coincidence of the resonances; (**b**) saturation of the basic resonance; and (**c**) the additional resonance

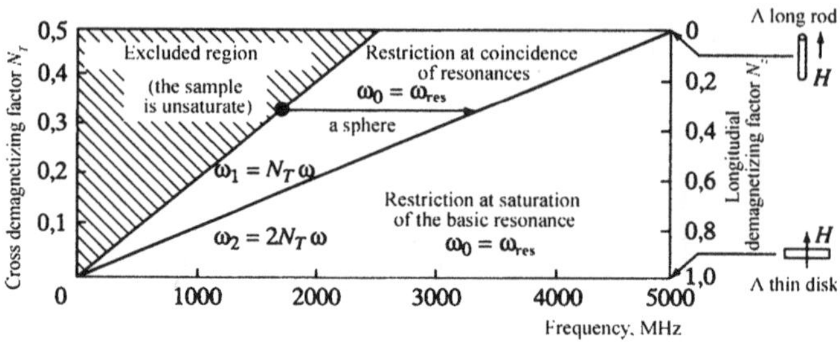

Fig. 15.8 The limiting levels of power in FMCR at coincidence of resonances and at saturation of the basic resonance

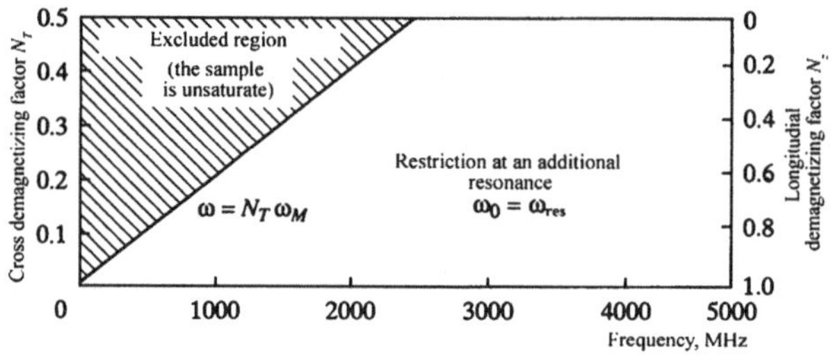

Fig. 15.9 The limiting levels of power in FMCR at an additional resonance

namely: a thin disk, a sphere, and a long rod (Figs. 15.8 and 15.9). So, for a YIG sphere (Fig. 15.8) restriction at coincidence of resonances takes place for a range of frequencies between the two straight lines $\omega_1 = N_T\omega_M$ and $\omega_2 = 2N_T\omega_M$, $N_T = 1/3$.

On frequencies lying above the straight line $\omega_2 = 2N_T\omega_M$, restriction takes place at saturation of the basic resonance at a higher, than in the first case, level of threshold power.

In the mode of an additional resonance, restriction comes for frequencies lying above the straight line $\omega_1 = N_T\omega_M$ as the area of spin waves is below $H_{add} < H_{res}$, and in this case

$$\omega_0 < \frac{\omega}{2} + N_T\omega, \quad \omega_0 = \omega_{res}.$$

From the diagrams in Figs. 15.8 and 15.9, it is obvious that high levels of power restriction in FMCR can be reached on low frequencies by the use of rare-earth-doped YIG monocrystals with a lowered saturation magnetization $4\pi M_s$. For a pure YIG monocrystal, high levels of power restriction can be realized in higher frequency ranges.

The mode of saturation of the basic resonance at FMR in a range of 1–4 GHz was investigated in MECE on crossed strip lines between which a ferrite sphere with Ø 0.4 mm (Fig. 15.10) was placed.

At the location of FMCR for creation of the maximal HF magnetic field, the strip lines were narrowed by three times (the thickness of the substrate was 500 μm,

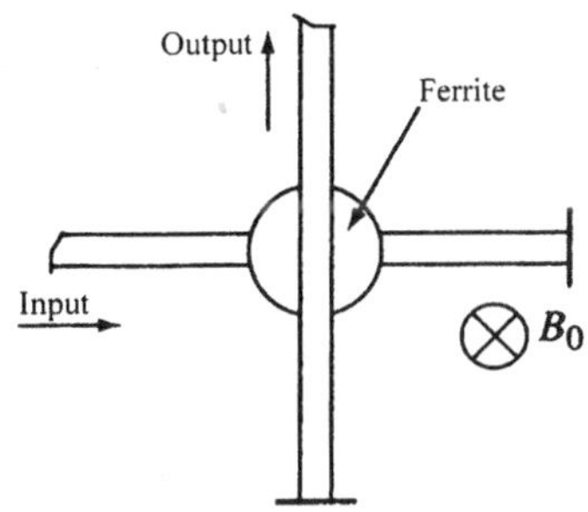

Fig. 15.10 MECE on crossed strip lines between which a ferrite sphere was placed

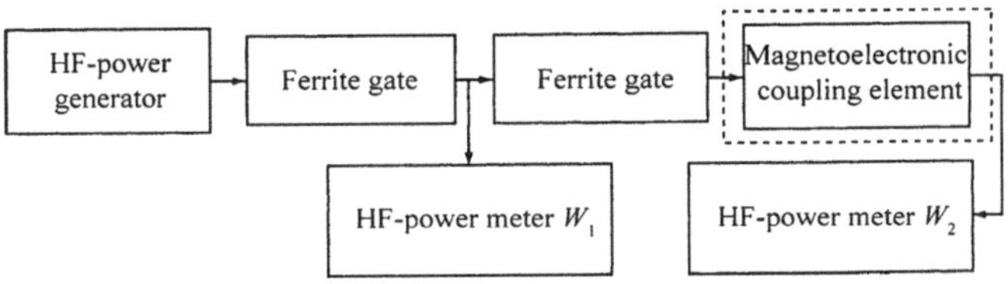

Fig. 15.11 The circuit of measurements

$\varepsilon = 10$). Unlike [38], FMCR samples with a low saturation magnetization (KG-35 with $4\pi M_s = 440$ mT; KG-65 with $4\pi M_s = 816$ mT) were used. The levels of threshold HF power in the mode of saturation of the basic resonance were determined at FMR.

The circuit of measurements is presented in Fig. 15.11.

To avoid the influence of MECE on the HF-power generator and the power meter W_1, ferrite gates were used. The frequency of oscillations f_0 was selected from the condition $f_0 > 2N_T f_M$ (see Fig. 15.8).

In Figs. 15.12 and 15.13, the peak characteristics of MECE containing ferrite samples KG-35 and KG-65 for two frequencies are presented: $f_{01} > 2N_T f_M$ and $f_{02} \gg 2N_T f_M$. In Fig. 15.12, the characteristics of MECE in the FMR mode are shown: $f_0 = 3{,}000$ MHz, a sphere with $d = 0.4$, KG-35, $4\pi M_s = 44.0$ mT, $2N_T \times 4\pi M_s \times 2.8 = 820.5$ MHz; a sphere with $d = 0.4$, KG-65, $4\pi M_s = 81.6$ mT, $2N_T \times 4\pi M_s \times 2.8 = 1{,}522.4$ MHz. Adjustment by magnetic induction B_0 at $P_{inp} = 29.1$ mW is made. In Fig. 15.13, the characteristic of MECE at $f_0 = 1{,}200$ MHz, KG-35, $4\pi M_s = 44.0$ mT, $f = 2N_T \times 4\pi M_s \times 2.8 = 820$ MHz $< f_0$ is shown. Adjustment by magnetic induction B_0 at $P_{out} = 100$ mW is made.

As is seen from the experimental dependences, in neither the first ($f_0 = 1{,}200$ MHz) nor the second ($f_0 = 3{,}000$ MHz) case, restriction of the output HF power was found; the constancy of the transfer factor in the whole range of change of output HF-power levels was kept.

From Figs. 15.12 and 15.13, it is obvious that at adjustment on FMR at the maximal output HF power, there was a significant reduction of the transfer factor on a low (below mW) output power.

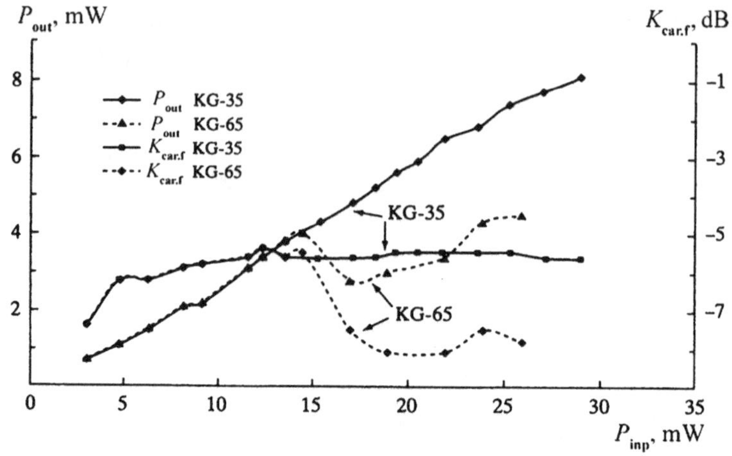

Fig. 15.12 The peak characteristics of MECE containing ferrite samples KG-35 and KG-65 for two frequencies: $f_{01} > 2N_T f_M$

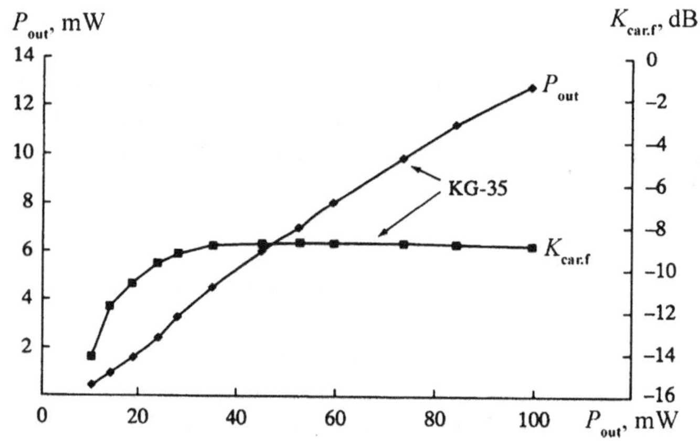

Fig. 15.13 The peak characteristics of MECE containing ferrite samples KG-35 and KG-65 for two frequencies: $f_{02} \gg 2N_T f_M$

Thus, the mode of saturation of the basic resonance at FMR provides significant levels of threshold HF power up to tens and hundreds mW that will allow using of MECE for control over the parameters of powerful (up to W and higher) HMTs.

Conclusion

Research on interactions of physical mechanisms in ferrite-semi-conductor structures leads to:

- On various types of bipolar and field transistors.
- At low, average, and high levels of capacity in continuous and pulse modes.
- In a wide frequencies range (VHF, UHF, MWF, EHF, HHF).
- For various kinds and spectra of signals (regular, pseudo-noise, noise, as multipurpose synthesizers of frequencies).
- For various types of ferrite microresonators and their orientations in an external magnetic field.
- For various modes of generation, amplification, multiplication, division and frequency modulation of a signal of the basic frequency.
- For registration of small values of a magnetic induction vector and a mechanical displacement vector.

Modern CAD means, original programs for calculation of parameters and characteristics of magnetotransistors various kinds and microsystems of active type are used.

The devices, which were considered in the book at the given stage of researches should be carried to hybrid-integrated microassemblies. FMCR integration of various kinds and types with transistor and diode (Hannah diodes, IMPATTD) devices and performance of magnetotransistors, magnetodiodes on a monolithic technology is a following stage in development of heteromagnetic microelectronics.

The fullest use of the advantages of heteromagnetic microsystems, in comparison with the known devices of microelectronics, will be demanded and will find wide practical application provided that management of their parameters and characteristics is possible; the sensor lead will require processing of response signals of vector character by means of microcontrollers and microprocessors.

Integration of heteromagnetic microsystems and microcontrollers is a following step in development of a discussed direction.

It is necessary to emphasize the opportunities of development of a new direction on researches and development of multipurpose operated semi-conductor lasers – magneto-lasers.

For development of various magnetotronics in radio- and optical ranges development of CAD heteromagnetic elements, devices, microsystems, adjustment of equivalent models, formation of help tables have basic value.

Many directions of results application of heteromagnetic microelectronics, such as signals coding and means of protection of the information, sensorics of a medical and biologic direction, defectoscopy, including complex, multiple-parameter monitoring systems of complex machines, monitoring systems of an environment on a key parameters (magnetosphere, seismic activity, including warning systems of earthquakes and a tsunami, mobile means of a tomography, system of borders protection and critical objects, communications and locations on a magnetic component, magnetic PAA, etc.), can be developed in the near future.

A new generation of radio components to form, at the level of a crystal, signals of various kinds and spectra, including signals of a complex shape, signals like those generated by multipurpose frequency synthesizers, and multiple-parameter vector sensors for control of mechanical, dynamic, and electrophysical (field) quantities will follow soon.

The following directions can be promising:

- Magneto-optics, multipurpose operated magneto-lasers
- Magnetotransistors in computer facilities
- Systems and means of testing of complex electronic schemes and modules
- Ferrointroscopy and ferrodefectoscopy
- Systems of the vector control

The stated bases of heteromagnetic microelectronics and microsystems of the active type, extensive results of experimental researches, approaches to physical modeling of such structures will not leave indifferent young inquisitive specialists and engineers–developers.

Authors hope that this book will serve as a starting point in the development of new directions of researches and applied development of the various microsystems meeting constantly increasing requirements of various market aspects.

References

1. Ignatiev, A.A., Magnetoelectronics of microwave and extremely high frequencies in ferrite films, Springer, NY, 2009, p. 333.
2. Meservey R., Tedrow P.W., Experimental test of the theory of high-field superconductivity, Phys. Rev. Lett., 1979, Vol. 43, p. 384–387.
3. Portier X., Tsymbal E.Yu., Petford-Long A.K., Correlation between domain structure and magnetoresistance in an active spin-valve element, Phys. Rev., 1998, Vol. 58, p. 591–594.
4. Levy P.V., Zhang Sh., Current opinion in solid state and materials science, 1999, Vol. 4, p. 223.
5. Gulyaev Yu. V., Zilberman P.E., Elliot R.J., Epshtein E.M., Movement of domain borders under the influence of spin-polarized current in magnetic junction, ZTF (Russ.), 2002, Vol. 72, p. 79–85.
6. Gulyaev Yu. V., Zilberman P.E., Epshtein E.M., Elliot R.J., Injection of spins by a current in magnetic metal-metal transitions, Radio Eng. Electron., 2003, Vol. 48 (9), p. 1030–1044.
7. Gulyaev Yu. V., Temiriazeva A.G., Tikhomirova M.P., Zilberman P.E., Nonlinear spin waves and resonances in non-uniform YIG films. Radio Eng. Electron., 2003, Vol. 48 (2), p. 161–168.
8. Gulyaev Yu. V., Nikitin S.A., Zhivotovskiy L.V., et al., Ferromagnetic films with periodic structures with a magnon forbidden band – magnon crystals, Letters to ZTF, Russ. J. Tech. Phys., 2003, Vol. 77, p. 670–674.
9. Gulyaev Yu. V., Nikitin S.A., Photon and magnetophoton crystals – a new medium for information transfer, Radio Eng., 2003 (8), p. 26–30.
10. Gulyaev Yu. V., Zilberman P.E., Epshtein E.M., Spin injection and magnetization reversal by a current in magnetic transitions (on the possibilities of design of nano-sized magnetic elements for high-density information recording), Radio Eng., 2003 (8), p. 31–35.
11. Gulyaev Yu. V., Zilberman P.E., Elliot R.J., Epshtein E.M., Spin injection mechanism of magnetization reversal and current hysteresis in magnetic transitions, Letters to ZTF, Russ. J. Tech. Phys., 2002, Vol. 76, p. 189–193.
12. Yakovlev J.M., Merkulov A.I., Magnetic semiconductors for devices of functional electronics, A review, Ser. 6 "Materials", 1989, Vol. 9 (985), p. 99.
13. Korutsenko I.V., Melkov G.A., Ukhanov S.A., Amplification of return bulk magnetostatic waves in YIG films, Kiev, Department in Ukraine central research engineering institute, 1988, p. 68.
14. Kazakov G.T., Sukharev A.G., Filimonov Yu. A., Magnetic "amplification" of MSW in YIG films, Digest Intermag'89: International Magnetic Conference, Washington, DC, March 28–31, New York, 1989, p. BP-7.
15. Melkov G.A., Sholom S.V., Amplification of surface magnetostatic waves by parametrical excitation, Russ. J. Tech. Phys., 1990, Vol. 60 (8), p. 118.
16. Kotsarenko N. Ya., Rapoport Yu. G., On the possibility of superheterodyn amplification of magnetostatic waves in ferromagnetics, Letters to Russ. J. Tech. Phys., 1984, Vol. 10 (14), p. 843.
17. Gurevich A.G., Magnetic resonance in ferrites and antiferromagnetics, Nauka, Moscow, 1973.
18. Gurevich A.G., Melkov A.G. Magnetic oscillations and waves, Nauka, Moscow, 1994.

19. Yakovlev Yu. M., Gendelev S. Sh., Ferrite monocrystals in radio electronics, Soviet Radio, Moscow, 1975.
20. Topical collection of reports and articles of the scientific and technical meeting "Heteromagnetic microelectronics", Vol. 1: Multipurpose complexification devices and systems of microwave and EHF ranges, Ed. Ignatiev A.A., Publishing house of Saratov State University, Saratov, 2004.
21. Topical collection of reports and articles of the II and III scientific and technical meetings 2004 "Heteromagnetic microelectronics", Vol. 2: Methods of designing of magnetoelectronic devices, Ed. Ignatiev A.A., Publishing house of Saratov State University, Saratov, 2005.
22. Physical optics (nomenclature), Collection of suggested terms, Vol. 74, Nauka, Moscow, 1968.
23. RF Patent No 2280917, application No 2005117825/28 (02081) of Ignatiev A.A., Lyashenko A.V. "A multifunctional integral magneto-semiconductor device", H01L27/14, priority of 09.06.2005, issued 27.07.06, Bul. No. 21.
24. USA Patent No 5, 313, 082. C1 *HO1L* 27/085 Feb. 16, 1993.
25. USA Patent No 5, 200, 713. C1 *HO3B* 5/18 May. 18, 1992.
26. Kalinin V.I., Zalogin N.M., Myasin E.A., Transition to chaos in a parametrical system with a nonlinear ferrite microresonator, Letters to ZTF (Russ.), 1984, Vol. 76 (21), p. 1311.
27. USA Patent No 5, 130, 670. C1 *HO3L* 7/07 Aug. 1, 1991.
28. USA Patent No 5, 146, 186. C1 *HO3L* 7/07 May. 13, 1991.
29. Bezborodov Yu. M., et al., Multipurpose generating microwave devices with dielectric resonators: A review on electronics, Ser.1 "Microwave electronics", 1983, Vol. 21.
30. Petrov T.V., et al., Low-noise autogenerators on field transistors with Schottky's barrier, Foreign Electron., 1983 (3), p. 24.
31. RU Patent No 2012102, C1 HO1L 29/812, 30.04.94, Bul. 8.
32. RU Patent No 2212090, C1 HO3B 5/18, 7/14, 25.04.95, Bul. 10.
33. Vikulin I.V., Glauberman M.A., et al., A magnetic-field sensor on the basis of a two-collector magnetotransistor, Appl. Tech. Electron., 1974 (5), p. 45.
34. RU Patent No 2055419, C1 6HO1L 29/82, 27.02.96, Bul. 6.
35. RU Patent No 2008748, C1 5HO1L 29/82, 28.02.94, Bul. 4.
36. RU Patent No 2097873, C1 6HO1L 29/82, 27.11.97. Bul. 33.
37. Bogdanov G.B., Basics of the theory and applications of ferrites in measurements and control, Soviet Radio, Moscow, 1967.
38. Lax B, Button K., Microwave ferrites and ferrimagnetics, McGraw-Hill Book Company, Inc. New York, San Francisco, Toronto, London, 1962.
39. Gummel H.K., Poon H.C., An integral charge control model of bipolar transistors, Bell Syst. Tech. J., May–June 1970, Vol. 49, p. 827–852.
40. Poon H.C., Modeling of bipolar transistor using charge control model with application to third-order distortion studies, IEEE Trans. Electron. Devices, 1972, Vol. ED-12 (6), p. 719.
41. Gummel H.K., A charge control relation bipolar transistor, Bell Syst. Tech. J., 1970, Vol. 49 (1), p. 115.
42. Khibnik A.I., Algorithms and FORTRAN programs, Pushino, 1979 (5).
43. Rizzoli V., Lipparini A., Computer-aided noise analysis of linear multiport networks of arbitrary topology, IEEE Trans. Microw. Theory Tech., 1985, Vol. MTT-33 (12), pp. 1507–1512.
44. Veselov G.I., Yegorov E.I., Alekhin Yu. I., et al., Microelectronic microwave devices, Moscow, 1988.
45. Tolstov Yu. G., Tevryukov A.A., Theory of electric circuits, Vishaya shkola, Moscow, 1971.
46. Landa P.S., Self-oscillations in systems with a finite number of degrees of freedom, Moscow, 1980.
47. Khibnik A.I., Algorithms and FORTRAN programs, Pushino, 1979 (1).
48. Kiyashko S.V., Pikovskiy A.S., Rabinovich M.I., An autogenerator of the radio range with stochastic behaviour, Radio Eng. Electron., 1980, Vol. 25, p. 336.
49. Kuznetsov S.P., Yerastova E.N., Feigenbaum theory: Lectures on microwave electronics and radiophysics, Saratov, 1983, Book 2, p. 3.
50. Feigenbaum M.J., Quantitative universality for a class of nonlinear transformations, J. Stat. Phys., 1978, Vol. 19 (1). p. 25.

51. Anishchenko V.S., Complex oscillations in simple systems, Nauka, Moscow, 1990.
52. Anishchenko V.S., Stochastic oscillations in radiophysical systems, Saratov State University, Saratov, 1986.
53. Samoldanov V.N., Lyashenko A.V., Ignatiev A.A., Sotov L.S., A program of calculation of Gummel-Poon's model parameters of a bipolar transistor: Reg. certificate No 20046110989 RU, application No 2004610417 of 26.01.2004, Date of registration 21.04.2004.
54. Pleshkov V.V., Lyashenko A.V., Ignatiev A.A., Khvalin A.L., A program of calculation of a field magnetoelectronic transistor: Reg. Certificate No 2004610992 RU, application No 2004610420 of 26.01.2004, Date of registration 21.04.2004.
55. Marinin A.V., Lyashenko A.V., Ignatiev A.A., Khvalin A.L., A program of calculation of magnetoelectronic coupling elements: Reg. Certificate No 2004610991 RU, application No 2004610419 of 26.02.2004, Date of registration 21.04.2004.
56. Samoldanov V.N., Lyashenko A.V., Ignatiev A.A., Khvalin A.L., A program of calculation of a powerful bipolar magnetoelectronic transistor: Reg. Certificate No 2004610987 RU, application No 2004610417 of 26.01.2004, Date of registration 21.04.2004.
57. Production catalogue of NIIET State Corp., Voronezh (RF), 2002.
58. Dul'nyov G.N., Semyashkin E.M., Heat exchange in radio-electronic devices, Energiya, Leningrad, 1968.
59. Rotkon L.L., Spokoinyi Yu.E., Maintenance of thermal modes at designing radio electronic devices, Soviet radio, Moscow, 1976.
60. Ignatiev A.A., Ovchinnikov S.V., Samoldanov V.N., Thermophysical analysis of a ferrite-semiconductor generator with heteromagnetic interaction, Prospects of the development of electronics and vacuum engineering for 2001–2006: proc. of scientific and technical conference on February 22–23, 2001, Saratov, p. 85.
61. Dul'nyov G.N., Parfyonov V.G., Sigalov A.V., Methods of calculation of thermal modes of devices, Radio and Svyaz, Moscow, 1990.
62. Ovchinnikov S.V., Vasiliev A.V., Lyashenko A.V., Solopov A.A., A program of calculation of a non-stationary temperature field of the semiconductor structure of a powerful magnetoelectronic transistor in a pulse mode: Reg. Certificate No 2004610985 RU, application No 2004610413 of 26.02.2004, Date of registration 21.04.2004.
63. Ovchinnikov S.V., Samoldanov V.N., Lyashenko A.V., Solopov A.A., A program of calculation of the stationary thermal resistance of the multilayered constructive elements of a powerful magnetoelectronic transistor of a rectangular shape: Reg. Certificate No 2004610993 RU, application No 2004610421 of 26.02.2004, Date of registration 21.04.2004.
64. Ovchinnikov S.V., Ignatiev A.A., Lyashenko A.V., Solopov A.A., A program of calculation of the stationary thermal resistance of the constructive elements of a powerful magnetoelectronic transistor in the form of a multilayered cylinder: Reg. Certificate No 2004610986 RU, application No 2004610414 of 26.02.2004, Date of registration 21.04.2004.
65. Berdichevskiy B.E., Question of reliability maintenance of radio-electronic equipment at its design, Soviet Radio, Moscow, 1977.
66. Kochetov V.T., Pavlenko A.D., Kochetov M.V., Materials strength, Rostov-on-Don, Phoenix, 2001.
67. Lykov A.V., Theory of heat conductivity, Vysshaya Shkola, Moscow, 1967.
68. Knyazev A.D., Elements of theory and practice of electromagnetic compatibility of radio-electronic means, Radio and Svyaz, Moscow, 1984.
69. Petrovskiy V.I., Sedel'nikov Yu. E., Electromagnetic compatibility of radio-electronic means: Manual for high schools, Radio and Svyaz, Moscow, 1986.
70. Kravchenko V.I., Bolotov E.A., Letunova N.I., Radio-electronic means and powerful electromagnetic interferences, Radio and Svyaz, Moscow, 1987.
71. Mal'tsev V.A., Myakin'kov V. Yu., Novoselets V.I., State-of-the-art of design of solid-state microwave low-power generators and devices on their basis, Radio Eng. (Russ.), 1999 (4), p. 44.
72. Rudiko F.L., DDS: direct digital synthesis of frequency, www.rf.atnn.ru/s6/DDS_1.htm.
73. Ferrite filters and their application in devices with magnetic frequency change: reviews on electronic engineering, Ser. 1 "Microwave electronics", 1982 (10), (914).

74. Ollivier P.M., Microwave YIG – tuned transistor oscillator amplifier design: application to C Band, IEEE J Solid State Circuits, 1972, Vol. Sc-7 (1), p. 54.
75. Stal'makhov V.S., Ignatiev A.A., Lectures on spin waves, Publishing house of State University, Saratov, 1983.
76. Vashkovskiy A.V., Stal'makhov V.S., Sharaevskiy Yu. P., Magnetostatic waves in ultrahigh frequency electronic engineering, Publishing house of Saratov State University, Saratov, 1993.
77. Vapne G.M., Microwave devices on magnetostatic waves: reviews on electronic engineering, Ser. 1 "Microwave electronics", Central Research institute "Electronics", Moscow, 1984, Vol. 8 (1060).
78. Gololobov V.P., Tureyeva O.V., Tsymbal V.I., Shelamov G.N., Ferrite-varactor microwave circuits, Proc. Inst. High. Educ. Phys., Ser. "Radio-electro-technics", 1983, Vol. 26 (8), p. 74.
79. Vaisblat A.V., Switching microwave devices on semiconductor devices, Moscow, 1987.
80. Gorodnitskiy A.M., Filin A.M., Studying of ocean bottom and search for metal objects by magnetic methods, Bull. Russ. Acad. Nat. Sci., 2001, Vol. 1, p. 21.
81. Krivoshey A.V., Bel'iev A.N., Soldering and welding in manufacture of radio-electronic equipment, Energy, Moscow, 1974.
82. Ultrahigh frequency engineering: a catalogue of production of Salute State Corp., Nizhni Novgorod, 2001.
83. Phillips E., Microwelding techniques, Microelectron. Reliab., 1966, Vol. 15 (3), p. 107.
84. Kolesova V.I., Analytical methods of magnetic cartography, Nauka, Moscow, 1985.
85. Belyaev I.I., Valyashko G.M., Ivanenko A.N., Magnetic field of the ocean, Nauka, Moscow, 1993, p. 299.
86. Yanovskiy B.M., Earth's magnetism, Leningrad, 1964.
87. Pleshkov V.V., Lyashenko A.V., Ignatiev A.A., Sotov L.S., A program of processing of the information signal of a magnetoelectronic sensor: Reg. certificate No 2004610990 RU, application No 2004610418 of 26.02.2004, Date of registration 21.04.2004.
88. Directory on radio-electronic device elements, Ed. Dulin V.N., Zhuk M.S., Moscow, 1977.
89. Devices and systems of prospecting geophysics, SOEAGO 2002, Vol. 2, p. 15.
90. Sivukhin D.V., A general course in physics, Vol. 1 "Mechanics", Nauka, Moscow, 2006.
91. Gofman Yu. V., Laws, formulas, problems of physics: a directory, Naukova Dumka, Kiev, 1977.
92. Korn G., Korn T., Mathematical handbook: For scientists and engineers, Dover Publications, Mineola, NY, 1968.
93. Monosov Ya. A., Non-linear ferromagnetic resonance, Nauka, Moscow, 1971.

Index

FSC
www.fsc.org
MIX
Papier | Fördert
gute Waldnutzung
FSC® C083411

Kolibri 360 GmbH
Ferdinand-Jühlke-Straße 7
99095 Erfurt, Deutschland
produktsicherheit@kolibri360.de